World Geomorphological Landscapes

Series Editor

Piotr Migoń, Institute of Geography and Regional Development, University of Wrocław, Wrocław, Poland

Geomorphology – 'the Science of Scenery' – is a part of Earth Sciences that focuses on the scientific study of landforms, their assemblages, and surface and subsurface processes that moulded them in the past and that change them today. Shapes of landforms and regularities of their spatial distribution, their origin, evolution, and ages are the subject of geomorphology. Geomorphology is also a science of considerable practical importance since many geomorphic processes occur so suddenly and unexpectedly, and with such a force, that they pose significant hazards to human populations. Landforms and landscapes vary enormously across the Earth, from high mountains to endless plains. At a smaller scale, Nature often surprises us creating shapes which look improbable. Many geomorphological landscapes are so immensely beautiful that they received the highest possible recognition – they hold the status of World Heritage properties. Apart from often being immensely scenic, landscapes tell stories which not uncommonly can be traced back in time for millions of years and include unique events. This international book series will be a scientific library of monographs that present and explain physical landscapes across the globe, focusing on both representative and uniquely spectacular examples. Each book contains details on geomorphology of a particular country (i.e. The Geomorphological Landscapes of France, The Geomorphological Landscapes of Italy, The Geomorphological Landscapes of India) or a geographically coherent region. The content is divided into two parts. Part one contains the necessary background about geology and tectonic framework, past and present climate, geographical regions, and long-term geomorphological history. The core of each book is however succinct presentation of key geomorphological localities (landscapes) and it is envisaged that the number of such studies will generally vary from 20 to 30. There is additional scope for discussing issues of geomorphological heritage and suggesting itineraries to visit the most important sites. The series provides a unique reference source not only for geomorphologists, but all Earth scientists, geographers, and conservationists. It complements the existing reference books in geomorphology which focus on specific themes rather than regions or localities and fills a growing gap between poorly accessible regional studies, often in national languages, and papers in international journals which put major emphasis on understanding processes rather than particular landscapes. The World Geomorphological Landscapes series is a peer-reviewed series which contains single and multi-authored books as well as edited volumes.

World Geomorphological Landscapes – now indexed in Scopus® !

More information about this series at http://www.springer.com/series/10852

Achim A. Beylich
Editor

Landscapes and Landforms of Norway

Springer

Editor
Achim A. Beylich
Geomorphological Field Laboratory
Selbustrand, Norway

ISSN 2213-2090 ISSN 2213-2104 (electronic)
World Geomorphological Landscapes
ISBN 978-3-030-52562-0 ISBN 978-3-030-52563-7 (eBook)
https://doi.org/10.1007/978-3-030-52563-7

This Springer imprint is published by the registered company Springer Nature Switzerland AG
The registered company address is: Gewerbestrasse 11, 6330 Cham, Switzerland

Solen/The Sun Edvard Munch 1910–13

Nord

Se oftere mot nord.
Gå mot vinden, du får rødere kinn.
Finn den ulente stien. Hold den.
Den er kortere.
Nord er best.
Vinterens flammehimmel, sommer-
nattens solmirakel.
Gå mot vinden. Klyv berg.
Se mot nord.
Oftere.
Det er langt dette landet.
Det meste er nord.

Rolf Jacobsen, Nattåpent, Gyldendal Oslo, 1985

North

Look more often towards the North.
Go against the wind, you get redder cheeks.
Find the rugged trail. Hold it.
It is shorter.
North is the best.
Winter's flaming sky, summer-
night sun miracle.
Go against the wind. Climb the mountain.
Look to the North.
More often.
This country is vast.
Most of it is North.

Rolf Jacobsen, Nattåpent, Gyldendal Oslo, 1985
(translated from Norwegian to English)

Preface

"Geomorphology is the interdisciplinary study of landforms, their landscapes and the earth surface processes that create and change them" (International Association of Geomorphologists 2019).

Mainland Norway stands for a great variety of beautiful and spectacular landscapes with these landscapes being a function of geological and geomorphological processes working over very long time spans and under varying climates. The outstanding beauty of the Norwegian nature has been an inspiration for famous Norwegian artists, e.g., the painter Edvard Munch (1863–1944), the poet Rolf Jacobsen (1907–1994) and the composer and pianist Edvard Grieg (1843–1907).

Building on the large body of existing literature on the long-term geological and Quaternary geological development of the Norwegian landscapes, the purpose of this book is to provide the first compilation of selected geomorphological review works and in-depth studies on relevant geomorphological earth surface processes and the resulting modification of landscapes and/or creation of new landforms and landscapes. The book shall contribute to filling a still existing gap regarding the in-depth understanding of Holocene, and particularly of contemporary geomorphological earth surface processes and how these processes change existing landforms and landscapes and/or create new landforms and landscapes. An improved scientific in-depth understanding of the mechanisms and drivers of geomorphological earth surface processes and their resulting landforms and/or landform and landscape changes is of utmost importance with respect to urgently needed qualified assessments of the possible geomorphological effects of ongoing and accelerated environmental changes, and in view of the increasing importance of the efficient development of geo-hazard assessment applications. Also, the status and value of geomorphological heritage are addressed with selected examples from different key landscapes in mainland Norway. All accepted chapters of this book are well illustrated with numerous figures and photographs and shall, also for non-academic readers, increase the awareness of the outstanding beauty, the increasing vulnerability and the hazardous potential of the various landscapes of mainland Norway.

The preparation of this book on Landscapes and Landforms of Norway would not have been possible without the very valuable work of the selected expert reviewers. The work and contributions of the invited chapter authors and the selected expert reviewers are greatly acknowledged. I also would like to thank the Book Series editor Piotr Migoń and Michael Leuchner, Robert Doe, Manjula Saravanan and Banu Dhayalan of Springer Nature Verlag for their support during the preparation and publishing process of this book.

Selbustrand, Norway
September 2019

Achim A. Beylich

Reference

International Association of Geomorphologists (2019) http://www.geomorph.org (September 2019)

Contents

Editor and Contributors

About the Editor

Achim A. Beylich (Geomorphological Field Laboratory (GFL), Sandviksgjerde, Strandvegen 484, 7584 Selbustrand, Norway) is a geomorphologist with more than 25 years of work experience in field- and laboratory-based quantitative process geomorphic research in various climatic environments and landscapes in Iceland, Sweden, Finland, Canada, Russia, Germany, Austria, Norway and Spain. Since 2004, he has been initiating and leading a number of large international and interdisciplinary research groups, networks and programs on geomorphologic earth surface processes and landscape development under ongoing or accelerated climate change and increasing anthropogenic impacts and pressures. During his scientific career, he has carried out research and has been working at research institutes and universities in Germany, Sweden, Iceland, Canada and Norway, and he is currently Head of Operations at the Geomorphological Field Laboratory (GFL) in Selbustrand, Norway. He is a senior scientist with more than 100 scientific publications in journals and books, numerous edited works and with formal full professor competence in geomorphology. He is Editor-in-Chief for the scientific journal *Geomorphology* (Elsevier), an Editorial Board Member for several international scientific journals, and serves frequently as a peer-reviewer for a number of international scientific journals and for various national and international funding bodies and agencies. He is the President of the IAG Geomorphological Research Group of Norway (IAG GeoNor), initiator and Chair of the Nordic IAG Network of National Geomorphology Groups from Norway, Sweden, Finland, Denmark and Iceland (IAG GeoNorth), and the Norwegian National Scientific Representative (IAG National Scientific Member Norway) for the International Association of Geomorphologists (IAG).

Contributors

Achim A. Beylich Geomorphological Field Laboratory (GFL), Sandviksgjerde, Selbustrand, Norway

Alastair M. Curry Department of Psychology, Sports Sciences and Geography, University of Hertfordshire, Hatfield, Hertfordshire, UK

Anika Donner Institute of Geology, University of Innsbruck, Innsbruck, Austria

Riwan Kerguillec University of Nantes, Nantes, France

Gunnar Ketzler Department of Physical Geography and Climatology, RWTH Aachen University, Aachen, Germany

Katja Laute Geomorphological Field Laboratory (GFL), Sandviksgjerde, Selbustrand, Norway

John A. Matthews Department of Geography, College of Science, Swansea University, Swansea, UK

Geraint Owen Department of Geography, College of Science, Swansea University, Swansea, UK

Lena Rubensdotter Geological Survey of Norway (NGU), Trondheim, Norway

Wolfgang Römer Department of Physical Geography and Climatology, RWTH Aachen University, Aachen, Germany

Gro Sandøy Geological Survey of Norway (NGU), Trondheim, Norway

Dominique Sellier University of Nantes, Nantes, France

Kari Sletten Geological Survey of Norway (NGU), Trondheim, Norway

Angela Tintrup gen. Suntrup Department of Geography and Geology, University of Würzburg, Würzburg, Germany

Stefan Winkler Department of Geography and Geology, University of Würzburg, Würzburg, Germany

Part I

Introduction to Landscapes and Landforms of Norway

Geomorphological Landscapes, Earth Surface Processes and Landforms in Norway

1

Achim A. Beylich

Abstract

Mainland Norway stands for a great variety of partly spectacular landscapes with these landscapes being a function of geological and geomorphological processes working over very long time spans. Building on the large body of existing literature on the long-term geological and Quaternary geological development of the Norwegian landscape, the purpose of this book is to provide the first compilation of selected review works and geomorphological in-depth studies on relevant geomorphological earth surface processes and the resulting modification of existing and/or creation of new landforms. The book shall contribute to filling a still existing gap regarding the in-depth understanding of Holocene, and particularly of contemporary geomorphological earth surface processes and how these processes change existing landscapes and landforms and/or create new landforms. The review works and geomorphological in-depth studies selected for this book cover a range of varied geomorphological key landscapes, earth surface processes and landforms in mainland Norway. An improved scientific in-depth understanding of the key drivers, mechanisms, spatiotemporal variability and quantitative rates of contemporary geomorphological earth surface processes and of their resulting landforms and/or landscape and landform changes is of utmost importance with respect to urgently needed qualified assessments of the possible geomorphological effects of ongoing and accelerated environmental changes and in view of the increasing importance of the efficient development of geo-hazard assessment applications. Also, the status and value of geomorphological heritage are addressed with selected examples from different key landscapes of mainland Norway.

Keywords

Landscapes • Landforms • Geomorphological earth surface processes • Environmental changes • Geo-hazards • Geomorphological heritage • Mainland Norway

1.1 Introduction

Norway stands for a great variety of partly spectacular landscapes with these landscapes being a function of geological and geomorphological processes working over very long lime spans. There is a large body of existing literature explaining in great detail the geological and Quaternary geological long-term development of the Norwegian landscapes, including the major key compilations on *The Making of a Land—Geology of Norway* edited by Ramberg et al. (2008) and *Quaternary Geology of Norway* edited by Olsen et al. (2013). Distinct first-order structures of the present Norwegian landscape can be traced back to ancient denudational processes, the Caledonian orogeny or break-up of the North Atlantic, whereas a large portion of today's large-, intermediate- and small-scale landforms in Norway were created by earth surface processes operating during the Quaternary time period, and mostly by the action of glaciers through numerous glaciations (e.g. Fredin et al. 2013; Olsen et al. 2013). The Quaternary time period, comprising the last ca. 3 million years, is characterized by cool and variable climate with temperatures oscillating between relative mildness to frigid ice-age conditions. Fredin et al. (2013) summarize that the numerous glaciations during the Quaternary "have had the most profound effect with the production of large U-shaped valleys, fjords and Alpine relief. On the other hand, interior and upland areas in Norway seem to be largely unaffected by glacial erosion and exhibit a possibly pre-Quaternary landscape with only some periglacial influence. The ice sheets in Scandinavia thus have

A. A. Beylich (✉)
Geomorphological Field Laboratory (GFL), Sandviksgjerde, Strandvegen 484, 7584 Selbustrand, Norway
e-mail: achim.beylich@geofieldlab.com

A. A. Beylich (ed.), *Landscapes and Landforms of Norway*, World Geomorphological Landscapes,
https://doi.org/10.1007/978-3-030-52563-7_1

redistributed rock mass and sediments in the landscape with the largest glaciogenic deposits being found on the continental shelf. Large Quaternary deposits and valley fills can also be found onshore and these are now valuable resources for aggregates, ground water and agriculture. Important Quaternary processes have also been acting along the Norwegian coast with denudation of the famous strandflat, where the formation processes are not fully understood. The isostatic depression of crust under the vast ice sheets have also lead to important consequences, with thick deposits of potentially unstable, fine-grained glaciomarine sediments in quite large areas of Norway" (Fredin et al. 2013).

1.2 Geomorphological Earth Surface Processes and Resulting Landforms

"Geomorphology is the interdisciplinary study of landforms, their landscapes and the earth surface processes that create and change them" (International Association of Geomorphologists 2020). Building on the large body of existing literature on the long-term geological and Quaternary geological development of the Norwegian landscape, the purpose of this book is to provide the first compilation of selected review works and geomorphological in-depth studies on relevant geomorphological earth surface processes and the resulting modification of existing and/or creation of new landforms. The various scientific contributions of this book include review works and geomorphological in-depth studies on glacial, periglacial, and denudational hill slope and fluvial processes, and explain in detail (i) the mechanisms, controls, quantitative rates as well as (ii) the modification of existing landscapes and landforms and the creation of new landforms resulting from these varied earth surface processes. While highlighting the advanced existing knowledge on the geology and Quaternary geology of mainland Norway, this book shall contribute to filling a still existing gap regarding the in-depth understanding of Holocene, and particularly of contemporary geomorphological earth surface processes and how these processes change existing landscapes and landforms and/or create new landforms and landscapes. An improved scientific in-depth understanding of the key drivers, mechanisms, spatiotemporal variability and quantitative rates of contemporary geomorphological earth surface processes and of their resulting landforms and/or landscape and landform changes is of utmost importance with respect to urgently needed qualified assessments of the possible geomorphological effects of ongoing and accelerated environmental changes and in view of the increasing importance of the efficient development of geo-hazard assessment applications.

1.3 Geomorphological Study Regions and Landscapes Presented in This Book

Following the introductory part of this book (book part I with this chapter and book chapter 2, including a brief general introduction to geomorphological landscapes, earth surface processes and landforms in Norway (this chapter), and a detailed overview of the climate of mainland Norway and geomorphologically relevant aspects of the contemporary climate (Ketzler et al. 2021), the geomorphological review works and in-depth studies selected for this book (book part II with book Chaps. 3–10) cover a range of varied geomorphological key landscapes, earth surface processes and landforms across mainland Norway (Fig. 1.1). In book part III (book Chaps. 11 and 12), the status and value of geomorphological heritage in Norway are addressed with selected examples from different key landscapes of mainland Norway (Fig. 1.1).

In book part II, the book Chaps. 3–7 focus on the magnificent Jostedalsbreen ice cap and the steep and spectacular fjord landscapelandscape in south-western Norway (Fig. 1.1). In book Chap. 3 (Winkler 2021), terminal moraine formation processes and the geomorphology of glacier forelands at selected outlet glaciers of the Jostedalsbreen ice cap are presented, whereas contemporary ice retreat and the associated formation and changes of proglacial lakes at this impressive ice cap are the topics of book Chap. 4 (Laute and Beylich 2021). In book Chap. 5 (Curry 2021), paraglacial rock-slope failures following deglaciation in rock-slope mountain landscapes in western Norway are discussed in detail. Snow-avalanche impact landforms in western Norway are explained and discussed in book Chap. 6 (Matthews and Owen 2021), whereas fluvial processes and contemporary fluvial denudation in the different glacierized and non-glacierized mountainous landscapes of western and central Norway (Fig. 1.1) are the topics of book Chap. 7 (Beylich and Laute 2021). Book Chap. 8 (Winkler et al. 2021) presents and discusses periglacial processes and landforms in the Alpine mountain landscape of Jotunheimen in central southern Norway (Fig. 1.1). The mountainous landscape of Rondane in south-central Norway (Fig. 1.1) is in the focus of book Chap. 9 (Sellier and Kerguillec 2021a) with the characterization and the explanation of the significance of scree slopes investigated in this region. The morphological description of erosional and depositional landforms created by debris flow processes is

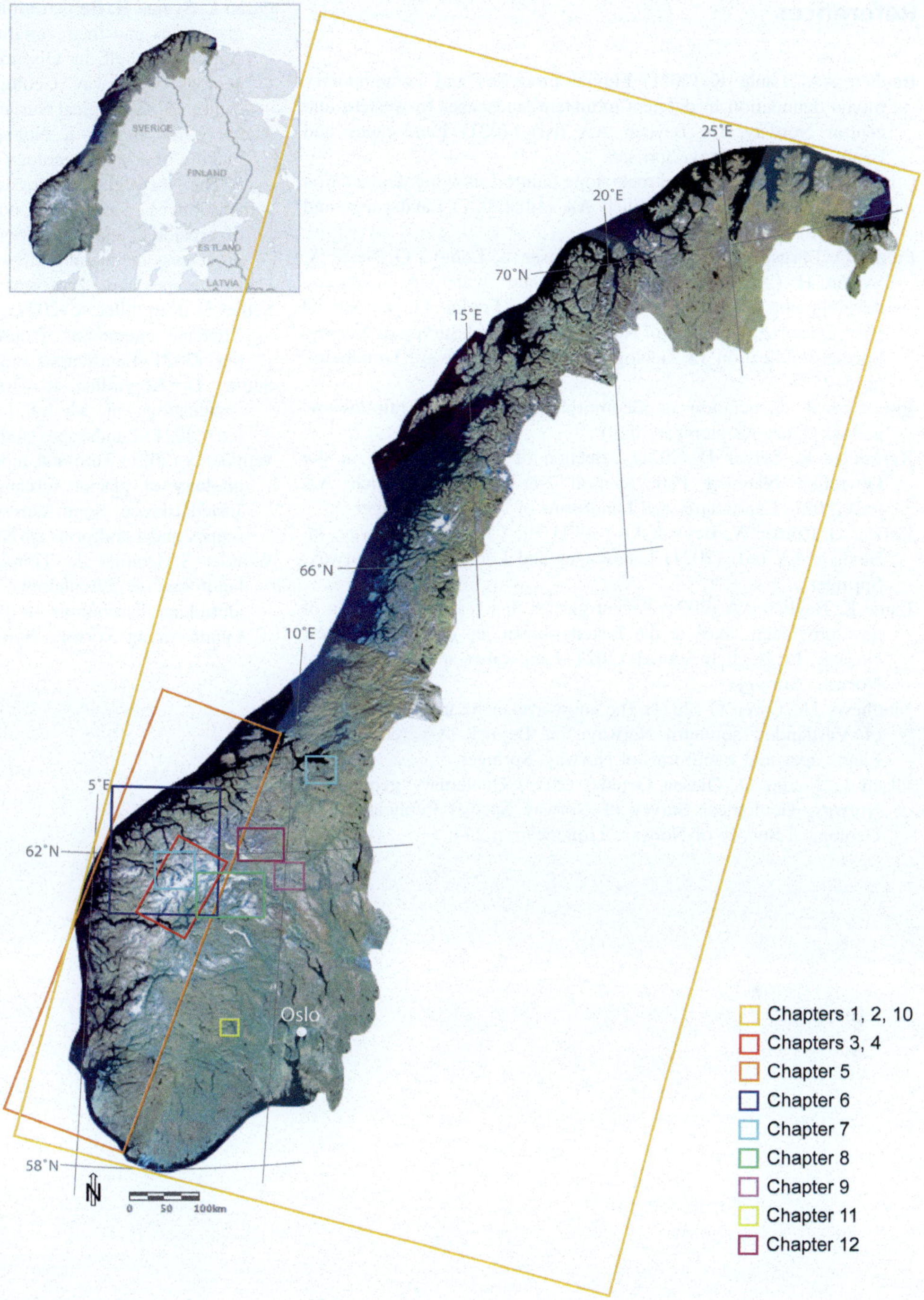

Fig. 1.1 Location of the selected study regions and geomorphological landscapes presented in the 12 chapters of this book. *Source* of Norway map and inset ©NVE Atlas 3.0 (Kartverket, Geovekst og kommuner-Geodata AS, NVE)

presented and discussed in book Chap. 10 (Rubensdotter et al. 2021) with examples from different selected mountain regions across mainland Norway (Fig. 1.1). In book part III, landforms and geomorphosite designation on Mount Gausta in Telemark (Fig. 1.1) are described and discussed in detail in book Chap. 11 (Sellier and Kerguillec 2021b), whereas the selection of geomorphosites in the mountainous upland landscape of the Dovrefjell and Sunndalsfjella National Park in central Norway (Fig. 1.1) is the focus of the final book Chap. 12 (Kerguillec and Sellier 2021).

References

Beylich AA, Laute K (2021) Fluvial processes and contemporary fluvial denudation in different mountain landscapes in western and central Norway. In: Beylich AA (ed) (2021) Landscapes and Landforms of Norway, Springer

Curry AM (2021) Paraglacial rock-slope failure following deglaciation in western Norway. In: Beylich AA (ed) (2021) Landscapes and Landforms of Norway, Springer

Fredin O, Bergstrøm B, Eilertsen R, Hansen L, Longva O, Nesje A, Sveian H (2013) Glacial landforms and quaternary landscape development in Norway. In: Olsen L, Fredin O, Olesen O (eds) Quaternary geology of Norway. Geological Survey of Norway Special Publication 13. Geological Survey of Norway, Trondheim, pp 5–25

International Association of Geomorphologists (2020) http://www.geomorph.org (September 2020)

Kerguillec R, Sellier D (2021) Selection of geomorphosites in the Dovrefjell National Park (central Norway). In: Beylich AA (ed) (2021) Landscapes and Landforms of Norway, Springer

Ketzler G, Römer W, Beylich AA (2021) The Climate of Norway. In: Beylich AA (ed) (2021) Landscapes and Landforms of Norway, Springer

Laute K, Beylich AA (2021) Recent glacier changes and formation of new proglacial lakes at the Jostedalsbreen ice cap in southwest Norway, In: Beylich AA (ed) (2021) Landscapes and Landforms of Norway. Springer

Matthews JA, Owen G (2021) The snow-avalanche impact landforms of Vestlandet, southern Norway. In: Beylich AA (ed) (2021) Landscapes and Landforms of Norway, Springer

Olsen L, Fredin O, Olesen O (eds) (2013) Quaternary geology of Norway. Geological Survey of Norway Special Publication 13. Geological Survey of Norway, Trondheim, p 174

Olsen L, Sveian H, Bergstrøm B, Ottesen D, Rise L (2013) Quaternary glaciations and their variations in Norway and on the Norwegian continental shelf. In: Olsen L, Fredin O, Olesen O (eds) Quaternary geology of Norway. Geological Survey of Norway Special Publication 13. Geological Survey of Norway, Trondheim, pp 27–78

Ramberg IB, Bryhni I, Nøttvedt A, Rangnes K (eds) (2008) The making of a land—geology of Norway. Norsk Geologisk Forening—The Norwegian Geological Association, Trondheim, p 624

Rubensdotter L, Sletten K, Sandøy G (2021) Morphological description of erosional and depositional landforms formed by debris flow processes in mainland Norway. In: Beylich AA (ed) (2021) Landscapes and Landforms of Norway, Springer

Sellier D, Kerguillec R (2021a) Characterization of scree slopes in the Rondane mountains (south-central Norway). In: Beylich AA (ed) (2021) Landscapes and Landforms of Norway, Springer

Sellier D, Kerguillec R (2021b) Landforms and geomorphosite designation on Mount Gausta (Telemark). In: Beylich AA (ed) (2021) Landscapes and Landforms of Norway, Springer

Winkler S (2021) Terminal moraine formation processes and geomorphology of glacier forelands at the selected outlet glaciers of Jostedalsbreen, South Norway. In: Beylich AA (ed) (2021) Landscapes and Landforms of Norway, Springer

Winkler S, Donner A, Tintrup gen. Suntrup A (2021) Periglacial landforms in Jotunheimen, central southern Norway, and their altitudinal distribution. In: Beylich AA (ed) (2021) Landscapes and Landforms of Norway, Springer

The Climate of Norway

2

Gunnar Ketzler, Wolfgang Römer, and Achim A. Beylich

Abstract

Mainland Norway shows a complex combination of physical factors leading to various climate settings. Due to the huge extension in north–south direction from 57° 58′ N to 71° 11′ N, Norway encompasses five climate zones according to the Köppen classification. Its location on the west side of the Scandinavian Peninsular close to the North Atlantic Current, however, shifts most climate effects to a more temperate level compared to what is to expect from the given geographical zone. Especially during the last two decades, a marked temperature increase is observed over the whole country. The close interlink with sea climates due to the very long coastline with many fjords and islands, effects of altitude as well as of luv and lee situations of different mountainous regions up to 2469 m a.s.l. and the general west–east gradient from maritime to continental climates result in various patterns of climate elements on a regional and local scale. Southwest Norwegian coastal lowlands have a quite temperate climate and the mountainous areas situated behind often show huge amounts of precipitation during all seasons including partly enormous snow accumulation in winter supplying numerous glaciers. The more continental areas of Eastern Norway are very dry, and the elevated mountain plateaus, especially in Northern Norway, are of subarctic appearance including phenomena of permafrost.

G. Ketzler (✉) · W. Römer
Department of Physical Geography and Climatology, RWTH Aachen University, Templergraben 55, 52062 Aachen, Germany
e-mail: gunnar.ketzler@geo.rwth-aachen.de

A. A. Beylich
Geomorphological Field Laboratory (GFL), Sandviksgjerde, Strandvegen 484, 7584 Selbustrand, Norway

Keywords

Solar radiation • Temperature • Precipitation • Wind • Topoclimate

2.1 Introduction

This chapter aims at giving an overview of the main climate factors influencing geomorphological processes and their geographical distribution. This also includes general statements, but it is not intended to outline a complete climatology of Norway but rather an overview of regional morphoclimatology.

By doing so, we generally follow the morphoclimatological approach of Ahnert (e.g. 1987). This approach has the intention to focus on those climatic conditions and processes relevant for morphological processes and to quantify as far as reasonable the relation between their magnitude and their frequency. An example of a field study from Norway based on morphoclimatic analysis is given by Beylich and Laute (2018). For the present—in a spatial and functional sense—more general study, extensive quantification, e.g. in the form of detailed analysis of frequency distributions, is not performed, but, however, it is intended to discuss the most important aspects of a morphoclimate on the basis of figures describing intensity and temporal dimension.

There are few surveys on the climate of Norway or Scandinavia in international publications. Williams (1901) already gives an overview of the main patterns caused by the North Atlantic Current and the country's situation in relation to mountain ranges, already taking into account the results of early meteorological measurements. Johanessen (1970) extensively documents and discusses the—at that time—already large extent of data from numerous stations in Norway with a scope on complete Scandinavia, beginning to analyse newer data series like solar radiation. Norseth (1987) continues this work focussing on the annual radiation deficit and

A. A. Beylich (ed.), *Landscapes and Landforms of Norway*, World Geomorphological Landscapes,
https://doi.org/10.1007/978-3-030-52563-7_2

its compensation by large-scale advection of (latent) heat including the role of atmospheric circulation types and interaction with the hydrological cycle. Tikkanen (2005), too, underlines the role of circulation types, introduces topoclimatological effects in the regional climatology of Norway and Scandinavia and—in this context—points to local climate extremes like extreme precipitation data from newer measurements and modelling activities.

After a short overview over the general setting and Norway's specific location, this chapter on the morphoclimate of Norway focusses on fundamental astronomical conditions for and virtual input of solar radiation, followed by a review on spatial distribution and frequency of average and extreme temperature and precipitation data including a closer view on effects of the adjacent marine areas on temperature and humidity. A section on wind conditions is added. Regional and topoclimate features and expected effects of climate change are discussed afterwards.

Norway's location on the west side of the Scandinavian Peninsula in north-western Europe between 57° 58′ N and 71° 11′ N respectively 4° 40′ E and 30° 58′ E (from Kartverket 2018) indicates a central position in the westerlies and a classification as part of both the temperate and polar climate zones. It forms an elongated territory on the west and partly on both sides of Scandes (see Fig. 2.1), with an extension of about 1700 km from the southernmost point Cape Lindesnes to the North Cape having a very long and structured coastline as well as mountain areas with an altitude of up to 2469 m a.s.l. (Galdhøppigen). The country has —in relation to other western European countries—a remarkable N-S-extension of 13° 13′ or 1470 km, but its E-W extension (26° 18′ = 1160 km based on the latitude of the northern polar circle) is also not negligible although this fact usually attracts little attention.

2.2 Solar Radiation

The energy for atmospheric and geomorphologic processes originally comes from solar irradiation. Being located between latitudes of 58.0°–71.2° N, Norway is characterized by marked differences in the solar climatic conditions. These differences are indicated in the strongly differing day length at the summer and winter solstices. At a latitude of 58° N, the longest and shortest days are experiencing 18:11–6:10 h daylight, at 60° N the daylight ranges from 18:53 to 5:52 h, at 63° N the daylight ranges from 20:19 to 4.42 h and at the Arctic Circle (66.5° N) the longest daytime is 24 h whilst the shortest day has 0 h, when twilight is excluded. During the summer, the northernmost regions experience a period of more than 2 months with daylight whilst in the winter a period of about 2 months with night (Tikkanen 2005). In the winter months, the high latitudes receive no direct solar radiation, though the astronomical twilight resulting from refraction and reflection of sunrays produces twilight until the sun is 18° below the horizon.

At the summer solstice at 58° N (e.g. Kristiansand), 60° N (e.g. Bergen and Gardermoen) and 63° N (e.g. Trondheim/Værnes), the sun's angle of incidence decreases from 55.5° to 53.5° and 50.5. At the Arctic Circle at the summer solstice, the sun's angle of incidence is never higher than 47.9° and at the northernmost point of Norway at the latitude of 71.2° N the maximum angle of incidence is 42.1°. This angle of solar incidence corresponds with an angle of incidence at the equinoxes at the latitude of 48°. For latitudes 71° N and 58° N, the ratio of the spread of solar radiation over a horizontal area is 0.81 (=sin 42.5°/sin 55.5°) at the summer solstice and appears to indicate only minor differences in the influx of solar energy. However, the ratio decreases markedly during the equinoxes to 0.61 and diminishes to zero in the winter months. When considering the influence of day length and the angle of incidence, the differences in the solar irradiation become even more pronounced. On 21 December, 21 March (23 Sept) and 21 June, the solar irradiation at 70° N attains 0 W m^{-2}, 149 W m^{-2} (147 W m^{-2}) and 492 W m^{-2} whilst at 50° N solar irradiation increases from 86 W m^{-2}, 280 W m^{-2} (276 W m^{-2}) to 482 W m^{-2}. Over a year, the summed solar energy attains 1768 kWh m^{-2} at 70° N, 2123 kWh m^{-2} at 60° N and 2559 kWh m^{-2} at 50° N (Weischet and Endlicher 2012). Thus, as a result of the northern location the intensity of solar radiation in Norway is never very high but in summer the sum of global radiation even in Northern Norway nearly equals the figure for 50° N.

However, on average solar intensity decreases towards the north as a function of decreasing angle of incidence, being modified by cloudiness, aerosols and relief. In the summer months, the duration of solar radiation increases northward and part of the energy is consumed for melting snow and ice and evaporating water. Although the astronomic sunshine hours increase from 4500 h a^{-1} at 62° N to 4600 h a^{-1} at 70° N, there is a decrease in the average daily global radiation of 8280 kJ m^{-2} at 63° N to about 6400 kJ m^{-2} at 69° N (Johanessen 1970). Accordingly, the energy deficit increases northwards from 1670 MJ m^{-2} at a latitude of 55° N to about 2800 MJ m^{-2} at a latitude of 70° N (Johanessen 1970: 25). The northward increase in the annual energy deficit is partly compensated by the release of latent heat by condensation and advection of air from southerly latitudes and by heat from the North Atlantic Ocean (Norseth 1987). However, the simple radiation tendencies are modified by several factors. The solar irradiance not only decreases northwards but also towards the west (Skartveit and Olseth 1986). As Norway is located in the zone of westerly winds, the increase in elevation along the Norwegian coast forces the rise of moist air masses. This results in an increase in cloudiness and precipitation along the coastal areas, which is indicated, in particular, at the

Fig. 2.1 Norway—general location and elevation; additionally weather station sites and other places referred to in the text (*Data source* NMI 2018a)

south-western coast of Norway, where the mean annual irradiance is markedly decreased by the high cloudiness (Skartveit and Olseth 1986).

The actual input of solar radiation (SR) at surface level is controlled by real combination of the factors mentioned above. As direct measurements of SR were generally rare during the last regular climatological normal period (1961–1990), data of sunshine duration (SD) were used here to analyse regional differences of such combinations. Nevertheless, these data give additional information for geomorphological processes as SD is defined as solar irradiance (SI) with an at least considerable intensity (minimum 120 W/m^2) and relative to a surface perpendicular to the beam of sunlight (WMO 2008). As horizontal surfaces, which are the reference for other SR measurements, are an exception in mountainous areas, SD data may give a realistic impression on the possible frequency of considerable solar energy input on mountain surfaces.

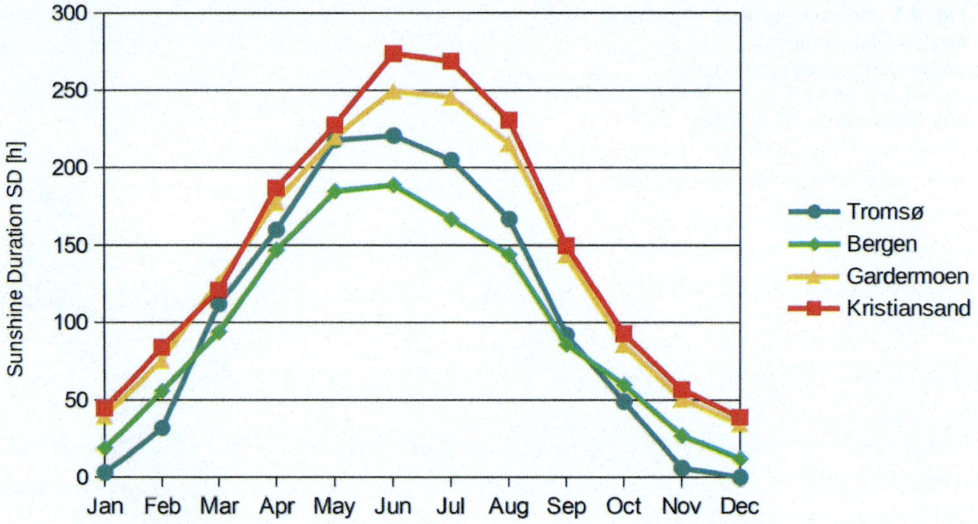

Fig. 2.2 Monthly sunshine duration at four Norwegian stations 1961–1990 (*Data source* NMI 2018b)

A general increase in sunshine duration according to latitude can be seen in Fig. 2.2. The monthly values of SD for the Southern Norwegian places Kristiansand (station Kjevik, 64° 73′ N; station locations see Fig. 2.1) and Gardermoen (66° 48′ N) are generally higher than for Tromsø (77° 19′ N; all station data from NMI 2018b). In the months around winter solstice (November, December and January), the places north of the arctic circle virtually receive no sunshine and only diffuse sky radiation according to the astronomical conditions. In the months of March to May, the SD values for North Norwegian Tromsø nearly equal those of Kristiansand (about 1400 km southward) or Gardermoen. Higher SD values in the north than in the south could be expected from the long potential sunshine duration in the Nordic summer (Tikkanen 2005) but are not recorded on an average base.

Differences between potential and actual SD are also due to the effects of topographic situation. A comparison of SD values for Bergen (Bergen-Florida, 67° 20′ N) and Gardermoen at nearly the same latitude along the relatively small W-E distance of 320 km shows differences comparable to those between Tromsø and Kristiansand. In western Norwegian Bergen, frequent precipitation—as discussed above—leads to increased cloudiness. Additionally, the station Bergen-Florida is situated close to the Byfjord at 45 m a.s.l. in a U-shaped glacial valley while there are surrounding mountains up to 673 m a.s.l. at a distance of 5 km, which leads to substantial horizon limitation and, thus, reduced actual sunshine duration (see Sect. 2.6 on Regional and topoclimates).

Solar radiation (SR) input in Norway is less dependent on anthropogenic aerosol as Northern Europe generally has lower particle concentrations with further decreasing concentrations at higher latitudes compared to Central Europe (Asmi et al. 2011). Nevertheless, the relatively low sun position in arctic summer causes long pathways of light through the atmosphere and, thus, considerable extinction even in air with lower aerosol content. Although aerosol contents in Norway are less than in more continental or highly industrialized regions, measurements indicate a general trend of decreasing irradiation; between 1950 and 2003 the annual sum of global radiation at a rural site near the Oslofjord is found to be reduced by 2.5% or 3.1% per decade (Grimenes and Thue-Hansen 2006).

2.3 Temperature

The general position of the Scandinavian Peninsula in relation to the Westerlies leads to very frequent advection of maritime air masses. Additionally, these air masses are strongly influenced by the temperature anomaly caused by the North Atlantic Current (Norseth 1987). 'Much of Norway's coastline lies within the Arctic region, but almost all of it remains free of ice and snow throughout the winter' (Alcamo and Olesen 2012).

This ocean current controls the climate of Norway as the oceanic heat content which is carried by the Gulf Stream, as the North Atlantic Current is also called, and its poleward extension, the Norwegian current. The North Atlantic current is part of the Atlantic Ocean's thermohaline circulation (THC). In this circulation, the heat of warm and saline Atlantic water is removed on its way to the Nordic seas by heat losses to the atmosphere and by freshwater inputs. The major drivers of this circulation pattern are differences in density resulting from salt content and temperature of the water (Blindheim 2004). In contrast to surface ocean currents, which are set in motion by winds, the THC provides an effective mechanism for exchanging substantial amounts of deep water across the equator which is also associated with various coupling mechanisms between sea surface waters and deep waters over both hemispheres (Wefer and Berger 2001). The THC in the Atlantic Ocean encompasses the production of the very dense Antarctic Bottom Water in Antarctic area, the travel across the equator, the formation of the warm and highly saline water of the Gulf Stream, the development of the North Atlantic Deep Water in the Nordic seas, including its southward flow, and the formation and circulation of various surface and intermediate water currents.

The Atlantic Bottom Water is formed in Nordic Seas. The Nordic Seas comprise the Norwegian Sea with the two deep basins, the Norwegian Basin and the Lofoten Basin, the Greenland Sea, and the Iceland Sea (Eldevik and Nilsen 2013). The basins are separated from the North Atlantic by the tectonic Greenland-Scotland Ridge. In the Nordic seas, the water temperature of the North Atlantic Current decreases and the highly saline water sinks down to the bottom. However, the southward travel of the bottom water is delayed by the barrier of Greenland-Scotland Ridge which limits the exchange of deep water as the water has to accumulate at the bottom until its thickness enables a flow over the ridge southwards into the northern Atlantic Ocean (Mauritzen 1996; Loeng and Drinkwater 2007). Estimates of the outflow of the Atlantic Bottom Water range from 15 to 20 Sv (Wefer and Berger 2001; SV = Sverdrup; 1 Sv = 10^6 m^{-3} s^{-1}). The overturning circulation resulting from the inflow of Atlantic water is compensated by fresh polar water at the surface. This water forms the Arctic surface layer above the Atlantic water and the main outflow and freshwater source for the Atlantic Ocean (Dickson et al. 2008; Isachsen et al. 2007). An additional source of surface water is provided by the Norwegian Coastal Current. This current is a continuation of the Baltic Sea outflow through the Skagerrak and carries low-salinity waters along the Norwegian coast to the Barents Sea (Christensen et al. 2018). The Norwegian Coastal Current forms the upper layer of the Norwegian Sea above and alongside the higher saline water of the Atlantic Ocean (Robert and Bousquet 2018).

The influences of the THC on the climatic conditions in Norway are indicated in the warm surface water entering the European Nordic Sea with a temperature of more than 8 °C (Blindheim 1989). With respect to the prevailing westerly winds, Norway lies downwind of the ocean having annual air temperatures nearly everywhere below 8 °C and the heat transferred from the ocean to the air is carried by the westerly winds. Figures on the heat loss of the North Atlantic Current in the Nordic Seas depend largely on the precise calculation of volume flux of the current. Mosby (1974) estimated that about 195.5 TW of the heat of the Nordic Sea are lost by evaporation, whilst 87.8 TW are lost by convection. More recent studies of the heat loss in the Nordic Seas range from 220 to 250 TW (Simonsen and Haugan 1996). In a detailed analysis of the mean volume and heat fluxes of the Atlantic water in the Svinøy section running north-westward from the Norwegian coast at a latitude of 62° N, Skagseth et al. (2008) estimated the volume flux and the heat loss of 4.3 Sv and 126 TW.

The North Atlantic current moderates the annual range of temperatures particularly along the coastal areas of Norway. During the winter months, the air temperature is usually lower than the water temperature of the North Atlantic Current which results in higher evaporation rates in the winter than in spring and summer and moist air masses are carried with the westerly winds. In summer and spring, the air masses are warmer than the ocean water, and there is a reverse effect, when the relatively warm air is advected over the cooler ocean water as the air transfers heat to ocean water and becomes cooler (Tikkanen 2005). The role of the North Atlantic current is indicated in the relatively narrow annual range of temperatures characterized by relatively mild winter temperatures and moderate summer temperatures and in the increase in seasonal differences in temperature from the coastal areas in the west towards the eastern hinterland (Tikkanen 2005). Direct consequences of the warm water of the North Atlantic Current are the ice-free harbours in the winter along the whole Norwegian coast which stand in striking contrast to the harbours in the northern Baltic Sea in the east.

The frequent advection of relatively warm and humid air masses towards the Norwegian mainland and its effects lead to considerable temperature variations (Norseth 1987). Especially the distance from the sea is an important factor, which modifies the general trend of increasing temperature level from north to south, but also location and altitude of mountain ranges have considerable effects. This can be seen from the station data of average temperatures in Fig. 2.3 which pairwise represent extreme places for North, Mid- and Southern Norway each and additionally a station near the southernmost point of Norway.

In the north, Finnmarksvidda (station Kautokeino, 307 m a.s.l.), an inland plateau about 150 km from the open sea, is 5.1 °C colder than Tromsø (100 m a.s.l.) on an island in the Tromsøy Sound about 30 km from open sea (3.8 °C after correction by an average lapse rate of 0.65 °C/100 m). The elevated inland plateau of Dovrefjell in central Norway (station Fokstugu, 973 m a.s.l., about 130 km from open sea) is 5.4 °C colder but 0.8 °C warmer—if altitude-corrected in the same way—than Værnes (12 m a.s.l.; both stations at about 63° N) at the open Trondheimsfjord with its undulating landscape. This indicates that the mountain station is too 'warm' in view of its altitude which can be explained by virtually lower lapse rates in maritime climates due to release of latent heat, which is also reported from other maritime mountain regions (Sattler et al. 2016). There are also seasonal variations; Skre (1971) calculates average

Fig. 2.3 Average annual air temperatures at Norwegian stations 1961–1990 (*Data source* NMI 2018b)

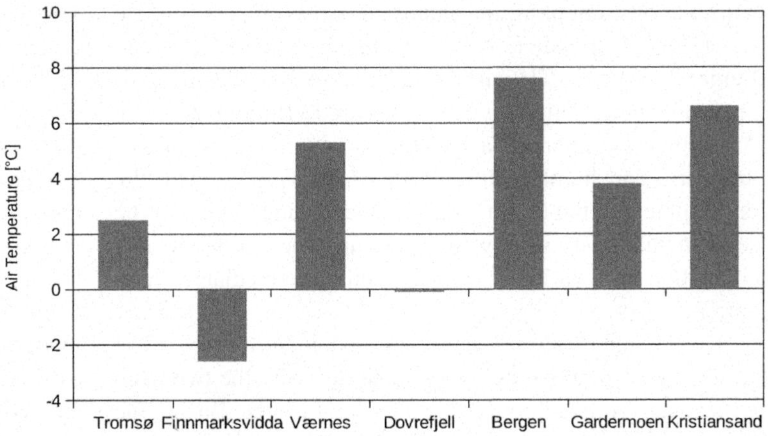

lapse rates between 0.4 °C/100 m (January) and 0.7 °C/100 m (April and July), the latter being the typical average lapse rate as a combination of times with dry and moist lapse rate and the former certainly including days with temperature inversions. The Norwegian west coast, here represented by the station of Bergen-Florida at 12 m a.s.l., which is situated at a fjord not too far from the open North Sea, shows the highest average temperatures and a difference of 3.8 °C (corrected 2.6 °C) to Gardermoen, an inland station at 202 m a.s.l. about 320 km from Bergen (both at about 60° N). This smaller corrected difference—compared to Tromsø-Kautokeino—does not match the larger distance to the coast; here, considerable foehn effects by the Southern Norwegian mountain range are concerned (Tikkanen 2005). Kristiansand with the southernmost weather station with long time series (at 58° N) is colder than Bergen although also situated not far from the sea and further southward. This region is not exposed to the open North Atlantic and continuous westerly winds but it is situated at the Skagerak which is a more continentally influenced marginal sea.

The coldest winter temperature in Norway was recorded in central Finnmark (Karasjok, −51.4 °C; Tikkanen 2005); inland Northern Norway shows extremely low minimum temperatures during nearly all months of the year, but in summer the lowest temperatures occur in the Jotunheimen Mountains in south-western Norway (Fannaråki weather station). The absolute minimum temperatures in the different regions are irregularly distributed with a wide range, which is due to distance to water bodies and more or less intense topoclimate effects of nocturnal cold air drainage. The highest summer extreme temperatures were generally measured in Eastern Norway, the region east of the southern Scandinavian Mountain Range at lower altitudes and greater distance to the northern Atlantic (maximum temperature T_{max} + 35.6 °C, Nesbyen, Buskerud region; Tikkanen 2005). But the range of summer maximum temperatures at low altitude stations in the Norwegian regions is very small, as the lowest maximum temperature is reported from the South Agder region with +32.6 °C and even Finnmark had an extreme heat event with up to 34.3 °C. Relatively high temperatures in the other seasons were reported from stations in various regions, especially where exposed to intense foehn effects normally on the east side of the Scandinavian Mountain Range (Tikkanen 2005), but in fact often also in the opposite direction as can be seen from extreme high winter temperature events at the west coast (18.9 °C on 23.2.1990 in Sunndalsøra; Tikkanen 2005).

At all stations the warmest month is July and the coldest is January (Fig. 2.4). All stations tend to an additional phase lag in their annual cycle with higher temperatures in August compared to June which has been described by Skre (1971), with the exception of the more continental stations Finnmarksvidda and Gardermoen. The annual temperature cycle as given by monthly air temperatures at coastal stations shows maritime effects with small amplitudes (Tikkanen 2005). The differences between coastal and inland stations in northern (Tromsø and Finnmarksvidda) and Southern Norway (Bergen and Gardermoen) are very small and even inverse in summer (−0.6 and −0.9 °C) but marked in winter (11.6 and 8.5 °C). For Værnes and Dovrefjell in central Norway, the differences do not fluctuate strongly throughout the year (5.6 °C in winter and 4.1 °C in summer). Kristiansand is the warmest of these stations in summer; in spring, it is hardly warmer than central-Norwegian Værnes. The negative summer differences between coast and inland are the result of low seawater temperature in the North Atlantic and marked warming of Scandinavian inland areas in summer, whereas shallow or calm fjord water—like the Trondheimsfjord near Værnes—may warm up at least superficially.

The number of freeze–thaw days (Fig. 2.5, left; days with $T_{max} > 0$ °C and $T_{min} < 0$ °C; calculated from NMI 2018b) shows no clear tendency, and the variation is small. Along the coast, Bergen (49d, all values rounded down) has the least number of freeze–thaw days but to the south the number increases (Kristiansand, 85d) as well as to the

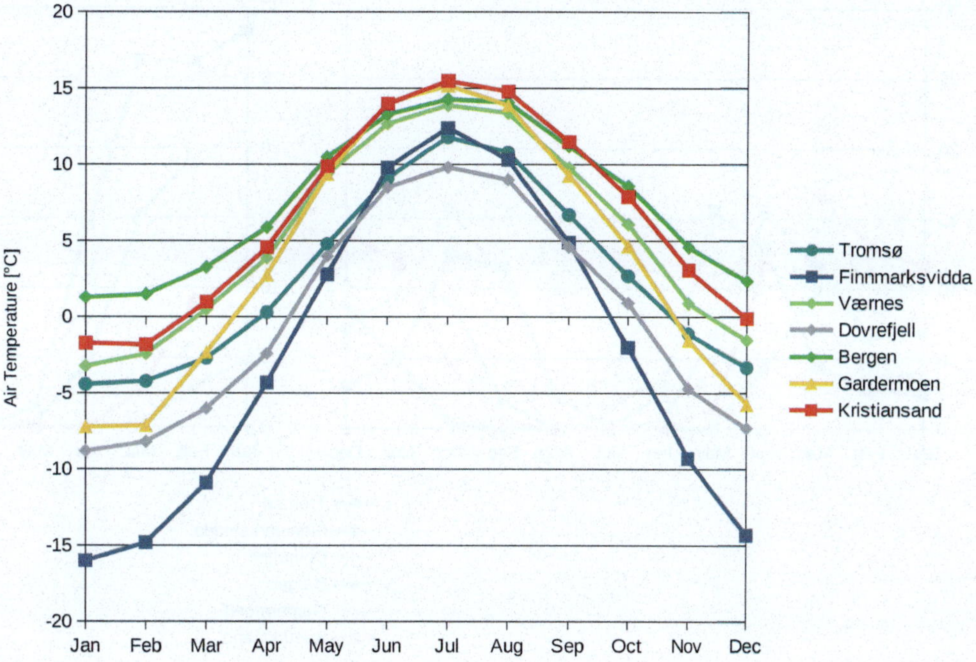

Fig. 2.4 Monthly average air temperatures at Norwegian weather stations 1961–1990 (*Data source* NMI 2018b)

north (Tromsø, 75d). The inland stations are also not systematically different from the coastal counterpart as Finnmarksvidda (75d) does not differ from Tromsø but Gardermoen (93d) does differ from Bergen. There is even no simple systematic difference to expect as the daily temperature cycle results from complex processes (Ketzler 2014) and its range intersects the freezing temperature of water in a complex way during the year.

In contrast to this, the number of ice days at the stations (Fig. 2.5, right; days with $T_{max} < 0$ °C and $T_{min} < 0$ °C; calculated from NMI 2018b) differs considerably and systematically at the stations. Ice days occur rarely along the South Norwegian coast (Bergen: 12d, Kristiansand: 33d, Værnes: 43d) but more frequently at southern inland station Gardermoen (79d). Northern coastal station Tromsø has more ice days (94d) than Bergen but much less than inland station Finnmarksvidda (159d). And—according to the tendencies in annual average temperature—Dovrefjell (133d) does not differ much from Finnmarksvidda but clearly from not very distant Værnes.

The annual cycles of freeze–thaw days for the stations (Fig. 2.6, left; calculated from NMI 2018b) show relatively small differences in amplitudes but a remarkable bimodal distribution and different compression. Finnmarksvidda (<1d) and Dovrefjell (1d) are the only stations with freeze–thaw days in all months; at the same time, they have the least number during winter. The two local maxima are closest to the summer also at these stations, for Bergen, Værnes and Kristiansand, they are closest to the winter months and they also have the greatest number of freeze–thaw days; for the other stations, the characteristics of the annual cycle are in between.

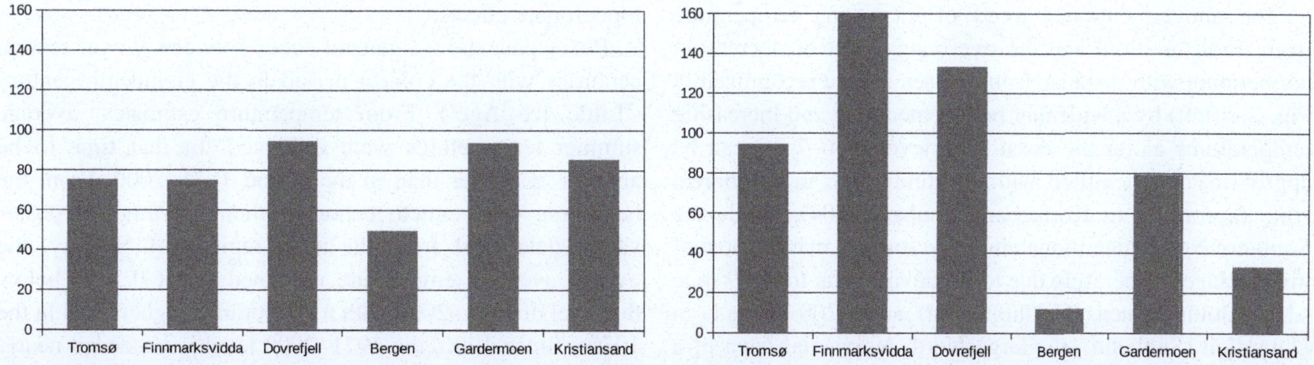

Fig. 2.5 Average annual numbers of freeze–thaw days (left) and ice days (right) at Norwegian weather stations 1961–1990 (calculated from NMI 2018b)

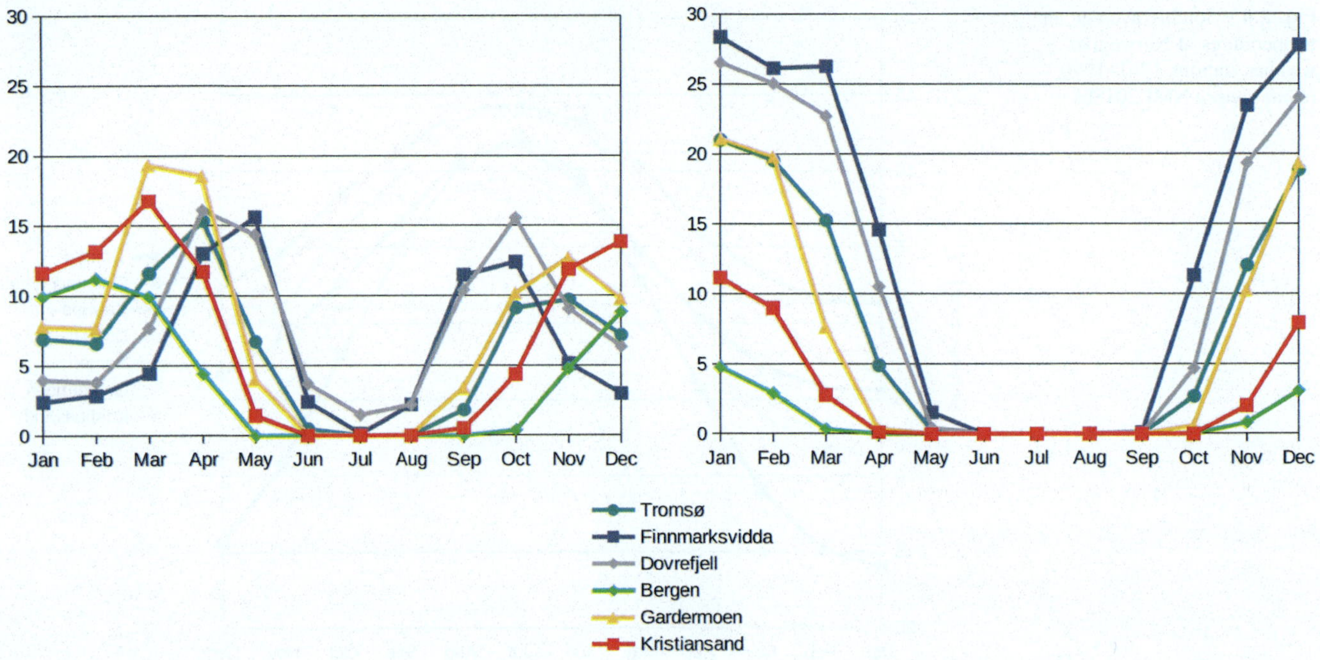

Fig. 2.6 Annual cycle of the monthly number of freeze–thaw days (left) and of ice days (right) at Norwegian weather stations 1961–1990 (*Data source* NMI 2018b)

The bimodal distribution of the freeze–thaw days is caused by the superimposed distribution of the ice days with its maximum in winter (Fig. 2.6, right; calculated from NMI 2018b). Besides the fact, that ice days do not occur in climatological summer at any of these stations, their annual distribution generally resembles the annual average. But it is to mention that the number of ice days in South Norwegian Gardermoen in January (19d) equals the number in North Norwegian Tromsø.

On the basis of climate data from geostatistical modelling by NMI (Lussana 2018) the average air temperature distribution in Norway and its variation can be calculated. In Fig. 2.7, the average air temperature (left) and the average absolute year-to-year temperature deviation for the period 1961–1990 are given on the base of these 1 km resolution data.

The above-mentioned trend of increasing temperature from north to south and the overlying effect of decreasing temperature with distance from the sea can be recognized in Fig. 2.7 (left) by a widening belt of moderate and increasing temperatures along the coast. The level of 6–7 °C can be approximately identified with the timberline in the boreal zone (according to Körner and Paulsen 2004). While the Southern Norwegian mountain range forms a marked area of low average temperature due to its elevation up to 2469 m a.s.l. in Jotunheimen (Galdhøppigen) and 2209 m a.s.l. in Dovrefjell (Snøhetta), the large North Norwegian area of a similar low-temperature level is situated in the low mountain area of Finnmarksvidda of only up to 600 m a.s.l. The undulating landscape around Central-Norwegian Trondheimsfjord and its hinterland along Stjørdalen (near Værnes) towards the, here, low-lying Scandinavian watershed situated far inland forms a region with a generally very moderate temperature level (see Fig. 2.8).

The average absolute year-to-year temperature deviation (Fig. 2.7, right) forms three different patterns: a more diffuse areal structure in the South-Eastern, Central- and North Norwegian inland, a more distinct microstructure mapping valleys and separated mountains, both showing a low absolute average temperature deviation, and also a more diffuse structure in some coast regions with small deviations. The former apparently represents regions of frequent high continental (probably in summer), the latter of constant maritime influence on the temperature regime. The areas of high deviation in microstructure indicate a role of special topoclimate effects.

Proxy data show a general cooling tendency over several centuries with the coldest period in the eighteenth century ('Little Ice Age'). From temperature estimates, average summer temperatures were estimated for that time to be about 1 °C lower than in the period 1971–2000. From the beginning of twentieth century, when instrumental observation data exist from the main regions of Norway, the overall average temperature remained about 0.5 °C below the level of 1971–2000 with a temporarily higher level in the 1930s. Since 1990, the 1971–2000 level is exceeded nearly constantly (Hanssen-Bauer et al. 2017). On the basis of data for those stations used in this paragraph which exist—with

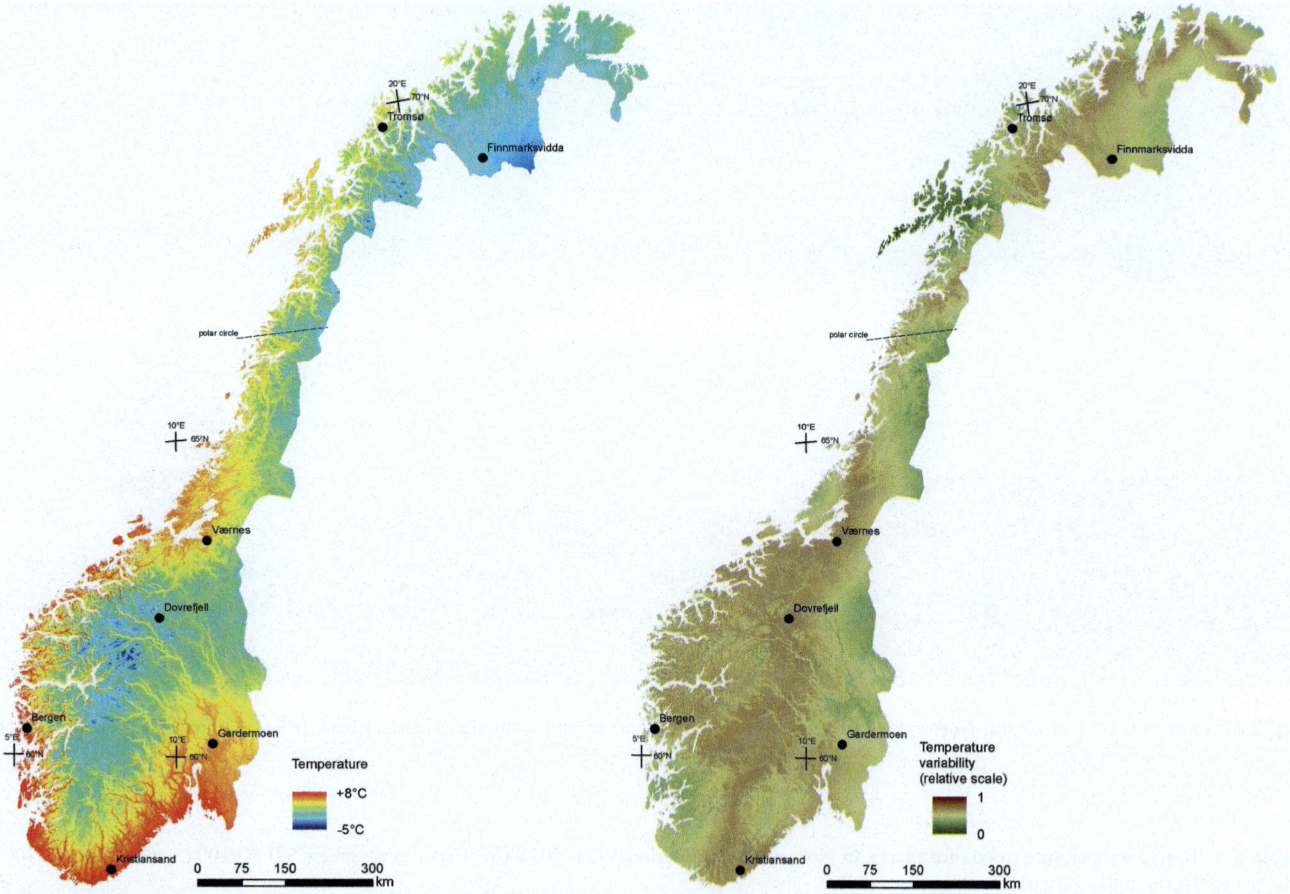

Fig. 2.7 Average air temperature (left) and the average absolute year-to-year temperature deviation (right) for the period 1961–1990 (*Data source* Lussana 2018)

partly minor changes of station location—for the last nearly 100 years, linear trends from regression analysis were computed: Kristiansand, Bergen, Dovrefjell (Fokstugu/Fokstua), Tromsø and Finnmarksvidda (data from NMI 2018b; a homogenisation for the whole period and all stations, as applied for parts and some stations in Lundstad and Tveito 2016, was not available).

As shown in Table 2.1, all of the time series have a positive trend on an annual base (average: 1.0 °C/100a) with the inland and more elevated places Dovrefjell and Finnmarksvidda both having the highest temperature increase of 1.6 °C/100a. All seasons have an average trend which is positive and in all seasons—except winter—all single stations also, with spring showing the strongest trend (1.5 °C/100a) and the weakest trend in summer (0.5 °C/100a). Moberg et al. (2005) cite a general trend for Fennoscandia since late nineteenth century of 0.6 °C. Winter is the only season with marked contrary trends: while there is a negative or weak positive trend at places at the coast, the inland and elevated places Dovrefjell and Finnmarksvidda have experienced a distinct warming. Here, the winter temperatures have increased with more than 1.2 °C since the middle of the last normal period 1961–1990.

A warming trend of less than 1.0 °C/100a (annual mean temperature) at coastal stations which is below the global average may be caused by the special sensible to latent energy flux relation over the oceans (Byrne and O'Gorman 2018) and may lead to a delay of global warming effects in these regions. However, the inland areas like Dovrefjell and Finnmarksvidda show a marked temperature increase, which is observed especially for the lasts decades in inland Scandinavia (Alcamo and Olesen 2012). This trend has severe consequences as these stations are situated in potential permafrost areas with an annual temperature level of 0 °C or below (marked by the blue colours in Fig. 2.7, left). An above-average warming (Table 2.1) and an increased frequency of extreme summer heat events (see Fig. 2.9) are expected to accelerate permafrost thaw with—inter alia—the effect of increased slope instability (e.g. Blikra and Christiansen 2014; Harbitz et al. 2014). In the period 1981–2010 in about 6% of mainland Norway's land area permafrost existed and, as arose from intermediate

Fig. 2.8 Summer aspect of central-Norwegian landscape towards the Skarvan and Roltdalen National Park (*Photo* Gunnar Ketzler, 26/7/2018)

Table 2.1 Linear temperature trends for places in Norway for the period 1921–2018 (in °C per year; green: <0 °C/100a, yellow: <1 °C/100a, orange: <2°/100a, red: ≥2°C/100a

	Spring	Summer	Autumn	Winter	Year
Kristiansand	0.016	0.01	0.004	-0.006	0.004
Bergen	0.009	0.008	0.011	0.008	0.009
Dovrefjell	0.017	0.002	0.013	0.029	0.016
Tromsö	0.016	0.001	0.007	-0.001	0.006
Finnmarksvid	0.017	0.004	0.013	0.029	0.016
Average	15	3.75	9.6	11.8	10.2

investigations, it already partly thawed and disappeared in the years after (Hanssen-Bauer et al. 2017).

2.4 Precipitation

If there is a clear general precipitation gradient in Norway, then it is directed from the coast to the inland areas (Fig. 2.10). In the North, the recordings show two and a half times as much annual precipitation 1961–1990 in maritime Tromsø (1031 mm) than in inland Finnmarksvidda (Kautokeino: 376 mm). In central and West Norway, the relation is similar: Værnes has twice as much precipitation (892 mm) as Dovrefjell (Fokstugu: 435 mm) and Bergen (2250 mm) two and a half times as much as the inland station Gardermoen (862 mm). And even in the coastal place Kristiansand, the southernmost station used in this study, the amount of precipitation is considerable (1299 mm) although, here, the westerlies are of less importance and only more southerly winds can contribute to this effect. These data also point to a certain gradient from north to south which is to be expected as a result of an increasing temperature level, but the complex relief situation is apparently of greater and clearer effect on precipitation.

The monthly distribution (Fig. 2.11) shows two distinct precipitation regimes: while the annual courses of precipitation at Finnmarksvidda and Dovrefjell have a marked maximum in summer (July) with a fast decrease in autumn, the precipitation maximum at the other stations is recorded in September or October (November). The latter can be

Fig. 2.9 Patterned grounds on Finnmarksvidda, Finnmark, 69.6° N, as a result of arctic climate conditions (Köppen type ET). The sky above with typical 'fair-weather clouds' of temperate environments (cumulus humilis, cu hu), indicating relatively high temperatures (the clouds completely consist of water droplets) and a marked positive energy balance at ground level recognizable by intense convection (summer heat period in 2018, *Photo* Gunnar Ketzler, 31/7/2018). An increased frequency of such summer heat events is expected to accelerate permafrost thawing and related geomorphological processes

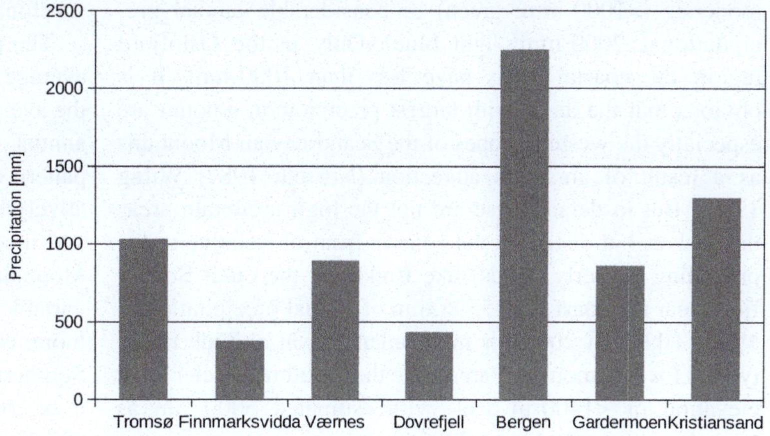

Fig. 2.10 Mean annual precipitation [mm] for Norwegian stations, precipitation normals 1961–1990 (*Data source* NMI 2018b)

associated with increased activity of the polar front system in autumn; a positive NAO phase is generally expected to correspond with higher precipitation normals in Scandinavia, especially along the western mountain slopes (Wibig 1999; Quante et al. 2016). The former is a typical inland regime with effects of convective precipitation events (Norseth 1987), which can be seen in inland mountain areas as well. All series have their minimum in April or May with Gardermoen and Værnes showing the smallest precipitation amounts. This is the inverse effect of the strong polar front system in autumn leading to a weaker activity of the westerlies in Norway in spring and more frequent stable anticyclones over Scandinavia as a ridge of an anticyclone in Russia (Johanessen 1970). The consequence is a dry and sunny weather type with southern or south-easterly winds in Norway including foehn effects on the west side of the Scandes.

The geographical distribution of precipitation in detail is given by results of geostatistical modelling using station data for the period 1961–1990 (Crespi et al. 2018; Fig. 2.12, left). Large areas along the South Norwegian west coast and along southern parts of the North Norwegian coast show remarkable amounts of annual precipitation (>3000 mm; dark blue). In some regions, the model data indicate results of >4000 mm—also in the northern Svartisen–Saltfjell region—and in a small area north of Bergen with the Ålfotbre Glacier even >5000mm (west of Jostedalsbre Glacier, the greatest plateau glacier in continental Europe and the

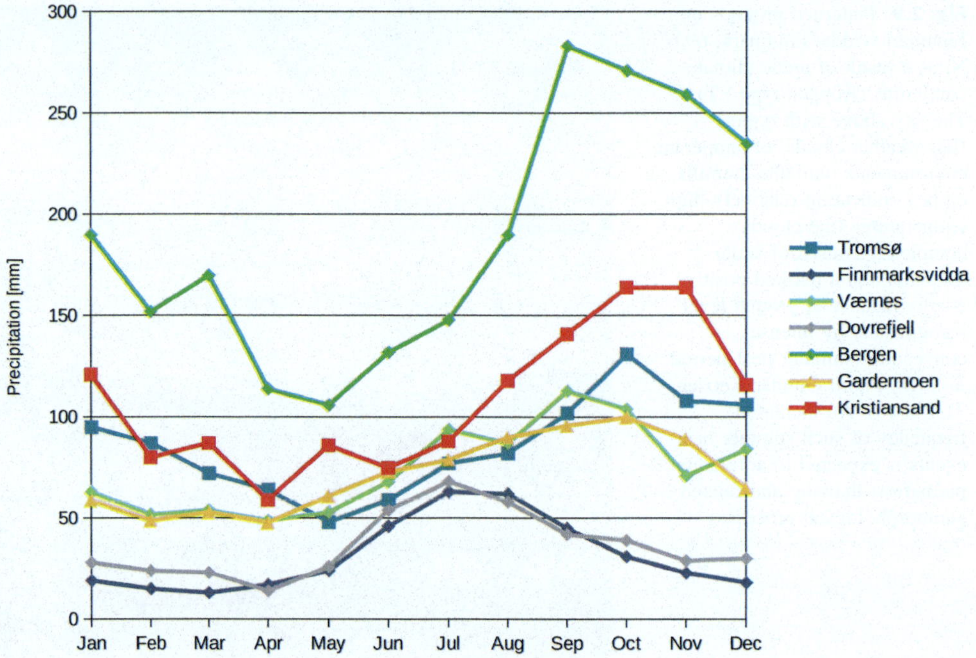

Fig. 2.11 Mean monthly precipitation [mm] for Norwegian weather stations, precipitation normals 1961–1990 (*Data source* NMI 2018b)

Jotunheimen Mountains with the highest summits of Scandinavia). Almost all coastal regions are marked as having moderate (>1000 mm; green) or considerable annual precipitation (>2000 mm; light blue). Only in the Oslofjord region the coastal areas have less than 1000 mm. It is obvious that the areas with largest precipitation amount are especially the western slopes of the Scandinavian Mountains as a result of air mass advection (Norseth 1987; Wibig 1999). But in detail, these are not the high mountain areas themselves but distinctly the zone in front of (relative to the prevailing westerly winds) like Brekke at the outer Sognefjord near the coast with 5596 mm of annual precipitation in 1990 (Tikkanen 2005) or more inland at an altitude rather typical for low mountain areas like the Ålfotbre Glacier at an elevation of <1400 m a.s.l. with estimated 5000 mm as annual average (Tikkanen 2005).

The locations of extreme daily precipitation events are concentrated in three areas at the windward coastal foreland of the Scandes, the distribution of which not showing exactly the same pattern as the annual extremes: the Norwegian south-west near Stavanger, the west side of the Jostedalsbre Glacier/Jotunheimen Mountains and the west side of the Saltfjell Mountains with the Svartisen Glacier. In Fig. 2.12 (left), the 10 most intensive events 1950–2008 are marked (Lupikasza 2016).

The inland regions are widespread characterized by relatively small amounts of precipitation (<600 mm; yellow). In Finnmark, greater areas have less than 500 mm. In some areas around Dovrefjell, an average annual precipitation of even 300 mm is not exceeded. Here, summers may be dry and agriculture, for which the greater valleys on the leeward side of the Jotunheimen Mountains with a foehn situation are very favourable due to relatively high temperatures and long sunshine duration is depending on irrigation.

The intra-annual variability of precipitation calculated as average absolute deviation of all monthly values relative to the average monthly precipitation on the basis of the average annual precipitation 1961–1990 shows a partly different pattern (calculated from NMI 2018b; Fig. 2.12; right). Relatively high values of variability (blue) show both mountain and inland areas. Along the watershed of the Scandinavian Mountains, this can be interpreted as an effect of a seasonal variable transport range of humid maritime air masses. In the more continental inland regions like the eastern part of Southern Norway or Finnmarksvidda, variability is expected to be connected to more or less intense high-pressure phases, which disconnect these areas completely from westerly winds (Johanessen 1970; Wibig 1999). Between these zones of great variability, a belt of small variability crosses great parts of the country from north to south including the Oslofjord and Trondheimsfjord region indicating a relatively even distribution of precipitation throughout the year.

The return periods (RI) for intense precipitation events (10 mm/d) in all months of the year cluster in two groups without any other apparent regional tendency as described in connection with the annual course (Fig. 2.13). Finnmarksvidda and Dovrefjell show an RI value of considerably more than 100 days, all other stations of about 10 days. For the former, this means an occurrence of these events of about twice a year, for the latter about once in a week or every second week. There is no massive difference between the data for the whole year and the months of March to

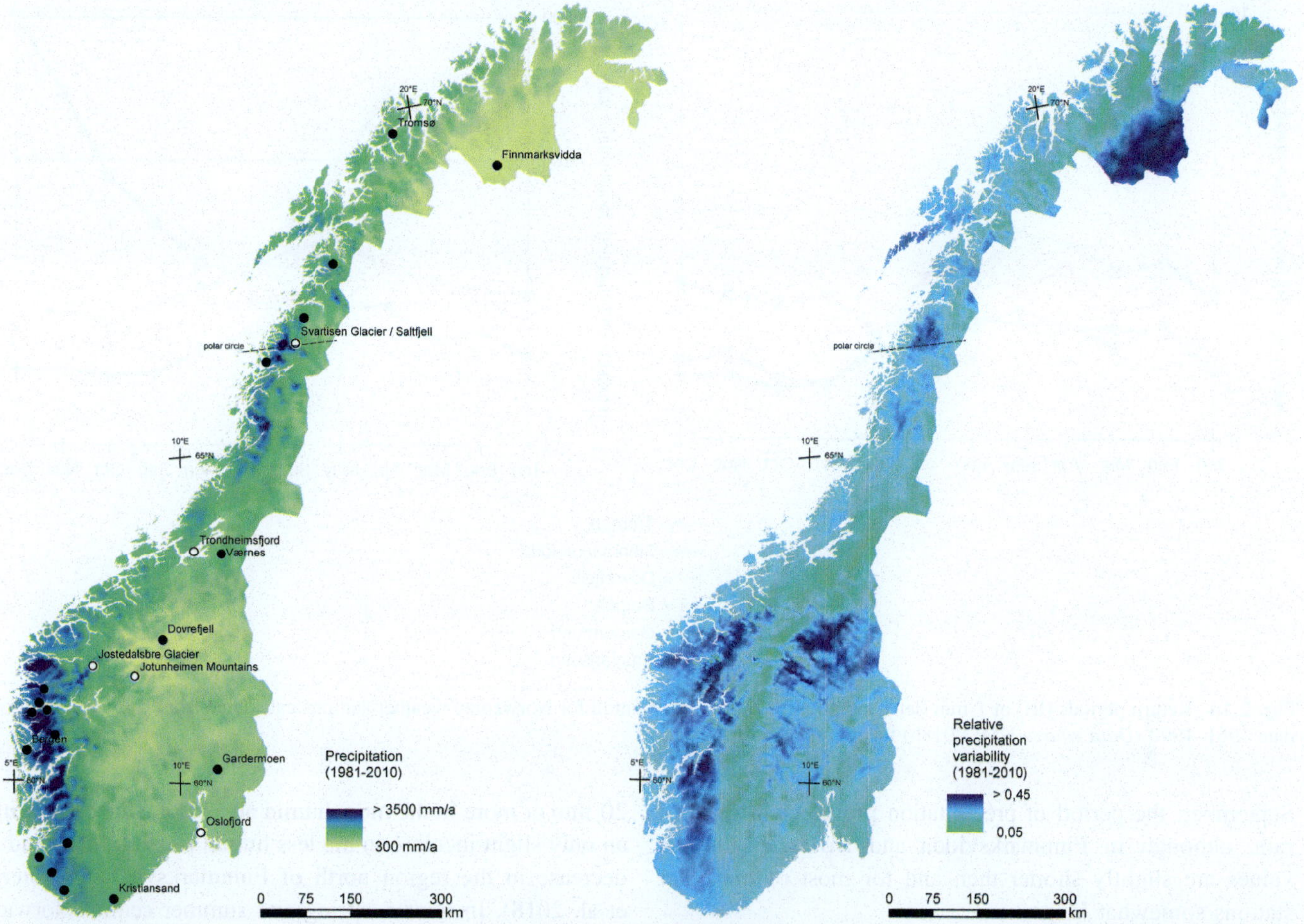

Fig. 2.12 Precipitation in Norway on the basis of station data and geostatistical modelling for the period 1961–1990 (*Data source* Crespi et al. 2018; Lussana and Crespi 2018)—average annual precipitation (left) and average relative month-to-month precipitation deviation calculated on the basis of monthly averages (right). Dots in Fig. 2.12 (left): locations of extreme daily rainfall events (after Lupikasza 2016)

Fig. 2.13 Return periods (RI) of 10 mm daily rainfall for Norwegian weather stations calculated from daily precipitation data 1961–1990 (calculated from NMI 2018b) for the whole year and the months March to November

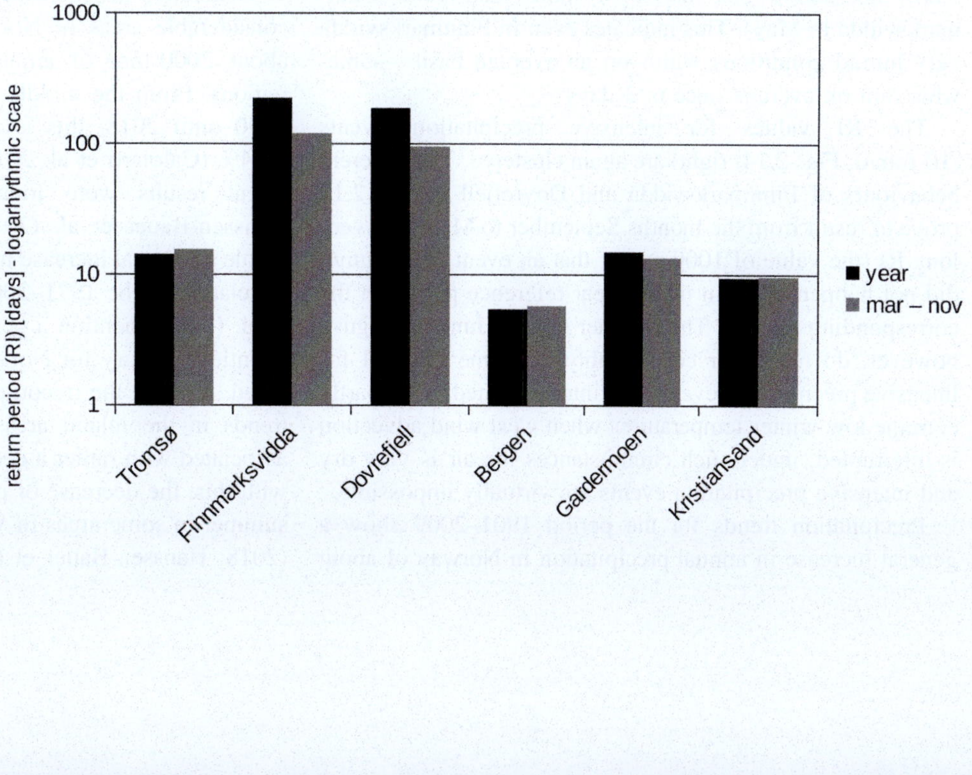

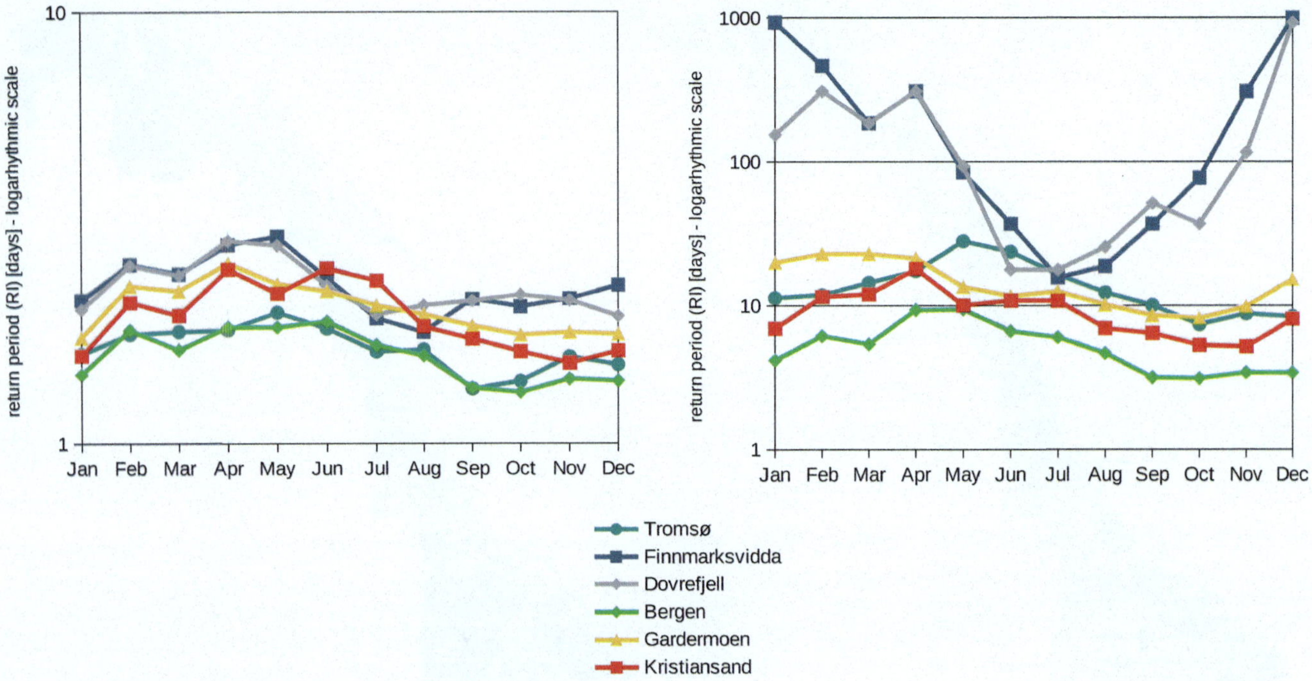

Fig. 2.14 Return periods (RI) of 1 mm (left) and 10 mm (right) daily rainfall for Norwegian weather stations calculated from daily precipitation data 1961–1990 (*Data source* NMI 2018b) for the months of the year

September, the period of precipitation prevalent in form of rain, although in Finnmarksvidda and Dovrefjell the RI values are slightly shorter then and for most of the other stations somewhat longer.

The annual cycles of RI values show some differences between stations (Fig. 2.14). The RI values for small precipitation events (1 mm/d; Fig. 2.14; left) are generally small and not apparently clustered. Although the very small values for Bergen, Værnes and Tromsø are somewhat detached (with an average frequency of such events of about 1.6d), no station has an RI of more than 3.0d (Finnmarksvidda in May). This indicates even in Finnmarksvidda very humid conditions with—on an average basis—somewhat rain on average once in 3 days.

The RI values for intensive precipitation events (10 mm/d; Fig. 2.14; right) are again clustered. The different behaviours of Finnmarksvidda and Dovrefjell in Fig. 2.13 prove to result from the months September to May with very long RI (the value of 1000 means that an event of 10 mm/d did not happen at all in the 30-year reference period in the corresponding month). The summer months June to August, however, do not differ considerably. Extreme long RI for intensive precipitation events in winter is related to the partly extreme low winter temperatures when west wind advection is interrupted; under such circumstances the air is very dry and intensive precipitation events are virtually impossible.

Precipitation trends for the period 1901–2009 show a general increase in annual precipitation in Norway of about 20 mm or more in the more humid areas along the coast with an only slight increase in the less humid (inland) areas and a decrease in the region north of Finnmarksvidda (Caloiero et al. 2018). In winter, spring and summer central Norway receives precipitation increased by 5–10 mm or more in each season, while there is a slight decline in the northernmost part of Norway and even in some places in West-Norway. In autumn, marked precipitation increases of 10 mm or more are recorded in West-Norway with a similar distribution in the rest of the country as in the other seasons.

Generally, an increase of about 20 mm—although in considerable areas of Norway—is moderate in relation to about 2000 mm of annual precipitation in these humid regions. From the middle of the last normal period 1961–1990 until 2018, this is an increase of about 87 mm or +4.4% (Caloiero et al. 2018). In other studies, slightly different results were found: for the period 1900–2014 Hanssen-Bauer et al. (2017) calculated for Norway as a whole an annual increase in precipitation of 1.6% per decade in relation to the 1971–2000 average precipitation. Moberg et al. (2005) mention a general trend of +5 to +20% in the twentieth century for Norway and Sweden. However, measured against the recorded marked positive temperature trends in the inland areas, the gains of precipitation are associated with rather a decrease of humidity. In connection with this, the decrease of precipitation in winter, spring and summer in some areas in West-Norway after Caloiero et al. (2018; Hanssen-Bauer et al. (2017) show a small increase instead in these areas) can indicate that the above-named dry and sunny weather type with southerly winds and foehn effects on the west side of the Scandes occurs more frequent now than before. Such a weather type continued in Scandinavia during whole summer 2018 and generally leads to extreme heat and drought including numerous wildland fires (Drobyshev et al. 2014).

In mainland Norway, partly huge amounts of precipitation are accumulated as snow in wintertime, where it does not immediately take part in runoff or infiltration; Tikkanen (2005) refers to a rate of 50% in Northern Fennoscandia. Hanssen-Bauer et al. (2017) note that in Southern, and partly in medium elevations of Central and Northern Norway the snow water equivalent has increased since 1931, while it has decreased below this altitude and that there is a trend towards a later snow accumulation and an earlier melt. But they found no historic trend of increased spring floods in connection with snowmelt beside an increased frequency of rain floods. Dyrrdal (2010) found a positive trend of maximum snow depth in colder regions. However, the question if winter precipitation is accumulated as snow or directly generates runoff plays a key role for present and changing sediment fluxes (e.g. Beylich et al. 2016).

Parts of the winter snow accumulation are stored in glaciers and, thus, not participating directly in the annual hydrological cycle which leads to substantial differences for the catchment systems concerned for sediment transport (Beylich and Laute 2015; Beylich et al. 2017; Chap. 7). Most of the glaciers in Norway had their maximum extent for the last 9000 years during the eighteenth century ('Little Ice Age'; Hanssen-Bauer et al. 2017). Since then, all glaciers have retreated, some of them stagnating since the 1950s or 1960s for a longer time and some few—especially outlet glaciers of Jostadalsbre plateau glacier and some other glaciers in very humid surroundings—had temporary advances with a peak around 2000 (Hanssen-Bauer et al. 2017) but are retreating again since then (see Fig. 2.15) (Chaps. 3–5). For several glaciers in humid regions, this extension phase was much more pronounced in terms of mass balance than glacier front movements indicated (UNEP 2009). Presently, all glaciers are retreating. In warm summers glacier melt is extreme; in 2018, some had extreme negative mass balances especially in Northern Norway (NVE 2019; see Fig. 2.16).

2.5 Wind

From mainland Norway's position in the westerlies, an extreme wind climate may be expected. But although the North Atlantic off the Norwegian coast often is a quite stormy sea, Norway generally has no harsh wind climate. Nevertheless, at coastal stations in Norway, considerable wind speeds are recorded more frequently than at inland stations: as an average over the year Tromsø and Bergen have about 25% of all 10 min intervals with a speed of 5 m/s or more, Kristiansand even more than 35%, while Finnmarksvidda and Gardermoen have about 10% and 15% (Fig. 2.17). These inland stations have a calm wind climate during most of the year. As the data for Tromsø, Dovrefjell and Bergen show, the winter situation along the coast or on mountains may be clearly more windy than the average over the whole year. In this regard, Tromsø is representative of the Northern Norwegian coast and other places exposed to wind in a similar way, where wind speeds of more than 5 m/s occur 50% more frequently in winter compared to the all-year average. And Tromsø—like Bergen, Kristiansand and many other coastal towns in Norway—is not even located directly at the open sea but sheltered by an offshore archipelago. Nevertheless, the highest maximum wind speed events can be expected on mountain tops (Dovrefjell). Near the open sea, high wind speeds are frequent and, if this is combined with extreme precipitation events or sea spray, considerable rainwash may occur (Fig. 2.18).

The main difference in wind climate between Northern and Southern Norway is the very windy winter in the northern coastal regions combined with a rather calm inland and less frequent windy weather in Southern Norway coastal areas combined with a smaller gradient towards the inland. But in general, for the period 1871–2008, an increase of daily maximum wind speed was observed, especially along coastal areas in Western Norway (Donat et al. 2011; Hanssen-Bauer et al. 2017 found a slight increase in higher wind speeds all over the county in the period 1961–2010). Moberg et al. (2005) discuss long-term changes of circulation patterns by historical time series of wind direction recordings which are available for a longer period than wind speed data.

2.6 Regional and Topoclimates

Some physical characteristics of Norway cause special mesoscale climate effects. These are especially the great valleys, the coastal regions with the fjords and the glaciers with their surroundings. Additionally, there are local and microclimate effects caused by special vegetation formations or surface cover. The processes behind such effects and their consequences for the climate conditions are typically well developed under calm and clear sky conditions; this occurs more frequently in areas with less precipitation and in more continental areas with less variability.

There are numerous large valleys in Norway like Gudbrandsdalen, Setesdalen or Østerdalen, which typically have a length of 100–200 km, a width of 10–20 km and a depth of 500–1000 m. These U-shaped valleys are relics of former glaciation, and they partly have lakes which are freshwater

Fig. 2.15 Bødalsbre in western Norway, an outlet glacier of Jostedalsbre plateau glacier. In the foreground: terminal moraines dating from Little Ice Age (Nesje 1992), in the background: Bødalsbre near its contemporary maximum stage. The area of completely bare bedrock and moraine around the glacier marks its maximal extension around 2000 (*Photo* Gunnar Ketzler, 20/8/2009)

fjords (they are also often named 'Fjord' in Norwegian); in many cases, they are eroded into plateau fjell areas; some are situated in alpine fjell.

One of the effects of their typical U shape is a reduction of solar radiation which partly is discussed above in connection with sunshine duration (see section 'Solar radiation'). As already described, there is a substantial horizon limitation at the bottom of U-shaped valleys which leads to reduced input of solar radiation. Sunshine duration in many Norwegian U-shaped valleys is estimated to be reduced to only 55% of the potential figure (Johanessen 1970).

Mountain breezes as another feature of valleys (Tikkanen 2005) have a different impact depending on the time of the day. Normally, daytime mountain breezes have minor effects as they surely redirect rising air warmed by radiative surface heating of slopes exposed to the sun in a direction parallel to slopes and valleys, force orographic cloud formation or induce thermal circulations between warmer and colder slopes but they do not cause massive temperature differences and rather reduce them. Entirely different from this, night-time mountain breezes lead to marked inverse lapse rates as the greater mass density of air cooled by nocturnal radiative thermal emission results in cold air accumulation in lower parts of the terrain. Typically, mountain breezes develop primarily along slopes which then converge in valleys to form slow but voluminous cold air drainage flows. Complete valleys can be filled up with cold air; the temperature differences between colder (lower) and warmer (upper) parts of a valley can exceed 10 °C (corrected for elevation) which is why cold air forming has great influence on frost distribution in the landscape. If similar frost events occur frequently, a lower treeline can be formed beside the upper tree line. However, the number of frost-change days is expected to be much greater here than outside. With wind speeds of only few meters per second, cold air drainage flow may transport air masses over a distance of some tens of kilometers per night (Mahrt 1986). Although the great Norwegian valleys have known effects on temperature, only very few topoclimatological investigations on cold air drainage have been published for this country. Besides mountain breeze or foehn effects, deep valleys tend to lower wind speeds because of decoupling from the free atmosphere winds (Troen and Lundtang Petersen 1989).

Some of the great valleys are clearly recognizable in Fig. 2.7 (left and right) because of their relatively high average temperature level and reduced year-to-year variability in a mountainous surrounding with lower temperatures and more variability. A higher average temperature level in the valleys is not completely self-evident although expectable due to lower altitude, because the night-time temperatures tend to be very low there despite adiabatic warming, and a high average temperature is, thus, only possible in combination with an above-average daytime temperature. Reduced solar radiation input as described above is another potential obstacle for higher temperatures. It is assumed that effective radiation and local cold air drainage explain 10–20% of the overall temperature variance in complex terrain (besides synoptic effects; Wolff et al. 2016).

Fig. 2.16 View of the Lyngen Alps, northern Norway, 69.4° N, in summer 2018. Some of the—mostly smaller—glaciers in northern Norway had extremely negative mass balances in the summer heat period that year with temperatures over 30 °C north of the polar circle (NVE 2019; *Photo* Gunnar Ketzler, 29/7/2018)

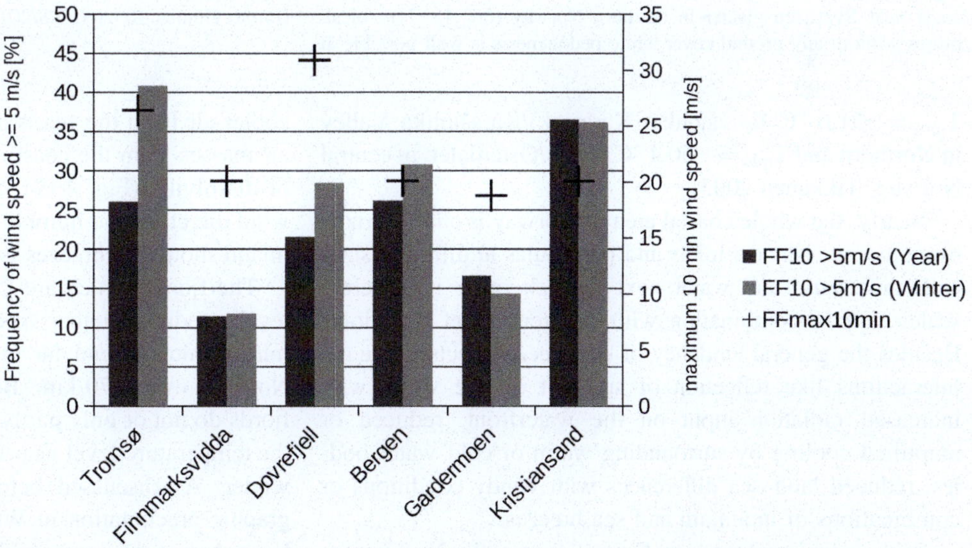

Fig. 2.17 Wind speed at Norwegian weather stations (*Data source* NMI 2018b)—frequency of wind speed events ≥ 5 m/s (year and winter, left scale) and maximum 10 min wind speed (crosses, right scale)

Finally, it is the specific combination of solar radiation input, cold air drainage, decoupling from the free atmosphere winds and actual land use which controls the resulting temperature level. Therefore, in such valleys both the highest summer and the coldest winter temperature in Norway were observed (T_{max} = 35.6 °C Nesbyen/Hallingdal in Southern Norway;

Fig. 2.18 Botnfjell Mountain a few kilometers inland of Helgeland coast near Svartisen glacier in northern Norway (66° 18′ N)—bleak ridges without any ground cover while pedogenesis is well possible at slightly sheltered places which is indicated by agricultural areas nearby (*Photo* Gunnar Ketzler, 5/8/2010)

T_{min} = −51.6 °C Kárášjohka (Karasjok)/Kárášjohka Valley in Northern and T_{min} = −50.4 °C Røros/Østerdalen in central Norway; Tikkanen 2005).

Nearly, the whole coastal area of Norway is characterized by numerous islands, inlets and peninsulas at different sizes and, thus, land and water next to each other in different scales and in combination with different relief situations. Besides the general tendency of sea breeze effects, there are interactions like reflection of sunlight on the water with increased radiation input on the waterfront, reduced or amplified cooling by surrounding warm or cold water bodies, reduced land–sea differences with windy conditions or combinations of mountain and sea breezes.

With calm weather, sea breeze events (in Norwegian: 'solgangsbris') are quite frequent along the Norwegian coast depending on season and latitude (Grønås and Sandvik 1998). During daytime, these events are driven by warmed-up land surfaces and, thus, rising warm air while cooler air from the sea is advected inland. This transport of air masses from the coast, which can cause cloud formation or dissolving (Fig. 2.19), can lead to temporary changes in wind direction and humidity conditions. A combination with inland mountain breezes is possible.

The fjords, which are U-shaped valleys partly below sea level, produce another special local climate. They represent inlets which extend far inland, the Sognefjord in Western Norway over 170 km. Because most of the Norwegian fjords do not or only partly and temporarily freeze in winter, the temperature level near the water is relatively high during winter. As discussed before, the zone of maximum orographic precipitation in Western and partly Mid-Norway is located near the coast, while the inland areas are dryer and partly sunny which is the case especially in some inner fjord regions. Here, major foehn weather situations over the Scandes caused by high pressures over the Baltic with south-easterly winds also have an impact leading to

Fig. 2.19 Sea breeze between the islands of Helgeland in the Nordland region in Northern Norway. Looking towards the open sea (behind the island in the centre of the image): shallow airflow drawn inland with water vapour condensing due to uplift over an island and cloud dissolving in the downstream flow field inland (*Photo* Gunnar Ketzler, 8/8/2010)

substantial warming or even drought phases in winter, spring and summer. But also local foehn winds may be induced with other wind directions as surrounding mountains often are high enough to cause orographic warming. Otherwise, wind speed in these inland areas is low. As a whole, in some inner fjord regions, winter temperatures remain mostly above 0 °C, spring temperatures can be extraordinarily high and summer temperatures moderate; additionally, the presence of water leads to humid conditions. This specific climate situation can be recognized by partly extensive fruit cultivation (if adequate soil conditions are given, e.g. on marine terraces and on glacial tills, at elevations lower than 150 m a. s. l.; Fezer 1966). Even around the polar circle, in the inner fjords a certain commercial cultivation of fruits and vegetable is possible (Fezer 1966; Gläßer 1993). Along humid and calm but cooler fjord and also valley parts in Southern Norway, a species-rich fern flora can be found (Fig. 2.20).

Inland freshwater fjords—as other lakes and shallow inlets in some coastal areas as well—are mostly covered with ice in winter and, therefore, interact differently with the atmosphere as the great coastal fjords, because their frozen surfaces react like land areas. In summer, the greater lakes (like Lake Mjøsa, 365 km^2, in Fig. 2.21 and some of the minor lakes) can generate local climate effects in their surroundings like flattening of the daily temperature cycle, increased humidity and sea breeze circulation.

The surroundings of the numerous glaciers in Norway can be affected by glacier winds. As the glacier surface is always colder than other surfaces at least during longer periods of the summer, a permanent down-valley wind is generated.

Typical vegetation cover in Norway is either boreal forest (see Fig. 2.8) in different varieties or heath-like dwarf-shrub vegetation (see Fig. 2.9). These plant formations both tend towards sparse plant cover and do not induce marked canopy

Fig. 2.20 Species-rich fern flora in cool and humid valley and fjord areas (Oppstrynsvatnet/Nordfjord; *Photo* Gunnar Ketzler, 23/8/2009)

climates as they surely are able to reduce wind speed near the ground but do not really decouple the understorey from the ambient atmosphere. But they often represent the environment for a species-rich and dense moss and lichen flora which forms a ground cover effectively reducing at least some atmospheric effects on the ground like SR or rain droplet impacts.

In some small areas in Southern Norway, beech forests similar to Central European forests are existing; in Western Norway, along the inner parts of fjords in central Norway and in some valleys inland mixed deciduous forest is irregularly distributed (Fig. 2.20), under more humid and colder conditions in the form of ravine forest (see Fig. 2.16). Within this vegetation type, an atmospheric canopy layer can develop with a wide range of climate effects like reduction of SR, flattening of the diurnal temperature cycle and increased humidity.

Anthropogenic climate canopies are rare in Norway as the population density is very small. Only few cities have a sufficient size for the formation of considerable urban climates (Oslo, Bergen, Trondheim, Stavanger).

In parts of Norway, the surface is not at all covered by vegetation and soil but it consists of block accumulations with a vertical extension of sometimes several meters with no finer material in between the blocks which are a product of in situ weathering or mass movements like stone runs. In these block accumulations, a special canopy climate is formed characterized by reduced solar radiation (SR) and wind speed as well as increased humidity. Depending on elevation, aspect and time of the year, this combination can lead to a higher or lower temperature level in the trapped air than in the ambient air. Cold air tends to flow down into the gaps between the blocks and replaces warm air, as it is known from caves. Thus, in the lower parts of such block accumulations, the temperature level is lower and hidden permafrost may be frequent (e.g. Blikra and Christiansen 2014). On the contrary, in upper parts of the cavities, SR may be less reduced than wind speed and then the temperature level can even be higher than at the surface; such effects in general are discussed in connection with microrefugia for the survival of certain thermophilous plants in glaciated Norway (Dobrowski 2010).

2.7 Expected Effects of Climate Change

The general climate trends expected can be outlined as follows (on the basis of the most relevant climate change scenarios for the period 2071–2100, i.e. in about 70 years and, thus, roughly two normal periods ahead; based on Hanssen-Bauer et al. 2017).

The overall temperature level in Norway will probably increase by 2.7–4.6 °C (this and all following pairs of values based on RCP4.5 and RCP8.5) from the average 1971–2000 to the period 2071–2100. Warming is supposed to be least in summer (1.5–3.8 °C) and most in winter (3.0–5.5 °C). The increase is expected to be less in Southern (2.3–2.5 °C) and greater in Northern Norway (3.7–6.2 °C). In the next 50 years, the seawater temperature is likely to rise by up to 1–2 °C in the northern Barents Sea and 3–4 °C in the Oslofjord, i.e. more than the inland temperature in the South and less in the North. The length of the growing season as a climate indicator for biological weathering in soils, which presently is less than 120 days per year in more than half of the country (Tveito et al. 2001), may extend by more than 60 days or even more than 90 days along the central and Northern Norwegian coast until the period 2071–2100 with an average increase in annual temperature of more than 6 °C (Wong et al. 2016; such environmental changes are considered to be so likely that new opportunities for agriculture are already planned following Uleberg et al. 2014, and

Fig. 2.21 Summer aspect of Lake Mjøsa in central eastern Norway—cumulus clouds (cu hu) indicate surface warming and convection over the inland areas and island Helgøya (left), while the air over the lake is cloud-free (as an effect of subsidence induced by relatively cold water) (*Photo* Gunnar Ketzler, 2/8/2010)

promising tests for the cultivation of fruit, e.g. strawberries, in Northern Norway are already being performed; Myrstad 2015).

The average annual precipitation in Norway is expected to increase by 8–18% (2071–2100 vs. 1971–2000). Heavy rain events, defined as 99.5th percentile for daily precipitation during the period 1971–2000, probably will occur 10–20% more frequent depending on RCP scenario and region. Precipitation events with a return period of 5 years are expected to increase by about 13–22%. However, the annual runoff is expected to gain no more than 5%, while summer runoff may decline by 20–30% (at the same time, soil moisture deficits may accumulate). Extreme flood events will probably have an intensity increased by more than 40% in the more humid coastal regions but in the dry inland areas of Southern Norway and Finnmarksvidda intensity may decrease by more than 40%. Only minor effects of sea-level rise (in some areas up to 0.6 m otherwise less) are expected for the Norwegian coast in the present century, as postglacial rebound will still compensate most of it.

The number of days with snow cover will probably decrease substantially in the period regarded, nearly everywhere in Norway and by up to 180 days per year. Glaciers are expected to be reduced by up to 1/3 of their area and volume, the remaining ones being much smaller and only being found at higher altitudes. From climate modelling results follows that permafrost may have completely thawed by 2050 in areas like Finnmarksvidda and that only some small permafrost areas in mountainous areas over 1800 m a.s.l. may remain by the end of the century.

No significant changes in wind speed are expected for the period 2071–2100.

Detailed descriptions and discussions of geomorphological effects of ongoing and accelerated climate change and connected changes of the cryosphere are included in the following book chapters.

References

Ahnert F (1987) An approach to the identification of morphoclimates. In: Gardiner V (ed) International geomorphology 1986. Proceedings of the 1st international conference on geomorphology [held at the University of Manchester from September 15–21, 1985], vol 2. Wiley, Chichester, pp 159–188 (A Wiley-interscience publication)

Alcamo J, Olesen JE (2012) Life in Europe under climate change. Wiley-Blackwell

Asmi A, Wiedensohler A, Laj P, Fjaeraa AM, Sellegri K, Birmili W, Weingartner E, Baltensperger U, Zdimal V, Zikova N, Putaud JP, Marinoni A, Tunved P, Hansson HC, Fiebig M, Kivekäs N, Lihavainen H, Asmi E, Ulevicius V, Aalto PP, Swietlicki E, Kristensson E, Mihalopoulos N, Kalivitis N, Kalapov I, Kiss G, de Leeuw G, Henzing B, Harrison RM, Beddows D, O'Dowd C, Jennings SG, Flentje H, Weinhold K, Meinhardt F, Ries L, Kulmala M (2011) Number size distributions and seasonality of submicron particles in Europe 2008–2009. Atmos Chem Phys 11 (11):5505–5538. https://doi.org/10.5194/acp-11-5505-2011

Beylich AA, Laute K (2015) Sediment sources, spatiotemporal variability and rates of fluvial bedload transport in glacier-connected steep mountain valleys in western Norway (Erdalen and Bødalen drainage basins). Geomorphology 228:552–567

Beylich AA, Laute K (2018) Morphoclimatic controls of contemporary chemical and mechanical denudation in a boreal-oceanic drainage basin system in central Norway (Homla drainage basin, Trøndelag). Geogr Ann A100(2):116–139. https://doi.org/10.1080/04353676.2017.1407219

Beylich AA, Dixon JC, Zwolinski Z (eds) (2016) Source-to-sink fluxes in undisturbed cold environments. Cambridge University Press, Cambridge, p 408

Beylich AA, Laute K, Storms JEA (2017) Contemporary suspended sediment dynamics within two partly glacierized mountain drainage basins in western Norway (Erdalen and Bødalen, inner Nordfjord). Geomorphology 287:126–143

Byrne MP, O'Gorman PA (2018) Trends in continental temperature and humidity directly linked to ocean warming; PNAS May 8, 2018 115(19):4863–4868; first published April 23, 2018 https://doi.org/10.1073/pnas.1722312115

Blikra LH, Christiansen HH (2014) A field-based model of permafrost-controlled rockslide deformation in northern Norway. Geomorphology 208:34–49

Blindheim J (1989) Ecological features of the Norwegian Sea. In: Rey L, Alexander V (eds) Proceedings of the sixth conference of the Comité Arctique International, 13–15 May 1985. Brill, Leiden, pp 366–401

Blindheim J (2004) Oceanography and climate. In: Skjoldal HR (ed) The Norwegian Sea ecosystem. Tapir Academic Press, Trondheim, pp 65–95

Caloiero T, Caloiero P, Frustaci F (2018) Long-term precipitation trend analysis in Europe and in the Mediterranean basin. Water Environ J 32(3):433–445. https://doi.org/10.1111/wej.12346

Christensen KH, Sperrevik AK, Broström G (2018) On the variability in the onset of the Norwegian Coastal Current. J Phys Oceanogr 48 (3):723–738. https://doi.org/10.1175/jpo-d-17-0117.1

Crespi A, Lussana C, Brunetti M, Dobler A, Maugeri M, Tveito OE (2018) High-resolution monthly precipitation climatologies over Norway: assessment of spatial interpolation methods. https://zenodo.org/record/1304042/files/hclimrk_prec_monthlyclim_1981_2010.nc?download=1. arXiv:1804.04867. https://doi.org/10.5281/zenodo.1304042

Dickson RR, Meincke J, Rhines P (eds) (2008) Arctic-Subarctic Ocean fluxes. Defining the role of the northern seas in climate. 1. Aufl. s.l.: Springer Netherlands. http://site.ebrary.com/lib/alltitles/docDetail.action?docID=10223402

Dobrowski SZ (2010) A climatic basis for microrefugia: the influence of terrain on climate. Glob Change Biol. https://doi.org/10.1111/j.1365-2486.2010.02263.x

Donat MG, Renggli D, Wild S, Alexander LV, Leckebusch GC, Ulbrich U (2011) Reanalysis suggests long-term upward trends in European storminess since 1871. Geophys Res Lett 38(14):n/a–n/a. https://doi.org/10.1029/2011gl047995

Drobyshev I, Granström A, Linderholm HW, Hellberg E, Bergeron Y, Niklasson M, McGlone M (2014) Multi-century reconstruction of fire activity in Northern European boreal forest suggests differences in regional fire regimes and their sensitivity to climate. J Ecol 102 (3):738–748. https://doi.org/10.1111/1365-2745.12235

Dyrrdal AV (2010) Analysis of past snow conditions in Norway—time periods 1931–60, 1961–90 and 1979–2008. In: MET report No. 10/2010 (Norwegian Meteorological Institute)

Eldevik T, Nilsen JEØ (2013) The Arctic-Atlantic Thermohaline circulation. J Climate 26(21):8698–8705. https://doi.org/10.1175/jcli-d-13-00305.1

Fezer F (1966) Vom Fjord zum Fjell – ein Profil aus Westnorwegen. Geographische Zeitschrift 54(2)

Gläßer E (1993) Norwegen, Wissenschaftliche Länderkunden, 14. Wissenschaftliche Buchgesellschaft, Darmstadt

Grimenes AA, Thue-Hansen V (2006) The reduction of global radiation in south-eastern Norway during the last 50 years. Theor Appl Climatol 85(1–2):37–40. https://doi.org/10.1007/s00704-005-0176-6

Grønås S, Sandvik AD (1998) Numerical simulations of sea and land breezes at high latitudes. Tellus A 50(4):468–489. https://doi.org/10.1034/j.1600-0870.1998.t01-3-00007.x

Hanssen-Bauer I, Førland EJ, Haddeland I, Hisdal H, Lawrence D, Mayer S, Nesje A, Nilsen JEØ, Sandven S, Sandø AB, Sorteberg A, Ådlandsvik B (2017) Climate in Norway 2100—a knowledge base for climate adaption. Unter Mitarbeit von Andreassen LM, Beldring S, Bjune A, Breili K, Dahl CA, Dyrrdal AV et al (eds), vol 1. Norwegian Centre for Climate Services NCCS. https://www.miljodirektoratet.no/globalassets/publikasjoner/m741/m741.pdf

Harbitz CB, Glimsdal S, Løvholt F, Kveldsvik V, Pedersen GK, Jensen A (2014) Rockslide tsunamis in complex fjords: from an unstable rock slope at Åkerneset to tsunami risk in western Norway. Coastal Eng 88

Isachsen PE, Mauritzen C, Svendsen H (2007) Dense water formation in the Nordic Seas diagnosed from sea surface buoyancy fluxes. Deep Sea Res Part I 54(1):22–41. https://doi.org/10.1016/j.dsr.2006.09.008

Johanessen TW (1970) The climate of Scandinavia. In: Wallén CC (eds) World survey of climatology, climates of Northern and Western Europe, vol 5

Kartverket (2018) norgeskart.no—detailed digital map of Norway. https://kartverket.no/en/data/Open-and-Free-geospatial-data-from-Norway/. Accessed 23 Dec 2018

Ketzler G (2014) The diurnal temperature cycle and its relation to boundary-layer structure during the morning transition. Boundary-Layer Meteorol 151(2):335–351. https://doi.org/10.1007/s10546-013-9898-7

Körner C, Paulsen J (2004) A world-wide study of high altitude treeline temperatures. J Biogeogr 31(5):713–732. https://doi.org/10.1111/j.1365-2699.2003.01043.x

Loeng H, Drinkwater K (2007) An overview of the ecosystems of the Barents and Norwegian Seas and their response to climate variability. Deep Sea Res Part II 54(23–26):2478–2500. https://doi.org/10.1016/j.dsr2.2007.08.013

Lundstad E, Tveito, OE (2016) Homogenization of daily mean temperature in Norway. MET report No. 6/2016 (Norwegian Meteorological Institute)

Lupikasza E (2016) Definitions and indices of precipitation extremes. In: Lupikasza E (eds) The climatology of air-mass and frontal extreme precipitation, pp 39–82

Lussana C (2018) seNorge_2018 daily mean temperature 1957–2017 (Version 18.12), data set. https://doi.org/10.5281/zenodo.2023997
Lussana C, Crespi A (2018) 1981–2010 monthly precipitation climatologies HCLIM-RK_1981_2010 (Version 1.0), data set. https://doi.org/10.5281/zenodo.1304042
Mahrt L (1986) Nocturnal topoclimatology, WCP, 117, TD, 132 World Meteorological Organization (WMO)
Mauritzen C (1996) Production of dense overflow waters feeding the North Atlantic across the Greenland-Scotland Ridge. Part 1. Evidence for a revised circulation scheme. Deep Sea Res Part I: Oceanogr Res Papers 43(6):769–806. https://doi.org/10.1016/0967-0637(96)00037-4
Moberg A, Tuomenvirta H, Nordli Ø (2005) Recent climate trends. In: Seppälä M (eds) The physical geography of Fennoscandia, Oxford regional environments series, vol 5. Oxford Univ. Press, New York, pp 113–133
Mosby H (1974) Die nordischen Meere. In: Sömme A (ed) Die Nordischen Länder, pp 25–34
Myrstad I (2015) Utprøving av jordbærsorter Sør-Troms, Målselv og Alta, Delrapport 2015. Landbrukstjenesten Midtre Hålogaland, Borkenes
Nesje A (1992) Younger Dryas and Holocene glacier fluctuations and equilibrium-line altitude variations in the Jostedalsbreen region, western Norway. Clim Dyn 6:221–227
NMI Norwegian Meteorological Institute (2018a) Digital elevation model. In: Norwegian gridded climate datasets (geoinfo seNorge2018_dem_UTM33.nc). https://thredds.met.no/thredds/catalog/senorge/geoinfo/catalog.html. Accessed 1 June 2018
NMI Norwegian Meteorological Institute (2018b) eKlima—free access to weather- and climate data from Norwegian Meteorological Institute; eklima.nmi.no 23.12.2018
Norseth K (1987) Climate and hydrology of Norden. In: Varjo U, Tietze W (eds) Norden, man and environment
NVE The Norwegian Water Resources and Energy Directorate (2019) Norske breer minket kraftig i 2018. https://www.nve.no/hydrologi/bre/brenyheter/norske-breer-minket-kraftig-i-2018/. Accessed 24 Feb 2019
Quante M, Colijn F, Bakker JP, Härdtle W, Heinrich HG, Lefebre C, Nöhren I, Olesen JE, Pohlmann T, Sterr H, Sündermann J, Tölle MH (2016) Introduction to the assessment-characteristics of the region. In: Quante M, Colijn F (eds) North sea region climate change assessment, pp 1–54
Robert C, Bousquet R (2018) Geowissenschaften. Die Dynamik des Systems Erde. Springer Berlin Heidelberg, Berlin, Heidelberg. Online verfügbar unter http://dx.doi.org/10.1007/978-3-662-50393-5
Sattler K, Anderson B, Mackintosh A, Norton K, de Róiste M (2016) Estimating permafrost distribution in the Maritime Southern Alps, New Zealand, based on climatic conditions at rock glacier sites. Front Earth Sci. https://doi.org/10.3389/feart.2016.00004
Simonsen K, Haugan PM (1996) Heat budgets of the Arctic Mediterranean and sea surface heat flux parameterizations for the Nordic Seas. J Geophys Res 101(C3):6553–6576. https://doi.org/10.1029/95jc03305
Skagseth Ø, Furevik T, Ingvaldsen R, Loeng H, Mork KA, Orvik KA, Ozhigin V (2008) Volume and heat transports to the Arctic Ocean via the Norwegian and Barents Seas. In: Dickson RR, Meincke J, Rhines P (eds) Arctic-Subarctic Ocean fluxes. Defining the role of the northern seas in climate, pp 45–64
Skartveit A, Olseth JA (1986) Modeling slope irradiance at high latitudes. Solar Energy 36(4):333–344. https://doi.org/10.1016/0038-092x(86)90151-9
Skre O (1971) Frequency distributions of monthly air temperature and their geographical distribution and seasonal variations in Northern Europe. In: Scientific Reports of the Agricultural College of Norway (50, 9)
Tikkanen M (2005) Climate. In: Seppälä M (eds) The physical geography of Fennoscandia. Oxford regional environments series, vol 5. Oxford Univ. Press, New York, pp 97–112
Troen I, Lundtang Petersen E (1989) European wind atlas. Risø National Laboratory, Roskilde
Tveito OE, Førland EJ, Alexandersson H, Drebs A, Jonsson T, Vaarby-Laursen E (2001) Nordic climate maps. DNMI—Report No. 06/01
Uleberg E, Hanssen-Bauer I, van Oort B, Dalmannsdottir S (2014) Impact of climate change on agriculture in Northern Norway and potential strategies for adaptation. Clim Change 122(1–2):27–39. https://doi.org/10.1007/s10584-013-0983-1
UNEP World Glacier Monitoring Service (2009) Global glacier changes: facts and figures. http://www.grid.unep.ch/glaciers/pdfs/glaciers.pdf. Accessed 1 March 2019
Wefer G, Berger WH (2001) Klima und Ozean. In: Huch M, Warnecke G, Germann K (eds) Klimazeugnisse der Erdgeschichte. Perspektiven für die Zukunft. Springer, Berlin, pp 51–107
Weischet W, Endlicher W (2012) Einführung in die Klimatologie, 8 ed
Wibig J (1999) Precipitation in Europe in relation to circulation patterns at the 500 hPa level. Int J Climatol 19(3):253–269. https://doi.org/10.1002/(sici)1097-0088(19990315)19:3%3c253:aid-joc366%3e3.0.co;2-0
Williams CT (1901) The climate of Norway and its factors. Q J R Meteorol Soc 27(118):105–116. https://doi.org/10.1002/qj.49702711802
Wolff M, Nygård H, van Nes A, Kielland G, Torfoss T, Kaukoranta JP, Santanen M, Huuskonen M, Haapa I, Wittskog C, Johansson P, Krabbi M, von Löwis of Menar S, Sigurðsson Á (2016) Temperature siting classification in Nordic Countries. Final report of NordObs activity on WMO-CIMO siting classification. Norwegian Meteorological Institute (17)
Wong WK, Haddeland I, Lawrence D, Beldringet S (2016) Gridded 1 x 1 km climate and hydrological projections for Norway. NVE Report 59/2016 (Norwegian Water Resources and Energy Directorate)
World Meteorological Organization WMO (2008) CIMO guide 7th edition 2008: Guide to Meteorological Instruments and Methods of Observation

Part II

Case Studies of Varied Landscapes, Geomorphological Processes and Landforms in Norway

3 Terminal Moraine Formation Processes and Geomorphology of Glacier Forelands at the Selected Outlet Glaciers of Jostedalsbreen, South Norway

Stefan Winkler

Abstract

Several outlet glaciers of Jostedalsbreen flow from the central plateau ice-cap into surrounding valleys. A well-established Holocene glacier chronology reveals that the 'Little Ice Age' with its culmination ~1750 CE was the most prominent neoglacial event in the region. Glacial and glaciofluvial landforms on the glacier forelands are, therefore, accurately determinable in their age and allow detailed interpretation of causal relationships between geomorphology and glacier dynamics. Glacier forelands are typically characterised by a high number of comparatively small moraine ridges and wide areas of glaciofluvial outwash and glaciofluvially overprinted subglacial till deposits. A recent regional advance during the 1990s offered the unique opportunity to study terminal moraine processes. Ice pushing/bulldozing of unconsolidated proglacial sediments by the advancing glacier tongues was the dominant process, whereas seasonal advances caused the formation of subsequent moraines during or shortly after culmination of this recent advance. Recent observations serve as analogues for previous advances and allow the interpretation that differences in moraine morphology of older moraines are mainly caused by different properties of the proglacial sediment in combination with detailed dynamics of the glacier fronts. The high number of terminal moraines reveals the efficiency of short-lived and seasonal advances for moraine formation following the 'Little Ice Age'-maximum.

Keywords

Glacier forelands • Terminal moraines • Moraine formation processes • Recent glacier advance • 'Littler Ice Age' (LIA) • Jostedalsbreen

3.1 Introduction

In addition to its world-famous fjord landscapes, Norway contains an important and diverse geomorphological heritage of glacial landforms. The Holocene glacier forelands of currently more than 2500 glaciers in Norway (Andreassen and Winsvold 2012) should be recognised as regionally important geomorphosites, with many exhibiting significant glacial features. This applies in particular to the outlet glaciers of the Jostedalsbreen plateau ice-cap, the largest glacier in Norway. Their glacier forelands constitute landform assemblages that are specific due to their general geological, glaciological, climatic and geomorphological framework. In comparison with typical glaciated valley landsystems in high mountains regions such as the European Alps, High Asian mountains, or the Southern Alps of New Zealand (Benn et al. 2005), significantly reduced to almost absent supraglacial debris supply due to a lack of strong neotectonic activity, less dynamic geomorphological process-systems (e.g. mass movements), and a specific glacier geometry are among the most prominent differences at Jostedalsbreen. Ice-marginal terrestrial landsystems at active outlet glaciers in Iceland and other parts of the subarctic (Evans 2005) are, on the other hand, exposed to colder climatic conditions and show less restrictions imposed by topography/gross morphology and its interaction with glacier dynamics.

An additional aspect enhancing the importance of glacier forelands as study objects is their potential in the context of palaeoclimatic research. Glacial landforms are utilised as the evidence of glacier variations and, hence, as valuable archives for any underlying climatic variability. The

S. Winkler (✉)
Department of Geography and Geology, University of Würzburg, Am Hubland, 97074 Würzburg, Germany
e-mail: stefan.winkler@uni-wuerzburg.de

A. A. Beylich (ed.), *Landscapes and Landforms of Norway*, World Geomorphological Landscapes,
https://doi.org/10.1007/978-3-030-52563-7_3

palaeoclimatic interpretation of moraines is, however, not a trivial task as Kirkbride and Winkler (2012) recently elaborated in their comprehensive conceptual review. Jostedalsbreen offers the rare opportunity of a detailed, well-established Holocene glacier chronology and climate history. Except few early Holocene moraine remnants outside the late Holocene glacier forelands, their landform assemblages solely represent the 'Little Ice Age' (LIA) and subsequent glacier variability because this was regionally the most extensive neoglacial event. With any older landforms having been destroyed, available detailed age constraints for those 'young' landforms on the forelands allow their interpretation and related analysis of relationships between glacier dynamics and moraine morphology or sedimentology.

An extremely valuable event for geomorphological research at Jostedalsbreen was the remarkable advance many outlets exhibited during the late twentieth century CE. It provided an opportunity, quite rare on a global scale, to observe and investigate recent moraine formation processes at temperate mountain glacier margins. These observations may serve as analogues for processes that occurred during the 'Little Ice Age', thus enabling to link moraine formation, glacial dynamics and palaeoclimatic interpretation in a more secure way. They have, furthermore, the potential to assess and enhance reliable application of dating techniques (e.g. lichenometry) and ultimately improve the palaeoclimatic analysis of Holocene glacier chronologies as requested by Kirkbride and Winkler (2012).

Based on the presentation of Jostedalsbreen's essential natural framework, the typical landform assemblage on glacier forelands at selected outlet glaciers will be described and interpreted in this chapter. Significant geomorphological features are highlighted before subsequently considerable space will be dedicated to the investigation of terminal moraine formation during the recent glacier advance. Apart from the aim to characterise the specific regional glacial process system of Jostedalsbreen and its landforms, there is also the intention to allow general conclusions regarding terminal moraine formation at modern temperate mountain glaciers.

3.2 Natural Environment

3.2.1 Geology

The Jostedalsbreen plateau ice-cap is located in the inner fjord region of western South Norway (Fig. 3.1). Southern and eastern parts of the Jostedalsbreen region belong to the major Sognefjorden system and northern and western parts to the inner Nordfjord system. This entire part of South Norway is characterised by deeply incised fjords, fjord valleys (i.e. terrestrial glacial valleys that at some stage during or after Deglaciation were inundated by the sea before glacio-isostatic uplift commenced and accelerated) and glacial valley with their tributary valleys, the latter are often hanging valleys. Between these valleys at higher altitudes, mostly flat or gently undulating remnants of the preglacial land surface (the so-called 'paleic surface'; Gjessing 1967) have survived Pleistocene glacial erosion

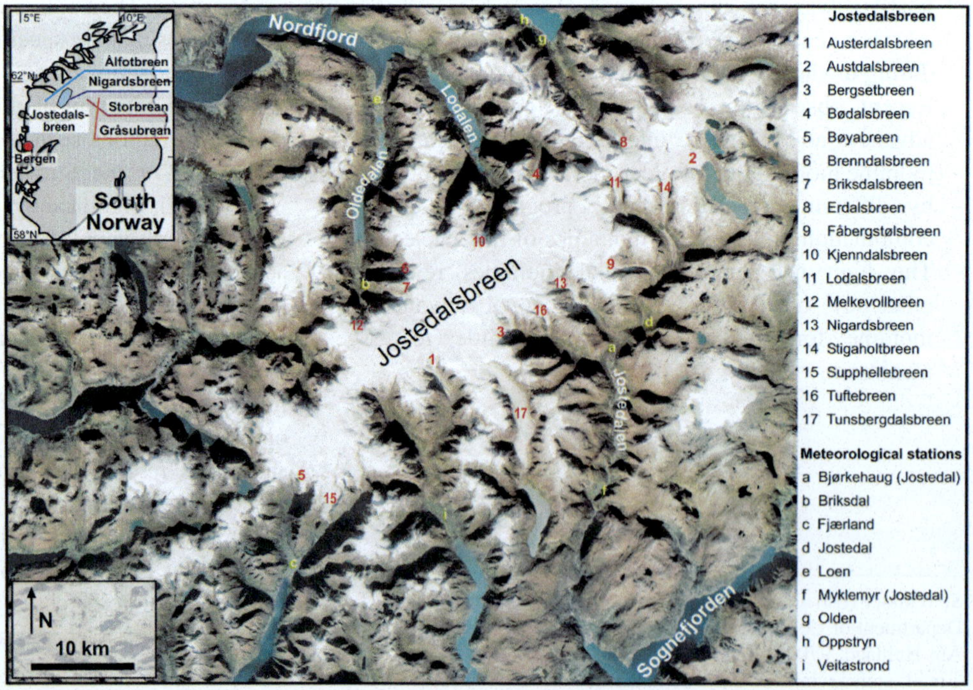

Fig. 3.1 Orthorectified aerial photo (2010) of Jostedalsbreen (adapted from www.norgeibilder.no, © Kartverket/www.kartverket.no) with its prominent outlets (cf. Table 3.1) and some meteorological stations (cf. Table 3.2) indicated. The inset depicts Jostedalsbreen's location within southern Norway and the location of those glacier mass balance records shown in Fig. 3.12a

Fig. 3.2 View from the icefall of Bergsetbreen eastwards into Krundalen, a tributary of Jostedalen (visible in the far distance; see Fig. 3.1). The rather flat and wide valley bottom and parts of the surrounding high-altitude land surfaces are clearly visible (*Photo* Stefan Winkler, 24.08.2004)

(Fig. 3.2). Glacial erosion concentrated its impact on (over) deepening the above-mentioned valleys as ice-streams of multiple Scandinavian Ice Sheets during the Pleistocene followed preexisting fluvial valleys that had developed previously during the Late Tertiary when the tectonic uplift of the western rim of Scandinavia commenced (Klemsdal and Sjulsen 1988; Nesje et al. 1992; Nesje and Sulebak 1994; Nesje and Whillans 1994; Lidmar-Bergström et al. 2000). Despite this development of a classic 'landscape of linear glacial erosion' sensu Sugden and John (1976), recent studies question the previously postulated limited effect of glacial erosion on the apparent relicts of the 'paleic surface' (Egholm et al. 2017; Andersen et al. 2018). Regardless of their detailed evolution, however, the mostly flat high-altitude surfaces constitute significant factors for both initial formation and morphology of the large plateau ice-caps typical for western South Norway. Even if radio echo-sounding (Sætrang and Wold 1986) reveals that the gentle surface of Jostedalsbreen's central ice accumulation does not accurately reflect its glacier bed and some larger valleys, especially in its south-eastern part, continue beneath the modern outlet glaciers up to the central plateau, the mostly low to moderately undulating glacier bed provided by the high-altitude surface supports the formation of a coherent plateau ice-cap.

The Jostedalsbreen region is geologically a part of the Norwegian basement with its Proterozoic rocks (Sigmond et al. 1984). It is part of the so-called Western Gneiss Region comprising almost the entire north-western part of South Norway west of the Caledonian Jotun Nappe Complex dominating central South Norway and forming the highest peaks of Norway in the Jotunheimen Mountains (see Chap. 8). The Jostedal Complex of the Western Gneiss Region is characterised by Precambrian migmatitic gneisses, granitic gneisses (both c. 1850–1500 Ma), and granites metamorphically overprinted to different degrees (1000–950 Ma; Nordgulen and Andresen 2008). The two latter bedrock types form the glacier bed of the central plateau ice-cap and the surrounding valleys where the glacier forelands of its outlets are located (Lutro and Tveten 1996). The granites become gradually more deformed and overprinted in the north-western part of the Jostedalen region and have partly developed into augen gneisses (Fig. 3.3e).

Under both Pleistocene cold and Holocene temperate climatic conditions, bedrock in the Jostedalsbreen region is

Fig. 3.3 **a** Brenndalen. A short tributary of Oldedalen in Inner Nordfjord occupied by Brenndalsbreen. **b** Exfoliation on the valley slope near Nigardsbreen. **c** View down the glacier tongue of Nigardsbreen in south-easterly direction depicting proglacial lake Nigardsvatnet (see Fig. 3.7). **d** Exfoliation and glacially/glaciofluvially modified bedrock in Brenndalen. **e** Polished specimen of typical augen gneiss from the inner Nordfjord, shown in the rock collection of Jostedalsbreen Nasjonalparksenteret/Oppstryn (*Photos* Stefan Winkler, **a** 24.06.2009, **b** 06.08.2014, **c** 24.08.2004, **d** 18.06.2007, **e** 02.09.2006)

comparatively resistant to both physical and chemical weathering. Due to its lithological properties, the degree of preservation of micro- and meso-scale bedrock features of glacial and glaciofluvial erosion is very high, not only within the Holocene glacier forelands. The hard and massive bedrock is, furthermore, one factor responsible for low frequencies and magnitudes of rock slope failures and mass movement processes in comparison to other mountain regions worldwide. Alongside the morphology of Jostedalsbreen and its outlets with few nunataks as potential debris sources, regional lithology thus additionally explains the absence of substantial supraglacial debris with resulting implications for moraine formation processes and glacial deposition. With their massive appearance and lack of prominent lithological structures such as bedding etc., granites and gneisses of the Jostedalsbreen region promote the development of 'typical' shapes of glacial macro-scale landforms like a glacial valley (Fig. 3.3a). The common weathering phenomenon of exfoliation helps to retain their initial shape even with ongoing weathering (Fig. 3.3b, d; Winkler 2009). Valleys surrounding the central plateau ice-cap are typically shaped glacial valleys with steep valley sides and mostly parabolic to trough-shaped cross sections (Baumhauer and Winkler 2014). There are, however, local differences that influence the configuration of glacier forelands. Most south-eastern valleys constitute tributaries of the long Jostedalen valley and the majority exhibits very low longitudinal slope gradients ($\leq 1°$) with considerable valley fills of mainly (Late)glacial and Holocene (glacio)fluvial sediment with some tills or slope/colluvial deposits (Fig. 3.2). Oldedalen and Lodalen and their tributaries valleys in inner Nordfjord are often narrower and small bedrock sills may occur within the Holocene glacier forelands. Basin sections between those are, however, comparatively flat. Modern proglacial lakes in front of some outlets occupy glacier-eroded depressions and potentially constitute pre-'Little Ice Age' features (or are supported by those; Fig. 3.3c).

3.2.2 Glaciology and Climate

The plateau ice-cap of Jostedalsbreen has an area of 473.75 km^2 (Andreassen and Winsvold 2012) and is the largest glacier in Norway. According to the current glacier inventory, Jostedalsbreen consists of 82 individual glacier units, around 20 of which constitute morphologically larger outlets that flow from the central accumulation area down into surrounding valleys (Fig. 3.1). Key glaciological data for selected Jostedalsbreen outlets are given in Table 3.1. It needs to be noted that the glacier tongues of most outlets have undergone significant changes since culmination of the recent advance around 2000 CE (Winkler et al. 2009; http://glacier.nve.no/Glacier/viewer/gpp/no/). Maximum thickness of the central plateau ice-cap reaches nearly 600 m (571 m, with thickness averaging 158 m average; Andreassen et al. 2015), but the glacier bed is far more undulating than the flat ice surface. During the recent glacier advance described in more detail below, considerable overall advance distances have been measured in particular at the short and steep outlets of Jostedalsbreen. Annual and partly seasonal length measurements during the twentieth century are available for most outlets investigated here in detail (Table 3.4). For more information about the ongoing monitoring of Jostedalsbreen and its outlets see Andreassen et al. (2005), Kjøllmoen (2017), or the regularly updated website of Norges Vassdragsog Energidirektoratet/NVE (https://www.nve.no/hydrologi/bre/bredata/).

Jostedalsbreen outlets are temperate and warm-based. Based on the mean annual air temperature normals of 1961–90 for Bjørkehaug in Jostedalen (3.7 °C, station 324 m a.s.l.) and Olden (5.9 °C, station 78 m a.s.l.; eKlima database by www.met.no; see Table 3.2), the mean air annual temperatures for the majority of glacier forelands experienced a fall within the range between 2.5 and 5.0 °C, if a standard mean adiabatic lapse rate of 0.6 °C/100 m is applied. The corresponding values for the few higher forelands (e.g. Tuftebreen) are estimated to c. 1.0 °C and higher, but local factors may cause the actual adiabatic lapse rate to deviate slightly, especially during winter (Winkler and Matthews 2010). Lowest monthly air temperatures (January, February) are accordingly estimated to be −2.0 °C (Kjenndalsbreen), −6.0 °C (Bergsetbreen) and −9.0 °C (Tuftebreen). These air temperatures are still comparatively mild for mid-latitude mountain environments due to the low elevation of the glacier tongues and the strong maritime climatic influence;

Table 3.1 Key glaciological data for selected Jostedalsbreen outlets (*Data source* see table; cf. Fig. 3.1)

Outlet	2000s[a]			1980/90s[b]		LIA[c]		
	Area (km^2)	Elevation (m a.s.l.)	Aspect	Length (km)	Front (m a.s.l.)	Length (km)	Front (m a.s.l.)	Maximum (CE)
Austerdalsbreen	19.85	424—1915	SE	8.5	390	11.5	310	1760/65 (l)
Bergsetbreen	11.15	854—1957	SE	4.8	540	7.7	410	1750/60 (l)
Bødalsbreen	8.10	654—1923	NW	6.5	650	8.2	590	1745/55 (l)
Bøyabreen	13.90	504—1733	S	5.7	145	8.4	120	c. 1750 (l)
Brenndalsbreen[d]	0.72 (20.12)	410—1368 (1108—1952)	W (W)	2.3	310	11.6	140	1743 (h)
Briksdalsbreen	11.73	349—1917	NW	6.0	350	7.4	290	c. 1760 (l)
Erdalsbreen	10.32	930—1906	NW	6.1	880	8.4	530	
Fåbergstølsbreen	20.29	720—1981	SE	7.0	760	9.5	550	1740/50 (l)
Kjenndalsbreen	19.07	489—1948	N	6.9	180	9.2	100	
Lodalsbreen	9.32	855—2004	E	6.0	860	10.2	580	1825/30? (l)
Melkevollbreen	6.58	620—1865	NW	4.3	410	5.6	260	
Nigardsbreen	42.02	345—1946	SE	9.6	350	13.9	260	1748 (h)
Stigaholtbreen	12.50	819—1773	S	7.7	880	9.8	600	1860? (l)
Supphellebreen	12.87	733—1734	S	8.4	720	10.4	40	c. 1750 (h)
Tuftebreen	6.79	799—1945	NE	6.5	820	8.6	400	1770/75 (l)
Tunsbergdalsbreen	47.64	656—1930	SE	19.1	590	21.8	460	c. 1750 (l)

[a]Data from the current glacier inventory (Andreassen and Winsvold 2012)

[b]Data representative for glaciers during the recent advance in the 1990s (Østrem et al. 1988; partly modified by Winkler 2001, 2002)

[c]Data representative for the LIA-maximum (Torsnes et al. 1993; Winkler 2001, 2002). Abbreviations stand for lichenometrically dated (l) and historically dated (h)

[d]The lower glacier tongue separated from the main glaciers during the first half of the twentieth century. The data given in brackets are for the main glacier as the sector of Jostedalsbreen's central ice accumulation

Table 3.2 (a) Climate normals (1961–90) of air temperature for selected meteorological stations around Jostedalsbreen (taken from Aune 1993; cf. Fig. 3.1). (b) Corresponding climate normals for precipitation (taken from Førland 1993)

(a)

Station	Altitude (m a.s.l.)	Jan (°C)	Feb	Mar	Apr	May	Jun	Jul	Aug	Sep	Oct	Nov	Dec	Year ∅
Bjørkehaug (Jost.)	324	−4.9	−4.6	−1.9	1.9	8.0	12.3	13.4	12.5	8.2	4.4	−1.1	−3.8	3.7
Fjærland	10	−3.3	−3.0	−0.1	3.7	9.6	13.3	14.3	13.3	9.3	5.7	0.6	−2.1	5.1
Loen	39	−1.4	−2.0	1.0	4.4	9.8	12.9	14.0	13.4	9.4	6.5	2.0	−0.4	5.8
Myklemyr (Jost.)	98	−7.1	−6.9	−2.5	2.4	8.7	12.9	14.0	12.9	8.6	4.5	−1.6	−5.3	3.4
Olden	78	−0.7	−1.2	1.2	4.2	9.6	12.8	13.7	13.0	9.2	6.4	2.2	0.0	5.9
Oppstryn	201	−1.0	−1.0	0.9	3.7	9.2	12.4	13.5	12.9	9.4	6.5	2.1	−0.2	5.7

(b)

Station	Altitude (m a.s.l.)	Jan (mm)	Feb	Mar	Apr	May	Jun	Jul	Aug	Sep	Oct	Nov	Dec	Year Σ
Bjørkehaug (Jost.)	324	148	97	113	52	61	69	73	97	162	181	157	170	1380
Briksdal	38	135	91	111	50	48	70	83	88	185	179	158	174	1372
Fjærland	10	198	138	158	82	71	88	102	129	238	245	217	239	1905
Jostedal	370	146	94	109	47	60	66	71	96	159	175	152	155	1330
Loen	39	96	82	87	46	41	59	81	77	129	135	114	128	1075
Myklemyr (Jost.)	98	140	88	107	44	55	71	82	99	166	180	153	165	1350
Olden	78	125	90	103	55	51	70	87	87	169	165	139	164	1305
Oppstryn	201	118	91	99	53	41	53	74	73	130	136	118	151	1137
Veitastrond	172	171	115	135	60	65	80	85	111	198	214	184	202	1620

the latter decreases slightly from the north-western towards the south-eastern glacier forelands. The comparatively mild air temperatures have considerable implications for potential processes of moraine formation. During both the 'Little Ice Age'-maximum and more recent time periods, neither permafrost nor deep penetrating winter frost is to be expected in the marginal zones of the glaciers. This needs to be considered with any comparison to studies of moraine formation in the Jotunheimen mountains east of Jostedalsbreen where a higher altitude allows a deeper penetration of winter frost and even the occurrence of permafrost on some glacier forelands (e.g. Matthews et al. 1995; Winkler and Matthews 2010; Matthews et al. 2014; Hiemstra et al. 2015; see Chap. 8). Annual precipitation around Jostedalsbreen exceeds 1500 mm on glacier forelands, but precise long-term local data for high elevations are not available. During the 1990s, some years showed record-high snow accumulation, but the snow cover on the glacier forelands usually disappears quickly in late spring/early summer (except for snow avalanche deposits and for a few years during the early 1990s CE).

3.3 Holocene Glacier Chronology

3.3.1 Early Holocene

Following the advances of a still impressive Scandinavian Ice Sheet during the Younger Dryas, ice-streams remaining in the deeply incised fjords and valley in the Jostedalsbreen region commenced a fast retreat. The general glacial dynamic was still affected by these ice-sheet remnants, but individual glaciers comparable to those of the late Holocene subsequently developed. There is evidence for early Holocene (Preboreal) advances of the latter in form of isolated terminal moraines outside the 'Little Ice Age'-maximum expansion at a few Jostedalsbreen outlets (Fig. 3.4b, c). Those are assigned to the so-called 'Erdalen Event' (Nesje et al. 1991; Nesje and Dahl 1993; Nesje et al. 2008a; Nesje 2009). The regionally two-phased advance is dated to 10,100–10,050 cal. a BP (Erdalen Event I) and c. 9700 cal a BP (Erdalen Event II; Dahl et al. 2002; Matthews et al. 2008). Apart from the type locality Erdalsbreen near

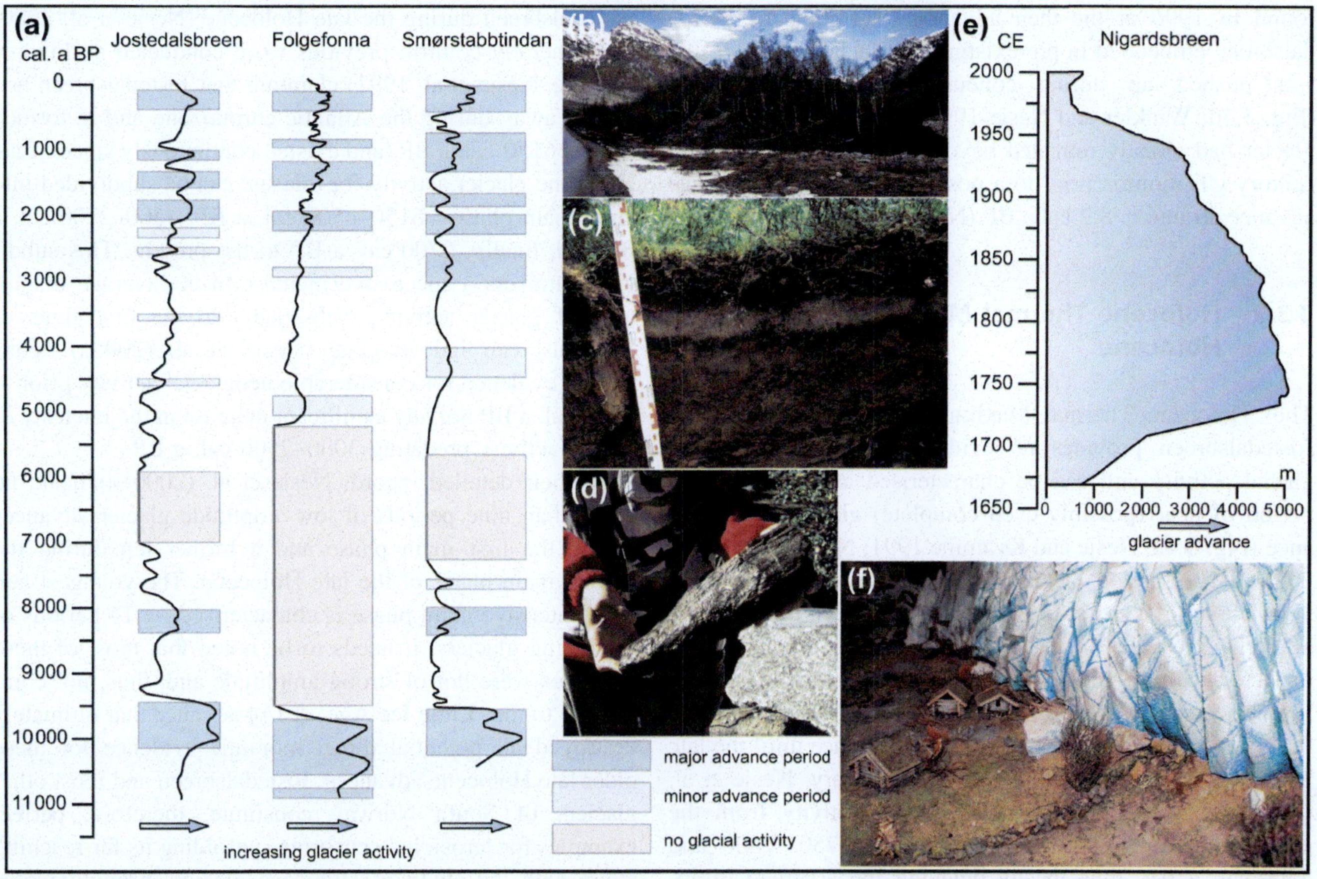

Fig. 3.4 **a** Holocene glacier chronology for Jostedalsbreen (after Nesje et al. 2001) compared with chronologies for Folgefonna in Southwest Norway (Bakke et al. 2005) and Smørstabbtindan in Jotunheimen (Matthews and Dresser 2008). **b** Early Holocene terminal moraine at Fåbergstølsbreen. **c** Podzolic soil on early Holocene terminal moraine at Fåbergstølsbreen (**b**). Despite the morphological similarity to local 'Little Ice Age'-moraines, the deeply developed soil and Schmidt-hammer measurements give evidence of its old age (Winkler 2001). **d** Fossil wood discovered at Briksdalsbreen (see text). **e** Glacier length variations of Nigardsbreen since the onset of the 'Little Ice Age' until culmination of the recent advance (modified after Fægri 1948 and Nesje et al. 2008a). **f** Artistic model of the advancing Nigardsbreen destroying the farm Nygaard during the 1740s CE, formerly display at Breheimsenteret/Jostedalen but destroyed by fire in 2011 (*Photos* Stefan Winkler, **b** 28.05.1998, **c** 25.08.1996, **d** 04.09.1996, **f** 17.08.1996)

Oppstryn in the north-western sector of Jostedalsbreen, early Holocene moraines individually dated or correlated to the Erdalen Event are reported from Bødalsbreen, Bøyabreen, Briksdalsbreen, Fåbergstølsbreen, Melkevollbreen and Nigardsbreen (Rye et al. 1997; Aa and Sjåstad 2000; Nesje 2009). Neither at Erdalsbreen nor at other Jostedalsbreen outlets evidence for earlier Preboreal advances corresponding to the 'Jondal Event I' (11.1 cal. ka BP) or 'Jondal Event II' (10.5 cal. ka BP) at Nordre Folgefonna south of the Jostedalsbreen region (Bakke et al. 2005) has been detected. The 'Jondal Event III' dated to c. 10 cal. ka BP seems, however, to correspond to the Erdalen Event (Fig. 3.4a).

In some valleys in the northwest and southern sectors of Jostedalsbreen the early Holocene experienced the sea inundating today's terrestrial fjord valleys up to 100 m a.s.l. (locally even slightly higher; Klakegg et al. 1989). This temporary high sea-level and an incomplete Deglaciation with remnants of ice-streams still occupying parts of some major valley may be responsible for the lack of evidence for early Holocene activity of individual glaciers prior to the Erdalen event around Jostedalsbreen. In Krundalen, a tributary valley to Jostedalen in the south-eastern sector of Jostedalsbreen (Fig. 3.2), Lukas (2007) links an older terminal moraine system downvalley supposedly Erdalen Event-moraines to the Jondal Event II. This correlation, however, is not constrained by numerical dating but by reconstructed equilibrium line altitude/ELA-lowering and afflicted with the uncertainty of isolated local evidence. There is only sole evidence for subsequent early Holocene glacier activity at Jostedalsbreen. It corresponds to the '8.2 ka BP-event' (Alley and Ágústsdóttir 2005) or 'Finse Event' regionally named after the type locality at Hardangerjøkulen (Dahl and Nesje 1994; Nesje 2009). Pieces of fossil wood (*salix spec.*) that predate the 'Finse Event' were found in 1996 at the then advancing terminus of Briksdalsbreen embedded in pre-existing proglacial sediment that was pushed up during concurrent moraine formation (Fig. 3.4d; Winkler and Nesje 1999). They suggest that the glacier had already retreated beyond modern mid-twentieth century CE minima prior to a possible final early Holocene advance around c. 8.2 cal a BP (Nesje 1997; Winkler 2001).

3.3.2 Holocene Thermal Maximum and Late Holocene

The 'Holocene Thermal Maximum' or 'Hypsithermal' at Jostedalsbreen provides no evidence for any substantial glacial activity and can be characterised as a prolongated period of near (possibly even complete) glacier disappearance (Fig. 3.4a; Nesje and Kvamme 1991; Nesje et al. 2000, 2008a). This period of climatic conditions that were unfavourable for glaciers seems to have commenced immediately after the Finse Event. Many small glaciers are, however, expected to have melted completely earlier, following the Erdalen Event, and possibly only re-formed briefly around 8200 cal a BP prior to their final disappearance until the late Holocene (Nesje et al. 2001). On the contrary, Nesje et al. (2000) report periods of minor glacier activity from the northern sector of Jostedalsbreen about 7750, 7600 and 7400 cal. a BP, thus locally delaying the assumed disappearance of the entire ice cap to c. 7350 cal a BP. Furthermore, they indicate minor interruptions of the Holocene Thermal Maximum during short intervals at 6900 and 6700 cal a BP. It remains unclear whether their findings can be interpreted as sufficient evidence to assume substantial re-formation of Jostedalsbreen at that stage. Summarising, following the Finse Event until at least around 6000 cal a BP, there is no clear and uncontested indication of any resuming glacier activity in the region.

As indicated by glaciofluvial sediments in lake and bog archives, the Holocene Thermal Maximum terminated with a stepwise re-formation of glaciers (Fig. 3.4a). This would correspond with the 'classic' concept of a 'neoglaciation' *sensu stricto*. For the large catchment of Oppstrynsvatnet in the northern sector of Jostedalsbreen, Vasskog et al. (2012) record a minimum of glacier-derived sediment input between 6700 and 5700 cal. a BP when the catchment was almost or completely free of glaciers. By contrast, the highest glacial input is dated to 600 and 200 cal. a BP indicating the local 'Little Ice Age'-maximum. Local 'neoglaciation' is assumed to have commenced at c. 4200 cal a BP and comprised five periods of retracted glaciers c. 3400, 3100–2700, 2100–2000, 1700–1500 and c. 900 cal. a BP. Lake sediment cores from nearby Sygneskardvatnet likewise indicate periods of increased or minor glacial activity related to the northern part of the Jostedalsbreen during the late Holocene (Nesje et al. 2000) and generally confirm previous work conducted at this key site (e.g. Nesje et al. 1991) claiming that Jostedalsbreen had melted away during the Atlantic chronozone and re-formed around 6150 cal. a BP (and existed continuously since). Late Holocene glacier activity for this site can be subdivided into three main phases: 6150–4500 cal. a BP, 4500–2700 cal. a BP and, finally, 2700 cal. a BP to the present. This subdivision mirrors a successive increase in the average magnitude of glacier activity. Subsequent to the first signs of Jostedalsbreen redeveloping, Nesje et al. (2008a) detect neither evidence for considerable neoglacial activity prior to 4500 cal. a BP nor any significant increase in the intensity of glacial activity predating 3000–2500 cal. a BP.

In their detailed record, Nesje et al. (2000) indicate no fewer than nine periods of low-amplitude glacier advances during the first main phase and a further ten during the second main phase of the late Holocene. The youngest and most intensive main phase is characterised by 14 periods of advancing glaciers. It needs to be noted that most of these advances were not of strong amplitude and, thus, not comparable to the 'Little Ice Age' (LIA)-advance that ultimately destroyed all potential direct morainic evidence for these older late Holocene advances. Jostedalsbreen and most other glaciers of South Norway constitute, therefore, perfect examples for 'erosional censoring' unfolding its far-reaching effect with the study of Holocene glacier chronologies as described by Kirkbride and Winkler (2012). Except for the above-mentioned few remnants of early Holocene moraines outside the LIA-maximum positions, glacier forelands contain only 'LIA'- and younger moraine sequences. Primarily due to the lack of substantial supraglacial debris supply, typical alpine-type lateral moraines (sensu Röthlisberger and Schneebeli 1979; Winkler and Hagedorn 1999) that potentially could reveal evidence for multiple (late) Holocene advances by their specific style of subsequent formation do not occur in the region. Hence, remarkable methodological progress has been achieved with the analysis of lake sediments to detect glaciofluvial sediment layers deposited by the inflow of glacier-sourced meltwater compensating for the lack of direct morainic evidence (see Matthews and Dresser 2008; Nesje 2009; Matthews 2013; Solomina et al. 2015 and sources within). By contrast, reliable age constraints are available for the 'LIA'-maximum positions that locally define the boundaries of the Holocene glacier forelands around Jostedalsbreen.

3.3.3 'Little Ice Age'

Unlike in the European Alps and other mountain regions where the 'Little Ice Age' (LIA) exhibited multiple major advances, at Jostedalsbreen it principally constituted a single prominent glacier advance (Fig. 3.4e). Culmination of this major advance was followed by an initially slow but consistent retreat, albeit frequently interrupted by stillstands or short-lived re-advances of a few years' duration only. Furthermore, LIA at Jostedalsbreen had a comparably late onset because documentary evidence of preceding cold climatic conditions lacks explicit indication of substantial glacier expansion (Grove 1972, 1985, 2001; Grove and Battagel 1983). The earliest reliable documentary evidence for glaciers advancing over mountain pastures dates from the last two decades of the seventeenth century CE (Rekstad 1904; Eide 1955; Hoel and Werenskiold 1962; Grove 2004). A general advance of Jostedalsbreen's outlet glaciers seems to have commenced during that time. It must have been substantial and rapid with calculated annual frontal advance distances of up to 100 m that were possibly even exceeded for several years. For example, Nigardsbreen advanced c. 2.8 km between 1710 and 1735 CE alone (Bogen et al. 1989; Fig. 3.4e). During the final phase of Nigardsbreen's major 'Little Ice Age'-advance, the farm Nygaard (subsequently serving as an eponym for the glacier) was abandoned and destroyed in 1742 (Fig. 3.4f). The glacier finally slowed down and reached its maximum extent in 1748 (Rekstad 1904; Fægri 1934). In Oldedalen, an ice avalanche from the advancing glacier tongue of Brenndalsbreen destroyed the farm Tungøyane in late 1743 (Rekstad 1901; Nesje 1994).

During the nineteenth-century CE historic evidence and documentation of individual glacier, terminus positions were recorded by an increasing number of scientific reports and maps (e.g. Forbes 1853), artistic visual displays (Grove 2004; Nussbaumer et al. 2011), and first photographs (cf. Winkler 1996a). This evidence provided reliable fixed points for glacier terminus positions and consequently facilitated the development of highly accurate lichenometric dating curves (Erikstad and Sollid 1986; Bickerton and Matthews 1992, 1993). Furthermore, lichenometry at Jostedalsbreen is substantially supported by well-developed and potentially complete moraine sequences in glacier forelands constituting typical landform assemblages (Fig. 3.7) and limited paraglacial modification of the original features. Lichenometric and historical dating reveal a fairly uniform image of the timing of the LIA-maximum at Jostedalsbreen with most individual outlets reaching their maximum extent between 1740 and c. 1760 CE (Table 3.1; Bickerton and Matthews 1993; Winkler 2002). At Jostedalsbreen, the LIA displays a unique character as the most important neoglacial event and is characterised by one major advance only. For these specific regional conditions its rather controversial name that does not describe its actual character in most mountain ranges worldwide very accurately seems even to be justified here. Investigations of related climatic causes and triggers (e.g. Nesje and Dahl 2003; Nesje et al. 2008b; Steiner et al. 2008; Imhof et al. 2011) reveal that increased winter precipitation has played a major role for the timing of the LIA-maximum advance at Jostedalsbreen. Despite its far minor magnitude, the recent advance during the 1990s can be considered as a modern analogue regarding its underlying mass balance perturbation and its climatic causes (Nesje and Matthews 2011).

Retreat from the LIA-maximum frontal positions at Jostedalsbreen during the second part of the eighteenth century CE was slow and continued through the entire nineteenth century. Bickerton and Matthews (1993) reveal about six clusters of simultaneous, widespread moraine formation at several Jostedalsbreen outlets and demonstrate their general concurrent response and comparable glacier front variability. Underlying mass balance perturbances (and thus their climatic triggers) cannot, however, be accurately analysed due to short duration and low magnitude of these individual advances. Nonetheless, the most distinct clusters of moraine formation coincide with two recorded advances of 6- to 7-year duration each in the early twentieth century (Bogen et al. 1989; Winkler 1996b; Andreassen et al. 2005). These advances culminated around 1910 and 1930, respectively. Whereas they have been recorded at all sensitively responding short outlets of Jostedalsbreen, the longer outlets merely exhibited slower retreat rates or stillstands at that time (Table 3.3). The different response can be explained by different terminus response times. Short and steep outlets respond in their majority only 3–4 years after a triggering climate/mass balance perturbation that may be insufficiently strong and/or long to cause the longer glaciers to commence advancing (Nesje 1989; Laumann and Nesje 2009). Analysis of meteorological data indicates that the first advance in the early twentieth century was mainly triggered by above-average winter precipitation, the second one by substantially below-average summer air temperatures (Winkler 2001).

Following the short-term advances during the early twentieth century CE, a widespread and substantial glacier retreat commenced at Jostedalsbreen (Fægri 1950; Hoel and Werenskiold 1962; Bogen et al. 1989; Andreassen et al. 2005). It is unanimously linked to warm summers and overall above-average air temperatures during the 1930s and 1940s (Nesje et al. 1995, 2008a). During the late 1940s and early 1950s, however, glacier fronts at some shorter outlets of Jostedalsbreen started to stabilise. The previous substantial retreat terminated and was followed by virtually stationary frontal positions displaying minor oscillations. By contrast, the longer outlets of Jostedalsbreen did not stabilise until later in the 1970s as a result of their longer terminus response times (Nesje 1989; Winkler 1996b). In a few cases, additional local factors such as calving into an emerging proglacial lake played an important role (e.g. at Nigardsbreen; Haakensen 1996). At several small outlets, glacier termini stabilisation was followed from the 1950s until the

Table 3.3 Advance distances at selected outlets of Jostedalsbreen during the two prominent re-advances in the first half of the twentieth century CE. Compare the advance distances with those recorded during the recent advance in the 1990s (see Table 3.4; raw data: www.nve.no)

Outlet	1st Advance	Distance (m)	2nd Advance	Distance (m)
Austerdalsbreen	(1905)[a]—09	62.5	No data	
Bergsetbreen	(1903)—10	135.0	1922—31	127.0
Bødalsbreen	(1900)—12	47.5	1922—30	62.0
Bøyabreen	(1904)—11	133.5	1921—31	164.0
Brenndalsbreen	1905—13	96.0	1922—32	73.0
Briksdalsbreen	1904—10	78.0	1921—31	66.5
Fåbergstølsbreen	(1907)—10	25.0	1922—30	112.5
Kjenndalsbreen	(1907)—09	44.0	1922—32	190.5
Lodalsbreen	No advance recorded		No advance recorded	
Melkevollbreen	1902—11	177.0	1922—29	208.0
Nigardsbreen	No advance recorded		1921—30	19.0
Stigaholtbreen	(1903)—11	36.5	1922—32	45.0
Supphellebreen	(1899)—12	95.5	1921—31	92.0
Tunsbergdalsbreen	No advance recorded		No advance recorded	

[a]Exact start of advance undetermined due to multi-year measurement interval

late 1980s by a slight overall advance, although this did not attract much attention and mass balance records demonstrate only a slight increase of the overall glacier mass in western South Norway during that time (Andreassen et al. 2005, 2016).

3.4 Glacier Forelands at Jostedalsbreen and Their Landform Assemblages

3.4.1 Landform Assemblages

Glacier forelands and their landform assemblages at Jostedalsbreen subsequently described here display some specific structures that deviate from common landform assemblages of the glaciated valley systems as outlined by Benn et al. (2005). Analysing the potential reasons for these differences, two partly interrelated facts need to be highlighted prior to any subsequent in-depth studies. There is no substantial supraglacial debris supply or debris transport at Jostedalsbreen impacting sediment transfer or geomorphological dynamics at its outlets. Even with the few exceptions from this general rule (e.g. Austerdalsbreen, Kjenndalsbreen), processes related to supraglacial debris are of little influence on the overall configuration of the glacier foreland. Glacier morphology (outlet glaciers from a central plateau ice-cap), a lack of nunataks and relatively high weathering resistance of the dominant bedrock are responsible for the lack of supraglacial debris that is otherwise an influential factor within the glacial process system of mountain regions (Benn et al. 2005; Baumhauer and Winkler 2014). A second fact that needs to be highlighted is the regional gross morphology determining generally very flat (1° and less) and wide glacial valley floors (Fig. 3.2). This geomorphological framework is (far) outside the dimension Holocene glaciers are able to actively modify; hence, gross morphology needs to be seen as an external factor for the configuration of glacier forelands and their landform assemblages. Additional to glacier front position changes themselves, the regional 'glaciological forcing' comprises a maritime glaciological regime exhibiting general high mass turnovers within the mass balance (Winkler 2009). This regime not only causes relatively sensitive responses of the glaciers to mass balance perturbations but also the high mass turnover creates regularly high meltwater discharge regardless of the overall mass balance trend. Finally, as empirically demonstrated during the recent glacier advance, the occurrence of seasonal glacier front changes is a typical phenomenon in maritime regions with high mass turnover and constitutes an important driver of moraine formation at Jostedalsbreen (see below).

The configuration of glacier forelands is decisive for their accurate chronological and palaeoclimatic interpretation. In this context, 'configuration' is understood as the relationship of number, dimension and position of glacial or glaciofluvial landforms and includes additional considerations of their respective individual genetic type. Applying this approach facilitates a solid geomorphological basis for the reconstruction of the temporal and spatial development of existing landform assemblages. Usually, the number of moraines and their morphology will particularly be in the focus of such analyses, but spatial distribution and overall areal coverage of glaciofluvial landforms and sediment are also regularly considered. Kirkbride and Winkler (2012) highlight, however, that the original configuration of a glacier foreland may be subject to erosional censoring with older landforms destroyed or modified by subsequent glacier activities.

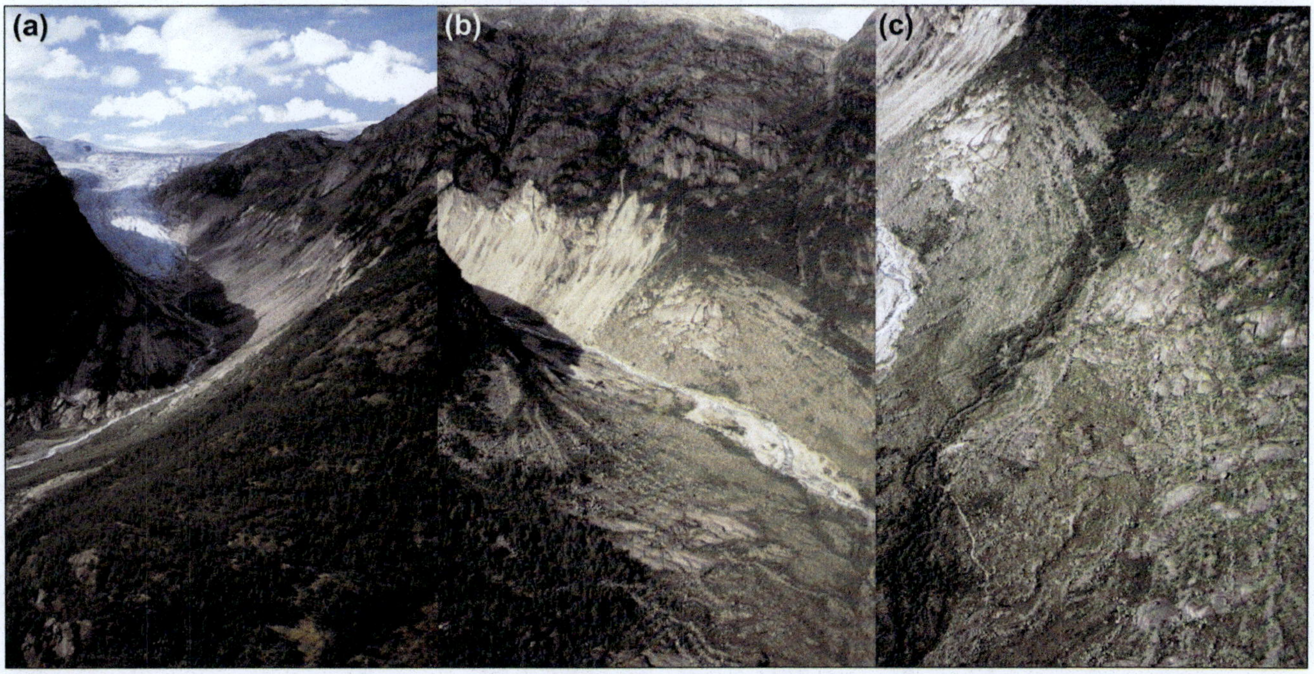

Fig. 3.5 Glacier foreland of Fåbergstølsbreen: **a** Fåbergstølsbreen currently retreated into a short western tributary valley of Jostedalen. The northern slope with old glacial sediments affected by paraglacial erosion is clearly visible in the centre of the image. **b** Detail of the glacier foreland of Fåbergstølsbreen showing the bedrock sill separating the inner and outer foreland. To the left (background), the paraglacially reworked sediment on the slope can be seen, and numerous (lateral) moraine ridges in the outer foreland. **c** Sequence of lateral moraine ridges in the northern outer foreland of Fåbergstølsbreen. Note the exposed bedrock indicating thin sediment cover in the right part of the image (*All photos* Stefan Winkler 24.08.2004)

At Jostedalsbreen, the well-established glacier chronology with all landforms on the glacier forelands dating from the LIA or later resolves problems that may emerge in cases of lacking detailed age constraints at some outlets comparatively easy by allowing relatively reliable correlations between individual outlets. Nevertheless, paraglacial processes and postdepositional erosion may inflict a modification of the original glacier foreland configuration as will geomorphological processes on the surrounding valley slopes such as mass movements and snow avalanches (Laute and Beylich 2013, 2014). Ballantyne (1995) investigated debris cone formation on deglaciated terrain on the foreland of Bergsetbreen (Fig. 3.2) and found that due to exhaustion of entrainable sediment on the upper slopes their formation had ceased before 1965 CE; they became inactive within 100–200 years of deglaciation. The bedrock resistance to weathering and resulting low production rates of regolith may contribute to relatively limited overprinting of most glacier forelands by paraglacial processes. A local exception is the inner foreland of Fåbergstølsbreen with its sediment-covered northern slope exposed to significant gullying (Ballantyne and Benn 1994; Curry 1999, 2000; Curry and Ballantyne 1999; Fig. 3.5). Glacial sediment is eroded from the upper slopes and deposited in the form of debris cones at their base by debris flows supported by snow avalanche activity. Gullies usually stabilised if the sediment becomes exhausted. Detailed sedimentological analysis (Curry and Ballantyne 1999; Curry 2000) has revealed that in this case, till deposited during the 'Little Ice Age' was overlying reworked early Holocene till indicating that paraglacial reworking on the steep slope had also occurred previously. Although mass movement processes (rockfalls, debris flows etc.) and snow avalanches have been observed quite frequently at other outlets, talus cones, alluvial fans or avalanche deposits do not substantially influence the general character of the glacier forelands.

In a number of glacier forelands around Jostedalsbreen the local bedrock is exposed or only thinly covered by sediment within the glacier forelands (Fig. 3.5c). The bedrock shows clear signs of glacial or glaciofluvial erosional overprinting (Figs. 3.3d and 3.6f). Micro-scale erosional landforms of glacial origin include not only striations but also other features belonging to the diverse group of friction cracks (e.g. crescent gouges; Fig. 3.6a–c). Together with the omnipresent glacially polished bedrock surfaces these features are evidence of a typical warm-based (wet-based) glacier regime at Jostedalsbreen. Larger areas of exposed bedrock have been modified into groups of meso-scale features like roches mountonées or rock drumlins (Fig. 3.6d). As pointed out by Bennett and

Fig. 3.6 Micro- and meso-scale glacial and glaciofluvial erosion features: **a** Glacial striae at Nigardsbreen, ice movement parallel to the ice axe. **b** Crescentic gouges at Tunsbergdalsbreen, ice movement from left to right. The total length of the sequence is c. 4 m. **c** Crescentic gouge at Tunsbergdalsbreen with bedrock particles resulting from fracturing caused by high clast-bedrock contact forces not been removed/entrained by the ice in the recently deglaciated area. **d** Rock drumlins at Nigardsbreen, ice movement from right to left. **e** Pothole at Tunsbergdalsbreen. **f** Inner foreland of Brenndalsbreen showing intense glacial and glaciofluvial erosion of the exposed bedrock almost devoid of any loose sediment (*Photos* Stefan Winkler, **a** 27.06.2006, **b/c** 01.09.2008, **d** 20.06.2007, **e** 06.09.2009, **f** 24.06.2009)

Glasser (2009) or Winkler (1996a, 2009), detailed shape and morphology of these features are primarily influenced by lithological factors like foliation or preferred orientation of cracks and not by glacier dynamics or glacial processes.

Meso-scale features occasionally show interrelation with glaciofluvial erosion channels (Baumhauer and Winkler 2014) that also occur alongside occasional potholes (Fig. 3.6e).

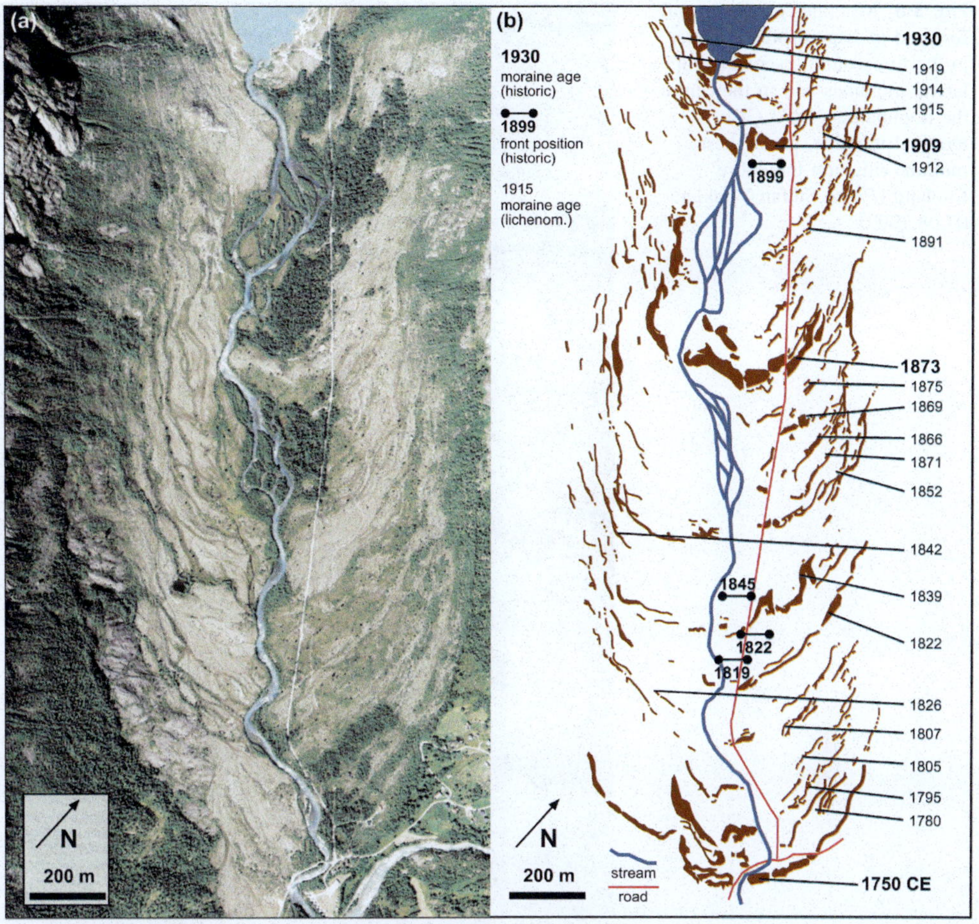

Fig. 3.7 **a** Orthorectified aerial photo (2012) of the glacier foreland of Nigardsbreen (adapted from www.norgeibilder.no, © Kartverket/www.kartverket.no). **b** Schematic map of moraine ridges on the glacier foreland of Nigardsbreen as historically/lichenometrically dated by Bickerton and Matthews (1992). Note that moraines and streams are shown as mapped by these authors. Predicted lichenometric ages may show inverse chronological order due to their uncertainty margins not displayed here in detail (modified after Bickerton and Matthews 1992)

3.4.2 Moraines

Possibly the most prominent characteristic of glacier forelands at Jostedalsbreen is their high number of moraine ridges, especially if one considers that their formation is chronologically restricted to the past c. 250 years since the LIA-maximum at around ~1750 CE (Fig. 3.7). Compared to moraines in other mountain regions, their size is mostly small to moderate and many moraine ridges do not exceed heights of 5 m (Figs. 3.8 and 3.9). Only a few reach crest heights of 10 m or more. On the flat forelands, they often appear as 'superimposed' on the otherwise flat terrain (Fig. 3.8). A distinct difference between terminal and lateral moraines in terms of both morphology and dimension that is generally considered characteristic for high mountain settings (Benn et al. 2005; Winkler 2009) does not occur at Jostedalsbreen. It can easily be explained because the typical huge 'alpine-type' lateral moraines sensu Winkler and Hagedorn (1999) are predominantly formed by dumping of supraglacial debris, the latter being significantly reduced or virtually absent in the region. Marginal moraines in lateral position at Jostedalsbreen are by contrast to those prominent lateral moraines in other high mountain regions generally of equal size to or smaller than terminal moraines on the glacier foreland. This fact points towards comparable processes of their respective formation. Lateral moraines are, however, more often bouldery and lack finer grain sizes (Fig. 3.9d, e), either due to primary differences in the preexisting sediment bulldozed up during their formation (see below) or secondary glaciofluvial erosion (winnowing) of finer grain sizes. An increase of subangular and angular boulders from the inner towards the outer lateral moraines has been observed, for example, at Nigardsbreen and Bødalsbreen, supporting the first explanation (Winkler 1996a). Where lateral moraines contain fine sediment, there are no differences in sedimentological composition with terminal moraines on the forelands. Unless glaciofluvial erosion has cut channels into the moraine ridges, they form continuous ridges comprised of terminal, latero-frontal and lateral sections reflecting the former ice-marginal positions. As a consequence of the sensitively responding outlets and efficient seasonal advances (see below), lateral moraine ridges occur in similarly high numbers as terminal moraines.

Prior to the recent advance during the 1990s, few studies focused on mechanisms of moraine formation at Jostedalsbreen. Andersen and Sollid (1971) investigated the most

Fig. 3.8 Moraines on the central foreland of Nigardsbreen. Notice in the foreground a sequence of lateral moraines and in the centre the dominating '1873 CE'-terminal moraine, the highest moraine ridge on the entire foreland (*Photo* Stefan Winkler 07.08.1992)

prominent terminal moraine on the Nigardsbreen foreland (Figs. 3.8 and 3.9c) formed during a short-term advance culminating in 1873 CE. They concluded that in addition to some bulldozing a large sediment slab frozen to a shear plane of the glacier and transported to its terminus was involved and responsible for the outstanding dimensions of that particular moraine. Its maximum crest height reaches c. 12 m, which is more than double the height of average moraine crests on the foreland. Although the occurrence of shear planes is documented on historic photographs (cf. Rekstad 1904; Fægri 1934; Winkler 1996a), its typical asymmetry (proximal slope 18°–20° and distal slope 24°–26°), a fluctuating height of the crest line and heterogenic sedimentology do not require such specific conditions but could well be explained by bulldozing alone. Shear planes have, furthermore, been observed during the recent advance (Fig. 3.13c) but did not or at least not substantially contribute to moraine formation or till deposition. By contrast, the famous "saw-tooth" terminal moraines at Bødalsbreen (Figs. 3.9a and 3.10; Fredin and Burki 2008) have been attributed by Matthews et al. (1979) and Lien and Rye (1988) to bulldozing processes and a (re-)advancing glacier snout deeply incised by radial crevasses.

Fig. 3.9 Marginal moraines on glacier forelands at Jostedalsbreen: **a** 'Saw-tooth' moraine at Bødalsbreen (formed c. 1815 CE). **b** Terminal moraine at Bergsetbreen (c. 1930). **c** Terminal moraine formed in 1873 (historical evidence) at Nigardsbreen. **d** Lateral moraine at Bergsetbreen (c. 1850). **e** Lateral moraine at Tuftebreen (c. 1800). **f** Small terminal moraine at Nigardsbreen (c. 1815). **g** Outermost terminal moraine at Nigardsbreen (historically dated to 1748; all lichenometric ages following Bickerton and Matthews 1992, 1993; *Photos* Stefan Winkler, **a** 03.09.1993, **b** 01.08.1992, **c/g** 25.08.2007, **d** 20.08.1995, **e** 25.07.2008, **f** 05.08.2018)

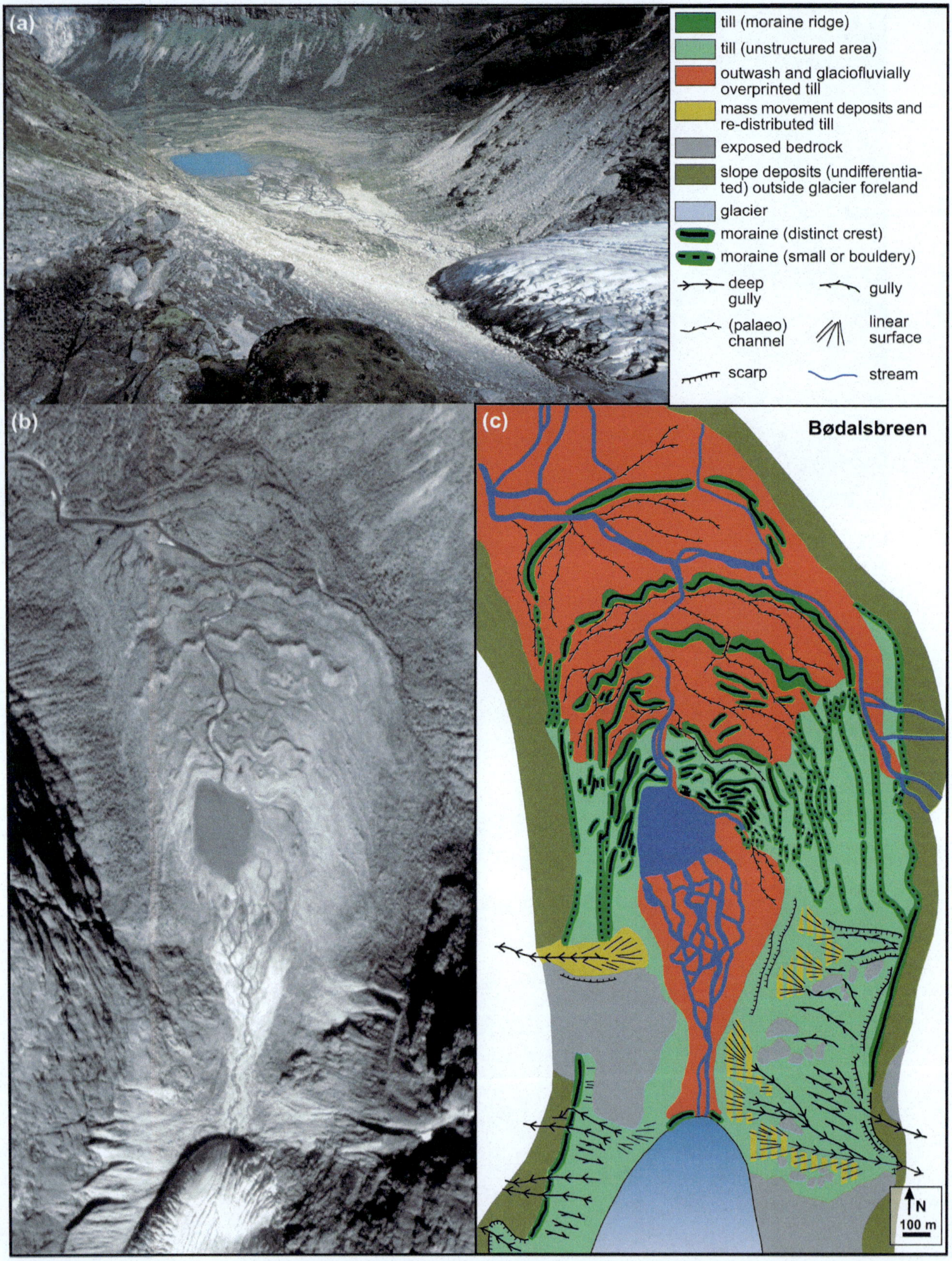

Fig. 3.10 Glacier foreland of Bødalsbreen: **a** View from the eastern lateral moraine of Bødalsbreen down the advancing glacier tongue into the inner foreland (*Photo* Stefan Winkler 26.08.1993). **b** Aerial photo of Bødalsbreen's glacier foreland taken on 08.09.1993 and showing the convex shape of the advancing glacier tongue (see **a**; © Fjellander Widerøe, modified). **c** Schematic geomorphological map of Bødalsbreen's glacier foreland following the mapping scheme for the Norwegian Quaternary geological maps (Kvartærgeologisk kart) and showing glacier position and proglacial lake outline around 2000 (modified after Winkler 2001)

Fig. 3.11 Glaciofluvial landforms: **a** Valley sandar/outwash plain at Bergsetbreen (deglaciated c. 1850–1880 CE). **b** Valley sandar and palaeochannels at Nigardsbreen (c. 1820–1840). **c** Central glacier foreland of Fåbergstølsbreen showing glaciofluvially overprinted till and glaciofluvial sediment (c. 1830–1870). **d** Typical vegetation of valley sandar at Begsetbreen (deglaciated c. 1850). **e** Active delta builds into the proglacial Sætrevatnet at Bødalsbreen (see Fig. 3.10). **f** Active valley sandar in Kjenndalen, an unusual narrow valley at Jostedalsbreen. The contemporary terminus of Kjenndalsbreen was just right off the image (all lichenometric ages following Bickerton and Matthews 1992, 1993; *Photos* Stefan Winkler, **a** 20.08.1995, **b** 07.08.1992, **c** 17.08.2006, **d** 23.08.1995, **e** 22.06.2009, **f** 23.08.1992)

The outermost LIA-moraine at Bødalsbreen does not display a comparable special outline or morphology (Fig. 3.10). It is also lower than the two major saw-tooth moraines with their crest heights of up to 8–9 m. Burki et al. (2009a) explain the difference between the arcuate shape of the outermost LIA-moraine at Bødalsbreen and the saw-tooth shape of the two large and subsequent smaller inner moraine ridges with the geometry of the Bødalen valley. Bødalen widens considerably downvalley of the modern glacier front and the proglacial Sætrevatnet lake (Fig. 3.10a) but narrows again in the outer part of the foreland at about Bødalsbreen's LIA-maximum position.

Detailed sedimentological studies that revealed no preferred clast orientation or dip of the moraine sediments allow

Burki et al. (2009a, b, 2010) to exclude any squeezing mechanism at Bødalsbreen. They interpret moraine sediment as reworked (recycled) proglacial sediment comprising mostly sandy diamicton (similar to the findings of Winkler 1996a; Fig. 3.18h). Unlike other studies, however, they consider some indications of limited thrusting of sediment frozen to the glacier sole by small thrust planes as described by Krüger (1993) in Iceland as a possible initial step of moraine formation prior to bulldozing developing its efficiency. They admit that such basal freezing would only have worked near the glacier margin and under thin ice and not in situations where the glacier underwent substantial advance displaying a convex-shaped, steep glacier front, as during the 1990s CE. Ultimately, observations during the recent glacier advance (see below) clearly revealed that bulldozing is the most prominent process of moraine formation at Jostedalsbreen, which also explains the high number of moraine ridges (Winkler and Matthews 2010).

3.4.3 Glaciofluvial Landforms and Sediments

The flat valley floors, typical of the glacier forelands around Jostedalsbreen, afford minimal topographic constraint to the dynamics of glaciofluvial erosion, transport or deposition. Because most moraine ridges are of small to moderate size, they also rarely act as major obstacles in this context and can, furthermore, be easily breached and eroded. With only a few forelands experiencing topographic control in the form of bedrock sills or narrow valleys (Fig. 3.11f), valley sandars (valley trains, glaciofluvial outwash plains) develop their typical dynamic behavior characterised by frequent changes of active channels and rapid development of sand or gravel point bars. The high glacier mass turnover causes large quantities of meltwater that are responsible for a high proportion of the entire sediment transfer of the glacial system being glaciofluvially controlled. The efficiency of sediment evacuation and subsequent deposition by meltwater is well demonstrated by the rate in which the proglacial lake Sætrevatnet at Bødalsbreen has shrunk by the growth of a delta (50% between 1966 and 2000 CE and c. 75% between its first appearance in the 1940s and 2005; Figs. 3.10c and 3.11e), or the ongoing growth of the glaciofluvial delta into Nigardsvatnet at Nigardsbreen (Haakensen 1996; Bogen et al. 2015; see also Chap. 4). Especially in the central and outer sections of glacier forelands, glaciofluvial features (mostly outwash plains) dominate and cover the majority of the surface, rather than moraine ridges or undisturbed subglacial till (Fig. 3.11a, b). Moreover, flat areas of initial subglacial till are frequently glaciofluvially overprinted (Fig. 3.11c; Winkler 1996a).

By applying lichenometry on the Austerdalsbreen glacier foreland, Maizels and Petch (1985) and Petch and Whittaker (1997) demonstrate that the dynamics of the outwash plains are not necessarily chronologically related to the adjacent moraine ridges. Instead, they have subsequently breached and intersected small moraine ridges formed during previous glacier re-advances. Glaciofluvial action is regarded as being responsible for the lack of any major imprint of supraglacial debris on the central and outer foreland of Austerdalsbreen. On most forelands at Jostedalsbreen outwash plains cover vast areas between the moraine ridges (Fig. 3.10c) and display a gravelly, stony surface that is easy to trace owing to a relatively sparse vegetation cover (Fig. 3.11b, d). In this regard, they contrast with moraine ridges that are often more densely vegetated, unless they are bouldery and lacking fine sediment. Except for areas affected by local stillwater sedimentation causing higher percentages of silt-rich fine sediment (Winkler 1996a), the majority of glaciofluvial outwash plains exhibit slightly convex-shaped surfaces structured by numerous small, shallow palaeochannels (usually 50–70 cm deep) as typical for braided rivers.

3.5 Observations of Recent Moraine Formation

3.5.1 Recent Glacier Advance

In response to several positive mass budget years without interruption commencing 1988/89 CE, some of them with unprecedented high mass gains, all short outlets of Jostedalsbreen started to advance during the early 1990s (Nesje 1995; Winkler et al. 1997, 2009). The originally published net balance record has since been adjusted after careful re-analysis of the direct glaciological long-term mass balance measurements based on geodetic measurements (Andreassen et al. 2016). Although this led to a slight reduction of the overall mass gain, the mass balance perturbation remains to be significant (Fig. 3.12a). It was, as far as can be judged, the strongest glacier mass increase during the twentieth century (Fig. 3.12c). Most long outlets (e.g. Nigardsbreen) showed a different response due to their longer terminus response times. Some did not show any strong advance comparable to Jostedalsbreen's short outlets, while others advanced likely because of a slight mass increase during the 1970s and not primarily related to the more recent mass gain (Table 3.4). As far as existing data allow for such comparison, the recent advance was the strongest advance since the LIA-maximum advance in the first half of the eighteenth century (Fig. 3.4e; see Sect. 3.3.3). It has been explicitly referred to as the 'Briksdalsbre Event' by Nesje and Matthews (2011) in the light of its significance.

Due to comprehensive glaciological and meteorological data, the recent advance provided a useful opportunity to

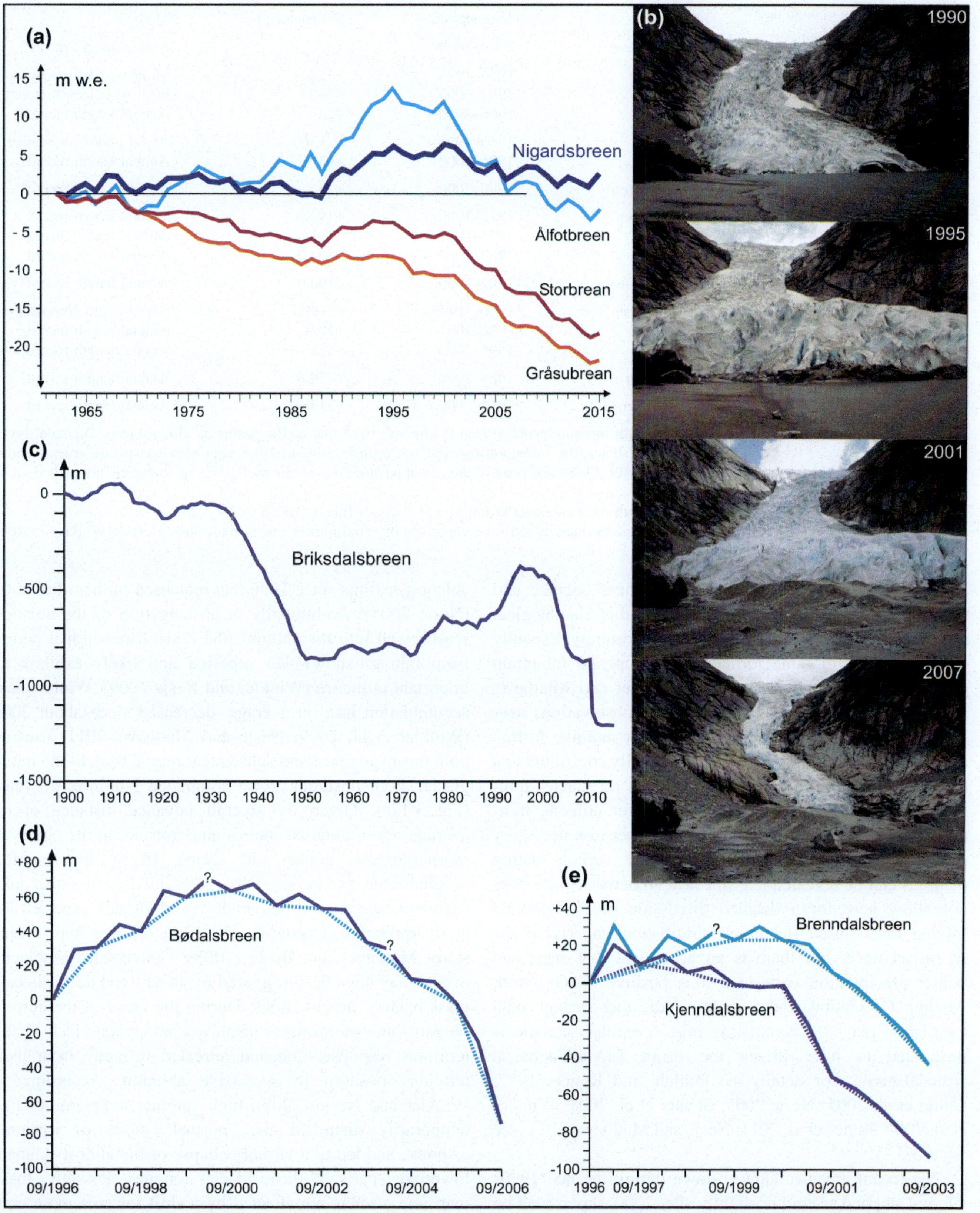

Fig. 3.12 **a** Cumulative glacier net balance of Nigardsbreen compared with the more maritime Ålfotbreen (outer Nordfjord), Storbrean and Gråsubrean (both in more continental Jotunheimen; data: www.nve.no; Fig. 3.1). **b** Image comparison of Briksdalsbreen (*Photos* Stefan Winkler, 08.08.1990, 05.09.1995, 05.09.2001, 30.08.2007). **c** Length change record for Briksdalsbreen, the only uninterrupted record for short outlet glaciers of Jostedalsbreen (data: www.nve.no). **d** Length changes recorded at Bødalsbreen as revealed by biannual measurements in June (06) and September (09). The dotted line represents the 'regular' annual length record and reveals that seasonal advances are not picked up (*Data* Stefan Winkler/www.nve.no). **e** Seasonal length changes at Brenndalsbreen and Kjenndalsbreen. The subsequent retreat commenced earlier than at Bødalsbreen (*Data* Stefan Winkler/www.nve.no)

Table 3.4 Length change data for selected outlets at Jostedalsbreen during the late twentieth century CE (*Data* www.nve.no, Winkler 1996s)

Outlet	Time period	Distance (m)	Method
Austerdalsbreen	1989–2000	+82	Annual length record[a]
Bergsetbreen	1966–1993 1993–2000 1996–2000	+412 +182 +25	Multi-year length record[b] Multi-year length record Annual length record
Bødalsbreen	1960–1995 1996–2000	c. +360 +63	Aerial photo comparison[c] Annual length record
Brenndalsbreen	1996–2000	+24	Annual length record
Briksdalsbreen	1973–1980 1973–1987 1987–1997	+199.0 +161.0 +304.0	Annual length record Annual length record Annual length record
Fåbergstølsbreen	1992–2000	+190.0	Annual length record
Kjenndalsbreen	1960s–1995 1996–1997 1996–2000	c. +900 +10.0 −2.0	Aerial photo comparison Annual length record Annual length record
Nigardsbreen	1989–2000	+259.0	Annual length record
Stigaholtbreen	1996–2000	+31.0	Annual length record

[a]Regular length measurements at annual interval. Note that at Bergsetbreen, Bødalsbreen, Brenndalsbreen and Kjenndalsbreen the measurements did not resume prior to 1996 after previous measurements ended during the 1940s. Onset and peak of the recent advance during the mid-1990s is, therefore, not picked up by the data

[b]Regular length measurements at multi-year intervals (breaks of up to 25 years)

[c]Length change estimate based on comparison of aerial photos, maps, and other sources (Winkler 1996a)

analyse the climatic forcing of glacier mass balance and subsequent length changes. Additional to this glaciological aspect, the advance offered the rare opportunity to study recent terminal moraine formation at temperate mountain glaciers (Winkler and Nesje 1999; Winkler and Matthews 2010). At least on a regional scale, these observations may serve as a model and modern analogue for moraine formation during the LIA and, therefore, potentially contribute to a better interpretation of the older moraines on glacier forelands. They may also have implications for utilising moraines as evidence for glacier chronologies because necessary precautions with sampling strategies for various dating methods can be revealed (e.g. for lichenometry). Space does not allow here for a detailed discussion of the climatic forcing that caused the mass perturbation underlying the recent advance, but there is an agreement that increased winter precipitation connected to a positive NAO (North Atlantic Oscillation) index regime indicating strong zonal circulation played a dominating role. A similar scenario is postulated to have driven the major LIA-advance at Jostedalsbreen (for details see Pohjola and Rogers 1997; Chinn et al. 2005; Nesje 2005; Steiner et al. 2008; Winkler et al. 2009; Imhof et al. 2011; Nesje and Matthews 2011; see Sect. 3.3.3).

The recent advance slowed down during the late 1990s CE and stopped around or slightly after 2000, superseded by a strong glacier retreat that accelerated until present (Fig. 3.12b and Table 3.4; Kjøllmoen 2017). The current retreat has been attributed to high air temperatures in the ablation seasons since 1996 that increased further after 2001 (Nesje 2005). Additionally, a prolongation of the ablation season well into the autumn (the critical transitional period from rain to snow) was reported and likely exhibits an important influence (Winkler and Nesje 2009). Winter snow accumulation has, on average, decreased since about 2000 (Winkler et al. 2009; Nesje and Matthews 2011) causing both recent advance and subsequent retreat back to the initial glacier front positions to be completed within c. 20 years (Fig. 3.12c). Given the overall advance distance of on average a few hundred metres and considering its observed morphological impact, it seems likely that similar decadal-scale events are also represented in regional lake sediment records and potentially yield signals expected for an interpretation as neoglacial or 'Little Ice Age'-type events sensu Matthews and Briffa (2005). Conversely, neoglacial events may have been triggered by short-lived decadal-scale mass balance perturbations. During the onset of the current retreat, some short outlets displayed an apparent lack of any terminus response time and retreated instantly from their terminus position by excessive ablation. According to Winkler and Nesje (2009), high summer temperatures had temporarily disturbed the 'normal' mode of dynamic response and led to a virtual collapse of the almost stagnant lowermost parts of those glacier tongues. Possibly, these maritime glaciers are affected by a shift towards more continental and ablation-season dominant glaciological regimes with a stronger impact of (currently high) summer air temperatures (Winkler et al. 2009).

During the culmination of the recent advance, considerable seasonal advances (winter advances) were recorded at some short outlets (Fig. 3.12d, e; Winkler 2008a). Based on length measurements conducted biannually in early and late summer they have been shown to generally range between 20 and 30 m, which is considerable in relation to the size of the glacier tongues. Seasonal length changes were most accentuated in years with small annual length change, that is, in the transitional period between the culmination of the advance and onset of the subsequent retreat. Seasonal advances were not monitored during the main phase of strong frontal advance and the accelerating retreat since the mid-2000s CE. Substantial seasonal length variations at maritime glaciers are causally linked to high glacier mass turnovers that result in efficient mass transfer. Whereas ice movement at the glacier front seasonally accelerates early in spring in response to, among others, basal hydrological conditions, ablation and frontal summer backmelting will peak later during the summer season to (over)compensate for the advance. This is the very reason annual length measurements are performed at the end of the summer season, preferably on similar dates. Any seasonal frontal variations are, therefore, generally not recorded. Recent observations of seasonal advances at Jostedalsbreen constituted rare opportunities and provided a good explanation for the high number of terminal moraines on glacier forelands, suggesting they regularly occurred during the retreat from the LIA-maximum position in the late eighteenth and nineteenth centuries.

3.5.2 Observations of Glacier Margins and Geomorphological Processes at Selected Outlets

The geomorphological situation at the glacier margins of several selected short and long outlet glaciers of Jostedalsbreen has been continuously monitored by the author at least once a year since the early 1990s throughout both recent advance and subsequent current retreat (Winkler 1996a, 2001, 2002, 2009; Winkler and Nesje 1999; Winkler et al. 2009, 2010; Winkler and Matthews 2010). This monitoring included a detailed observation of (i) the position and state of the glacier tongue, (ii) the occurrence of subglacial debris in marginal cavities or englacial debris connected to shear planes, (iii) the extent of supraglacial debris cover on the lower glacier tongue and any related depositional processes and (iv) the characteristics of moraine ridges (morphometry and sedimentology) during their ongoing formation process at the glacier fronts. Annual photo series documenting the morphology of the lower glacier tongues were initiated during this time as a complement to the standard annual length measurements to support the analysis of related glacier dynamics (Winkler 2008b; Fig. 3.12b). Geomorphological fieldwork was conducted at some outlets with special attention to mapping, morphometric measurements (e.g. cross sections), clast roundness and grain size analyses potentially indicating debris origin and, hence, modes and mechanics of moraine formation (cf. Winkler 1996a, 2001; Winkler and Nesje 1999; Winkler and Matthews 2010 for methodological details).

Briksdalsbreen has likely been selected as ‘type locality’ for the recent glacier advance (Nesje and Matthews 2011) because it is the only short outlet whose annual length changes were consecutively monitored during the twentieth century CE without interruption (Fig. 3.12c). At other short outlets measurements were either not performed annually (Bergsetbreen) or resumed after an extended break during the mid-twentieth century as late as 1996 when the advance already approached its culmination (Bødalsbreen, Brendalsbreen, Kjenndalsbreen; Table 3.4). Comparison of maps and aerial photos, however, facilitates estimation of their advance since they experienced their minimum positions during the 1950s or 1960s. On that basis, their length change can be assumed similar to Briksdalsbreen (Winkler 1996a). Between 1987 and 1997, Briksdalsbreen advanced 304 m and exceeded annual advance distances of 50 m; similar advance distances can be deduced for the other short outlets based on annual observations. At some longer outlets, considerable advance distances have been recorded as well, for example, at Fåbergstølsbreen and Nigardsbreen (Table 3.4). Although their advances were recorded parallel to that at short outlets, it needs to be noted that Briksdalsbreen advanced 199 m between 1973 and 1980 before it experienced a period of moderate retreat (38 m) until 1987. Given the different response times and the regionally representative net balance record (Fig. 3.12a), the slowly commencing advance at some long outlets is likely linked to that earlier mass balance perturbation. During the recent advance, all outlets of Jostedalsbreen advanced over proglacial areas of either preexisting sediment or exposed bedrock. The only exception was Briksdalsbreen, which advanced into a proglacial lake that fills an erosional bedrock basin (McManus and Duck 1988; Haakensen 1996) until it was almost completely covered by the glacier in 1997 (Fig. 3.12b).

After reaching their maximum extension between 1997 and 2000, the glacier fronts of all short outlets monitored in detail were stationary for two or three years prior to showing any significant retreat. Whereas the glaciers are listed as stationary in their annual length change records, they experienced the abovementioned seasonal glacier front variability with seasonal advances during spring (ironically also called ‘winter’ advances) that dominated the prevailing glacier dynamics. Moraine ridges that existed along the glacier margins were either modified (Briksdalsbreen, Bødalsbreen)

Fig. 3.13 Recent glacier advance during the 1990s CE at Jostedalsbreen: **a** Glacier front of Bergsetbreen during the peak of the recent advance. **b** Advancing glacier front of Bødalsbreen showing small radial crevasses (see Fig. 3.10). **c** Shear plane and melt-out of englacial debris at the advancing glacier front of Bergsetbreen. **d** Shear planes and melt-out of englacial debris at Austerdalsbreen. It is subsequently deposited like flow till over a large marginal snow patch in the foreground. **e** Birch on an area deglaciated during the 1950/60s becoming overridden during the recent advance of Briksdalsbreen. **f** Birch at Briksdalsbreen destroyed by the advancing glacier pushing up sediment. **g** Birches and other plants destroyed by recent moraine formation at Briksdalsbreen incorporated into the moraine ridge (*Photos* Stefan Winkler, **a** 20.08.1995, **b** 01.09.1996, **c** 30.08.1991, **d** 12.08.1993, **e** 04.09.1996, **f** 11.09.1997, **g** 19.09.1997)

or small subsequent moraines formed immediately proximal to them (Bergsetbreen, Kjenndalsbreen; see below and Fig. 3.18). Once the outlets entered the current, ongoing phase of accelerated retreat and left the narrow marginal zone previously affected by moraine formation during the culmination of the recent advance and the subsequent stationary period, only active glaciofluvial processes could be observed at the glacier margins. Coarse glaciofluvial outwash was deposited immediately proximal to the recent moraine ridge at Bødalsbreen caused by a temporarily restricted meltwater outflow (Fig. 3.17b). The area proximal to the recent moraines at most outlets is strongly modified by meltwater activity. The lower the topography of the inner forelands that restricts (glacio)fluvial dynamics the larger the area covered by outwash plains or glaciofluvially overprinted subglacial till deposits. The current situation of glaciofluvial dynamics dominating the time period after an advance has terminated is clearly analogue to the outer LIA glacier forelands where extensive valley sandars give evidence of identical processes dominating the time periods between re-advances, which are represented by moraine ridges (Figs. 3.10 and 3.11).

Like other short outlets, Bergsetbreen was characterised by steep glacier fronts with convex profiles during the advance in the 1990s CE (Fig. 3.13a). Bødalsbreen additionally displayed small radial crevasses (Figs. 3.10 and 3.13b) that, despite their minor size, may serve as modern analogues for re-advances during the nineteenth and early twentieth centuries and support the conclusions on how the 'saw-tooth' moraines at Bødalsbreen were formed (see Sect. 3.4.2). When Bergsetbreen was still close to the glacier's impressive icefall in the early 1990s shear planes became visible, indicated by thick englacial debris bands (Fig. 3.13c). These shear planes resulted from compressional ice flow at the foot of the icefall. Only sparse accumulation of flow till originating from this englacial debris took place during the early 1990s and never played a substantial role. Shear planes and flow till also occurred at Austerdalsbreen (Fig. 3.13d), the only outlet with considerable supraglacial debris cover. Despite Austerdalsbreen experiencing a recent advance followed by a stationary glacier front for several years the recorded, the overall advance distance was comparably small due to specific local conditions (Table 3.4). During the preceding retreat in the 1970s and 1980s, a massive complex of debris-covered dead-ice had developed immediately in front of the active glacier tongue. This complex constituted a prominent morphological feature and major obstacle when the active part of the glacier tongue commenced its advance. By 1995, the ablation moraine already had been compressed to 50% of its original length by the pressure of the advancing ice. Whereas the part in contact with the active glacier underwent ongoing modification, the downvalley part escaped rather unaffected and successive melting of dead-ice caused the development of kettle holes and successive downwasting (for details see Winkler 1996b, 2001).

Due to specific morphology and topography, two different sections of the glacier forland need to be differentiated for the investigation of moraine formation at Briksdalsbreen (Winkler and Nesje 1999). The flat central area of the foreland occupied by the proglacial lake possesses a specific sedimentology in the form of water-saturated glaciolimnic sediment dominated by silt with only a few large boulders along the shore and minor quantities of sand and gravel in a few places (Winkler 1996a). By contrast, the lateral and latero-frontal parts of the foreland at the foot of the steep valley show slope angles of up to 25°–30° increasing towards the upper parts of the slope. The latter is solid bedrock but its lower sections and foot slopes are covered by coarse, partly bouldery diamicton, predominately till deposited during the previous expansion period following the LIA-maximum until the mid-twentieth century CE when the area was successively exposed during glacier retreat. Birch vegetation and dense scrub had been established prior to the recent advance commencing. This resulted in impressive scenes of trees and scrubs being overridden by the steep, advancing lateral/latero-frontal ice-front (Fig. 3.13e–g). The differing terrain and sedimentology within the foreland of Briksdalsbreen subsequently offered a good opportunity to investigate the influence of these two factors for moraine formation processes in comparison to other small outlets.

No unambiguously distinct moraine ridges were observed during the peak advance in the mid-1990s CE when annual advance distances of several tens of metres were typical. At Bødalsbreen linear accumulations of large boulders with a total height of less than 2 m were present during some years (Fig. 3.14a, b, d). They were temporary only and were subsequently overridden and destroyed. In 1996, however, a cluster of very large boulders was rotated and pushed together to form a small moraine ridge at the eastern latero-frontal section along with the glacier tongue (Fig. 3.14c). Overriding and destruction of small temporary moraine ridges along the glacier margins were also observed at other outlets, for example, at Bergsetbreen. Here, a small moraine comprising plant remains, eroded soil and large boulders had been pushed up by the advancing glacier front. By 1996, it had developed into a continuous terminal moraine with a measured height of 2–3 m along the entire glacier margin. An interesting discovery was boulders with growing lichens of >30 mm individual thallus diameter incorporated within the distal slopes of the moraine (Fig. 3.14f; similar boulders were discovered at Bødalsbreen and Tuftebreen). According to lichenometric growth curves for the *Rhizocarpon* subgenus (Bickerton and Matthews 1993), these lichens would have colonised the boulders

Fig. 3.14 Recent glacier advance during the 1990s CE at Jostedalsbreen: **a** Advancing Bødalsbreen with a steep and convex-shaped front profile and a small accumulation of boulders at its margin. **b** Boulder concentration at the lateral glacier margin of Bødalsbreen. **c** Large boulder and small moraine ridged pushed up at the lateral margin of Bødalsbreen. **d** Small (temporary) terminal moraine at the advancing Bødalsbreen. **e** Recent terminal moraine at Bødalsbreen showing a dislocated boulder with former mark (circled—see text). **f** Lichen-covered boulder on the distal slope of the recent latero-frontal at Bergsetbreen (*Photos* Stefan Winkler, **a** 26.08.1993, **b** 25.08.1992, **c** 01.09.1996, **d** 28.08.1994, **e** 29.08.2006, **f** 25.07.2008)

during the 1960s. These lichen-covered boulders provide evidence for both (i) the incorporation of proglacial boulders into a young moraine by bulldozing and (ii) limited rotation and displacement of some boulders during culmination of the glacier advance and related moraine formation. Lichens would not survive significant rotation or any subglacial displacement. Evidence that such displacement actually occurred at some glaciers was discovered at Bødalsbreen when a coloured marker painted on a large proglacial boulder used for annual length measurements in September 1996 was re-discovered when the glacier commenced its retreat. The dislocated boulder was embedded in the recent moraine ridge and based on calculations using the remaining other markers it had moved downvalley subglacially a distance of $\sim$10 m between 1999 and 2001 and slightly changed its orientation (Fig. 3.14e).

At Briksdalsbreen, very large boulders (up to 100 m^3) came in contact with the advancing ice margin. Some of these originated from previous rock falls and others were deeply embedded in the proglacial sediment, exhibiting glacial micro-erosion features that indicated a subglacial origin. In particular, those deeply embedded boulders with their long axis orientated parallel to the direction of the advancing glacier front were simply overridden and remained virtually undisturbed (Fig. 3.15a). By contrast, other large boulders were rotated in contact with the ice (Fig. 3.15c, d), in a few cases with supportive ‘wedges’ of glacier ice pushing underneath them (see Winkler and Nesje

Fig. 3.15 Boulders at the margins of Briksdalsbreen during the recent advance: **a** Boulder embedded in old till overridden by the advancing Briksdalsbreen without dislocation. **b** Small 'bulge' in front of a boulder pushed forward by the advancing Briksdalsbreen. **c** Boulder in contact with the latero-frontal margin of Briksdalsbreen; the boulder was subsequently rotated. d Large rotated boulder at Briksdalsbreen. During the preceding 3 years the boulder was rotated by a total of >140° (see Winkler and Nesje 1999; *Photos* Stefan Winkler, **a** 05.09.1998, **b** 22.06.1996, **c** 19.04.1997, **d** 05.05.1999)

1999 for details). A few boulders were pushed forward into the proglacial sediment without rotational movement and created 'bulges' of soft sediments a few tens of centimetre high pushed up in front of them (Fig. 3.15b). Several factors including the degree of embedding in the proglacial sediment, orientation in relation to the advancing ice, the slope angle of the proglacial area and weight may have influenced the different way boulders were affected by the advance of Briksdalsbreen. Smaller boulders were, however, usually incorporated in the moraine ridges and pushed up alongside plant remnants of the overridden birch trees, moss and grass mats and upper organic soil layers (Figs. 3.14g and 3.16a, b).

Whereas in the lateral section of Briksdalsbreen a bouldery moraine of 1–2 m height was formed, towards the latero-frontal section in less steep terrain where proglacial sediment had higher quantities of sand and gravel the recent moraine ridge reached up to 6–7 m height. The lateral expansion during the recent advance was substantially more restricted than marginal advance at the latero-frontal and frontal sections of the glacier tongue. This in combination with higher slope gradients and coarser sediment (old till and rockfall deposits/colluvium) exerted a higher shear resistance towards bulldozing/pushing. In the latero-frontal section old till and glaciofluvial sediments displayed lower shear resistance causing higher moraine ridges (Winkler and Nesje 1999). In lateral and latero-frontal sections the moraine exhibited steep (>30°) distal and often even steeper proximal (ice-contact) slopes. The latter were prone to collapse promoted by summer backmelting of the glacier margin leaving a steep ice-erosional scarp with occasionally exposed ice-cores (Fig. 3.16e). Those ice-cores represented glacier ice that formerly protruded directly into the proglacial sediment (Fig. 3.16a, b) but subsequently had become separated from the active ice by a combination of summer backmelting, subsequent burial by sediment and plant remains, development of shear planes due to increased friction or glaciofluvial action of en- or subglacial meltwater channels. Until the recent advance reached its culmination, the moraine was successively modified because summer

Fig. 3.16 Recent moraine formation at Briksdalsbreen: **a/b** 'Wedge' of advancing Briksdalsbreen protruding into preexisting, proglacial sediment. **c/d** Frontal section of Briksdalsbreen pushing up fine glaciolimnic sediment. **e** Temporary core of dead glacier ice separated from the active tongue of Briksdalsbreen exposed in the proximal slope of the recent latero-frontal moraine. **f** Central terminal moraine at Briksdalsbreen; the glacier front is a few metres to the left. **g** Sediment pushed up between the advancing Fåbergstølsbreen (background left) and a prominent marginal snow patch (*Photos* Stefan Winkler, **b** 05.09.1995, **d** 31.08.1994, **e** 19.09.1997, **f** 15.09.1997, **g** 23.07.1999)

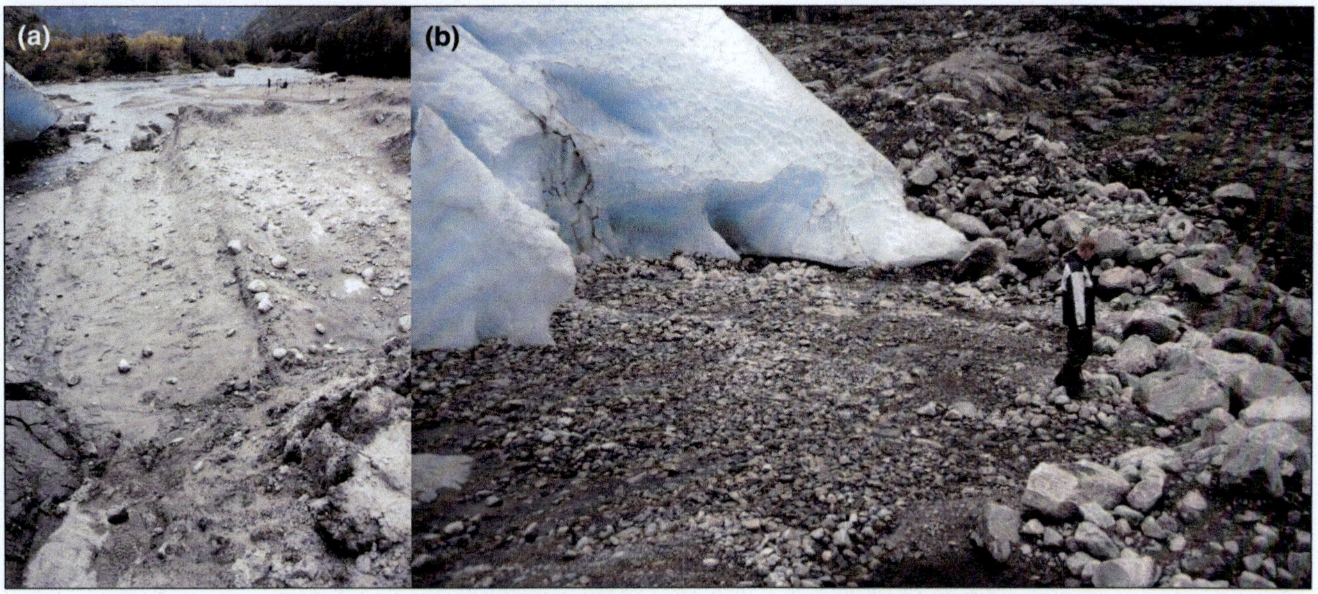

Fig. 3.17 **a** Subglacially formed micro-scale flutings exposed by summer backmelting at Briksdalsbreen. **b** Glaciofluvial sediment deposited between the recent terminal moraine (right part of image) and the retreating glacier margin at Bødalsbreen (*Photos* Stefan Winkler, **a** 22.09.1998, **b** 04.09.2001)

backmelting was restricted to a few metres. That was, however, sufficient to cause separation of the abovementioned protruding ice-wedges and the active glacier.

The central section of Briksdalsbreen's tongue was still located within the deeper parts of the proglacial lake in the early years of the advance. After it approached the shallower parts of the lake Briksdalsvatnet in summer 1994 CE a small moraine ridge consisting of water-saturated glaciolimnic sediment (silt-rich sand, Winkler 1996a) became visible (Fig. 3.16c, d). It showed deep extension cracks on its distal slope and contained an ice core. This moraine was destroyed during the ongoing advance and, at its culmination, a complex terminal moraine had formed at the frontal glacier margin (Fig. 3.16f). In contrast to the latero-frontal sections of the terminal moraine, the central section contained fine-grained glaciofluvial and glaciolimnic sediments. It was consistently affected by glaciofluvial reworking at its distal slope by marginal meltwater streams and by melting of incorporated ice-cores on its proximal slope. It was, furthermore, subsequently modified by seasonal advances during the period Briksdalsbreen remained in a stationary position. In a few years, squeezing in combination with bulldozing was observed but restricted to the silty, water-saturated glaciolimnic sediments. Bulldozing was, however, the dominant process also in the central section and heavily influenced by the properties of the proglacial sediment, the latter clearly revealed by comparison with other sections of the glacier margin. Detailed studies of subglacial sediment deformation by Hart (2006) and Hart et al. (2009) documented different types of micro-flutings and related processes (Fig. 3.17a) and concluded that till deformation required ice/sediment coupling and was, therefore, restricted to summer and autumn conditions. Briksdalsbreen retreated dramatically during the 2000s, accelerating due to calving over the proglacial Briksdalsvatnet causing a break-up of its lower tongue (Hart et al. 2011).

Kjenndalsbreen has a steep icefall that descends into a very narrow valley. This glacier experienced its greatest extension during the early summer 1997 CE, but the terminal moraine was modified by seasonal advances of the oscillating glacier terminus until 2000. These oscillations resulted in a multi-ridged moraine complex of up to 20 m base width with the outermost ridge (M1) being formed in 1997, the subsequent ridges M2–M4 in 1998–2000, respectively (Fig. 3.18c, d). A comparable complex morphology of the recent terminal moraine had formed in parts of the northern foreland at Bergsetbreen (Fig. 3.18a, b). The initial moraine ridge modified during 1998 and 1999 by seasonal advances was in some sections overridden in 2000 when the glacier experienced its last (seasonal) frontal advance. This resulted in a double-ridged moraine in a latero-frontal position. This example demonstrates that very local factors such as glacier motion trajectories, courses of existing sub- or proglacial meltwater streams, properties of proglacial sediments or local topography of the glacier foreland influenced moraine formation during the recent advance and the following years of stationary glacier front positions. All multi-ridged moraines at Jostedalsbreen have in common that they are clearly related to strong seasonal advances. Because the latter are

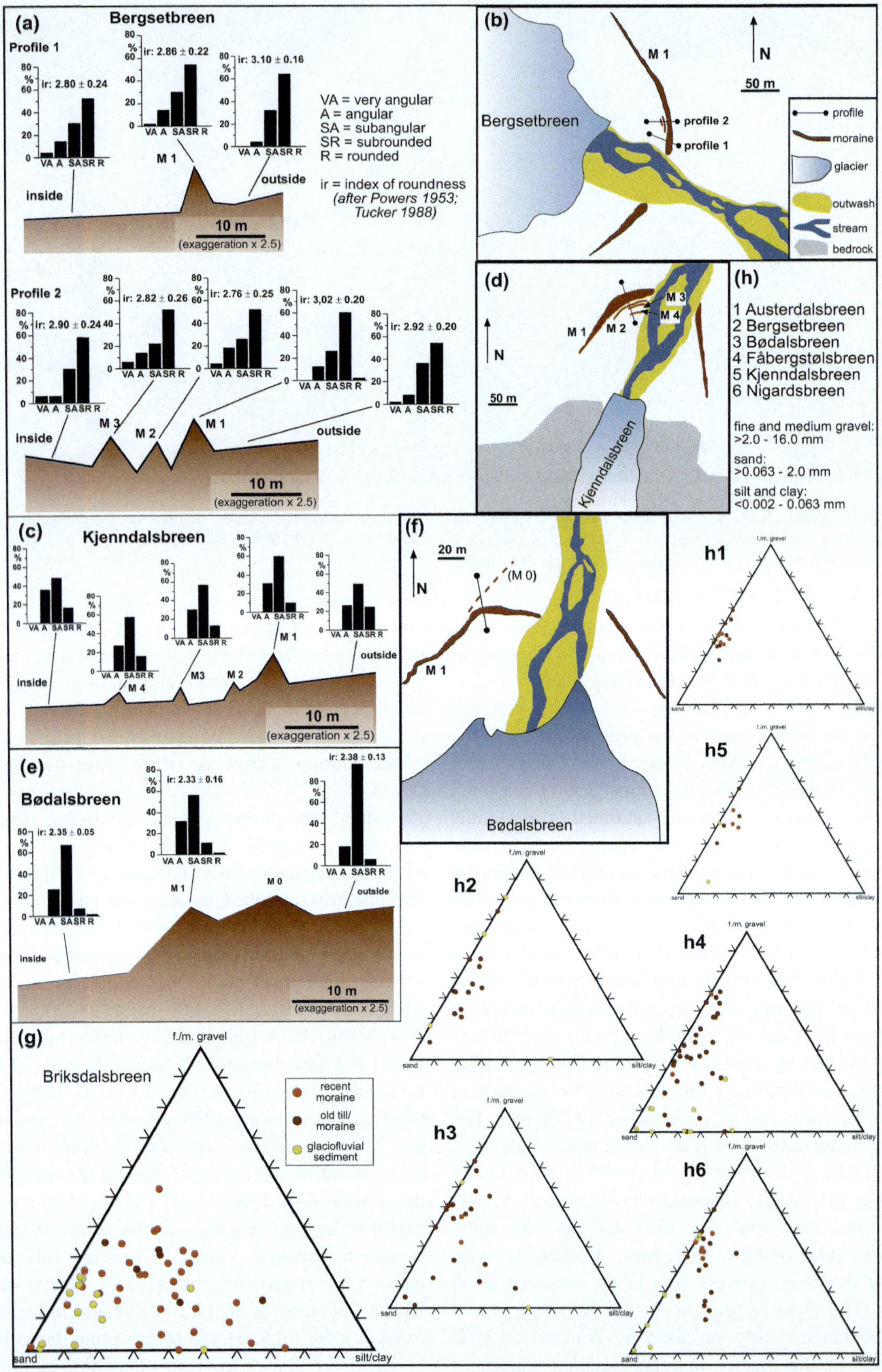

Fig. 3.18 **a–f** Cross-profiles and locations of the recent moraine(s) at Bergsetbreen, Kjenndalsbreen and Bødalsbreen with roundness measurements (cf. Winkler and Matthews 2010 for details on methodology). M0 on (**e**) is an older moraine formed during the 1960s or 1970s CE (Winkler 1996a). **g** Comparison of sediment samples (grain size analysis for grain sizes <16 mm) taken on recent moraines, older moraines/till deposits and glaciofluvial sediment deposits on the glacier foreland of Briksdalsbreen. Note the considerable variability and the lack of obvious characteristic clusters (cf. Winkler 1996a, 2001 for details on methodology). **h** Comparison of sediment samples from six selected Jostedalsbreen outlets (same presentation as on **g**)

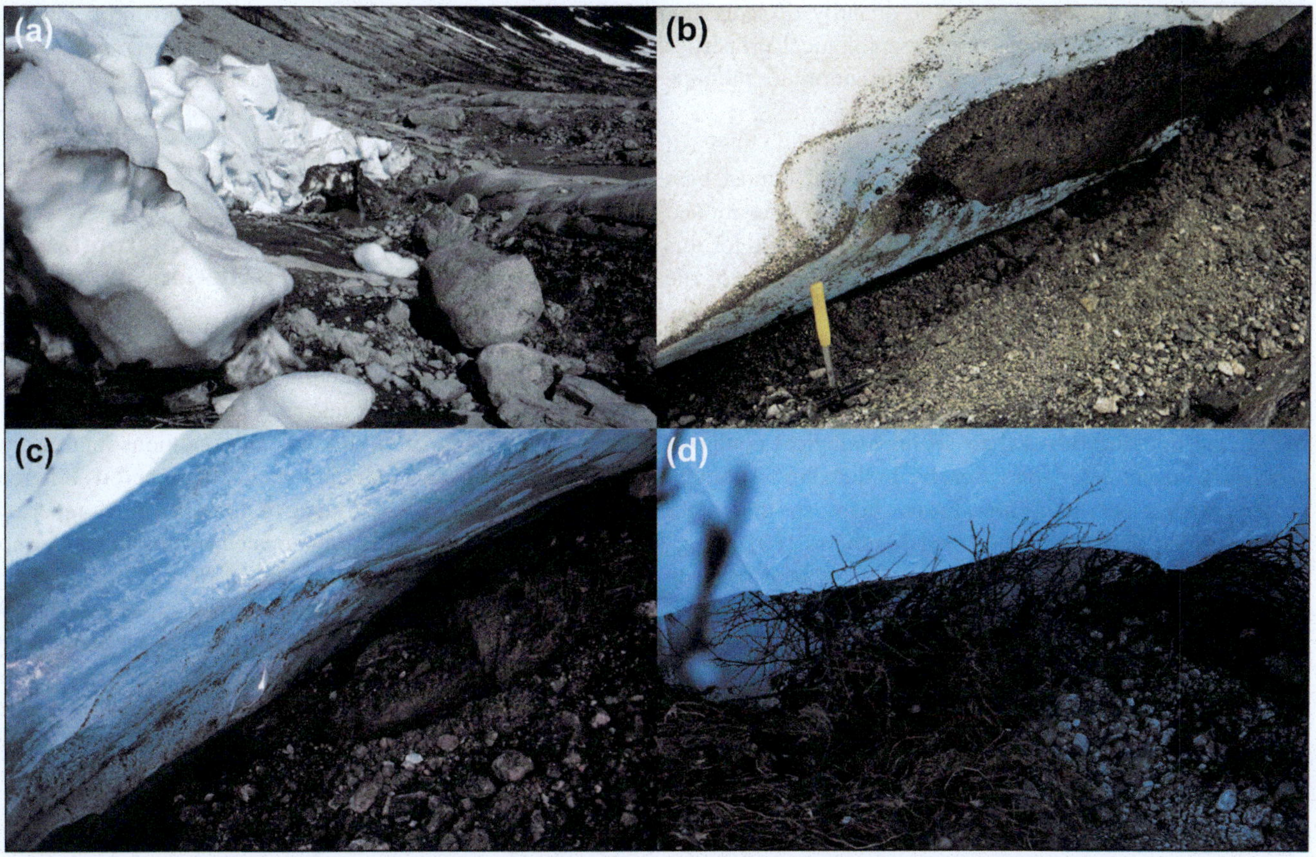

Fig. 3.19 **a** Boulders and sediment pushed up by the advancing Nigardsbreen in a small area where the exposed bedrock dominating the inner foreland was covered by loose sediments (see text). **b** Patch of sediment frozen to the base of Fåbergstølsbreen. **c** Almost debris-free glacier sole exposed in a marginal cavity at Kjenndalsbreen. **d** Debris-free glacier sole of Briksdalsbreen overriding proglacial scrubs (*Photos* Stefan Winkler, **a** 29.05.1998, **b** 07.08.1992, **c** 12.09.1997, **d** 19.04.1997)

usually not recorded with (annual) length measurements and, as a consequence, easily neglected, these observations during the recent advance carry important wider implications.

A special characteristic of Kjenndalsbreen is the presence of some supraglacial debris on its lowermost glacier tongue as a result of frequent rockfalls and snow avalanches from the steep valley sides. But dumping of supraglacial debris could neither be identified as a significant contribution to any sedimentation process nor was it involved with the recent moraine formation. Any supraglacial debris deposited by dumping along the glacier margins from the steep, convex glacier front was subsequently affected by ice-pushing/bulldozing. Only the slightly higher percentage of angular clasts compared to those at Bergsetbreen and Bødalsbreen reveals the presence of supraglacial debris (Fig. 3.18a, c, e; Winkler and Matthews 2010). In line with the results at other outlets, roundness measurements in different parts of the foreland of Kjenndalsbreen show very similar results without significant differences and indicate that sedimentological characteristics and processes during the recent advance were not different to previous ones.

The result that at all outlets no significant difference in roundness between the recent moraine ridge (and individual ridges of a multiple-ridge moraine complex) and either distal or proximal sites was detected supports the conclusion that ice-pushing/bulldozing of preexisting proglacial sediment was the dominant and decisive process of moraine formation (Winkler 1996a; Winkler and Matthews 2010). Sedimentological characteristics of grain sizes finer than coarse gravel were comparable between outlets (Fig. 3.18g, h; Winkler 1996a). Any variability in moraine sediment composition between individual sections of moraines or between different outlets unrelated to initial preexisting proglacial sediments is the result of different degrees of postdepositional glaciofluvial modification. Significant subglacial debris has never been observed in marginal subglacial cavities. Basal ice was mostly completely free of any debris and at Briksdalsbreen, remnants of scrubs and trees directly overridden by the glacier and subsequently exposed by summer backmelting remained uncovered by any debris (Fig. 3.19c, d). At Fåbergstølsbreen, a few patches of sediment frozen onto the glacier sole likely during the previous winter (Fig. 3.19b) were detected in the mid-1990s. The quantity of sediment, however, was insignificant and sedimentological properties suggest local transport only. No contribution to moraine formation took place.

The terminus of Tuftebreen in a tributary valley of Krundalen is located on a flat valley floor of almost bare bedrock and at a higher altitude than the sites mentioned above. No detailed length change record is available for this glacier and recent moraine ridges in the eastern part of the foreland cannot be dated with precision. But if its dynamics are assumed similar to the neighbouring Bergsetbreen, the outermost recent ridge at Tuftebreen likely dates from around 1995 or 1996 CE. Despite large differences in the size of the moraine ridges their sedimentological composition is consistent with formation by bulldozing. Although winter air temperatures are lower than at the other Jostedalsbreen outlets, Winkler and Matthews (2010) suggested that sediment slabs frozen-on to the glacier by deep penetrating winter frost have not been involved, given that moraine cross-sections do not exhibit the characteristic asymmetry of moraines composed of frozen-on sediment slabs (a steep distal slope and shallow proximal slope) as described from recent terminal moraine formation in Jotunheimen. Additionally, the general morphology of the glacier tongue (thick and convex-shaped) in the 1990s would hardly allow significant penetration of the winter cold wave (Winkler 2009).

In contrast to other outlets, geomorphological activity at Brenndalsbreen was very restricted during the recent advance and temporary 'moraines' of 0.4 to less than 1 m height were the only features occasionally observed in front of the central glacier snout. The reason is that the main part of the lower glacier snout and marginal glacier foreland constitutes glacially modified, exposed bedrock (Figs. 3.3d and 3.6f). Only in a small part of the northern foreland did loose sediment (in the form of old till and colluvium) facilitate moraine formation by bulldozing. Here, a moraine ridge of several metres height developed during the recent advance. The causal relationship between the occurrence of deformable sediment and moraine formation is not only demonstrated by the foreland of Brenndalsbreen, but also at Nigardsbreen. Small temporary moraine ridges were only formed where the glacier advanced across bedrock that supported small pockets of loose sediments in the proglacial zone (Fig. 3.19a). They were ultimately overridden and destroyed during the ongoing advance. Where the glacier margins of Nigardsbreen rested on exposed bedrock surfaces, no moraine formation or related geomorphological activity was observed. At Fåbergstølsbreen that similarly mainly advanced across exposed bedrock, temporary moraines or similar features were only observed at a few places where a thin cover of glaciofluvial sediment became affected by the glacier advance (Winkler 1996a). However, in early 1999 CE, a ridge of barely a few tens of centimetres had been pushed up between the glacier front and an old marginal snow patch acting as an obstacle (Fig. 3.16g). Similar processes have been observed at other outlets (e.g. Bergsetbreen and Bødalsbreen).

3.5.3 Analysis and Comparative Interpretation of Recent Moraine Formation

The literature on processes of moraine formation and their resulting morphology and sedimentology is comprehensive. It needs, however, to be taken into account that the considerable diversity of environmental conditions and glaciological characteristics naturally results in numerous different genetic and morphological types of moraines being reported. An often inconsistent use of the related terminology does not help in this context (Boulton and Eyles 1979; Hambrey 1994; Bennett 2001; Evans 2005; Bennett and Glasser 2009; Benn and Evans 2010). Even if restricted to present-day glaciers and terminal moraines in terrestrial temperate or subpolar environments, the picture remains complex. A considerable number of studies investigate moraine formation associated with thrusting and glaciotectonic processes at polythermal glaciers, partly during glacier surges (e.g. Eybergen 1987; Van der Wateren 1995; Bennett 2001; Benediktsson et al. 2008, 2010; Benn and Evans 2010). The processes described in these studies are connected to the presence of permafrost at the glacier margins of advancing glaciers. Such conditions are different from temperate, wet-based glaciers in mid-latitude maritime mountains like the Jostedalsbreen area and, as a result, are not transferrable. A depositional process highlighted in several studies on moraine formation in mountainous regions is marginal dumping. Because it predominately occurs in lateral positions and requires substantial supraglacial debris cover and a 'typical' alpine relief (Small 1983, 1987; Winkler and Hagedorn 1999), it similarly lacks relevance to Jostedalsbreen. Freezing-on of subglacial sediment slabs and the successive build-up of recent moraines at stationary glacier margins have been regularly reported during the past 3 decades, for example from Iceland (e.g. Krüger 1985, 1993, 1996; Evans and Hiemstra 2005; Benn and Evans 2010). This process of terminal formation by an annual freezing-on of overridden proglacial sediment slaps and their subsequent melt-out is described from Jotunheimem by Matthews et al. (1995), Winkler (1996a), Winkler and Matthews (2010) or Hiemstra et al. (2015), and requires deep seasonally frozen ground and occurred in locations close to the lower limit of alpine permafrost. Burki et al. (2009a) and Andersen and Sollid (1971) mention the possible occurrence of similar processes during the formation of LIA-moraines at Bødalsbreen and Nigardsbreen, respectively. The climatic

Table 3.5 Summary of characteristics related to recent terminal moraines formation at selected outlets of Jostedalsbreen during the 1990s CE (modified from Winkler and Matthews 2010)

	Bergsetbreen	Bødalsbreen	Kjenndalsbreen	Tuftebreen
Number of moraine ridges	3	1	4	6
Complex moraine system	Yes, in parts	No	No	Yes. In parts
Older moraine incorporated in recent moraine	No	No	No	Possibly, in parts
Temporary ice core	No	No	No	No
Dumping of supraglacial debris	No	No	Yes (limited)	No
Melt-out of englacial debris along shear planes	Yes (limited)	No	Yes (very limited)	No
Subglacial debris in the basal transportation zone	No	No	No	No
Freezing-on of subglacial sediment	No	No	No	No
Snow-foot push over marginal snow beds in early 1990s	Yes	Yes	Yes	Unknown
Bulldozing of proglacial sediments	Yes (dominant)	Yes (dominant)	Yes (dominant)	Yes (dominant)
Rotated/relocated boulders	Yes	Yes	Yes	No
Incorporated boulders with large lichens (distal slope)	Yes	Yes	Yes	Yes

conditions at Jostedalsbreen, however, cannot fully be compared to Jotunheimen or Iceland. Doubts are, therefore, legitimate and the extensive observations during the recent advance do not support suggestions of any relevant contribution of freezing-on of subglacial sediment slabs to moraine development. In particular, two factors need to be considered: (i) the stronger glacial dynamics at the outlets of Jostedalsbreen with higher rates of advance and subsequent retreat would prohibit the glacier front to remain stationary over periods longer than 2 or 3 years and (ii) the milder winter temperatures in combination with massive and convex lower glacier snouts would hardly allow winter frost to penetrate sufficiently deep into subglacial sediment near the glacier margin to create substantial freezing-on of sediment slabs.

According to the widely accepted current terminal moraine typology by Benn and Evans (2010), the recent terminal moraines at Jostedalsbreen can be classified as ‘push and squeeze moraines’ with bulldozing/pushing as the dominant process. Other processes included within this category are: squeezing, freeze-on, melt-out and mass flowage of sediments, though they only have been occasionally observed at Jostedalsbreen and seem negligible as they never played any substantial role (Table 3.5). Benn and Evans’s (2010) definition of ice pushing/bulldozing of ice-marginal sediment by an advancing glacier perfectly matches the situation at Jostedalsbreen. They state that during major advances proglacial sediment will usually be overridden and incorporated in the subglacial environment. The lack of moraine ridges or occurrence of merely small temporary accumulations of boulders observed during most of 1994–1996 CE when outlets reached their annual peak advances perfectly supports their statement. Successive seasonal advances in an order of up to 20–30 m as they have been measured post-1996 would, by contrast, develop sufficient pressure for efficient bulldozing (see below).

Moraine formation by bulldozing associated with glacier advances requires considerable high pressure by the glacier terminus ploughing into the preexisting proglacial soft sediment of the glacier bed to finally push it up to form moraine ridges. If the overall advance is too rapid, the glacier tongue tends to override the proglacial foreland without significant moraine formation because the slightly upward-moving advancing ice mass exerts less overburden friction/shear stress. There is, however, slight disagreement of observations at Jostedalsbreen with what Benn and Evans (2010) postulate as the most efficient period of ice push for terminal moraine formation. Whereas they see an ice-marginal push by seasonal/winter advances during glacier retreat as most efficient, most recent moraines at Jostedalsbreen were actually formed during the culmination of the recent advance when glaciers actually slowed down. Subsequent proximal moraines of minor dimensions were formed during the first years after this culmination when the glaciers remained stationary prior to their initial retreat. In some cases, these seasonal advances modified (single) recent terminal moraines, but at those outlets where subsequent moraine ridges had been formed proximal to the outermost recent ridge, they exhibit minor dimensions. After the retreat finally had commenced, substantial seasonal advances no longer occurred and no moraine formation took place. Nevertheless, the recent advance at Jostedalsbreen confirms and underlines the importance of seasonal advances for terminal moraine formation under specific environmental and glaciological conditions.

The outlines of recent push moraines at Jostedalsbreen are mostly arcuate, that is, their shape typically reflects the morphology of the glacier margin. Only minor radial crevasses occurred as a consequence of the topographically restricted width of the inner forelands that the outlets had retreated into during the preceding decades of the twentieth century CE. But despite their small size, they are plausible analogues for larger radial crevasses that occurred during re-advances interrupting the retreat following the LIA-maximum when glaciers exhibited positions further downvalley. They can, therefore, be linked to the formation of the unique 'saw-tooth' terminal moraines at Bødalsbreen. Sedimentological studies undertaken at the recent terminal moraines at Jostedalsbreen (Winkler 1996a; Winkler and Nesje 1999; Winkler and Matthews 2010) confirm Benn and Evans's (2010) statement that no general sedimentological characteristics for push moraine can be outlined. Their sedimentology highly or almost entirely depends on the properties of the preexisting proglacial sediment affected by the bulldozing process. Only small quantities of sediment have actually been glacially deposited in the course of the recent advance. With the exception of Austerdalsbreen (and to some extent Kjenndalsbreen), supraglacial debris was absent and melt-out of englacial debris at shear planes was only observed in single years without gaining importance. Consequently, hardly any dumping of debris occurred along the glacier margins. All subglacial observations showed no or limited subglacial debris in the basal transportation zones of the outlets. Based on these observations, melt-out of subglacial debris can be assumed to be limited (at least in the marginal zones).

A process similar to the 'snow-foot push mechanism' introduced by Birnie (1977) was observed at some outlets (Bergsetbreen, Bødalsbreen and Kjenndalsbreen). It generally describes stress developed by the advancing glacier transmitted by marginal snow patches and finally causing a moraine ridge to form without contact with the actual glacier. Although this particular case was not observed at Jostedalsbreen, an alternative mode mentioned by Birnie (1977) in which sediments buckled up between the advancing glacier front and marginal snow patches occurred at Jostedalsbreen during the early 1990s CE, a period characterised by record-high snow accumulation. Large quantities of winter snow remained at the glacier margins in late spring and early summer when frontal glacier movement peaked. After 1996 when the overall advance slowed down and winters were no longer extremely snow-rich, glacier margins only experienced small marginal snow patches except when major snow avalanches had covered entire parts of the lower glacier tongues and their surroundings until late summer (e.g. at Bødalsbreen and Kjenndalsbreen). Ice-marginal squeezing is another process generally linked to push moraine formation, and requires very wet marginal environments to enable the glacier to squeeze up sediment purely by its static loading. At the central frontal margin of Briksdalsbreen with its water-saturated glaciolimnic sediments such squeezing was observed, but only in combination with ice pushing (Winkler and Nesje 1999). This supports the statement of Benn and Evans (2010) that both processes often appear in combination.

There is little doubt that bulldozing was the dominant process for terminal moraine formation during the recent glacier advance at Jostedalsbreen (Winkler and Nesje 1999; Winkler and Matthews 2010). This supports conclusions drawn from investigations of older terminal moraines on the glacier forelands around Jostedalsbreen (Matthews et al. 1979; Lien and Rye 1988; Winkler 1996a; Burki et al. 2009a, b). Similar moraine formation processes by bulldozing have been described from other temperate glaciers (e.g. Rabassa et al. 1979; Krüger et al. 2002; Benn and Evans 2010). Extrusion/squeezing of water-saturated subglacial till in combination with bulldozing has been reported from a number of temperate glaciers (Price 1970; Worsley 1974; Sharp 1984), but the sedimentological properties only supported such a mechanism at Briksdalsbreen (Winkler and Nesje 1999; Hart 2006). The proglacial sediments at most outlets lack the fine grain sizes that facilitate squeezing (Fig. 3.18g, h). Apart from direct observations of glacier ice directly protruding/ploughing into proglacial sediments at various outlets, there is plenty of sedimentological evidence for bulldozing, including relocation and rotation of boulders during moraine formation, irrefutably indicated by lichen-covered boulders on the distal slopes of the outermost recent moraine ridges. Clast roundness measurements invariably revealed that the sediments in the moraine ridges are similar to those on the proglacial areas into which the glaciers advanced (Winkler and Matthews 2010). Grain size analyses are fully consistent with this interpretation (Winkler 1996a). Finally, the fact that no moraine formation took place at those outlets of Jostedalsbreen resting on bare bedrock during the advance demonstrates the dependency of bulldozing on the availability of suitable sediment.

With bulldozing identified as the decisive factor for moraine formation, the dimension of moraines ridges is causally related to the availability and properties of the proglacial sediment rather than to particular glaciological characteristics such as annual advance distance. This is clearly demonstrated at Briksdalsbreen where the highest and most impressive moraine ridge corresponds with the appearance of sediment with a high content of fines, particularly silt, in water-saturated conditions (Winkler and Nesje 1999).

Because high pore-water pressure reduces the shear strength of silty sediments (see Banham 1975; Boulton 1979; Sladen and Wrigley 1983; Alley 1989; Hart and Boulton 1991; Benn and Evans 2010) less pressure is required to push them up by the advancing ice. By contrast, the coarser the sediment and the higher the percentage of angular clasts, the higher is the shear strength and the pressure necessary to form moraine ridges. The lack of causal links between the dimension (crest height/base width) of moraines ridges and duration or distance of the related glacier advance is typical for push moraines and transferrable to older terminal moraines at Jostedalsbreen. The terminal moraine formed in 1873 CE at Nigardsbreen (Figs. 3.8 and 3.9c) exhibits by far the highest moraine crest on the whole foreland. It represents, however, a minor re-advance of only a few years (Andersen and Sollid 1971). In contrast, the outermost moraine at Nigardsbreen (Fig. 3.9g) formed during the culmination of the major LIA-advance that lasted several decades with >3 km total advance distance has a crest of less than half that size (Winkler 1996a).

3.6 Summary and Conclusions

The glacial geomorphological process-system of Jostedalsbreen is largely determined by its specific natural environment. Resistant bedrock, insignificant neotectonic activity, a central high-altitude plateau surrounded by deeply incised glacial valleys that determine Jostedalsbreen's distinctive glacier morphology, very limited supply of supraglacial debris, a typical maritime glaciological regime and a special Holocene glacier chronology dominated by the outstanding 'Little Ice Age' with its single major advance is the main factors responsible for landform assemblages on glacier forelands that show considerable differences to other mountain regions in Norway and elsewhere. Terminal and lateral moraine sequences with high numbers of individual ridges, small to moderate dimensions of those marginal moraines and consequential absence of huge alpine-style lateral moraine ridges, wide and flat valley sandars accounting for large portions of the glacier forelands and noticeable surface areas of glacially overprinted bedrock are characteristic for glacier forelands at Jostedalsbreen's outlet glaciers. Ultimately, the regionally specific attributes of glacier forelands at Jostedalsbreen highlight a general diversity among glaciated valley systems of temperate mountain glaciers.

Analysis of morphological and sedimentological data combined with detailed field observations, photographic evidence and length change records reveals that moraine formation at Jostedalsbreen's outlet glaciers during a recent advance in the 1990s CE was unambiguously dominated by bulldozing/ice-pushing and various associated mechanisms. Morphology and sedimentology of individual push-moraine ridges depend mainly on the properties of the preexisting proglacial sediment affected by the glacier advance. No moraine formation took place where bedrock dominated the proglacial area indicating the dominance of bulldozing just like other evidence such as lichen-covered boulders incorporated into the distal slopes of several recent moraines or large rotated boulders. Terminal moraine formation at Jostedalsbreen is partly connected to seasonal advances. Even if local individuality has to be taken into account and temporal/spatial diversity needs to be observed, comparison with older moraines formed during or after the 'Little Ice Age' on the glacier forelands reveals that most have likely been formed by identical processes. In particular, the high number of moraine ridges representing ice-marginal positions since the 'Little Ice Age' maximum-position ($\sim$1750 CE) can only be explained in terms of efficient moraine formation during either frequent short-term re-advances or seasonal advances during stationary periods or slow retreat. Large proportions of most glacier forelands are, however, occupied by flat glaciofluvial outwash or glaciofluvially overprinted subglacial till.

References

Aa AR, Sjåstad JA (2000) Schmidt hammer age evaluation of the moraine sequence in front of Bøyabreen, western Norway. Norsk Geol Tidsskr 80:27–32

Alley RB (1989) Water-pressure coupling of sliding and bed deformation: I. Water system. J Glaciol 35:108–119

Alley RB, Ágústsdóttir AM (2005) The 8 k event: cause and consequences of a major Holocene abrupt climate change. Quat Sci Rev 24:1123–1149

Andersen JL, Sollid JL (1971) Glacial chronology and glacial geomorphology in the marginal zones of the glaciers, Midtdalsbreen and Nigardsbreen, South Norway. Norsk Geogr Tidsskr 25:1–38

Andersen JL, Egholm DL, Knudsen MF. Linge H, Jansen JD, Pedersen VK, Nielsen SB, Tikhomirov D, Olsen J, Fabel D, Xu S (2018) Widespread erosion on high plateaus during recent glaciations in Scandinavia. Nat Commun 9, Article 830

Andreassen LM, Winsvold SH (eds) (2012) Inventory of Norwegian Glaciers. Norwegian Water Resources and Energy Directorate (NVE), Oslo, Rapport 38-2012

Andreassen LM, Elvehøy H, Kjøllmoen B, Engeset RV, Haakensen N (2005) Glacier mass balance and length variations in Norway. Ann Glaciol 42:317–325

Andreassen LM, Huss M, Melvold K, Elvehøy H, Winsvold SH (2015) Ice thickness measurements and volume estimates for glaciers in Norway. J Glaciol 61:763–775

Andreassen LM, Elvehøy H, Kjøllmoen B, Engeset RV (2016) Reanalysis of long-term series of glaciological and geodetic mass balance for 10 Norwegian glaciers. Cryosphere 10:535–552

Aune B (1993) Temperaturnormaler normalperiode 1961–1990. Oslo, DNMI Rapport 02/93 Klima

Bakke J, Dahl SO, Nesje A (2005) Lateglacial and early Holocene palaeoclimatic reconstruction based on glacier fluctuations and equilibrium-line altitudes at northern Folgefonna, Hardanger, western Norway. J Quat Sci 20:179–198

Ballantyne CK (1995) Paraglacial debris-cone formation on recently deglaciated terrain, western Norway. Holocene 5:25–33

Ballantyne CK, Benn DI (1994) Paraglacial slope adjustment and resedimentation following glacier retreat, Fåbergstølsdalen, Norway. Arctic Alpine Res 26:255–269

Banham PH (1975) Glaciotectonic structures: a general discussion with particular reference to the contorted drift of Norfolk. In: Wright AF, Mosely AF (eds) Ice ages: ancient and modern. Sea House Press, London, pp 69–96

Baumhauer R, Winkler S (2014) Glazialgeomorphologie – Formung der Landoberfläche durch Gletscher. Bornträger, Stuttgart

Benediktsson IÖ, Möller P, Ingólfsson Ó, Van der Meer JJM, Kjær KH, Krüger J (2008) Instantaneous and moraine and sediment wedge formation during the 1890 glacier surge of Brúarjökull, Iceland. Quat Sci Rev 27:209–234

Benediktsson IÖ, Schomaker A, Lokrantz H, Ingólfsson Ó (2010) The 1890 surge end moraine at Eyjabakkajökull, Iceland: a re-assessment of a classic glaciotectonic locality. Quat Sci Rev 29:484–506

Benn DI, Evans DJA (2010) Glaciers and glaciations, 2nd edn. Hodder, London

Benn DI, Kirkbride MP, Owen LA, Brazier V (2005) Glaciated valley systems. In: Evans DJA (ed) Glacial landsystems. Hodder-Arnold, London, pp 372–406

Bennett MR (2001) The morphology, structural evolution and significance of push moraines. Earth Sci Rev 53:197–236

Bennett MR, Glasser NF (2009) Glacial geology, 2nd edn. Wiley, Chichester

Bickerton RW, Matthews JA (1992) On the accuracy of lichenometric dates: an assessment based on the 'Little Ice Age' moraine sequence of Nigardsbreen, southern Norway. Holocene 2:227–237

Bickerton RW, Matthews JA (1993) 'Little Ice Age' variations of outlet glaciers from the Jostedalsbreen ice-cap, southern Norway: a regional lichenometric-dating study of ice-marginal moraine sequences and their climatic significance. J Quat Sci 8:45–66

Birnie RV (1977) A snow-bank push mechanism for the formation of some "annual" moraine ridges. J Glaciol 18:77–85

Bogen J, Wold B, Østrem G (1989) Historic glacier variations in Scandinavia. In: Oerlemans J (ed) Glacier fluctuations and climatic change. Reidel, Dordrecht, pp 109–128

Bogen J, Xu M, Kennie P (2015) The impact of pro-glacial lakes on downstream sediment delivery in Norway. Earth Surf Proc Land 40:942–952

Boulton GS (1979) Processes of glacier erosion on different substrata. J Glaciol 23:15–38

Boulton GS, Eyles N (1979) Sedimentation by valley glaciers; a model and genetic classification. In: Schlüchter C (ed) Moraines and varves. Balkema, Rotterdam, pp 11–23

Burki V, Larsen E, Fredin O, Margreth A (2009a) The formation of sawtooth moraine ridges in Bødalen, western Norway. Geomorphology 105:182–192

Burki V, Larsen E, Fredin O, Nesje A (2009b) Glacial remobilisation cycles as revealed by lateral moraine sediment, Bødalsbreen glacier foreland, western Norway. Holocene 19:415–426

Burki V, Hansen L, Fredin O, Andersen TA, Beylich AA, Jaboyedoff M, Larsen E, Tønnesen J-F (2010) Little Ice Age advane and retreat sediment budgets for an outlet glacier in western Norway. Boreas 39:551–566

Chinn TJH, Winkler S, Salinger MJ, Haakensen N (2005) Recent glacier advances in Norway and New Zealand—a comparison for their glaciological and meteorological causes. Geogr Ann A 87:141–157

Curry AM (1999) Paraglacial modification of slope form. Earth Surf Proc Land 24:1213–1228

Curry AM (2000) Observations on the distribution of paraglacial reworking of glacigenic drift in western Norway. Norsk Geogr Tidsskr 54:139–147

Curry AM, Ballantyne CK (1999) Paraglacial modification of glacigenic sediment. Geogr Ann A 81:409–419

Dahl SO, Nesje A (1994) Holocene glacier fluctuations at Hardangerjøkulen, central-southern Norway: a high-resolution composite chronology from lacustrine and terrestrial deposits. Holocene 4:269–277

Dahl SO, Nesje A, Lie Ø, Fjordheim K, Matthews JA (2002) Timing, equilibrium-line altitudes and climatic implications of two early-Holocene glacier glacier readvances during the Erdalen event at Jostedalsbreen, western Norway. Holocene 12:17–25

Egholm DL, Jansen JD, Brædstrup CF, Pedersen VK, Andersen JL, Ugelvig SV, Larsen NK, Knudsen MF (2017) Formation of plateau landscapes on glaciated continental margins. Nat Geosci 10:592–599

Eide TO (1955) Breden og bygda. Norveg Tidsskrift folgelivsgransking 5:1–42

Erikstad L, Sollid JL (1986) Neoglaciation in South Norway using lichenometric methods. Norsk Geogr Tidsskr 40:85–105

Evans DJA (2005) Ice-marginal terrestrial landsystems: active temperate glacier margins. In: Evans (ed) Glacial landsystems. Hodder-Arnold, London, pp 12–43

Evans DJA, Hiemstra JF (2005) Till deposition by glacier submarginal, incremental thickening. Earth Surf Proc Land 30:1633–1662

Eybergen FA (1987) Glacier snout dynamics and contemporary push moraine formation at the Trutmannglacier, Wallis, Switzerland. In: Van der Meer JJM (ed) Tills and glaciotectonics. Balkema, Rotterdam, pp 217–231

Fægri K (1934) Über die Längenvariationen einiger Gletscher des Jostedalsbre und die dadurch bedingten Pflanzensukzessionen. Bergen, Bergens Museum, Årbok 1933 Naturvidenskapelig rekke 7

Fægri K (1948) Brevariasjoner i Vestnorge i de siste 200 år. Naturen 72:230–243

Fægri K (1950) On the variations of western Norwegian glaciers during the last 200 years. International Association of Scientific Hydrology General Assembly 1948. Oslo 2:293–303

Forbes JD (1853) Norway and its glaciers—visited in 1851. Black, Edinburgh

Førland EJ (1993) Nedbørnormaler nomalperiode 1961–1990. Oslo, DNMI Rapport 39/93 Klima

Fredin O, Burki V (2008) A photogrammetric digital elevation model of the Bødalen valley saw-tooth moraine complex, western Norway. J Maps 4:63–70

Gjessing J (1967) Norway´s paleic surface. Norsk Geogr Tidsskr 21:69–132

Grove JM (1972) The incidence of landslides, avalanches and floods in western Norway. Arctic Alpine Res 4:131–138

Grove JM (1985) The timing of the Little Ice Age in Scandinavia. In: Tooley MJ, Sheail GM (eds) The climatic scene. Allen & Unwin, London, pp 132–153

Grove JM (2001) The initiation of the "Little Ice Age" in regions round the North Atlantic. Clim Change 46:53–82

Grove JM (2004) Little Ice Ages, 2 volumes. Routledge, London

Grove JM, Battagel A (1983) Tax records from western Norway, as an index of Little Ice Age environmental and economic deterioration. Clim Change 5:265–282

Haakensen N (ed) (1996) Excursion guide for IGS-symposium "changing glaciers" hold at Fjærland June 1996. NVE, Oslo

Hambrey MJ (1994) Glacial environments. UCL Press, London

Hart JK (2006) An investigation of subglacial processes at the microscale from Briksdalsbreen, Norway. Sedimentology 53:125–146

Hart JK, Boulton GS (1991) The interrelation of glaciotectonics and glacio-depositional processes within the glacial environment. Quat Sci Rev 10:335–350

Hart JK, Rose KC, Martinez K, Ong R (2009) Subglacial clast behaviour and its implication for till fabric development: new results derived from wireless subglacial probe experiments. Quatern Sci Rev 28:597–607

Hart JK, Rose KC, Waller RI, Vaughan-Hirsch D, Martinez K (2011) Assessing the catastrophic break-up of Briksdalsbreen, Norway, associated with rapid climate change. J Geol Soc London 168:673–688

Hiemstra JF, Matthews JW, Evans DJA, Owen G (2015) Sediment fingerprinting and the mode of singular and composite annual moraine ridges at two glacier margins, Jotunheimen, southern Norway. Holocene 25:1772–1785

Hoel A, Werenskiold W (1962) Glaciers and snowfields in Norway. Norsk Polarinstitut Skrifter, Oslo, p 114

Imhof P, Nesje A, Nussbaumer SU (2011) Climate and glacier fluctuations at Jostedalsbreen and Folgefonna, southwestern Norway and in the western Alps from the 'Little Ice Age' until the present: the influence of the North Atlantic Oscillation. Holocene 22:235–247

Kirkbride MP, Winkler S (2012) Timescales of climate variability, glacier response, and chronological resolution: issues for correlation of Late Quaternary moraines. Quat Sci Rev 46:1–29

Kjøllmoen B (ed) (2017) Glaciological investigations in Norway 2016. Oslo, NVE Rapport 76/2017

Klakegg O, Nordahl-Olsen T, Sønstegaard E, Aa AR (1989) Sogn og Fjordane fylke. Kvartærgeologisk kart M 1:250.000. Trondheim, NGU

Klemsdal T, Sjulsen E (1988) The Norwegian macro-landforms: definition, distribution and system of evolution. Norsk Geogr Tidsskr 42:133–147

Krüger J (1985) Formation of a push moraine at the margin of Höfdabrekkujökull, South Iceland. Geogr Ann A 67:199–212

Krüger J (1993) Moraine-ridge formation along a stationary ice front in Iceland. Boreas 22:101–109

Krüger J (1996) Moraine ridges formed from subglacial frozen-on sediment slabs and their differentiation from push moraines. Boreas 25:57–63

Krüger J, Kjær KH, Van der Meer JJM (2002) From push moraine to single-crested dump moraine during a sustained glacier advance. Norsk Geogr Tidsskr 56:87–95

Laumann T, Nesje A (2009) A simple method of simulating the future frontal position of Briksdalsbreen, western Norway. Holocene 19:221–228

Laute K, Beylich AA (2013) Holocene hillslope development in glacially formed valley systems in Nordfjord, western Norway. Geomorphology 188:12–30

Laute K, Beylich AA (2014) Morphometric and meteorological controls on recent snow avalanche distribution and activity at hillslopes in steep mountain valleys in western Norway. Geomorphology 218:16–34

Lidmar-Bergström K, Ollier CD, Sulebak JR (2000) Landforms and uplift history of southern Norway. Global Planet Change 24:211–231

Lien R, Rye N (1988) Formation of saw-toothed moraines in front of the Bødalsbreen glacier, western Norway. Norsk Geol Tidsskr 68:21–30

Lukas S (2007) Early-Holocene glacier fluctuations in Krundalen, south central Norway: palaeoglacier dynamics and palaeoclimate. Holocene 17:585–598

Lutro O, Tveten E (1996) Geologisk kart over Norge, berggrunnskart Årdal M 1:250.000 Årdal. Trondheim, Norges Geologiske Undersøkelse

Maizels JK, Petch JR (1985) Age determination of intermoraine areas, Austerdalen, southern Norway. Boreas 14:51–65

Matthews JA (2013) Neoglaciation in Europe. In: Elias SA (ed) Encyclopedia of quaternary sciences, 2nd edn. Elsevier, Rotterdam, pp 257–268

Matthews JA, Briffa K (2005) The 'Little Ice Age': re-evaluation of an evolving concept. Geogr Ann A 87:17–36

Matthews JA, Dresser PQ (2008) Holocene glacier variation chronology of the Smørstabbtindan massif, Jotunheimen, southern Norway, and the recognition of century- to millennial-scale European Neoglacial events. Holocene 18:181–201

Matthews JA, Cornish R, Shakesby RA (1979) "Saw-tooth" moraines in front of Bødalsbreen, southern Norway. J Glaciol 22:535–546

Matthews JA, McCarroll D, Shakesby RA (1995) Contemporary terminal-moraine ridge formation at a temperate glacier: Styggedalsbreen, Jotunheimen, southern Norway. Boreas 24:129–139

Matthews JA, Shakesby RA, Schnabel C, Freeman S (2008) Cosmogenic ^{10}Be and ^{26}Al ages of Holocene moraines in southern Norway I: testing the method and confirmation of the date of the Erdalen event (c. 10 ka) at its type-site. Holocene 18:1155–1164

Matthews JA, Winkler S, Wilson P (2014) Age and origin of ice-cored moraines in Jotunheimen and Breheimen, Southern Norway: insights from Schmidt-hammer exposure-age dating. Geogr Ann A 96:531–548

McManus J, Duck RW (1988) Localised enhanced sedimentation from icebergs in a proglacial Lake in Briksdal, Norway. Geogr Ann A 70:215–223

Nesje A (1989) Glacier-front variations at the outlet glaciers from Jostedalsbreen and climate in the Jostedalsbre region of western Norway in the period 1901–1980. Norsk Geogr Tidsskr 43:3–17

Nesje A (1994) A gloomy 250 year memory: the glacier destruction of the Tungøyane farm in Oldedalen, western Norway, 12 December 1743. Norsk Geogr Tidsskr 48:133–135

Nesje A (1995) Breene I Vest-Norge vokser med rekordfart. Naturen 119:7–10

Nesje A (1997) 8 400 år gamal seljestokk frå Briksdalsbreen. Naturen 121:3–4

Nesje A (2005) Briksdalsbreen in western Norway: AD 1900–2004 frontal fluctuations as a combined effect of variations in winter precipitation and summer temperature. Holocene 15:1245–1252

Nesje A (2009) Late Pleistocene and Holocene alpine glacier fluctuation in Scandinavia. Quat Sci Rev 28:2119–2136

Nesje A, Dahl SO (1993) Lateglacial and Holocene glacier fluctuations and climate variations in we-stern Norway: a review. Quat Sci Rev 12:255–261

Nesje A, Dahl SO (2003) The 'Little Ice Age'—only temperature? Holocene 13:139–145

Nesje A, Kvamme M (1991) Holocene glacier and climate variations in western Norway: evidence for early Holocene glacier demise and multiple Neoglacial events. Geology 19:610–612

Nesje A, Matthews JA (2011) The Briksdalsbre event: a winter precipitation-induced decadal-scale glacial advance in southern Norway in the AD 1990s and its implications. Holocene 22:249–261

Nesje A, Sulebak JR (1994) Quantification of late Cenozoic erosion and denudation in the Sognefjord drainage basin, western Norway. Norsk Geogr Tidsskr 48:85–92

Nesje A, Whillans IM (1994) Erosion of Sognefjord, Norway. Geomorphology 9:33–45

Nesje A, Kvamme M, Rye N, Løvlie R (1991) Holocene glacier and climate history of the Jostedalsbreen region, western Norway; evidence from lake sediments and terrestrial deposits. Quat Sci Rev 10:87–114

Nesje A, Dahl SO, Valen V, Ovstedal J (1992) Quaternary erosion in the Sognefjord drainage basin, western Norway. Geomorphology 5:511–520
Nesje A, Johannesen T, Birks HJB (1995) Briksdalsbreen, western Norway: climatic effects on the terminal response of a temperate glacier between AD 1901 and 1994. Holocene 5:343–347
Nesje A, Dahl SO, Andersson C, Matthews JA (2000) The lacustrine sequence in Sygneskardvatnet, western Norway: a continuous, high-resolution record of the Jostedalsbreen ice cap during the Holocene. Quat Sci Rev 19:1047–1065
Nesje A, Matthews JA, Dahl SO, Berrisford MS, Andersson C (2001) Holocene glacier fluctuations of Flatebreen and winter-precipitation changes in the Jostedalsbreen region, western Norway, based on glaciolacustrine sediment records. Holocene 11:267–280
Nesje A, Bakke J, Dahl SO, Lie Ø, Matthews JA (2008a) Norwegian mountain glaciers in the past, present and future. Global Planet Change 60:10–27
Nesje A, Dahl SO, Thun T, Nordli Ø (2008b) The "Little Ice Age" glacial expansion in western Scandinavia: summer temperature or winter precipitation? Clim Dynam 30:789–801
Nordgulen Ø, Andresen A (2008) The Precambrian. In: Ramberg I, Bryhni I, Nøttvedt A, Rangnes K (eds) The making of a land—geology of Norway. NGF, Trondheim, pp 62–119
Nussbaumer SU, Nesje A, Zumbühl HJ (2011) Historical glacier fluctuations of Jostedalsbreen and Folgefonna (southern Norway) reassessed by new pictorial and written evidence. Holocene 21:455–471
Østrem G, Dale Selvig K, Tandberg K (1988) Atlas over breer i Sør-Norge. Oslo, NVE Meddelelser fra Hydrologisk Avdeling 61
Petch JR, Whittaker RJ (1997) Chronology of the Austerdalen glacier foreland. Z Geomorphol 41:309–317
Pohjola VA, Rogers JC (1997) Atmospheric circulation and variations in Scandinavian glacier mass balance. Quat Res 47:29–36
Powers MC (1953) A new roundness scale for sedimentary particles. J Sediment Petrol 23:117–119
Price RJ (1970) Moraines at Fjallsjökell, Iceland. Arctic Alpine Res 2:27–42
Rabassa J, Rubulis S, Suárez J (1979) Rate of formation and sedimentology of (1976–1978) Push-Moraines, Frias Glacier, Mount Tronador (41°10′ S; 71° 53′ W), Argentina. In: Schlüchter C (ed) Moraines and varves. Balkema, Rotterdam, pp 65–79
Rekstad J (1901) Iagttagelser fra bræer i Sogn og Nordfjord. Kristiania, Norges Geologiske Undersøkelse, p 34
Rekstad J (1904) Fra Jostedalsbræen. Bergen, Bergens Museums Aarbog 1904 No. 1
Röthlisberger F, Schneebeli W (1979) Genesis of lateral moraine complexes, demonstrated by fossil soils and trunks: indicators of postglacial climatic fluctuations. In: Schlüchter C (ed) Moraines and varves. Balkema, Rotterdam, pp 387–419
Rye N, Nesje A, Lien R, Blikra LH, Eikenæs O, Hole PA, Torsnes I (1997) Glacial geology and deglaciation chronology of the area between Nordfjord and Jostedalsbreen—Strynefjellet, western Norway. Norsk Geol Tidsskr 77:51–63
Sætrang AC, Wold B (1986) Results from radio echo-sounding on parts of the Jostedalsbreen ice cap, Norway. Ann Glaciol 8:156–158
Sigmond EMO, Gustavson M, Roberts D (1984) Berggrunnskart over Norge 1:1 milion. Norges Geologiske Undersøkelse, Trondheim
Sladen JA, Wrigley W (1983) Geotechnical properties of lodgement till. In: Eyles N (ed) Glacial geology. Pergamon, Oxford, pp 184–212
Sharp M (1984) Annual moraine ridges at Skálafellsjökull, South-East Iceland. J Glaciol 30:82–93
Small RJ (1983) Lateral moraines of Glacier de Tsidjiore Nouveau: forms, development and implications. J Glaciol 29:250–259
Small RJ (1987) Englacial and supraglacial sediment: transport and deposition. In: Gurnell AM, Clark MJ (eds) Glacio-fluvial sediment transfer. Wiley, Chichester, pp 111–145
Solomina ON, Bradley RS, Hodgson DA, Ivy-Ochs S, Jomelli V, Mackintosh AN, Nesje A, Owen LA, Wanner H, Wiles GC, Young NE (2015) Holocene glacier fluctuations. Quat Sci Rev 111:9–34
Steiner D, Pauling A, Nussbaumer SU, Nesje A, Luterbacher J, Wanner H, Zumbühl HJ (2008) Sensitivity of European glaciers to precipitation and temperature—two case studies. Clim Change 90:413–441
Sugden DE, John BS (1976) Glaciers and landscape. Arnold, London
Torsnes I, Rye N, Nesje A (1993) Modern and Little Ice Age equilibrium-line altitudes on outlet valley glaciers from Jostedalsbreen, western Norway: an evaluation of different approaches to their calcu-lation. Arctic Alp Res 25:106–116
Tucker M (1988) Techniques in sedimentology. Blackwell, London
Van der Wateren FM (1995) Processes of glaciotectonism. In: Menzies J (ed) Modern glacial environments. Butterworth-Heinemann, Oxford, pp 309–335
Vasskog K, Paasche Ø, Nesje A, Boyle JF, Birks HJB (2012) A new approach for reconstructing glacier variability based on lake sediments recording input from more than one glacier. Quat Res 77:192–204
Winkler S (1996a) Frührezente und rezente Gletscherstandsschwankungen in Ostalpen und West-/Zentralnorwegen. Trier, Trierer Geographische Studien 15
Winkler S (1996b) Front variations of outlet glaciers from Jostedalsbreen, western Norway, during the twentieth century. Norg Geol Unders B 431:33–47
Winkler S (2001) Untersuchungen zur Klima- und Morphodynamik in skandinavischen Gebirgsregionen während des Holozän - ein Vergleich ihrer Wechselwirkungen und Prozeßsysteme im überregionalen Kontext kaltgemäßigter maritimer Gebirgsregionen. Habilitation thesis (unpublished), University of Trier
Winkler S (2002) Von der "Kleinen Eiszeit" zum "globalen Gletscherrückzug" - eignen sich Gletscher als Klimazeugen? Mainz, Akademie der Wissenschaften und der Literatur Abhandlungen der Mathematisch-naturwissenschaftlichen Klasse 3
Winkler S (2008a) 11 years of bi-annual glacial length measurements. In: Kjøllmoen B (ed) Glaciological investigation in Norway 2007. Oslo, NVE Rapport 2008/1, p 88
Winkler S (2008b) Photographic documentation of front position change at Jostedalsbreen. In: Kjøllmoen B (ed) Glaciological investigation in Norway 2007. Oslo, NVE Rapport 2008/1, pp 86–87
Winkler S (2009) Gletscher und ihre Landschaften. Wissenschaftliche Buchgesellschaft, Darmstadt
Winkler S, Hagedorn H (1999) Lateralmoränen - Morphologie, Genese und Beziehung zu Gletscherstandsschwankungen (Beispiele aus Ostalpen und West-/Zentralnorwegen). Z Geomorphol Supp 113:69–84
Winkler S, Matthews JA (2010) Observations on terminal moraine-ridge formation during recent advances of southern Norwegian glaciers. Geomorphology 116:87–106
Winkler S, Nesje A (1999) Moraine formation at an advancing temperate glacier, Brigsdalsbreen, western Norway. Geogr Ann A 81:17–30
Winkler S, Nesje A (2009) Perturbation of climatic response at maritime glaciers? Erdkunde 63:229–244
Winkler S, Haakensen N, Nesje A, Rye N (1997) Glaziale Dynamik in Westnorwegen - Ablauf und Ursachen des aktuellen Gletschervorstoßes am Jostedalsbreen. Petermann Geogr Mitt 141:43–63

Winkler S, Elvehøy H, Nesje A (2009) Glacier fluctuations of Jostedalsbreen, western Norway, during the past 20 years: the sensitive response of maritime mountain glaciers. Holocene 19:389–408

Winkler S, Chinn T, Gärtner-Roer I, Nussbaumer SU, Zemp M, Zumbühl HJ (2010) An introduction to mountain glaciers as climate indicators with spatial and temporal diversity. Erdkunde 64:97–118

Worsley P (1974) Recent "annual" moraine ridges at Austre Okstindbreen, Okstindan, North Norway. J Glaciol 13:265–277

www.met.no—open portal of Meteorologisk insitutt/www.met.no. Last accessed 19 Dec 2018

www.norgeibilder.no—open portal of Kartverket/www.kartverket.no. Last accessed 23 Dec 2018

www.nve.no—open portal of Norges Vassdrags og Energidirektoratet (NVE)/www.nve.no. Last accessed 27 May 2019

Recent Glacier Changes and Formation of New Proglacial Lakes at the Jostedalsbreen Ice Cap in Southwest Norway

4

Katja Laute and Achim A. Beylich

Abstract

At present, glaciated mountain environments are among the most dynamic geomorphic systems as they are exposed to various climatic and environmental changes. Climate-induced widespread glacier retreat and thinning lead to a gradual enlargement of formerly glaciated terrain. Due to a continuing increase of summer temperatures predicted for western Norway until the end of this century, it is likely that the current trend of the accelerated mass loss of Norwegian glaciers will continue. As one consequence of this development, new lakes will emerge within the formerly glaciated and newly exposed terrain. Because glaciers and glacier-fed streams in mainland Norway have a high importance for hydropower production, tourism and climate research, it is essential to gain an improved understanding of the possible environmental impacts of proglacial lakes in order to being prepared for advantages and challenges connected to these newly emerging landscape elements. This chapter highlights the significant transformation of the Norwegian glacial landscape by illustrating recent glacier changes and the formation of new proglacial lakes at the Jostedalsbreen ice cap in southwest Norway. The created lake inventory contributes to the global set of worldwide existing glacier lake inventories and serves as a baseline study for possible future comparisons at regional and global scales.

Keywords

Recent glacier changes • Proglacial lakes • Lake inventory • Lake type and distribution • Glacial lake outburst flood • Jostedalsbreen ice cap

K. Laute (✉) · A. A. Beylich
Geomorphological Field Laboratory (GFL), Strandvegen 484, 7584 Selbustrand, Norway
e-mail: katja.laute@geofieldlab.com

4.1 Introduction

At present, glaciated mountain environments are among the most dynamic geomorphic systems as they are exposed to various climatic and environmental changes. Worldwide glacier retreat causes the most visible changes but high-altitude and high-latitude environments are as well affected by, e.g. permafrost degradation and an increased frequency of intense rainfall events (e.g. Haeberli and Beniston 1998; Beniston 2000; Harris et al. 2003; Gruber and Haeberli 2007; Krautblatter et al. 2013; Hanssen-Bauer et al. 2017). Climate-induced widespread glacier retreat and thinning lead to a gradual enlargement of formerly glaciated terrain. For instance, for the entire European Alps an area of about 2200 km^2 has transformed from being glacier-covered to an ice-free area within 150 years (from the 1850 century CE to 2000 CE) as based on the extrapolation of reconstructed Little Ice Age (LIA) maximum glacier extents (Zemp et al. 2008). As a consequence, these newly exposed proglacial areas are then exposed to subaerial conditions, with implications for hydrological, geomorphic and ecological processes (Heckmann et al. 2019). The term 'proglacial' is used here in the sense of the description and clarification given in Heckmann et al. (2019), whereas the termino-lateral moraines formed by many glaciers during the end of the LIA and the present-day glacier extent (of a mountain glacier terminus, ice cap margin or ice sheet margin) represent the boundaries of the proglacial area. This definition enables quantitative analyses and large-scale comparisons as the proglacial area becomes linked to glacier recession within a defined time frame depending on the time when the LIA reached its culmination (e.g. ~1850 CE in the European Alps and ~1750 CE in Norway). Proglacial environments undergo a transition from glacial to non-glacial conditions (Johnson et al. 2002) and comparatively high geomorphic process intensities have been observed on recently deglaciated areas (Carrivik and Heckmann 2017) often associated with the paraglacial concept

A. A. Beylich (ed.), *Landscapes and Landforms of Norway*, World Geomorphological Landscapes,
https://doi.org/10.1007/978-3-030-52563-7_4

(e.g. Church and Ryder 1972; Ballantyne 2002). Proglacial systems are characterised by a sediment cascade where corresponding sediment transfer pathways link sediment sources to storage landforms and to the river channel network (Carrivik and Heckmann 2017). With respect to sediment connectivity and the sediment budget system of proglacial areas, proglacial lakes play a geomorphic key role by being important first-order sediment sinks which effectively disconnect downstream valley sections from upstream sediment source input (Liermann et al. 2012; Geilhausen et al. 2013). Following the explanation for the term 'proglacial lake' given by Carrivick and Tweed (2013), we refer in this chapter to all lakes that are, or have been, directly influenced by either a glacier margin or subaerial glacial meltwater.

In recent years, the number and size of proglacial lakes in mountain regions have increased worldwide which is associated to the documented climate-induced glacier retreat and thinning (e.g. Paul et al. 2007; Carrivick and Quincey 2014; Wang et al. 2014; Buckel et al. 2018; Wilson et al. 2018). By buffering meltwater discharge dynamics and acting as effective sediment traps, the formation of proglacial lakes affects the geomorphological activity in proglacial systems and beyond (e.g. Carrivick and Tweed 2013; Carrivik and Heckmann 2017). In addition, proglacial lakes also have diverse regional and global socio-economic implications. Such lakes create new opportunities and challenges related to freshwater supply, hydropower production, touristic development, hazard prevention and landscape protection (Haeberli et al. 2016). At the same time, new lakes may also constitute an increased hazard potential for mountain and foreland communities, e.g. due to the possible formation of glacial lake outburst floods (GLOFs) (Carrivick and Tweed 2016).

Presently, the development of new proglacial lakes receives an increasing awareness by both scientists and the public within the context of potential consequences of ongoing and future climate change in mountain environments. This fact is reflected in, e.g. the growing number of scientific publications which cover case studies, review articles and book chapters (e.g. Liermann et al. 2012; Linsbauer et al. 2012; Carrivick and Tweed 2013; Carrivick and Quincey 2014; Wang et al. 2014; Bogen et al. 2015; Haeberli et al. 2016; Buckel et al. 2018; Wilson et al. 2018; Otto et al. 2019). A comprehensive review paper about the character, behaviour and geological importance of proglacial lakes is provided by Carrivick and Tweed (2013). A first comprehensive global assessment of the societal impacts of glacier outburst floods has been published by the same authors in 2016. Buckel et al. (2018) present a detailed inventory of glacial lakes for the Austrian Alps combined with a brief review of glacial lake formation studies and worldwide existing glacial lake inventories.

It has been shown that high-latitude and high-altitude areas are highly sensitive to and significantly affected by recent climate change (e.g. Beniston 2003; Rangwala and Miller 2012; IPCC 2013). Meteorological records for mainland Norway show the general trend that the last 100 years, especially, the last three decades have been warmer and wetter than the time periods before (see also Ketzler et al. 2021, Chap. 2). In the Northern Hemisphere, 1983–2013 was likely the warmest 30-year period of the last 1400 years (IPCC 2013). For mainland Norway, the annual air temperature has increased by c. 1 °C from 1900 to 2014. Mean annual precipitation has increased by c. 18% since 1900 (Hanssen-Bauer et al. 2017). The documented retreat of Norwegian glacier termini since the LIA glacial maximum has facilitated as well the enlargement of proglacial areas and the formation of proglacial lakes (e.g. Burki et al. 2010; Liermann et al. 2012; Laute and Beylich 2014a, b; Bogen et al. 2015; Winkler 2021, Chap. 3). Within the last two decades, the Jostedalsbreen ice cap (the largest ice mass in continental Europe) has experienced a recognisable ice mass deficit and most of its outlet glaciers display a major frontal retreat (e.g. Nesje et al. 2008a; Winkler et al. 2009; Andreassen and Winsvold 2012; Winsvold et al. 2014; Kjøllmoen et al. 2018). Long-term forecasts for western Norway predict an increase in summer temperature of 2.3 °C until 2070–2100 CE which would result in a glacier area reduction of 34% by 2100 (Nesje et al. 2008a). Hence, the sediment delivery from the Jostedalsbreen outlet glaciers will most likely be altered in the near future with consequences for, e.g. stream hydrology and ecology as well as hydropower production. However, it is likely that a recognisable share of the future suspended sediment load from the retreating glaciers will probably be buffered by newly formed proglacial lakes (e.g. Liermann et al. 2012; Geilhausen et al. 2013; Bogen et al. 2015).

Because glaciers and glacier-fed streams in mainland Norway have a high importance for hydropower production, tourism and climate research, it is essential to gain a better understanding of the environmental impact of proglacial lakes in order to being prepared for the possible advantages and challenges connected to these newly emerging landscape elements.

The objectives of our study are: (1) to produce an up-to-date glacier area outline for the entire Jostedalsbreen ice cap; (2) to compile the first inventory of proglacial lakes which were formed within the newly exposed ice-free area since the 1952–1985 mapped glacier area outline at the Jostedalsbreen ice cap; (3) to investigate the proglacial lake characteristics and types; (4) to analyse their spatial distribution and hazard potential and (5) to discuss their geomorphic importance and future development.

This chapter highlights the significant transformation of the Norwegian glacial landscape by illustrating recent glacier changes and the formation of new proglacial lakes. The created lake inventory contributes to the global set of worldwide existing glacier lake inventories and serves as a baseline study for possible future comparisons at regional and global scales.

4.2 The Jostedalsbreen Ice Cap

4.2.1 Characteristics of the Jostedalsbreen Ice Cap

In mainland Norway, a total area of 2692 km^2 is covered by glaciers, whereof 57% of this glacier-covered area is located in southern Norway (Andreassen and Winsvold 2012). The main proportion of this southern area is made up of the Jostedalsbreen ('breen' = glacier) ice cap which represents the largest glacier in mainland Norway and the largest ice mass in continental Europe. Located between the Nordfjord and the Sognefjord, the Jostdalsbreen ice cap belongs to the maritime western glaciers in southern Norway (Fig. 4.1). Presently, the ice cap stretches over a distance of ca. 60 km in NE-SW direction and encompasses a width of ca. 15 km in NW–SE direction (Fig. 4.1). The ice cap covers an area of 474 km^2 according to the latest glacier inventory based on Landsat imagery from 1999–2006 (Paul et al. 2011). A new glacier area outline is presented in Fig. 4.1 as based on Landsat imagery from 2017/2018 which results in an area of 464 km^2 covered by the ice cap in 2017/2018. The Jostedalsbreen ice cap is a plateau glacier which encompasses about 20 major outlet glaciers of different size and morphology. From the ice mass located on top of the bedrock plateau, these outlet glaciers descend downwards in all directions and terminate within the adjacent valleys (Fig. 4.1) (see also Winkler 2021, Chap. 3; Beylich and Laute 2021, Chap. 7). The outlet glaciers are temperate and warm-based. A maximum ice thickness of 571 m and an average ice thickness of 158 m for the central Jostedalsbreen ice cap are stated in Andreassen et al. (2015). One distinct characteristic of plateau glaciers is that their glacial morphology is strongly influenced by the underlying topography (Baumhauer and Winkler 2014). Situated within the Scandinavian mountain range, the ice cap currently ranges in elevation from ca. 400 m a.s.l. to ca. 1950 m a.s.l. The Lodalskåpa, as an exposed nunatak, is the highest peak of the Jostedalsbreen region (2083 m a.s.l.; Fig. 4.1). Directly within the Jostedalsbreen ice cap 'Høgste Breakulen' (the highest glacier point) is regarded as the highest point with an elevation of 1957 m a.s.l. The predominant bedrock which exists below the ice cap and which covers the surrounding valleys is rather homogenous and belongs to the Jostedal Complex as part of the Western Gneiss Region. The Jostedal Complex consists largely of Precambrian migmatites, orthogneisses and granites (Nordgulen and Andresen 2013). These predominant bedrock types are, in general, relatively weathering resistant (with respect to physical and chemical weathering) resulting in comparatively thin thicknesses of debris and unconsolidated sediments. As a consequence, the modification of the inherited glacial relief since the Last Glacial Maximum (LGM) around the Jostedalsbreen ice cap is altogether minor which is mainly due to the generally rather low intensities of denudational surface processes in comparison to other mountain areas (Laute and Beylich 2014a; Laute et al. 2016). A detailed description of the geology of the Jostedalsbreen region is provided by Winkler (2021, Chap. 3).

4.2.2 Climatic Setting

The climate along the Norwegian coast is strongly influenced by large-scale circulation patterns (see also Ketzler et al. 2021, Chap. 2). The mild and wet winters in western Norway are primarily caused by two factors: firstly, low pressure cells which cross the North Atlantic Ocean and transport warm and humid air to the north, and secondly, the North Atlantic Current (NAC) as the northern extension of the Gulf Stream which transports warm seawater from the Mexican Gulf to the Norwegian coast (e.g. Glässer 1993). Along the western side of the Jostedalsbreen, frontal and orographic precipitation prevails (e.g. Glässer 1993). The orographic effect forces the warm and moist air to rise above the mountain range causing at some locations very high amounts of precipitation in western Norway. Precipitation maxima usually occur during autumn and winter due to cyclonic activity (Beylich and Laute 2012). A large gradient from west to east exists for both precipitation and air temperature in southern Norway. With increasing distance from the west coast precipitation amounts decrease due to the shading effect of the mountain range. The local climate at the Jostedalsbreen is characterised by its proximity to the maritime west coast and the existing orographic effect. Already a small difference along a linear distance reveals a high spatial variability in the air temperature and precipitation regimes (Laute and Beylich 2018). Mainly due to local topographic effects, mean annual precipitation sums can vary, for example, between 1433 mm in Briksdal (40 m a.s.l., 1896–2015; Laute and Beylich 2018) and 3400 mm at the Nigardsbreen glacier (1961–2012; Engelhardt et al. 2015).

The mass balances of Scandinavian glaciers are mainly influenced by winter precipitation and summer temperatures (e.g. Trachsel and Nesje 2015). A large proportion of the variability in winter precipitation may be attributed to variations in the northern hemisphere atmospheric circulation described by the North Atlantic Oscillation (NAO). Strong positive phases of the NAO are associated with stronger than average westerlies

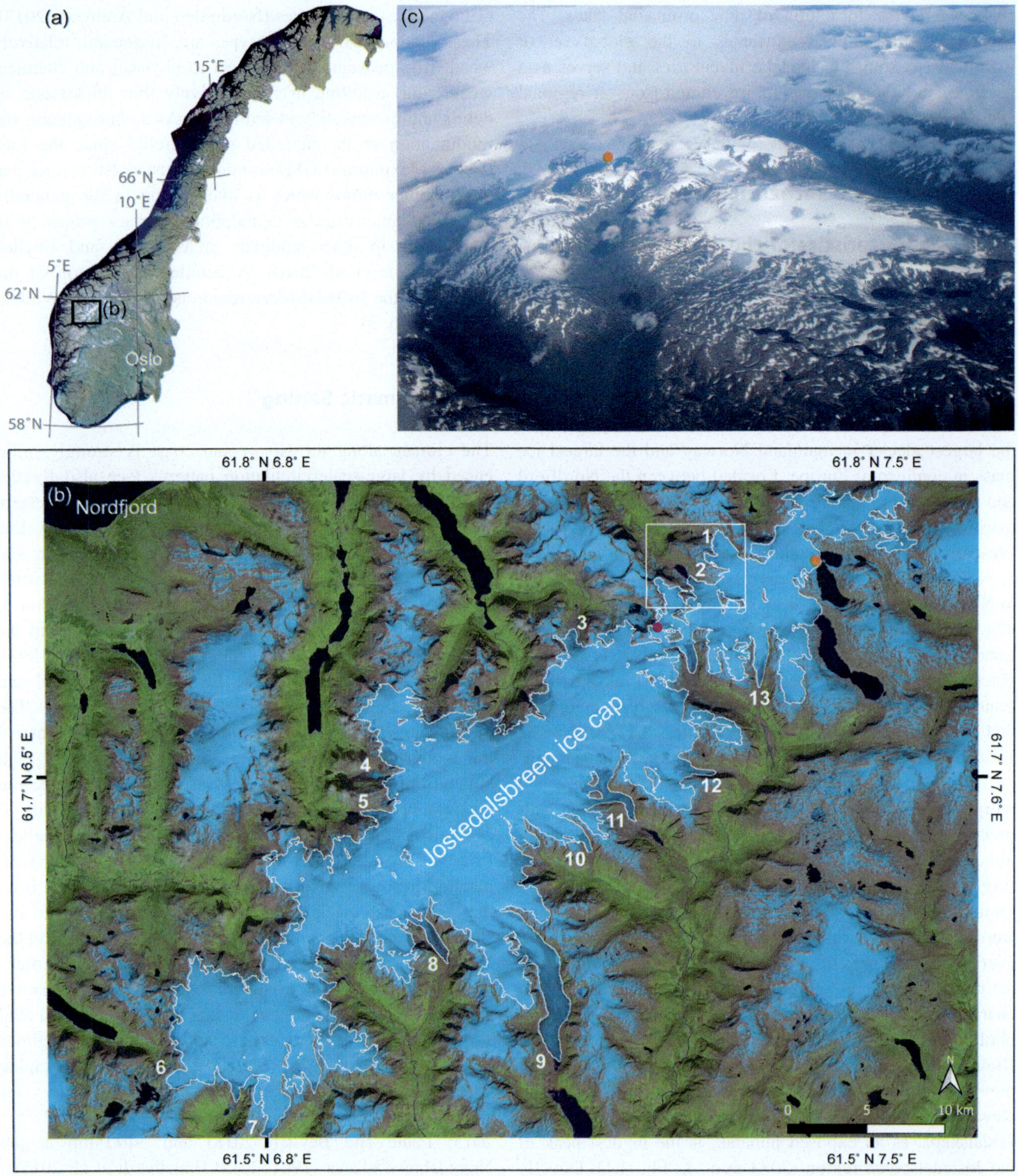

Fig. 4.1 **a** Location of the Jostedalsbreen ice cap in southwest Norway. *Source* of Norway map © NVE Atlas 3.0 (Kartverket, Geovekst og kommuner- Geodata AS, NVE). **b** Landsat 8 satellite image (29.06.2018) displaying the current glacier area extent of the Jostedalsbreen ice cap. *Source* © USGS Earth Explorer (earthexplorer.usgs.gov). The white delineation represents the glacier area outline for 2017/2018. Numbers 1–13 display the exact location of all mentioned outlet glaciers in the text from north to south. The numbers correspond to the following outlet glaciers: 1-Vesledalsbreen, 2-Erdalsbreen, 3-Bødalsbreen, 4-Brenndalsbreen, 5-Briksdalsbreen, 6-Marabreen, 7-Supphellebreen/Flatbreen, 8-Austerdalsbreen, 9-Tunsbergdalsbreen, 10-Tuftebreen, 11-Nigardsbreen, 12-Fåbergstølsbreen, 13-Stigaholtbreen. The purple point denotes the location of the Lodalskapå, the highest peak in the Jostedalsbreen region. The white square delineates the location of Fig. 4.2. **c** The northern tip of the Jostedalsbreen ice cap (for orientation see location of orange points in **b** and **c**). *Photograph* © K. Laute.

over the middle latitudes and more intense weather systems over the North Atlantic supporting mild, stormy and wet winter conditions in Scandinavia. Reversed patterns are observed during strong negative NAO phases with a weaker than usual difference in air pressure between the two regions causing cold, calm and dry winters (Hurrell 1995; Greatbatch 2000; Hurrell et al. 2003). Hence, positive NAO phases cause a comparably higher winter precipitation on the glaciers whereas less winter precipitation on the glaciers occurs during negative NAO phases (e.g. Hanssen-Bauer and Førland 1998; Nesje et al. 2000; Trachsel and Nesje 2015). A recent study by Trachsel and Nesje (2015) stresses the importance of a possible interplay between the NAO and the Atlantic Multi-decadal Oscillation (AMO). The AMO is based on the average anomalies of sea surface temperatures in the North Atlantic Basin (Trenberth et al. 2019). According to Trachsel and Nesje (2015), the relative importance of precipitation and temperature for glacier mass balances in the recent past was probably influenced by the state of the AMO and the NAO, as these two indexes are associated with changes in summer temperature (AMO) and winter precipitation (NAO).

4.2.3 Holocene and Recent Glacier Fluctuations

Reconstructions of Holocene glacier variations of the Jostedalsbreen ice cap are primarily based on several lacustrine and terrestrial sedimentary records (e.g. Nesje et al. 2001; Nesje 2005). The size of the Jostedalsbreen ice cap has fluctuated throughout the entire Holocene but most important are three main glacier fluctuations. During the early Holocene, the final deglaciation was interrupted by a recognisable glacial re-advance (the so-called Erdalen-Event), leaving a distinctive terminal moraine ridge in one of the glacial valleys (Erdalen) behind (Nesje 1984; Matthews et al. 2008). After this event, the deglaciation continued until the Jostedalsbreen ice cap completely disappeared during the Holocene thermal optimum between 7300 and 6100 cal. a BP (Nesje 2009). From approximately 6000 to 2000 cal. a BP the Jostedalsbreen ice cap was reformed (e.g. Nesje and Kvamme 1991; Nesje et al. 2001, 2008a; Nesje 2009) and the outlet glaciers from the Jostedalsbreen ice cap experienced their maximum neoglacial position during the 'Little Ice Age' (LIA) period (regionally around 1750 CE), as almost all Norwegian glaciers (e.g. Bickerton and Matthews 1993; Grove 2004; Nesje et al. 2008a). The term LIA is used to describe a period in which many glacier regions on a global scale experienced one or multiple major glacier advances between the twelfth and the late nineteenth centuries (Grove 1988, 2004; Luckman 2000; Matthews and Briffa 2005). The LIA glacial maximum extent is defined as the furthest downvalley extent of a glacier during the last few centuries (Luckman 2000). For instance, glaciers in the European Alps reached their maximum LIA glacial advances during the fourteenth, seventeenth and nineteenth centuries, with the majority of the glaciers attaining their greatest LIA extent in the final 1850/1860 advance (e.g. Ivy-Ochs et al. 2009; Nussbaumer et al. 2011). In contrast, the LIA period in Norway is usually defined from 1500/1650 to 1920 (e.g. Nesje 1995; Nesje et al. 2008a) with most of the glaciers attaining their maximum LIA glacial extent during the mid-eighteenth century (Nesje 2009). Based on high-resolution lichenometric dating of moraine ridges (Bickerton and Matthews 1993) and historical data, the outlet glaciers of the Jostedalsbreen ice cap reached their greatest LIA extent around 1750 (e.g. Nesje and Kvamme 1991; Bickerton and Matthews 1993; Nussbaumer et al. 2011). The rapid glacier advance during the mid-eighteenth century in southern Norway is primarily attributed to mild and wet winters as a consequence of a positive NAO index in the first half of the eighteenth century (e.g. Nesje and Dahl 2003; Nesje et al. 2008a, b; Rasmussen et al. 2010). A comprehensive summary about the Holocene glacier chronology of the Jostedalsbreen region is given by Winkler (2021, Chap. 3).

Since the LIA glacial maximum advance, the outlet glaciers of the Jostedalsbreen have been retreated to their recent positions, but with several still-stands or partly remarkable re-advances (e.g. Bickerton and Matthews 1993; Winkler et al. 2009; Winkler and Matthews 2010). At the late 1980s, the Jostedalsbreen experienced its latest substantial ice mass increase, resulting in strong frontal advances of many outlet glaciers (Winkler et al. 1997; Nesje 2005; Winkler 2021, Chap. 3). This advance is primarily attributed to an increase in winter precipitation and above-average winter balances due to a positive NAO phase which caused a period of predominantly strong zonal atmospheric circulation (Andreassen et al. 2005; Chinn et al. 2005). Since the year 2000, the Jostedalsbreen ice cap has experienced a recognisable ice mass deficit and most of its outlet glaciers display a major frontal retreat, and in some cases, the lowermost glacier tongues have, in fact, been completely separated (e.g. Nesje et al. 2008a; Winkler et al. 2009; Andreassen and Winsvold 2012; Winsvold et al. 2014; Kjøllmoen et al. 2018).

An increase of high air temperatures since the last 20 years, mainly within the elevation range of the outlet glaciers and especially in late summer and autumn, seems to be the most important factor for the current glacier retreat (Nesje 2005; Winkler and Nesje 2009). Besides the currently high summer air temperatures also the average winter snow accumulation has decreased since about 2000 (Winkler et al. 2009; Nesje and Matthews 2011). Additional factors are a prolongation of the ablation season along the lowermost glacier termini due to the fact that frost and snowfall occur later in the year (Winkler and Nesje 2009) and partly a locally increased ablation due to a combination of a shrunken ice thickness of the outlets and an increased thermal reflection at the bedrock valley sides. However, it is important to note that the changes recorded for the actual glacier mass balances do not reflect the magnitude and the morphological change which is currently visible at the retreating glacier fronts (Winkler et al. 2009; Winkler 2021, Chap. 3). As long-term forecasts for western Norway point towards higher summer temperatures until the end of this century (Nesje et al. 2008a), it can be assumed that the current trend of the accelerated mass loss of Norwegian glaciers will continue.

4.2.4 Glacier Length Changes and Formation of New Proglacial Lakes

Norway has a long history in glacier monitoring and systematic observations are dating back to the beginning of the last century. Glacier length measurements at the Jostedalsbreen ice cap were initiated by J. Rekstad and started around 1900 CE (e.g. Rekstad 1902; Øyen 1906). Since the early 1960s, long-term investigations of glacier length changes and glacier mass balances have been conducted by the Norwegian Water Resources and Energy Directorate (NVE). Examples of annual and cumulative glacier length changes measured at six outlet glaciers of the Jostedalsbreen are presented in Table 4.1. The Table includes four outlet glaciers with continuous records since around 1900 whereas the time series for Briksdalsbreen had to be stopped in 2015 as the lower part of the terminus was completely detached from the upper part. Annual glacier length change measured in 2017 for the five outlet glaciers varied from 0 to −54 m whereas for the year 2018 it varied from +2 to −81 m. The strongest terminus retreat was recorded in both years at the Nigardsbreen outlet glacier related to the presented examples in Table 4.1. The cumulative length change varies from −1203 to −2860 m for a time period of approximately 118 years.

The ongoing glacier dynamics and fluctuations of the Jostedalsbreen ice cap have obvious environmental effects on the glacier forefields (e.g. Winkler and Matthews 2010; Andreassen and Winsvold 2012; Bogen et al. 2015; Laute and Beylich 2014a, b; Laute et al. 2016; Winkler 2021, Chap. 3). One aspect is the formation of new proglacial lakes within the significantly enlarged ice-free glacier forelands as a consequence of glacier retreat (e.g. Buckel et al. 2018; Otto et al. 2019). A number of new proglacial lakes have formed within the LIA maximum advance of several outlet glaciers of the Jostedalsbreen at different times. Some of the lakes have been covered and uncovered again during short-term glacier advances and retreats (e.g. Briksdalsvatn). Based on the comparison of the glacier area outline from 1952–1985 to 2017/2018 of the Jostedalsbreen ice cap, a number of 57 new lakes have formed within the newly exposed ice-free area.

4.3 Data and Methods

4.3.1 Delineation of the 2017/2018 Glacier Area Outline

NVE has created several inventories of Norwegian Glaciers with the latest being published in 2012 (Andreassen and Winsvold 2012). Glacier area outlines (GAO) for the Jostedalsbreen ice cap are available for the periods 1952–1985 and 1999–2006 (Andreassen and Winsvold 2012; Winsvold et al. 2014). Topographic map information (N50) and aerial photographs have been used to delineate the GAO for mainland Norway for the period 1952–1985 (Table 4.2). For the latest GAO produced by NVE, Landsat TH/ETM + satellite images with a 30 m resolution from the period 1999–2006 in combination with a semi-automatic band ratio method were applied (for technical details see Andreassen et al. 2008; Paul et al. 2011; Andreassen and Winsvold 2012). Within this latest 'Inventory of Norwegian Glaciers' (Andreassen and Winsvold 2012) the Jostdalsbreen ice cap is defined as a glacier complex which is divided into 82 glacier units using drainage divides. The NVE glacier identification and delineation follow the GLIMS (Global Land Ice Measurements from Space) definition of a 'glacier' which is conforming to the standards used by the World Glacier Monitoring Service (WGMS) (Andreassen and Winsvold 2012). The definition is as

Table 4.1 Cumulative and current annual length change data for selected outlet glaciers of the Jostedalsbreen ice cap

Glacier	Annual glacier length Change (m) 2017	Annual glacier length Change (m) 2018	Cumulative length Change (m)	Measuring Period(s)
Fåbergstølsbreen	–27	1	–2700	1899–
Nigardsbreen	–54	–81	–2860	1900–
Briksdalsbreen	Nm	Nm	–1203	1900–2015
Stigaholtbreen	–17	2	–1269	1903–
Austerdalsbreen	0	–35	–1721	1905–19, 1933–
Tuftebreen	–23	–33	–209	2007–

Data source NVE/Glacier data, www.nve.no/hydrology/glaciers
Nm = not measured

follows: 'A glacier or perennial snow mass, identified by a single GLIMS glacier ID, consists of a body of ice and snow that is observed at the end of the melt season, or, in the case of tropical glaciers, after transient snow melts. This includes, at a minimum, all tributaries and connected feeders that contribute ice to the main glacier, plus all debris covered parts of it. Excluded is all exposed ground, including nunataks' (Raup and Khalsa 2007; Racoviteanu et al. 2009). Based on that definition NVE includes both glaciers and perennial snowfields in their inventory.

The delineation of the 2017/2018 GAO produced by the authors is based on two georectified Landsat 8 satellite images with a multi-spectral resolution of 30 m which cover the entire Jostedalsbreen ice cap (Fig. 4.1). Both Landsat images were obtained from the United States Geological Survey's (USGS) Earth Explorer interface (Table 4.2). The two selected satellite images represent the best possible image quality with respect to remaining seasonal snow cover, cloud cover and mountain shadowing. However, the satellite image from 2018, taken in the end of June, exhibits partly remaining seasonal snow which covers primarily the higher margins and parts of the glacier. Unfortunately, all images recorded later in summer 2018 are useless due to a too extensive cloud cover. Due to this fact, the delineation of the glacier area outline is primarily based on the satellite image 2017 (taken in the end of August) in combination with the 2018 satellite image and the available digital orthophotos 'Sogn 2017' and 'Sogn 2015' (Table 4.2). Based on different spectral band combinations, colour-infrared (CIR) and false-colour images (FCI) are created to enable a better distinction of snow and ice as well as a better identification of proglacial lakes (Fig. 4.2). The open-source Geographic Information System QGIS 3.4 is used for all spatial analysis of this study. A manual and restrictive mapping approach is applied with respect to the GAO from 1999–2006 meaning that the delineation from 1999–2006 is taken over in cases of uncertainties. Also the existing division of the Jostedalsbreen glacier complex into the 82 glacier units following the drainage divides is kept for the new 2017/2018 GAO. The authors also follow the GLIMS glacier definition but with the main emphasis on the criteria that all delineated

Table 4.2 Overview of used Landsat satellite imagery, digital orthophotos, topographic maps and available glacier area outlines (GAOs) of the Jostedalsbreen ice cap, their specifications and sources

	Label	Date of acquisition	Resolution/scale	Specifications	Source
Satellite image	LC08_L1TP_201017_20180629_20180716_01_T1	29.06.2018	30 m	Landsat 8, OLI/TIRS C1 Level-2	USGS EarthExplorer (earthexplorer.usgs.gov)
	LC08_L1TP_200017_20170822_20170911_01_T1	22.08.2017	30 m	Landsat 8, OLI/TIRS C1 Level-2	USGS EarthExplorer (earthexplorer.usgs.gov)
Digital orthophoto	Sogn 2017	25.09.2017	0.25 m	Digital sensor, colour	Kartverket/Norge i bilder (www.kartverket.no)
	Sogn 2015	21.09.2015	0.25 m	Digital sensor, colour	Kartverket/Norge i bilder (www.kartverket.no)
	Sogn 2010	29.09.2010	0.50 m	Digital sensor, colour	Kartverket/Norge i bilder (www.kartverket.no)
	Stryn 2006	15.07.2006	0.20 m	Undefined, colour	Kartverket/Norge i bilder (www.kartverket.no)
	Skjåk Stryn 2005	10.09.2005	0.50 m	Undefined, colour	Kartverket/Norge i bilder (www.kartverket.no)
	Jotunheimen 2004	12.08.2004	0.50 m	Undefined, colour	Kartverket/Norge i bilder (www.kartverket.no)
	Hornindal-Gloppen-Stryn 1967	20.07.1967	0.20 m	Analogue sensor, Panchromatic	Kartverket/Norge i bilder (www.kartverket.no)
	Luster 1964/1965	01.07.1965	0.20 m	Analogue sensor, Panchromatic	Kartverket/Norge i bilder (www.kartverket.no)
Topographical map	N50 Kartdata	09.01.2019*	1:50,000	N50	Kartverket/Geonorge (www.kartverket.no)
Glacier Area Outline	Cryoclim_GAO_NO_1952_1985	1952–1985	–	N50 Topographic maps/air photos	NVE (www.nve.no/hydrology/glaciers)
	Cryoclim_GAO_NO_1999_2006	1999–2006	30 m	Landsat 5 TM	NVE (www.nve.no/hydrology/glaciers)

*Date of download

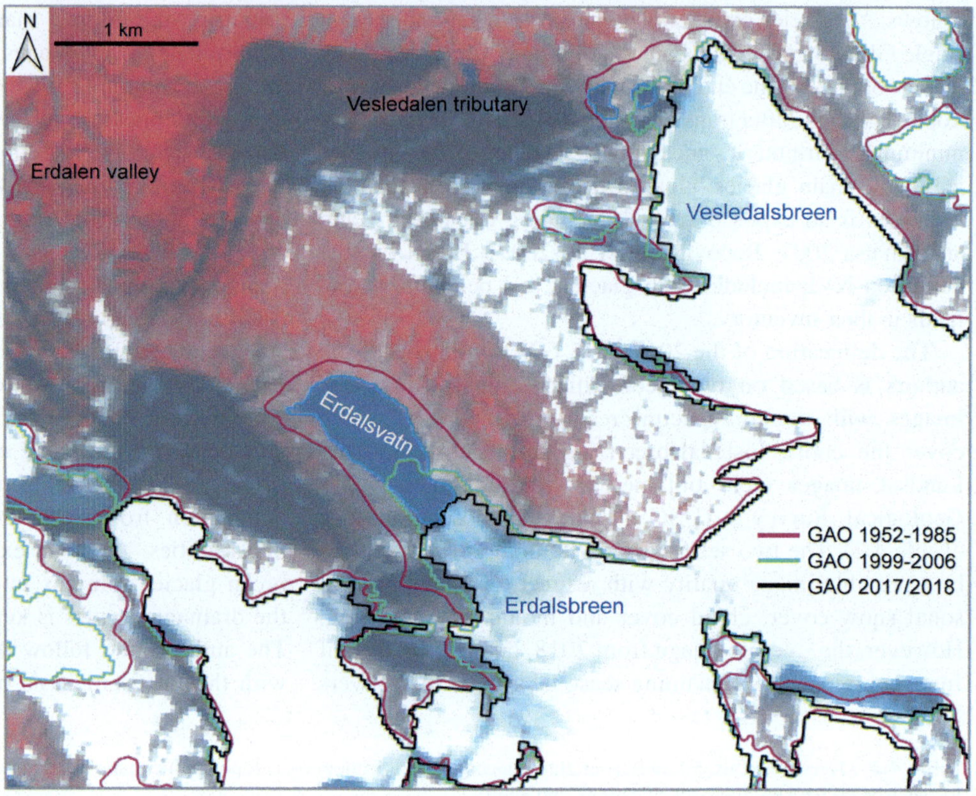

Fig. 4.2 Colour-infrared (CIR) image created from the Landsat 8 satellite scene (22.08.2017) with the two available glacier area outlines (GAOs) for the periods 1952–1985, 1999–2006 (*Data source* NVE/Glacier data, www.nve.no/hydrology/glaciers; Andreassen and Winsvold 2012; Winsvold et al. 2014) and the new 2017/2018 GAO produced by the authors. *Source* Landsat 8 satellite scene © USGS Earth Explorer (earthexplorer.usgs.gov). The retreat of both outlet glaciers has exposed four glacial lakes since the period 1952–1985. For location see white square in Fig. 4.1

parts have to be clearly connected to the main glacier complex of the Jostedalsbreen ice cap. Uncertain cases (due to the 30 m resolution of the Landsat images) are cross-checked and verified with the available orthophotos. Although the manual approach is more labour-intensive and partly subjective, it ensures a consistent examination and a high-quality outcome particularly with respect to image quality issues. However, this method is also subject to errors and uncertainties. NVE estimates the accuracy of their mapped glacier areas in regions that do not require manual correction as better than 3% (Andreassen and Winsvold 2012). As the authors use similar satellite images with the same resolution as NVE, and due to the described restrictive mapping approach, a comparable accuracy for the new 2017/2018 GAO is assumed. Figure 4.2 displays an example of the three different GAOs mapped along the Erdalsbreen and Vesledalsbreen outlet glaciers located on the northern side of the Jostedalsbreen ice cap (Fig. 4.1). The retreat of both outlet glaciers has exposed four glacial lakes since the period 1952–1985 (Fig. 4.2).

4.3.2 Compilation of the Proglacial Lake Inventory

The conducted identification and mapping of lakes which were recently formed within the newly exposed ice-free area following the ice retreat since the 1952–1985 glacier area extent/outline are based on the existing topographical map information in combination with the digital orthophotos and the two satellite images (Table 4.2). The majority of the identified lakes were already mapped in the N50 topographic map of Norway. Additional new lakes are identified by using the digital orthophotos with their given high resolution. Only lakes with a minimum lake area of 1000 m^2 are included in the inventory in order to avoid lakes which only persist periodically or intermittently after, e.g. heavy rainfall or snowmelt (see Buckel et al. 2018). The borders of lakes which are not yet marked in the topographic map are manually delineated. In cases where the lake area has recognisably changed in comparison to the N50 topographic map data, the new lake area was also manually mapped as far as this was possible due to sometimes still existing seasonal lake ice and/or snow cover on the respective orthophoto. A unique identification number is assigned to each lake and a number of different attributes are recorded for all lakes including surface area of the lake, type of lake damming, characteristics of local sedimentary covers, existence/non-existence of direct glacier contact, elevation above sea level and given orientation of the lake (located at the north or south side of the Jostedalsbreen ice cap). Lakes which already existed before the 1952–1985 time period but obviously increased in surface area, primarily due to the retreat of the glacier terminus, are included but listed separately in the inventory. The total recorded number of the new proglacial lakes represents a minimum number due to image quality issues with respect to remaining seasonal lake ice and/or snow cover on some of the orthophotos and the comparably rough resolution of the satellite images.

4.4 Results

4.4.1 Recent Changes of the Spatial Glacier Area of the Jostedalsbreen Ice Cap and Examples of Newly Formed Proglacial Lakes

Referring to the study by Winsvold et al. (2014), the glacierised areas in mainland Norway decreased by −326 km^2 (which corresponds to 11%) within the time period from 1947 to 2006. An average glacier length change of −240 m was determined for the same period. The study of Winsvold et al. (2014) indicates that glaciers in western Norway are affected by a larger glacier shrinkage as compared to glaciers in the eastern parts, and that northern glaciers have retreated more than southern glaciers. They relate the observed spatial trends in their glacier change analysis to a combination of several factors including glacier geometry, elevation and different climatic aspects.

Figure 4.3 presents a comparison of the delineated glacier area outlines from 1952–1985 and 2017/2018 for the entire Jostedalsbreen ice cap. The created overlay displays impressively the change that has taken place. The entire glacier area of the Jostdalsbreen ice cap has experienced a loss of 79 km^2 which corresponds to 14.6% within the period from 1952–1985 to 2017/2018 (Table 4.3). Since the last GAO from 1999–2006, a glacier area reduction of 10 km^2 or 2.1% has occurred (Table 4.3). Based on the newly created 2017/2018 GAO the Jostedalsbreen ice cap covers currently a total surface area of 464 km^2. Especially since the year 2000 the Jostedalsbreen has experienced a recognisable ice mass deficit (e.g. Nesje et al. 2008a, Winkler et al. 2009; Andreassen and Winsvold 2012; Winsvold et al. 2014; Kjøllmoen et al. 2018). A clear change in the glacier geometry of the Jostedalsbreen is visible in Fig. 4.3, including emerging and enlarging rock outcrops and comparably large ice patches becoming separated from the main ice mass. The observed enlargement of rock outcrops occurs due to an almost self-reinforcing process. A locally lower ice thickness along the glacier margin and an increased thermal reflection of the newly exposed bedrock cause an increasing

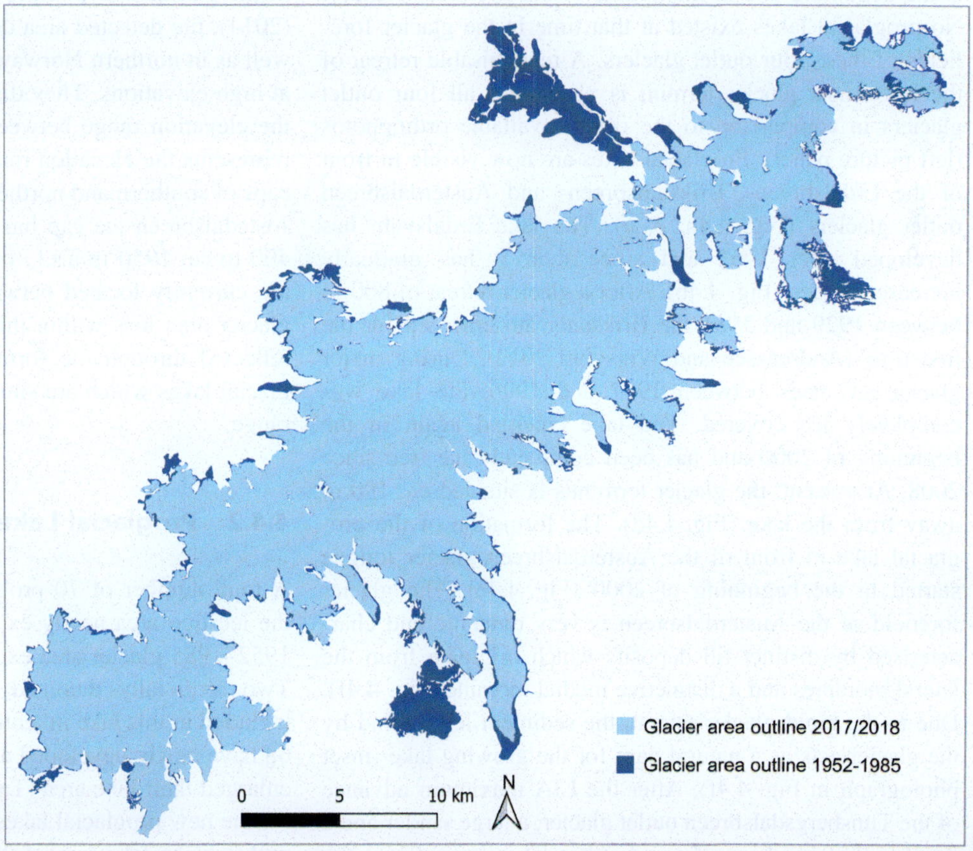

Fig. 4.3 Comparison of the delineated glacier area outlines from 1952–1985 and 2017/2018 for the entire Jostedalsbreen ice cap. *Source* GAO 1952–1985 © NVE/Glacier data, www.nve.no/hydrology/glaciers; Andreassen and Winsvold (2012); Winsvold et al. (2014)

Table 4.3 Change in glacier-covered area of the Jostedalsbreen ice cap based on the different glacier area outlines (GAO)

	GAO 1952–1985* (km^2)	GAO 1999–2006* (km^2)	GAO 2017/2018 (km^2)
Glacier-covered area	543	474	464
Newly exposed ice-free area		69	10

**Data source* NVE/Glacier data, www.nve.no/hydrology/glaciers; Andreassen and Winsvold (2012); Winsvold et al. (2014)

ablation and therefore a continuing expansion of the ice-free surface area. The enlargement of rock outcrops occurs currently, for example, in the upper parts of the outlet glaciers Nigardsbreen, Tunsbergsdalsbreen and Austerdalsbreen (Figs. 4.1 and 4.3). The observed significant thinning and retreat of the glacier margins cause in addition the separation of partly very large ice patches from the main ice cap, as formerly existing very narrow ice connections disappear (Fig. 4.3). The probably most obvious change is the significant retreat of lower situated glacier termini (Figs. 4.2, 4.3 and 4.4). At several glacier termini, the lowermost parts have even detached completely from the upper parts of the outlet glaciers (e.g. Brenndalsbreen and Briksdalsbreen (Fig. 4.4 d)). Figure 4.4 shows recent changes of four outlet glaciers in connection with the formation of new proglacial lakes. The oldest available glacier photographs (black and white inset photographs) display the position of the glacier termini at the beginning and in the middle of the twentieth century. No proglacial lakes existed at that time in the glacier forefields of these four outlet glaciers. A recognisable retreat of the lowermost glacier termini is visible for all four outlet glaciers in comparison to the oldest available orthophotos (left picture panel). Proglacial lakes are now visible in front of the Erdalsbreen-, Briksdalsbreen- and Austerdalsbreen outlet glaciers (Fig. 4.4a, c, e). The lake Erdalsvatn has developed after 1952 and since then it has gradually increased in size (Fig. 4.4b). After a glacier retreat of 800 m between 1929 and 1951 the Briksdalsvatn emerged for the first time (Andreassen and Winsvold 2012). During major glacier advances between 1987 and 1996, the lake was completely ice covered. The lake emerged again in the beginning of 2000 and has been completely ice free since 2008. At present, the glacier terminus is situated ca. 190 m away from the lake (Fig. 4.4d). The formation of the proglacial lake in front of the Austerdalsbreen glacier tongue started in the beginning of 2000 (Fig. 4.4e). The glacier forefield of the Austerdalsbreen is very dynamic and characterised by distinct till deposits which originate from the lateral moraines and a distinctive medial moraine (Fig. 4.4f). Due to the rapid glacier retreat, the sediment left behind by the glacier acts as a natural dam for the growing lake (inset photograph in Fig. 4.4f). After the LIA maximum advance of the Tunsbergsdalsbreen outlet glacier, a large sander and a proglacial lake have formed (the lake is outside the orthophoto in Fig. 4.4g). In 1978, this lake was transformed into an artificial reservoir for hydropower operations. Another new proglacial lake, about 500 m long and 9 m deep, was formed by glacier recession in 1987–2006 in front of the glacier (Bogen et al. 2015; Fig. 4.4h).

These significant changes can be partly related to the glacier geometry of the Jostedalsbreen ice cap. Ice caps in Norway including the Jostedalsbreen are comparably flat, and a high fraction of their surface remains close to the modern equilibrium line. In the case that the equilibrium line rises, large parts of the accumulation area turn into the ablation area leading to the situation that the glacier mass balance becomes negative (Winsvold et al. 2014). The described characteristic is one fact which is considered to make ice caps particularly sensitive to climatic changes (e.g. Nesje et al. 2008a), for instance, in comparison to steep mountain glaciers. Another key factor is the given elevation of the glacier above sea level. According to Winsvold et al. (2014), the detected area decrease of glaciers in southern as well as in northern Norway is larger at lower elevations than at high elevations. They discovered a large total area loss in the elevation range between 1000 and 1700 m a.s.l. which represents the elevation range where many of the largest ice caps of southern and northern Norway are located. While the Jostedalsbreen ice cap has a total elevation range from ca. 400 to ca. 1950 m a.s.l., the majority of its glacier margins are currently located between 1200 and 1700 m a.s.l. The glacier area loss within this elevation range is, in addition, reflected through the formation of a number of new proglacial lakes which are situated exactly within this elevation range.

4.4.2 Proglacial Lake Characteristics and Types

A total number of 70 proglacial lakes are identified within the ice-free area newly exposed by the ice retreat since the 1952–1985 glacier area extent of the Jostedalsbreen ice cap. Two artificially dammed lakes (reservoir lakes) are not included in this lake inventory. 57 lakes out of the remaining 68 have newly developed and 11 already existing lakes have enlarged their lake area (Table 4.4). The individual lake area for the new proglacial lakes varies from 1016 to 447,241 m^2 with a mean lake area of 28,308 m^2. The largest lake of the

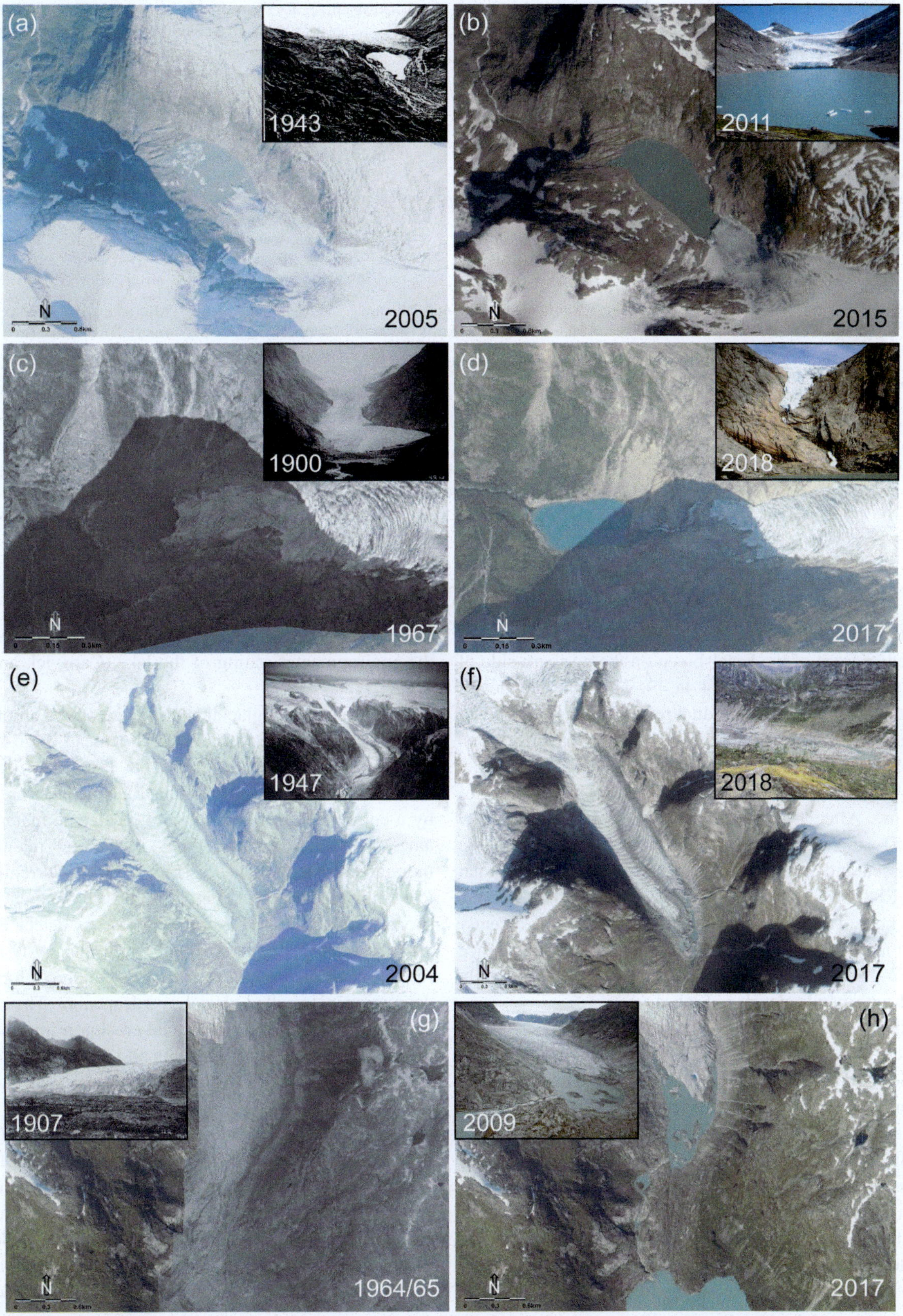
(a)
1943
2005
(b)
2011
2015
(c)
1900
1967
(d)
2018
2017
(e)
1947
2004
(f)
2018
2017
(g)
1907
1964/65
(h)
2009
2017

◀ **Fig. 4.4** Recent glacier length changes of selected outlet glaciers of the Jostedalsbreen in connection with the formation of new proglacial lakes. The oldest available glacier photographs (black and white inset photographs) display the position of the glacier termini at the beginning and in the middle of the twentieth century. The coloured inset photographs give an impression of the present-day situation. *Source* Digital orthophotos (a–h) © Kartverket, www.kartverket.no. **a** and **b** Position of Erdalsbreen glacier terminus in 2005 and 2015 in connection with the enlargement of the proglacial lake Erdalsvatn. *Source* inset photograph 1943 © Nesje (1984); inset photograph 2011 © K. Laute. **c** and **d** Position of Briksdalsbreen glacier terminus in 1967 an 2017 in connection with the enlargement of the proglacial lake Briksdalsvatn. *Source* inset photograph 1900 © J. B. Rekstad (NVE/Glacier data, www.nve.no/hydrology/glaciers); inset photograph 2018 © S. Winkler. **e** and **f** Position of Austerdalsbreen glacier terminus in 2004 and 2017 in connection with the development of the proglacial lake. *Source* inset photograph 1947 © Widerøe Flyselskap (NVE/Glacier data, www.nve.no/hydrology/glaciers); inset photograph 2018 © P. Solnes (NVE/Glacier data, www.nve.no/hydrology/glaciers). **g** and **h** Position of Tunsbergdalsbreen glacier terminus in 1964/65 and 2017 in connection with the development of the proglacial lake. *Source* inset photograph 1907 © J.B. Rekstad (NVE/Glacier data, www.nve.no/hydrology/glaciers); inset photograph 2009 © H. Elvehøy (NVE/Glacier data, www.nve.no/hydrology/glaciers)

Table 4.4 Characteristics of the proglacial lakes included in the created lake inventory for the Jostedalsbreen ice cap

	Number of lakes	Total lake area (km^2)	Mean lake area (m^2)	Min lake area (m^2)	Max lake area (m^2)
New lakes	57	1.56	28,308	1016	447,241
Enlarged lakes	11	0.33	29,787	2772	83,191
Sum	68	1.89			
		Mean elevation (m a.s.l.)	Min elevation (m a.s.l.)	Max elevation (m a.s.l.)	Number of ice-contact lakes
All lakes	68	1360	347	1724	36

newly developed lakes is the Erdalsvatn (447,241 m^2) which is located in front of the Erdalsbreen on the northern side of the Jostedalsbreen (Figs. 4.1, 4.4a, b). This lake has developed after 1952 due to the gradual retreat of the Erdalsbreen outlet glacier. Within the group of already existing lakes, Briksdalsvatn located in front of the Briksdalsbreen outlet glacier (Fig. 4.1) is the largest lake covering an area of 83,191 m^2. After a glacier retreat of 800 m between 1929 and 1951, the Briksdalsvatn emerged for the first time (Andreassen and Winsvold 2012). All 68 lakes cover a total area of 1.88 km^2 whereof only the new lakes cover an area of 1.56 km^2 (Table 4.4). Related to the 79 km^2 of recently exposed ice-free surface areas following the ice retreat since the 1952–1985 GAO, two percent of this area is currently covered with newly developed lakes.

Proglacial lake formation is strongly linked to glacier dynamics and the surrounding environmental conditions (Carrivick and Tweed 2013). For instance, retreating glaciers can expose a topographic bedrock depression or a space behind a dam that retard runoff and provoke storage of water and sediment causing a situation where these newly created lakes become first-order sediment sinks (Otto et al. 2019). According to the most commonly used classification, proglacial lakes can be impounded by ice, glacigenic material (moraines), landslide debris or bedrock (Costa and Schuster 1988). The lake character, evolution and drainage are depended on the respective type of dam in combination with the environmental setting of the lake (Carrivick and Tweed 2013).

Four types of proglacial lakes are identified among the 68 detected lakes at the Jostedalsbreen ice cap (Fig. 4.5a). The majority (52 lakes) of these 68 lakes are defined as bedrock-dammed lakes (Fig. 4.5a, b). Nine lakes are dammed by a combination of bedrock and glacigenic material (moraines) (Fig. 4.5c), and three lakes are identified as moraine-dammed lakes (Fig. 4.5d). Currently, two ice-dammed lakes exist (Fig. 4.5e) but due to their generally short-lived nature it is more difficult to detect them. No landslide-dammed lake is found within the study area. The formation of *bedrock-dammed lakes* requires a topographic depression, also called glacial overdeepening. These bedrock depressions are formed by glacial erosion primarily through the processes of glacial quarrying and glacial abrasion (Benn and Evans 2010; Cook and Swift 2012). They are common in mountainous regions, along ice cap fringes and in ice sheet outlet troughs (Carrivick and Tweed 2013). The high share of bedrock surfaces and consequently the low share of surfaces with local sedimentary covers in the immediate vicinity of the Jostedalsbreen ice cap (situated on top of a bedrock plateau) clearly favour the formation of bedrock-dammed lakes. Most of the proglacial lakes situated within the higher elevated parts of the glacially carved valleys have developed within

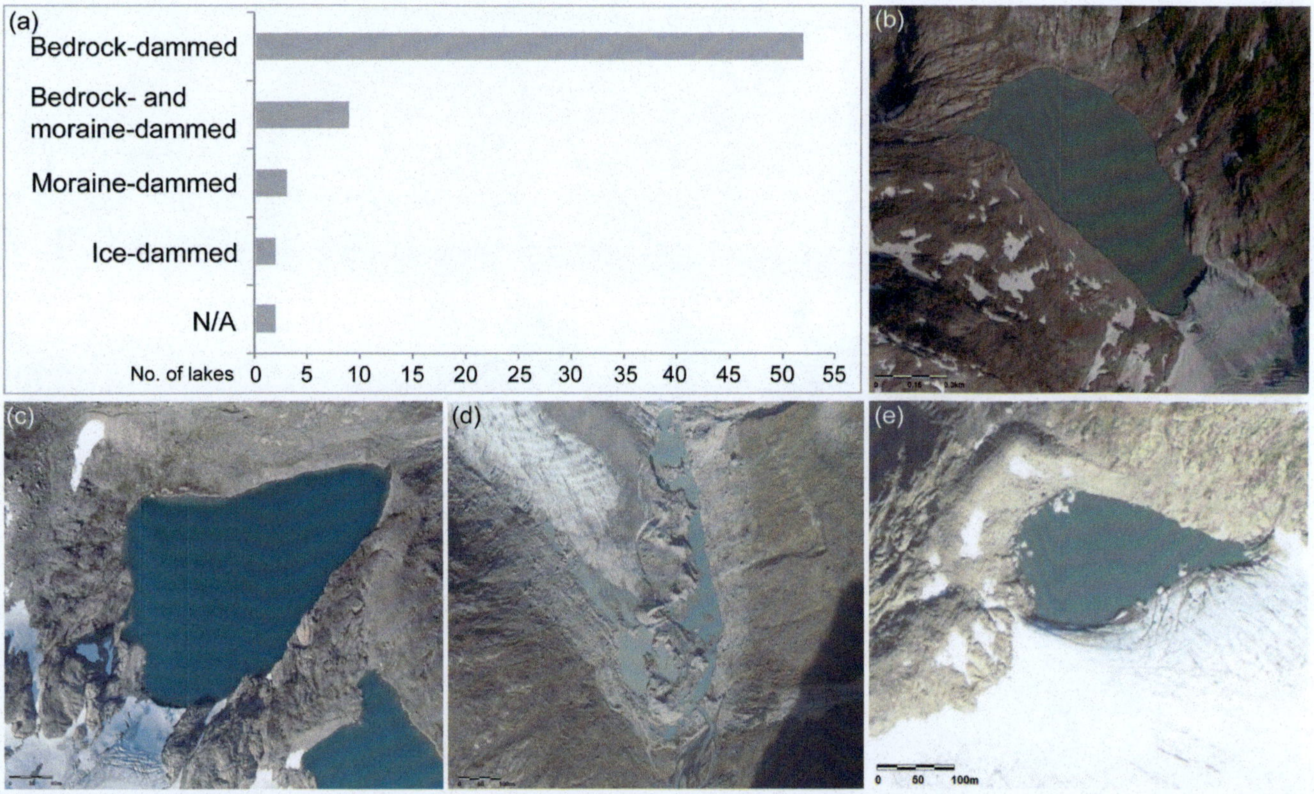

Fig. 4.5 Distribution **a** and examples of identified proglacial lake types. **b** Bedrock-dammed lake, **c** bedrock- and moraine-dammed lake, **d** moraine-dammed lake and **e** ice-dammed lake. *Source* Digital orthophotos (b–e) © Kartverket, www.kartverket.no

bowl-shaped depressions located behind distinct bedrock bars (Fig. 4.5b). In some of the adjacent valleys, multiple bedrock depressions with associated multiple lakes are found (Fig. 4.6). Three lakes have developed since the LIA glacier maximum extent of the Vesledalsbreen in Vesledalen (a tributary valley in Erdalen) and a fourth lake is currently in the process of its formation (Fig. 4.6a). For most of the existing bedrock depressions, a maximum depth of several metres can be assumed. In contrast, a number of very young lakes which are situated along the present glacier margins on top of the bedrock plateau have developed in comparably shallow bedrock depressions. The formation of *moraine-dammed lakes* is linked to periods of glacier retreat following an earlier period of glacier advance. Due to the glacier retreat the deposited sediment left behind by the glacier as terminal and lateral moraines acts as a natural dam for the emerging lake (Fig. 4.5d). Moraine-dammed lakes develop in topographic depressions which were formerly occupied by ice or melt-water channels (Carrivick and Tweed 2013). Lake formation can also be related to buried bedrock overdeepenings (Otto et al. 2019) or depressions formed by the melting of large masses of dead ice (e.g. Kääb and Haeberli 2001). The low number of moraine-dammed lakes within the recently exposed and now ice-free surface areas around the Jostedalsbreen ice cap is again related to a low availability of sediments and, connected to this, a low share of surface areas with debris covers and unconsolidated sediments. The third identified lake type is impounded by a combination of the two previously described dam types, and is termed *bedrock- and moraine-dammed lake* (Fig. 4.5c) in this study. For the lakes which fall into this category, it was not possible to confirm if the damming of the lake is solely to either a bedrock bar or to glacigenic material (moraines). In most of these cases, it may actually be a combination of an existing bedrock depression and glacigenic material deposited as a terminal moraine which reinforces the damming effect. One important characteristic is that the moraines of this lake type are always recognisably smaller as compared to the moraines of the moraine-dammed lake type. This lake type exists more often than moraine-dammed lakes (Fig. 4.5a) due to larger ice-free areas which exhibit thin and discontinuous covers of mostly glacigenic sediments especially in higher elevations. The formation of the two *ice-dammed lakes* results from the direct retreat of the glacier termini. The melt water is impounded between the glacier termini and a lateral or terminal moraine (Fig. 4.5e). Ice-dammed lakes are generally characterised by a shorter lifespan as they are strongly connected to the glacier behaviour and as their damming material (ice) is less stable than for instance bedrock or moraines. The present drainage of the ice-dammed lakes occurs usually through subglacial

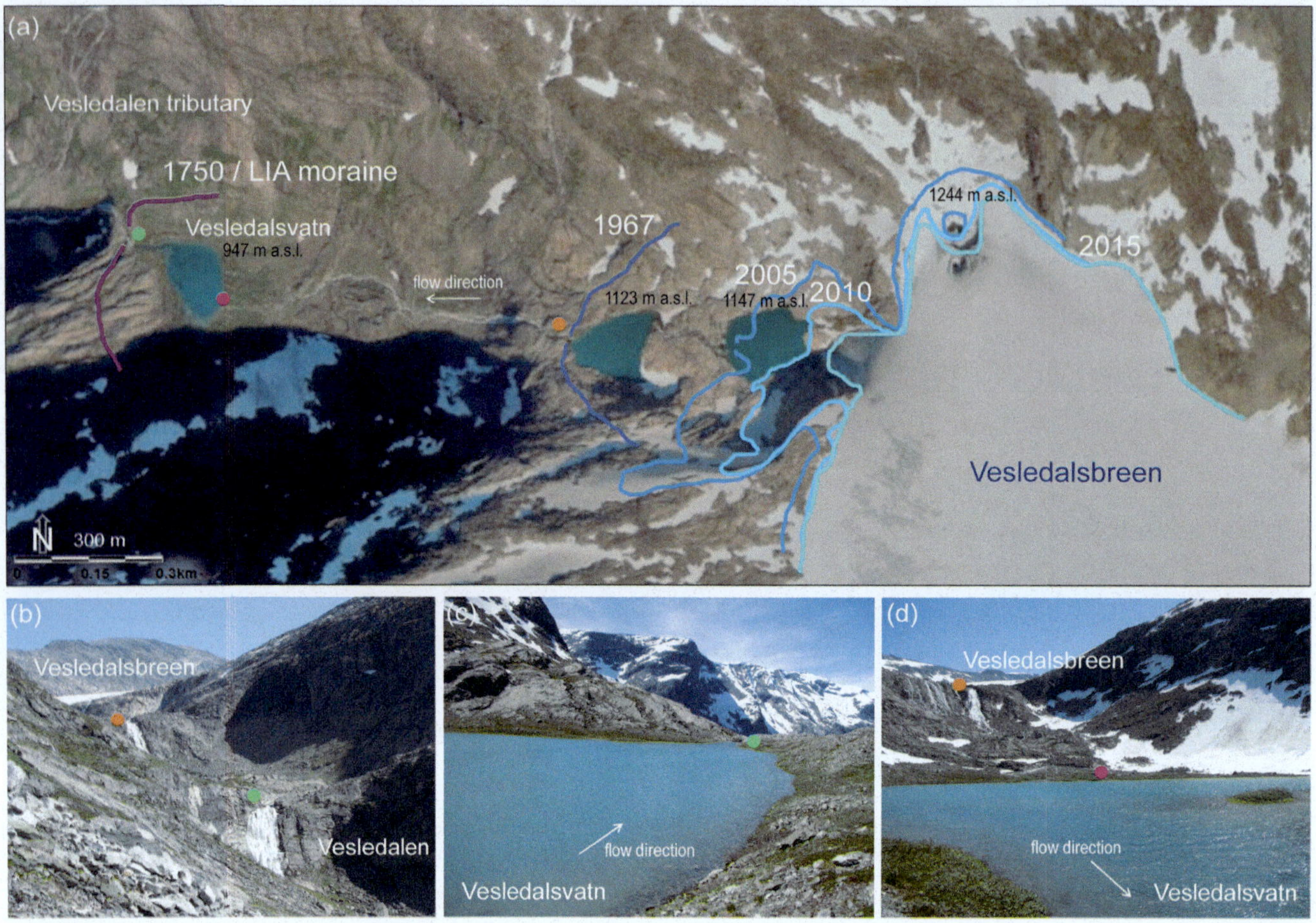

Fig. 4.6 **a** Example of the retreating Vesledalsbreen outlet glacier which has gradually exposed multiple bedrock depressions enabling the formation of multiple bedrock-dammed lakes. *Source* Digital orthophoto © Kartverket, www.kartverket.no. **b** View upvalley towards the Vesledalsbreen glacier showing three distinct bedrock bars. **c** and **d** Impressions of the first lake (Vesledalsvatn) looking down- and upvalley. For a better orientation, the location of the green and orange-coloured points marks the position of the distinct bedrock bars and the location of the purple point marks the position of the lake inflow in the orthophoto (**a**) and the three photos (**b**–**d**). *Photographs* © K. Laute.

drainage paths following the topographic downhill gradient. Especially ice-dammed and moraine-dammed lakes can be the source of potentially catastrophic GLOFs. In the case of a partial or complete failure of the dam, GLOFs have the ability to travel long distances from their source and to affect downstream areas by inundating them with large amounts of reworked sediment and debris (e.g. Clague and Evans 2000; Breien et al. 2008).

From the 68 detected lakes are currently 36 in direct contact with glaciers (Table 4.4, Fig. 4.4). So-called ice-contact lakes can directly affect the glacier system resulting in, e.g. increased ice velocities, changed mass balances and enhanced melting (Otto et al. 2019). Ice-contact lakes can partly decouple the glacier behaviour from climatic perturbations (Carrivick and Tweed 2013; Sutherland et al. 2019) leading to a potentially increased ice mass loss of glaciers terminating in lakes as compared to glaciers without ice-contact lakes situated within the same region.

4.4.3 Spatial Distribution of Proglacial Lakes

The detected 68 proglacial lakes are situated between 347 m a.s.l. and 1724 m a.s.l. with a computed average lake elevation of 1360 m a.s.l. (Table 4.4 and Fig. 4.7). Lakes situated in lower elevations (green elevation range in Fig. 4.7) represent proglacial and partly ice-contact lakes of the lowermost outlets of the Jostedalsbreen, as for example Tunsbergsdalsbreen, Austerdalsbreen and Briksdalsbreen (Figs. 4.1 and 4.4). The elevation range from 800 to 1200 m a.s.l. includes lakes connected to shorter outlets as, for example, the outlets of Erdalsbreen, Vesledalsbreen and Flatbreen/Supphellebreen (Figs. 4.1, 4.4 and 4.6). The density of lakes steadily increases with increasing elevation, except between 600 and 800 m a.s.l. where no lake exists, and clearly peaks between 1400 and 1600 m a.s.l. (light blue elevation range in Fig. 4.7). Only a few lakes exist in elevations above 1600 m a.s.l. Almost 74% of the 68 lakes are

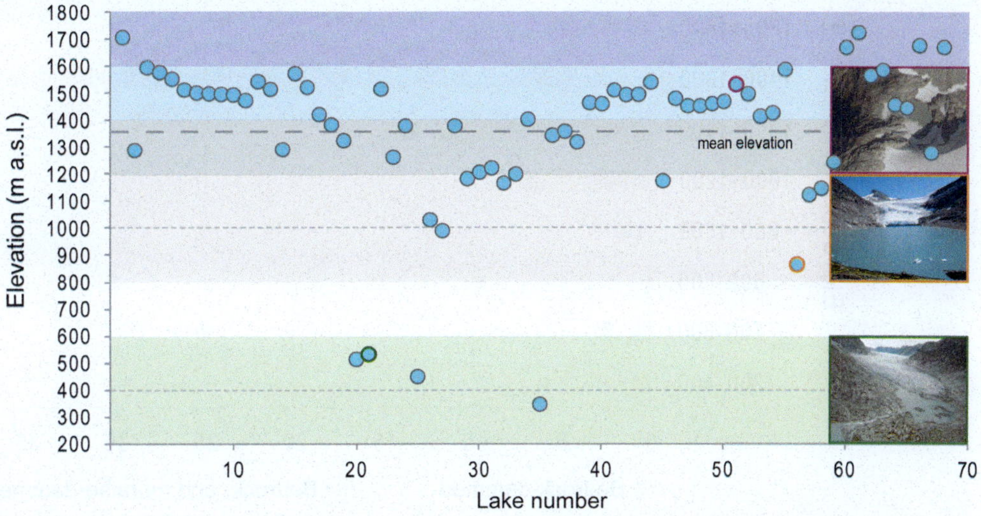

Fig. 4.7 Distribution of lake number over elevation for all lakes included in the inventory. The coloured margins of the circles correspond to the examples of proglacial lakes shown in the photographs. *Photograph-green margin* © H. Elvehøy (NVE/Glacier data, www.nve.no/hydrology/glaciers); *Photograph-orange margin* © K. Laute; *Digital orthophoto-purple margin* © Kartverket, www.kartverket.no

situated between 1200 and 1600 m a.s.l. This comparably high density of lakes clearly reflects the present mean elevation of the glacier forefields caused by the ongoing loss in glacier area.

The altitudinal distribution of lakes also varies with lake type and lake surface area (Fig. 4.8), primarily indicating differences in bedrock topography, damming conditions and prevalent surrounding material. Bedrock-dammed lakes represent the most common lake type and also have a widespread altitudinal distribution with an apparent dominance in higher elevations from 1200 m a.s.l. upwards (Fig. 4.8a). The type of bedrock- and moraine-dammed lakes exists in almost all elevations up to 1600 m a.s.l., whereas the type of moraine-dammed lakes is found up to 1400 m a.s.l. No ice-dammed lakes are found below 800 m a.s.l. (Fig. 4.8a). Interestingly, the percentage share of lake surface area per elevation range deviates from the number of lakes per elevation range (Fig. 4.8b). Although the majority of lakes are situated between 1400 and 1600 m a.s.l., the highest percentage of total lake surface area is found between 800 and 1000 m a.s.l. 48.9% of the total lake surface area is made up of only 8.8% of the total number of lakes located between 200 and 1000 m a.s.l. This elucidates that, so far, larger proglacial lakes have only developed within the adjacent valley systems and not on top of the bedrock plateau. In contrast, lakes developed in higher elevations are obviously smaller in size but more numerous. Also with regard to percentage share of lake surface area, the lake type *bedrock-dammed* is again dominant between 1200 and 1600 m a.s.l. (Fig. 4.8b). In addition, it is dominant between 800 and 1000 m a.s.l. but not in elevations below 600 m a.s.l. Ice-dammed lakes have the lowest share connected to lake surface area in comparison to all other lake types. The two lake types bedrock- and moraine-dammed and moraine-dammed have the highest shares of lake surface area between 1000 and 1200 m a.s.l. and at elevations below 600 m a.s.l. (Fig. 4.8b). The altitudinal distribution of these two lake types is strongly controlled by the availability of sediments.

From the 68 lakes, 37 are located along the northern side and 31 along the southern side of the Jostedalsbreen ice cap (Fig. 4.8c). In elevations above 1000 m a.s.l., a slightly higher number of lakes is situated on the northern side compared to the southern side. However, no clear trend can be observed regarding the distribution of the number of lakes over elevation with respect to lake orientation.

4.4.4 Glacier Lake Outburst Floods (GLOFs) / Jøkulhlaups

Glacier lake outburst floods (GLOFs) or jøkulhlaups (from Icelandic *jøkull* = glacier and *hlaup* = flood) are sudden releases of large amounts of water from a glacier (Carrivick and Tweed 2016). These floods are often initiated by failure of ice-, moraine- or landslide-dammed glacial lakes (Tweed and Russell 1999). On a global perspective, ice-dammed lakes are the most common source of glacier outburst floods (Carrivick and Tweed 2016). GLOFs are dependent on climatic conditions as the formation and evolution of ice- and moraine-dammed lakes are related to environmental factors (Carrivick and Tweed 2013). Due to their catastrophic and far-reaching effects, GLOFs pose a significant natural hazard in mountainous areas and can cause loss of human life as well as damage to infrastructure and property.

According to the inventory of glacier-related hazardous events in Norway (Jackson and Ragulina 2014), the most common hazardous events are jøkulhlaups/GLOFs, ice avalanches and incidents related to glacier length changes. Around the Jostedalsbreen ice cap several historical but also

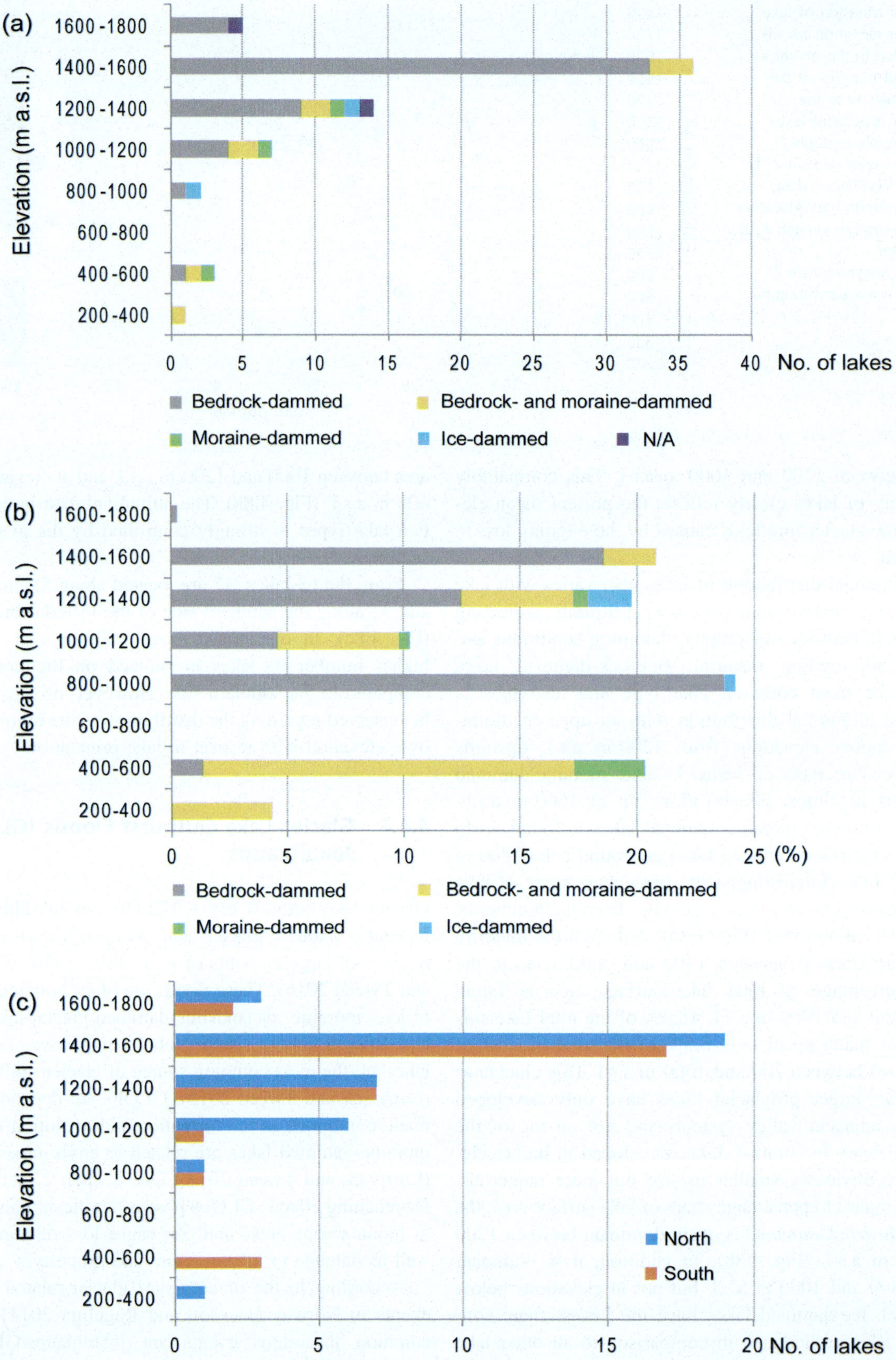

Fig. 4.8 **a** Altitudinal distribution of the numbers of lakes classified in the four different lake types. **b** Distribution of the percentage share of lake surface area classified in the four different lake types over elevation. **c** Distribution of numbers of lakes over elevation with respect to lake orientation (location of the lake along the northern or southern side of the Jostedalsbreen).

Table 4.5 Summary of documented historical and recent glacier lake outburst floods at the Jostedalsbreen ice cap

Glacier name	Date (month/year)	Reported damage	Reference
Brenndalsbreen	01/1720	Farmland destroyed	Grove (1988)
	01/1734	Farmland flooded and destroyed	Grove (1988)
	01/1743	Farmland flooded and destroyed	Jackson and Ragulina (2014)
	01/1760	–	Grove (1988)
Marabreen	07/2004	No damage	Jackson and Ragulina (2014)
	09/2015	No damage	Jackson and Ragulina (2014)
	08/2018	No damage	Jackson, NVE database 2018
Nigardsbreen	08/1998	–	Norges Klatreforbund
Supphellebreen/ Flatbreen	01/1760	Flood damage, debris flow	Breien et al. (2008)
	01/1924	Flood damage, debris flow	Breien et al. (2008)
	11/1947	Flood, debris flow, bridge/road damaged	Breien et al. (2008)
	05/2004	Farmland flooded and destroyed	Breien et al. (2008)
Stigagholtbreen	08/2017	No damage	Kjøllmoen et al. (2018)
Tunsbergdalsbreen	07/1896	Flood damage	Mottershead and Collin (1976)
	07/1897	Flood damage	Mottershead and Collin (1976)
	07/1898	Flood damage	Mottershead and Collin (1976)
	07/1899	Flood damage	Mottershead and Collin (1976)
	08/1900	Flood damage, bridge destroyed	Liestøl (1956)
	08/1903	Flood damage	Liestøl (1956)
	08/1926	Farmland flooded and destroyed	Liestøl (1956)
	06/1970	Flood damage	Mottershead and Collin (1976)
	08/1973	Flood damage	Mottershead and Collin (1976)
	01/1999	Flood damage	Kjøllmoen et al. (2000)

Data source NVE/Glacier data, www.nve.no/hydrology/glaciers

recent events are documented (Table 4.5). The majority of the events caused partly severe damage to farmland and infrastructure but fortunately no people have been harmed by today. A comparably high frequency of GLOF events is recorded at Tunsbergdalsbreen (Fig. 4.1), the largest and longest outlet glacier in southern Norway. All these events originated from a former ice-dammed lake (called Brimkjelen) which was located along the western side of the glacier tongue. Since 1999 this former ice-dammed lake is empty and no further GLOF events have been observed (Jackson and Ragulina 2014).

On 8 May 2004, a glacial lake outburst flood occurred on the western side of the lowermost glacier terminus of Flatbreen (which belongs to Supphellebreen) (Fig. 4.1). This GLOF was caused by a rapid failure of the neoglacial moraine ridge located at 1000 m a.s.l. which led to a sudden drainage of the ice-dammed lake in front of Flatbreen (Fig. 4.9a; Breien et al. 2008). The resulting outburst flood developed quickly into a debris flow which caused significant erosion along its 3-km-long path downwards (Fig. 4.9a). Breien et al. (2008) report that the erosion by the debris flow was tenfold, and the flow volume increased from around 25,000 up to 240,000 m^3. At the main valley bottom a 75,000 m^2 large debris fan was deposited by the debris flow and floodwater and fine material inundated 250,000 m^2 of farmland (Fig. 4.9a). Field and historic evidence revealed that at

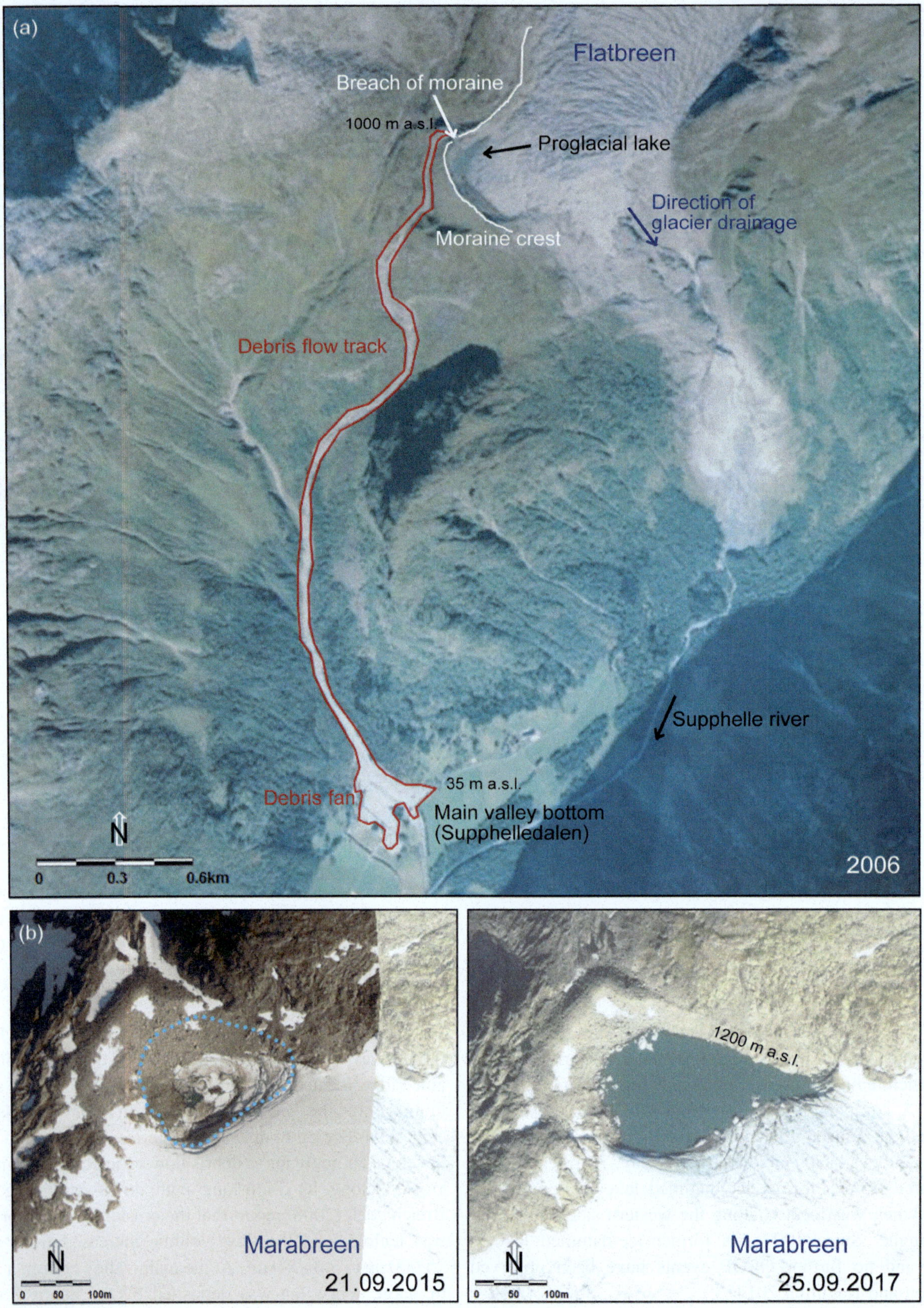
(a)
Flatbreen
Breach of moraine
1000 m a.s.l.
Proglacial lake
Direction of
glacier drainage
Moraine crest
Debris flow track
Supphelle river
35 m a.s.l.
Debris fan
Main valley bottom
(Supphelledalen)
N
0 0.3 0.6km
2006
(b)
Marabreen
N
0 50 100m
21.09.2015
1200 m a.s.l.
Marabreen
N
0 50 100m
25.09.2017

◀ **Fig. 4.9** **a** Example of the impact of a glacial lake outburst flood which occurred on 8 May 2004 at the glacier terminus of the Flatbreen/Supphellebreen outlet glacier. A rapid failure of the moraine ridge led to a sudden drainage of the ice-dammed lake. The resulting outburst flood quickly developed into a debris flow which caused significant erosion along its 3-km-long path downwards. *Source* Digital orthophoto © Kartverket, www.kartverket.no. **b** A recent glacier lake outburst flood occurred most likely in September 2015 on the northwestern side of Marabreen. The blue dotted line indicates the likely former lake level. The drainage of the lake occurs most likely through a subglacial drainage path. The second orthophoto shows that the ice-dammed lake has completely filled up again in September 2017. *Source* Digital orthophoto © Kartverket, www.kartverket.no

least two smaller debris flow events must have occurred earlier in the same gully path (Breien et al. 2008). A recent example of a GLOF occurred most likely in September 2015 on the northwestern side of Marabreen (Fig. 4.1). The estimated size of the full ice-dammed lake is about 30,000 m^2 which relates to a volume of ca. 100,000 m^3 (Jackson and Ragulina 2014). Figure 4.9b shows an orthophoto from 21 September 2015 where the lake has completely drained. The blue dotted line indicates the assumed former lake level. The drainage of the lake occurs most likely through a subglacial drainage path orientated towards the southwestern terminus of Marabreen which drains into the lake Trollavatnet. As the Trollavatnet lake is used as a hydropower reservoir and is therefore regulated, no damage originating from GLOFs has occurred so far. The second orthophoto in Fig. 4.9b shows that the ice-dammed lake has been completely filled up with water again in September 2017.

4.5 Discussion

4.5.1 Geomorphic Importance of Proglacial Lakes

In recent years, the number and size of proglacial lakes in mountain regions have increased worldwide associated to the climate-induced glacier retreat and thinning (e.g. Paul et al. 2007; Carrivick and Quincey 2014; Wang et al. 2014; Buckel et al. 2018; Wilson et al. 2018). Proglacial lakes represent a significant component of many glaciated environments and play a geomorphic key role with respect to the sediment connectivity, sediment storage and the sedimentary budgets of proglacial systems. By buffering meltwater discharge dynamics and acting as effective sediment traps, proglacial lakes affect the geomorphological activity in proglacial systems and beyond (e.g. Carrivick and Tweed 2013; Carrivick and Heckmann 2017). The formation of proglacial lakes reduces significantly sediment delivery from glaciated mountain drainage basins to the lowlands by trapping efficiently coarse and partly suspended sediment loads. The stored sediment within proglacial lakes represents an important geochronological and paleoenvironmental archive of glacier-derived meltwater fluctuations and can be used as proxy for reconstructing glacier history and climatic conditions in the past (e.g. Larsen et al. 2011; Vasskog et al. 2012; Bilt et al. 2016).

Although the current development of proglacial lakes receives an increasing awareness, comparably little information on the actual trapping efficiency of proglacial lakes is available in contrast, for instance, to well-studied lake reservoirs. A study by Geilhausen et al. (2013) showed that the connectivity between glacial sediment production and downstream sediment fluxes of a retreating Alpine glacier is significantly reduced by the development of proglacial lakes. Based on a 20-month monitoring period including a systematic up- and downstreaming sampling, the study reports that the proglacial lake reduced suspended sediment concentrations by 88–95% related to an upstream catchment of ca. 18.7 km^2 of the study area. However, the study also emphasises that the lake can change from a sink to a temporal source 'when the lake's suspended sediment budget is elevated by external forcing (e.g. rainfall induced hillslope sediment supply)' (Geilhausen et al. 2013).

Bogen et al. (2015) compare the effects of larger proglacial lakes developed in bedrock overdeepenings on fluvial sediment transport from three different Norwegian glaciers including Nigardsbreen and Tunsbergdalsbreen, two outlets from the Jostedalsbreen ice cap (Fig. 4.1). By measuring the sediment in- and output of the lake a trapping efficiency of 80% has been estimated for the 1.8-km-long and on average 15-m-deep lake Nigardsvatn (not included in the lake inventory). In contrast, a trapping efficiency of 36% has been estimated for the newly formed proglacial lake (about 500 m long and 9 m deep) in front of Tunsbergdalsbreen (Fig. 4.4h; included in the lake inventory).

A similar study conducted by Liermann et al. (2012) highlights, in particular, the importance of small ephemeral proglacial lakes as storage sites within larger valley-fjord sediment routing systems. Contemporary suspended sediment transfer and accumulation processes are investigated in the small proglacial Sætrevatnet sub-catchment located in Bødalen along the northwestern side of the Jostedalsbreen (Figs. 4.1, 4.10a, b). The moraine-dammed lake Sætrevatnet was formed after the retreat of the Bødalsbreen glacier in 1930 (note that this lake is not included in the lake inventory as it is located outside of the 1952–1985 GAO). Based on the combination of sediment in- and outflow data and delta sediment accumulation rates a trapping efficiency of about 80–85% for the sediments delivered from the upstream-located glacial system (17.5 km^2; Fig. 10a) was calculated for 2010. Recorded delta accumulation rates ranging from 4 cm $year^{-1}$ (2009) to 3 cm $year^{-1}$ (2010) and

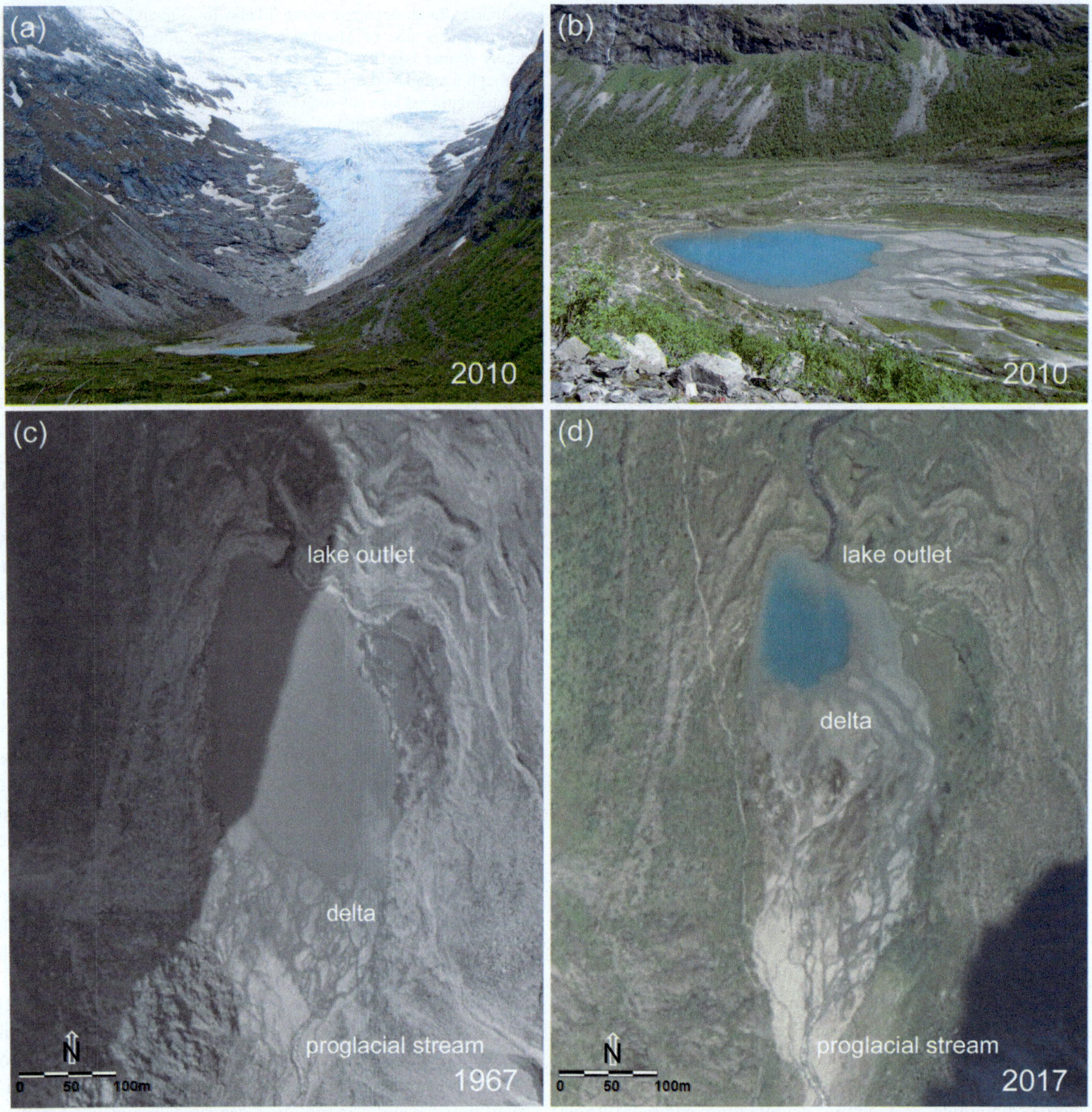

Fig. 4.10 **a** and **b** Impressions of the proglacial Sætrevatnet sub-catchment located in Bødalen. **a** View upvalley towards the Bødalsbreen outlet glacier and **b** view downvalley with the moraine-dammed lake Sætrevatnet and its delta in the foreground. *Photographs* © K. Laute. **c** and **d** Change in lake size of lake Sætrevatnet over a period of 50 years. *Source* Digital orthophotos © Kartverket, www.kartverket.no

the average annual delta advance ranging from 3.8 to 4.6 m are in agreement with the calculated high annual lake sedimentation rate of 1.7 cm $year^{-1}$ in 2010. Figure 4.10c, d illustrate the change in lake size over a period of 50 years. It is conceivable that the lake will be filled up with sediments in the near future. In this case, it is possible that the lake transforms from a sediment sink to a new temporal sediment source of erodible sediments.

The lifespan of proglacial lakes is highly variable and depends primarily on the type of lake dam, lake size and lake depth as well as on the sediment input rate. Once formed, proglacial lakes can be persistent elements of a landscape, but they can also be transitory and dynamic (Carrivick and Tweed 2013). The lifespan can range from a few years to several thousand years (Buckel et al. 2018).

With respect to sediment connectivity in proglacial environments it is important to distinguish erosion-resistant lake damming types (e.g. bedrock-dammed) from failure-prone lake damming types (e.g. moraine- or landslide-dammed) (Cavalli et al. 2019). For instance, bedrock-dammed lakes represent comparably long-lasting water bodies which are able to disconnect the upper part of the catchment from downstream sediment transfer until the lake is completely filled up with sediment. But even after the lake is filled, the resulting low-slope alluvial deposit still effectively impedes the downstream sediment flux from the upper located source areas (Cavalli et al. 2019). In contrast, lakes impounded by failure-prone dames can lead to a sudden release of a large water volume which may transform into a debris flow or debris flood, depending on the slope steepness of the downstream flow path, which can deliver huge amounts of sediment downwards through the catchment (Breien et al. 2008; Cavalli et al. 2019).

4.5.2 Future Development of Proglacial Lakes and Possible Implications

It is generally expected that the number of proglacial lakes in mountain landscapes will further increase in the near future. Several techniques are currently used to identify potential subglacial bedrock depression which may become future proglacial lakes (a short overview is given in Otto et al. 2019). A common approach includes the modelling of the ice thickness distribution of glaciers based on digitised glacier inventories, digital elevation models (DEMs) and simplified glaciological principles (e.g. Clarke et al. 2009; Linsbauer et al. 2009; Paul and Linsbauer 2012). The subtraction of the modelled ice thickness from a surface DEM results in a DEM without glaciers reflecting an estimate of the subglacial topography (e.g. Binder et al. 2009; Linsbauer et al. 2009, 2012). Modelled results are typically validated with measured ice thickness data obtained from ground penetrating radar (GPR) or seismic measurements.

Such an approach has been applied, for instance, in the Swiss Alps and in the Himalaya–Karakorum region (Linsbauer et al. 2012, 2016). Linsbauer et al. (2012) present a modelled ice thickness distribution for all Swiss glaciers and analyse the characteristics of the resulting glacier bed topography in terms of potential future lake formation. They identified a considerable number of (partly large) overdeepenings in the modelled glacier beds with a total area of about 50–60 km^2. These overdeepenings can be seen as sites of potential future lake formation.

Such a comprehensive modelling approach has not yet been conducted for the Jostedalsbreen ice cap but the subglacial topography was partly mapped during a ground penetrating radar survey in the 1980s carried out by NVE (Sætrang and Wold 1986). Burki et al. (2009) constructed a DEM of the subglacial topography of the Bødalsbreen glacier (Fig. 4.1) based on these GPR data. Two semicircular subglacial depressions with diameters in the order of 0.6–0.8 km and a depth of more than 50 m are identified near the edge of the plateau within the Bødalsbreen catchment. According to the authors, it is possible that the two depressions have stored large quantities of sediments which would be exposed and made available in case of a persistent glacier retreat. Hence, it is very likely that both depressions will develop into proglacial lakes. Based on the same subglacial topography data from NVE, Bogen et al. (2015) present two maps of the subglacial morphology of the Nigardsbreen and Tunsbergdalsbreen (Fig. 4.1) highlighting the development of potential future lakes in both glacier catchments. A 500-m-long subglacial overdeepening was mapped in the upper part of the Nigardsbeen outlet (at 900 m a.s.l.) which potentially will become a small proglacial lake and would act as sediment trap. Several bedrock depressions are found beneath the Tunsbergdalsbreen, with the largest one being 4 km long and the deepest area exceeding 100 m. Even if it is still unclear how many new lakes will develop at the Jostedalsbreen ice cap in the near future, it is often assumed that a significant share of the future suspended sediment load from the retreating glaciers could be trapped in these newly emerging proglacial lakes (Liermann et al. 2012; Geilhausen et al. 2013; Bogen et al. 2015).

The development of new proglacial lakes affects the hydrology, geomorphological activity and sediment dynamics in proglacial systems, and has also diverse regionally and globally socio-economic implications. Newly developed proglacial lakes create, for example, new opportunities and challenges related to freshwater supply, hydropower production, touristic development, hazard prevention and landscape protection (for a more detailed review see Haeberli et al. 2016). Especially in Norway, where glacier meltwater is an important resource for hydropower, new proglacial lakes can act either as important sediment traps for already existing lake reservoirs further downstream or they constitute possible new reservoirs for expanding the hydropower production. Also in intensely exploited regions such as the European Alps, high-mountain water reservoirs can play a key role in future multi-complex energy supply systems (Haeberli et al. 2016). The current drastic changes in mountain environments also have impacts on the touristic development. While the outlet glaciers are increasingly losing their aesthetic appeal, newly formed lakes offer new attractive destinations for tourists. At the same time, new lakes may also constitute an increased hazard potential for tourists as well as for mountain and foreland communities. A recent and comprehensive global assessment of the societal impacts of GLOFs has been published by Carrivick and Tweed (2016). Their study presents data compiled from 20 countries comprising 1348 glacier outburst floods from 332 sites whereof 36% of these sites have recorded societal impacts. Multipurpose projects that, e.g. combine flood retention, freshwater supply and hydropower production may be a first step towards a future-orientated and integrative management of potentials and challenges arising from the formation of the new lakes (Haeberli et al. 2016).

4.6 Summary

The glacial landscape of the Jostedalsbreen ice cap in southwestern Norway is currently undergoing significant changes reflected by progressing glacier length changes of the outlet glaciers and the formation of new glacial lakes within the recently exposed glacier forefields.

This chapter has presented a newly produced glacier area outline for 2017/2018 for the Jostedalsbreen ice cap facilitating the latest comparison of the glacial area. For the period from 1952–1985 to 2017/2018, the entire glacier area of the Jostdalsbreen ice cap experienced a loss of 79 km^2. A glacier area reduction of 10 km^2 occurred since 1999–2006. Two percent of the recently exposed surface area (since 1952–1985) is currently covered with newly developed lakes corresponding to a total number of 57 lakes. In addition, 11 lakes that already existed have enlarged in size. Four types of proglacial lakes are identified including bedrock-dammed, bedrock- and moraine-dammed, moraine-dammed and ice-dammed lakes. Bedrock-dammed lakes represent the most common lake type and have a widespread altitudinal distribution. The density of lakes steadily increases with increasing elevation and clearly peaks between 1400 and 1600 m a.s.l. This comparably high density of lakes clearly reflects the present mean elevation of the glacier forefields caused by the ongoing loss in glacier area. Although the present number of ice-dammed lakes is very low, they can be the source of potentially catastrophic glacier lake outburst floods (GLOFs). Several historical and also some recent GLOF events are documented around the Jostedalsbreen ice cap.

Proglacial lakes play a geomorphic key role with respect to sediment connectivity and the sedimentary budgets of proglacial areas. The formation of proglacial lakes significantly reduces sediment delivery from glaciated mountain drainage basins to the downstream valley sections by trapping efficiently coarse and partly suspended sediment loads. However, proglacial lakes can also change from a sink to a temporal sediment source in specific weather conditions. In addition, a completely filled up lake can develop into a new source of erodible sediments. The development of new proglacial lakes affects the geomorphological activity in proglacial systems and also has diverse regional and global socio-economic implications. Especially in mainland Norway, where glaciers and glacier-fed streams have a high importance for hydropower production, tourism and climate research, it is essential to gain a better understanding of the possible impacts of proglacial lakes for being prepared for advantages and challenges arising from these newly emerging landscape elements. Due to the predicted increase in summer temperatures for western Norway until the end of this century, it is very likely that the current trend of an accelerated mass loss of Norwegian glaciers will continue. As one consequence of this development, further new lakes will emerge within the newly exposed terrain.

For the first time, a detailed glacial lake inventory for the Jostedalsbreen ice cap in mainland Norway has been created and is presented in this chapter. This new lake inventory contributes to the set of worldwide existing glacier lake inventories and serves as a baseline study for possible future comparisons at regional and global scales.

Acknowledgements The authors thank Stefan Winkler for informative and constructive comments on Section 4.2. The revised and final version of this chapter was sent to Springer in August 2019.

References

Andreassen LM, Winsvold SH (eds) (2012) Inventory of Norwegian glaciers. Norwegian Water Resources and Energy Directorate (NVE), Oslo, Report p 38

Andreassen LM, Elvehøy H, Kjøllmoen B, Engeset RV, Haakensen N (2005) Glacier mass balance and length variations in Norway. Ann Glaciol 42:317–325

Andreassen LM, Paul F, Kääb A, Hausberg JE (2008) Landsat-derived glacier inventory for Jotunheimen, Norway, and deduced glacier changes since the 1930s. The Cryosphere 2:131–145

Andreassen LM, Huss M, Melvold K, Elvehøy H, Winsvold SH (2015) Ice thickness measurements and volume estimates for glaciers in Norway. J Glaciol 61:763–775

Ballantyne CK (2002) Paraglacial geomorphology. Quat Sci Rev 21:1935–2017

Baumhauer R, Winkler S (2014) Glazialgeomorphologie—Formung der Landoberfläche durch Gletscher. Bornträger, Stuttgart

Beniston M (2000) Environmental change in mountains and uplands. Taylor & Francis Ltd., London

Beniston M (2003) Climatic change in mountain regions: a review of possible impacts. Clim Change 59:5–31

Benn DI, Evans DJA (2010) Glaciers and glaciation. Routledge, London

Beylich AA, Laute K (2012) Spatial variations of surface water chemistry and chemical denudation in the Erdalen drainage basin, Nordfjord, western Norway. Geomorphology 167–168:77–90

Beylich AA, Laute K (2021) Fluvial processes and contemporary fluvial denudation in different mountain landscapes in western and central Norway. In: Beylich AA (ed) Landscapes and Landforms of Norway. Springer, Dordrecht

Bickerton RW, Matthews JA (1993) Little Ice Age' variations of outlet glaciers from the Jostedalsbreen ice-cap, southern Norway: a

regional lichenometric-dating study of ice-marginal moraine sequences and their climatic significance. J Quat Sci 8:45–66
Bilt WGMvd, Bakke J, Vasskog K, Røthe T, Støren EWN (2016) Glacier-fed lakes as palaeoenvironmental archives. Geol Today 32:213–218
Binder D, Brückl E, Roch K, Behm M, Schöner W, Hynek B (2009) Determination of total ice volume and ice-thickness distribution of two glaciers in the Hohe Tauern region, Eastern Alps, from GPR data. Ann Glaciol 50:71–79
Bogen J, Xu M, Kennie P (2015) The impact of pro-glacial lakes on downstream sediment delivery in Norway. Earth Surf Proc Land 40:942–952
Breien H, De Blasio FV, Elverhøi A, Høeg K (2008) Erosion and morphology of a debris flow caused by a glacial lake outburst flood, Western Norway. Landslides 5:271–280
Buckel J, Otto JC, Prasicek G, Keuschnig M (2018) Glacial lakes in Austria—distribution and formation since the Little Ice Age. Global Planet Change 164:39–51
Burki V, Larsen E, Fredin O, Nesje A (2009) Glacial remobilization cycles as revealed by lateral moraine sediment, Bødalsbreen glacier foreland, western Norway. The Holocene 19:415–426
Burki V, Hansen L, Fredin O, Andersen TA, Beylich AA, Jaboyedoff M, Larsen E, Tønnesen JF (2010) Little Ice Age advance and retreat sediment budgets for an outlet glacier in western Norway. Boreas 39:551–566
Carrivik JL, Heckmann T (2017) Short-term geomorphological evolution of proglacial systems. Geomorphology 287:3–28
Carrivick JL, Quincey DJ (2014) Progressive increase in number and volume of icemarginal lakes on the western margin of the Greenland Ice Sheet. Glob Planet Change 116:156–163
Carrivick JL, Tweed FS (2013) Proglacial lakes: character, behaviour and geological importance. Quat Sci Rev 78:34–52
Carrivick JL, Tweed FS (2016) A global assessment of the societal impacts of glacier outburst floods. Glob Planet Change 144:1–16
Cavalli M, Heckmann T, Marchi L (2019) Sediment connectivity in proglacial areas. In: Heckmann T, Morche D (eds) Geomorphology of proglacial systems: landform and sediment dynamics in recently deglaciated alpine landscapes. Springer, Basel, pp 271–287
Chinn TJH, Winkler S, Salinger MJ, Haakensen N (2005) Recent glacier advances in Norway and New Zealand: a comparison for their glaciological and meteorological causes. Geografiska Annaler A 87:141–157
Church M, Ryder JM (1972) Paraglacial sedimentation: a consideration of fluvial processes conditioned by glaciation. Geol Soc Am Bull 83:3059–3071
Clague JJ, Evans SG (2000) A review of catastrophic drainage of moraine-dammed lakes in British Columbia. Quat Sci Rev 19:1763–1783
Clarke GKC, Berthier E, Schoof CG, Jarosch AH (2009) Neural networks applied to estimating subglacial topography and glacier volume. J Clim 22:2146–2160
Cook SJ, Swift DA (2012) Subglacial basins: their origin and importance in glacial systems and landscapes. Earth-Sci Rev 115(4):332–372
Costa JE, Schuster RL (1988) The formation and failure of natural dams. Geol Soc Am Bull 100:1054–1068
Engelhardt M, Schuler TV, Andreassen LM (2015) Sensitivities of glacier mass balance and runoff to climate perturbations in Norway. Ann Glaciol 56(70):79–88
Geilhausen M, Morche D, Otto JC, Schrott L (2013) Sediment discharge from the proglacial zone of a retreating Alpine glacier. Zeitschrift für Geomorphologie Supplementary Issues 57(3):29–53
Glässer E (1993) Wissenschaftliche Länderkunden. Bd.14: Norwegen. Wissenschaftliche Buchgesellschaft, Darmstadt
Greatbatch RJ (2000) The north atlantic oscillation. Res Risk Assess 14:0213–0242
Grove JM (1988) The little ice age. Methuen, London
Grove JM (2004) Little ice ages, 2 volumes. Routledge, London
Gruber S, Haeberli W (2007) Permafrost in steep bedrock slopes and its temperature-related destabilization following climate change. J Geophys Res 112:F02S18
Haeberli W, Beniston M (1998) Climate change and its impacts on glaciers and permafrost in the Alps. Ambio 27(4):258–265
Haeberli W, Buetler M, Huggel C, Lehmann Friedli T, Schaub Y, Schleiss AJ (2016) New lakes in deglaciating high-mountain regions—opportunities and risks. Climatic Change 1–14
Hanssen-Bauer I, Førland EJ (1998) Annual and seasonal precipitation trends in Norway 1896–1997. DNMI report 27/98
Hanssen-Bauer I, Førland EJ, Haddeland I, Hisdal H, Mayer S, Nesje A, Nilsen JEØ, Sandven S, Sandø AB, Sorteberg A, Ådlandsvik B (2017) Climate in Norway 2100: a knowledge base for climate adaptation. NCCS report 1/2017
Harris C, Von der Mühll D, Isaksen K, Haeberli W, Sollid JL, King L, Holmlund P, Dramis F, Guglielmin M, Palacios D (2003) Warming permafrost in European mountains. Glob Planet Change 39(3–4):215–225
Heckmann T, Morche D, Becht M (2019) Introduction. In: Heckmann T, Morche D (eds) Geomorphology of proglacial systems: landform and sediment dynamics in recently deglaciated alpine landscapes. Springer, Basel, pp 1–19
Hurrell JW (1995) Decadal trends in the North Atlantic Oscillation: regional temperatures and precipitation. Science 269:676–679
Hurrell JW, Kushni Y, Ottersen G, Visbeck M (2003) An overview of the North Atlantic Oscillation. The North Atlantic Oscillation: climatic significance and environmental impact. Geophys Monogr Ser 134:1–35
IPCC (2013) Summary for policymakers. In: Stocker TF, Qin D, Plattner GK, Tignor M, Allen SK, Boschung J, Nauels A, Xia Y, Bex V, Midgley PM (eds) Climate change 2013: the physical science basis. Contribution of working group I to the fifth assessment report of the intergovernmental panel on climate change. Cambridge University Press, Cambridge
Ivy-Ochs S, Kerschner H, Maisch M, Christl M, Kubik PW, Schlüchter C (2009) Latest Pleistocene and Holocene glacier variations in the European Alps. Quatern Sci Rev 28:2137–2149
Jackson M, Ragulina, G (2014) Inventory of glacier-related hazardous events in Norway. Oslo, NVE Report 83/2014
Johnson PG (2002) Proglacial and paraglacial fluvial and lacustrine environments in transition. In: Hewitt K, Byrne M-L, English M, Young G (eds) Landscapes of Transition. Springer, Dordrecht, pp 43–62
Kääb A, Haeberli W (2001) Evolution of a high-mountain thermokarst lake in the Swiss Alps. Arct Antarct Alp Res 33(4):385–390
Ketzler G, Römer W, Beylich AA (2021) The Climate of Norway. In: Beylich AA (ed) Landscapes and Landforms of Norway. Springer, Dordrecht
Kjøllmoen B, Andreassen LM, Elvehøy H, Haakensen N (2000) Glasiologiske undersøkelser i Norge 1999. Oslo, NVE Rapport 2/2000
Kjøllmoen B, Andreassen LM, Elvehøy H, Jackson M (2018) Glaciological investigations in Norway 2017. Oslo, NVE Report 82/2018
Krautblatter M, Funk D, Günzel FK (2013) Why permafrost rocks become unstable: a rock–ice-mechanical model in time and space. Earth Surf Proc Land 38:876–887
Larsen DJ, Miller GH, Geirsdóttir Á, Thordarson T (2011) A 3000-year varved record of glacier activity and climate change from the proglacial lake Hvitárvatn, Iceland. Quat Sci Rev 30:2715–2731
Laute K, Beylich AA (2014a) Environmental controls, rates and mass transfers of contemporary hillslope processes in the headwaters of two glacier-connected drainage basins in western Norway. Geomorphology 216:93–113

Laute K, Beylich AA (2014b) Environmental controls and geomorphic importance of a high-magnitude/low-frequency snow avalanche event in Bødalen, Nordfjord, western Norway. Geografiska Annaler A 96A:465–484

Laute K, Beylich AA (2016) Sediment delivery from headwater slope systems and relief development in steep mountain valleys in western Norway. In: Beylich AA, Dixon JC, Zwolinski Z (eds) Source-to-sink-fluxes in undisturbed cold environments. Cambridge University Press, Cambridge, pp 293–312

Laute K, Beylich AA (2018) Potential effects of climate change on future snow avalanche activity in western Norway deduced from meteorological data. Geografiska Annaler A 100(2):163–184

Liermann S, Beylich AA, van Welden A (2012) Contemporary suspended sediment transfer and accumulation processes in the small proglacial Sætrevatnet subcatchment, Bødalen, western Norway. Geomorphology 167–168:91–101

Liestøl O (1956) Glacier dammed lakes in Norway. Nor Geogr Tidsskr 15(3/4):121–149

Linsbauer A, Paul, F, Hoelzle M, Frey H, Haeberli W (2009) The Swiss Alps without glaciers: a GIS-based modelling approach for reconstruction of glacier beds. In: Purves R, Gruber S, Straumann R, Hengle T (eds) Geomorphometry. Zürich, pp 243–247

Linsbauer A, Paul F, Haeberli W (2012) Modeling glacier thickness distribution and bed topography over entire mountain ranges with GlabTop: application of a fast and robust approach. J Geophys Res 117:F03007

Linsbauer A, Frey H, Haeberli W, Machguth H, Azam MF, Allen S (2016) Modelling glacier-bed overdeepenings and possible future lakes for the glaciers in the Himalaya-Karakoram region. Ann Glaciol 567(71):119–130

Luckman BH (2000) The little ice age in the Canadian rockies. Geomorphology 32:357–384

Matthews JA, Briffa K (2005) The 'Little Ice Age': re-evaluation of an evolving concept. Geografiska Annaler A 87:17–36

Matthews JA, Shakesby RA, Schnabel C, Freeman S (2008) Cosmogenic 10Be and 26Al ages of Holocene moraines in southern Norway I: testing the method and confirmation of the date of the Erdalen event (c. 10 ka) at its type-site. The Holocene 18:1155–1164

Mottershead DN, Collin RL (1976) A study of glacier-dammed lakes over 75 years: Brimkjelen southern Norway. J Glaciol 17(77):491–505

Nesje A (1984) Kvartærgeologiske undersøkningar i Erdalen, Stryn, Sogn og Fjordane. M.S. Thesis. University of Bergen, p 201

Nesje A (1995) Breene I Vest-Norge vokser med rekordfart. Naturen 119:7–10

Nesje A (2005) Briksdalsbreen in western Norway: AD 1900–2004 frontal fluctuations as a combined effect of variations in winter precipitation and summer temperature. The Holocene 15:1245–1252

Nesje A (2009) Latest Pleistocene and Holocene alpine glacier fluctuations in Scandinavia. Quatern Sci Rev 28:2119–2136

Nesje A, Dahl SO (2003) The 'Little Ice Age'—only temperature? Holocene 13:139–145

Nesje A, Kvamme M (1991) Holocene glacier and climate variations in western Norway: evidence for early Holocene glacier demise and multiple Neoglacial events. Geology 19:610–612

Nesje A, Matthews JA (2011) The Briksdalsbre event: a winter precipitation-induced decadalscale glacial advance in southern Norway in the AD 1990s and its implications. The Holocene 22:249–261

Nesje A, Lie Ø, Dahl SO (2000) Is the North Atlantic Oscillation reflected in Scandinavian glacier mass balance records? J Quat Sci 15:587–601

Nesje A, Matthews JA, Dahl SO, Berrisford MS, Andersson C (2001) Holocene glacier fluctuations of Flatebreen and winter-precipitation changes in the Jostedalsbreen region, western Norway, based on glaciolacustrine sediment records. The Holocene 11(3):267–280

Nesje A, Bakke J, Dahl SO, Lie Ø, Matthews JA (2008a) Norwegian mountain glaciers in the past, present and future. Glob Planet Change 60:10–27

Nesje A, Dahl SO, Thun T, Nordli Ø (2008b) The "Little Ice Age" glacial expansion in western Scandinavia: summer temperature or winter precipitation? Clim Dyn 30:789–801

Nordgulen Ø, Andresen A (2013) De eldeste bergartene dannes, Jordas Urtid; 4600–850 Ma. In: Ramberg IB, Bryhni I, Nøttvedt A, Rangnes K (eds) Landet blir til. Norges geologi. 2. utgave. Norsk Geologisk Forening, Trondheim, Norway, pp 62–119

Nussbaumer SU, Nesje A, Zumbühl HJ (2011) Historical glacier fluctuations of Jostedalsbreen and Folgefonna (southern Norway) reassessed by new pictorial and written evidence. The Holocene 21:455–471

Otto JC (2019) Proglacial lakes in high mountain environments. In: Heckmann T, Morche D (eds) Geomorphology of proglacial systems: landform and sediment dynamics in recently deglaciated alpine landscapes. Springer, Basel, pp 231–247

Øyen PA (1906) Klima und Gletscherschwankungen in Norwegen. Zeitschrift für Gletscherkunde 1:46–61

Paul F, Linsbauer A (2012) Modeling of glacier bed topography from glacier outlines, central branch lines and a DEM. Int J Geograph Inform Sci 26(7):1173–1190

Paul F, Kääb A, Haeberli W (2007) Recent glacier changes in the Alps observed by satellite: consequences for future monitoring strategies. Glob Planet Change 56(1–2):111–122

Paul F, Andreassen LM, Winsvold SH (2011) A new glacier inventory for the Jostedalsbreen region, Norway, from Landsat TM scenes of 2006 and changes since 1966. Ann Glaciol 52(59):153–162

Racoviteanu AE, Paul F, Raup B, Khalsa SJS, Armstrong R (2009) Challenges and recommendations in mapping of glacier parameters from space: results of the 2008 Global Land Ice Measurements from Space (GLIMS) workshop. Boulder, Colorado, USA. Ann Glaciol 50(53)

Rangwala I, Miller JR (2012) Climate change in mountains: a review of elevation-dependent warming and its possible causes. Clim Change 114:527–547

Rasmussen LA, Andreassen LM, Baumann S, Conway H (2010) Little Ice Age precipitation in Jotunheimen, southern Norway. The Holocene 20(7):1039–1045

Raup B, Khalsa SJS (2007) GLIMS Analysis Tutorial. Digital Media: https://www.glims.org/MapsAndDocs/assets/GLIMS_Analysis_Tutorial_a4.pdf

Rekstad J (1902) Iakttagelser fra bræer i Sogn og Nordfjord. Norges geologiske undersøkelse, Aarbog 1902(3):1–48

Sætrang AC, Wold B (1986) Results from the radio-echo sounding on parts of the Jostedalsbreen ice cap, Norway. Ann Glaciol 8:156–158

Sutherland JL, Carrivick JL, Shulmeister J, Quincey DJ, James WHM (2019) Ice-contact proglacial lakes associated with the Last Glacial Maximum across the Southern Alps, New Zealand. Quat Sci Rev 213:67–92

Trachsel M, Nesje A (2015) Modelling annual mass balances of eight Scandinavian glaciers using statistical models. The Cryosphere 9:1401–1414

Trenberth K, Zhang, R, National Center for Atmospheric Research Staff (eds) (2019) The climate data guide: Atlantic Multi-decadal Oscillation (AMO). https://climatedataguide.ucar.edu/climate-data/atlantic-multi-decadal-oscillation-amo. Accessed 10 May 2019

Tweed FS, Russell AJ (1999) Controls on the formation and sudden drainage of glacier-impounded lakes: implications for jökulhlaup characteristics. Prog Phys Geogr 23:79–110

Vasskog K, Paasche Ø, Nesje A, Boyle JF, Birks HJB (2012) A new approach for reconstructing glacier variability based on lake sediments recording input from more than one glacier. Quat Res 77:192–204

Wang W, Xiang Y, Gao Y, Lu A, Yao T (2014) Rapid expansion of glacial lakes caused by climate and glacier retreat in the Central Himalayas. Hydrol Process 29(6):859–874

Wilson R, Glasser NF, Reynolds JM, Harrison S, Iribarren Anacona P, Schaefer M, Shannon S (2018) Glacial lakes of the Central and Patagonian Andes. Glob Planet Change 162:275–291

Winkler S (2021) Terminal moraine formation processes and geomorphology of glacier forelands at the selected outlet glaciers of Jostedalsbreen, South Norway. In: Beylich AA (ed) Landscapes and Landforms of Norway. Springer, Dordrecht

Winkler S, Matthews JA (2010) Observations on terminal moraine-ridge formation during recent advances of southern Norwegian glaciers. Geomorphology 116:87–106

Winkler S, Nesje A (2009) Perturbation of climatic response at maritime glaciers? Erdkunde 63:229–244

Winkler S, Haakensen N, Nesje A, Rye N (1997) Glaziale Dynamik in Westnorwegen—Ablauf und Ursachen des aktuellen Gletschervorstoßes am Jostedalsbreen. Petermanns Geogr Mitt 141:43–63

Winkler S, Elvehøy H, Nesje A (2009) Glacier fluctuations of Jostedalsbreen, western Norway, during the past 20 years: the sensitive response of maritime mountain glaciers. Holocene 19:389–408

Winsvold SH, Andreassen LM, Kienholz C (2014) Glacier area and length changes in Norway from repeat inventories. The Cryosphere 8:1885–1903

Zemp M, Paul F, Hoelzle M, Haeberli W (2008) Alpine glacier fluctuations 1850–2000: an overview and spatio-temporal analysis of available data and its representativity. In: Orlove B, Luckman B, Wiegandt E (eds) Darkening peaks: glacier retreat, science, and society. University of California Press, Berkeley and Los Angeles, pp 152–167

Paraglacial Rock-Slope Failure Following Deglaciation in Western Norway

5

Alastair M. Curry

Abstract

The paraglacial framework describes the geomorphological response to glaciation and deglaciation, whereby non-renewable, metastable, glacially conditioned sediment sources are progressively released by a range of nonglacial processes. These include slope failures that directly modify the bedrock topography of mountain landscapes. This chapter synthesises recent research on the paraglacial evolution of western Norway's mountain rock-slopes, and evaluates the importance of glaciation, deglaciation and associated climatic and non-climatic processes. Following an introduction to the concept of paraglacial landscape change, current understanding of rock-slope responses to deglaciation is outlined, focusing on the spatial distribution, timing, duration and causes of rock-slope failure activity. Preliminary analysis of an inventory of published ages for 49 prehistoric, moderate-large ($>10^3$ m^3) rock-slope failures (RSFs) indicates that the great majority occurred in the Late Weichselian/Early Holocene transition ($\sim$13–9 ka), within 2 ka of deglaciation. Subsequent RSFs were much smaller, though event frequency increased again at 8–7 ka and 5–4 ka BP. The majority of dated RSFs were not directly triggered by deglaciation (debuttressing) but were preconditioned for more than 1000 years after ice withdrawal, until slopes collapsed. It is proposed that the primary causes of failure within 2 ka of ice retreat were stress redistribution, subcritical fracture propagation, with some events possibly triggered by seismic activity. While earthquakes may have triggered renewed failure of rock-slopes in the Late Holocene, it seems likely that permafrost degradation and water supply were locally important. Priority avenues for further research are briefly identified.

Keywords

Paraglacial • Rock-slope failure • Glaciation • Deglaciation • Rockfall • Rock avalanche • Stress release

5.1 Introduction

Rock-slope failures (RSFs) are often located close to retreating mountain glaciers, dominating bedrock erosion and sediment distributions in alpine landscapes, and representing a hazard to mountain communities and resources. Understanding the distribution, timing and duration of rock-slope activity is critically important for accurately reconstructing Quaternary landscape evolution and sediment flux, and managing hazard risks in glaciated areas.

Definition and scope

This chapter reviews current understanding of the legacy of glaciation on the failure of rock-slopes in western Norway's mountain landscape. The act of failure usually involves the initial formation of a fully developed rupture surface as a displacement or strain discontinuity (Hungr et al. 2014), though this may be preceded by rock-slope deformation (rock-slope instability). Both phenomena are on a continuum, and many RSFs exhibit both. Here, 'rock-slope failure' is used as an overall umbrella term to refer to 'any substantial rock-mass exposed to slope gravitational processes which has lost structural integrity, regardless of its degree of disintegration or distance travelled' (Jarman and Harrison 2019, p. 202). Accordingly, RSFs reflect a range of mechanisms that directly displace in situ bedrock, and include rock slides (that may develop into rock avalanches), rock-slope deformations, rock topples and falls (McColl 2012).

A. M. Curry (✉)
Department of Psychology, Sports Sciences and Geography, University of Hertfordshire, College Lane, Hatfield, Hertfordshire AL10 9AB, UK
e-mail: a.curry4@herts.ac.uk

A. A. Beylich (ed.), *Landscapes and Landforms of Norway*, World Geomorphological Landscapes,
https://doi.org/10.1007/978-3-030-52563-7_5

After introducing the reader to the concept of paraglacial landscape adjustment and the study area, following sections highlight the most important characteristics with reference to deglaciated rock-slopes in western Norway, including the processes, spatial distribution, timing, periodicity and causes of rock-slope failure in the region. Promising avenues for future research are briefly highlighted, and principal findings are summarised.

Unless otherwise indicated, radiometric ^{14}C ages have been calibrated to calendar (cal.) years before present using the IntCal13 dataset (Reimer et al. 2013), and average ^{10}Be exposure ages are also given in years before present.

Paraglacial landscape change

While geomorphologists have long recognised how landscapes may be influenced by glaciation and deglaciation, specific consideration of the transitional, 'paraglacial' adjustment of a landscape to nonglacial conditions is relatively new (Ryder 1971). The term 'paraglacial' highlights 'non-glacial processes that are directly conditioned by glaciation' (Church and Ryder 1972, p. 3059) in proglacial and ice-marginal settings. Early use of the paraglacial concept focused primarily on the abrupt and radical change in terrestrial fluvial entrainment and sedimentation associated with Late Pleistocene or Early Holocene deglaciation, whereby vast quantities of unconsolidated glacigenic detritus became liable to enhanced erosion, reworking and redeposition by rivers and debris flows, manifest in the accumulation of impressive fans and valley fills within millennia of deglaciation.

More recently, however, the notion of paraglacial landscape change has been applied to a wide range of nonglacial processes, landforms, landsystems and deposits conditioned by both Pleistocene and present-day glaciation and deglaciation within diverse geomorphological contexts over a range of process and spatial scales. In particular, since the mid-1980s, the concept of paraglacial landscape adjustment (or relaxation) from a glacially conditioned state to nonglacial conditions has been increasingly recognised as being of critical importance in understanding postglacial landscape evolution and its theoretical framework, as well as predicting landform response to current and future environmental change.

Following widespread deglaciation, the sediment budgets of a variety of landscape subsystems may be characterised by a state of disequilibrium. In rock-slope landsystems, for example, rockwalls may progressively weaken and eventually fail in response to glacier thinning and locally induced stress changes, variations in temperature and moisture, or as a result of seismicity associated with glacio-isostatic adjustment. Release of such sediments may be conceptualised as a sediment cascade (Ballantyne 2002a) in which transport of sediment from non-renewable, glacially conditioned metastable sources is stored for varying timescales in a range of landforms and deposits such as talus accumulations, landslide debris and fjord deposits (referred collectively as accommodation space (Brierley 2010)).

While there is nothing unique about the paraglacial environment or paraglacial processes per se, one of the most fruitful aspects of this research lies in recognising and interpreting the 'paraglacial period'. This temporal adjustment is characterised by high rates of glacially conditioned sediment release that peak soon after the land surface emerges from retreating glacier ice, when unstable or metastable sediment stores are exposed and subsequently depleted by a wide range of processes (Fig. 5.1). The idea of paraglacial landscape response was first considered as a unique episode of rapid readjustment in the evolution of formerly glacierised landscapes (André 2009; Slaymaker 2009); however, it has subsequently developed into a unifying, dynamic systems concept, illustrated, for example, by steady-state and exhaustion models of sediment release and storage (Cruden and Hu 1993; Ballantyne 2002a, b; Cossart and Fort 2008; Ballantyne and Stone 2013). During deglaciation, primary sediment flux rates are predicted to increase rapidly, then decline towards background nonglacial denudation rates, the rate of decline (and duration of the paraglacial period) being controlled by sediment availability and stability.

Such an orderly, monotonic evolution is commonly disrupted, however, by extrinsic perturbations, transient storage, lags and feedbacks in the sediment transport system and reworking of secondary paraglacial stores, leading to delayed, renewed and rejuvenated reworking of glacially conditioned sediment (Ballantyne 2003; Knight and Harrison 2018).

Accordingly, different geomorphic systems exhibit paraglacial relaxation over widely differing timescales, with the duration of rock-slope system responses to deglaciation measured in terms of millennia. Moreover, the trajectory and rate of response and recovery are dependent on the pre-existing system state, the extent of the glacial disturbance and spatial scale (Church and Slaymaker 1989; Slaymaker 2011), implying that individual rock-slope subsystems may attain adjustment with their nonglacial environment while mountain range systems are still in recovery mode.

This persistence of landscape memory is widely evident. Many areas deglaciated in the Late Pleistocene can be regarded as having not yet fully adjusted to nonglacial conditions, at least in terms of glacially conditioned sediment supply, and many landforms in these areas are out of equilibrium with both former glacial and contemporary nonglacial conditions. If this is the case, then the development of most formerly glaciated landscapes can be regarded as transitional or transient, rather than linear (Church 2002; Hewitt 2002; Slaymaker 2009, 2011). Paraglaciation (the collective and cumulative effects of paraglacial rates of

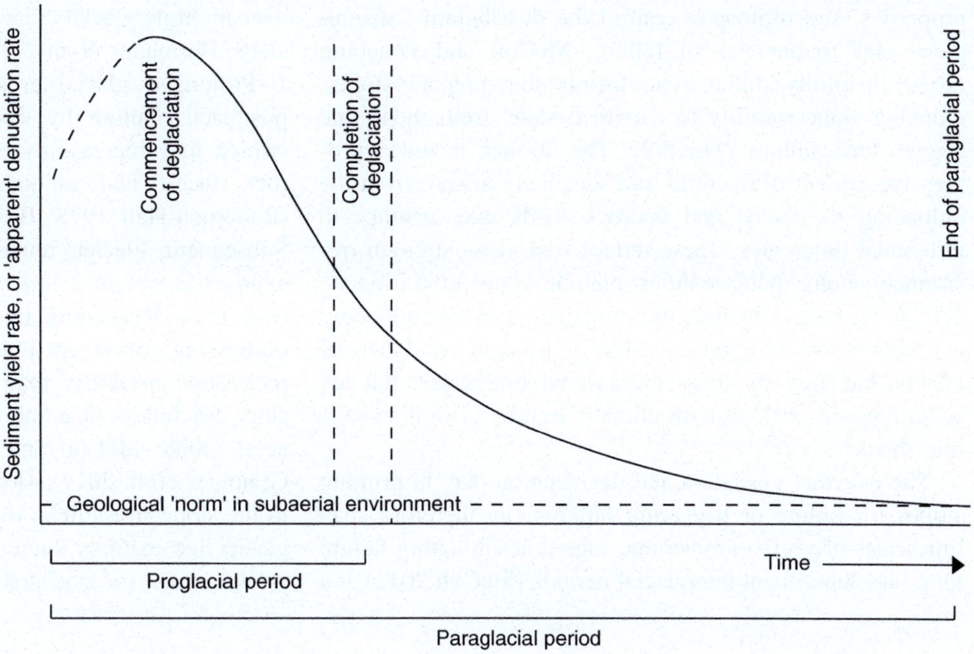

Fig. 5.1 The Paraglacial period: conceptual representation of the pattern of glacially conditioned sediment release and reworking, as envisaged by Church and Ryder (1972). Deglaciation marks the onset of this period of enhanced sediment yield which terminates when sediment yield has declined to long-term subaerial denudation norms

activity in modifying the landscape) hereby represents system recovery following glacial disturbance (Hewitt 2006; Slaymaker 2009), where a transitional, disturbance regime landscape persists until glacially conditioned sediment sources either become exhausted or stable.

The temporal pattern of paraglacial recovery is further complicated by Neoglaciation, the waxing and waning of Holocene glaciers in upper source areas, as periods of localised glacial retreat result in the exposure of fresh metastable sediments. Rapid shrinkage of mountain and arctic glaciers during the twentieth and twenty-first centuries has also resulted in widespread exposure of deglaciated terrain.

While no preserved analogues exist for the climatic contexts envisaged under future, anthropogenically enhanced global warming, the paraglacial response to recent deglaciation arguably provides a valuable template for predicting how contemporary paraglacial systems in sensitive glaciated mountain environments may respond to future changes. As deglaciation gathers pace in high-altitude and high-latitude regions presently experiencing increased sediment flux, a major priority is to better understand the trajectory and behaviour of paraglaciation. Indeed, Knight and Harrison (2014, p. 256) claim that the increasing dominance of the paraglacial process domain in mid- to high-latitude glaciated mountains this century will represent 'the most significant and fastest change to take place in mountain cryosystems in at least the last 9000 years'.

The following section focuses specifically on the response of rock-slope systems to glaciation and deglaciation.

5.2 Paraglacial Rock-Slope Adjustment: Overview

Many authors have proposed a causal connection between deglaciation of steep, glacially modified rockwalls and subsequent RSF activity, associated with both recent glacier retreat (McSaveney 1993; Evans and Clague 1994; Holm et al. 2004; Fischer et al. 2006; Kos et al. 2016) and the demise of the Late Pleistocene ice sheets (Gardner 1980; Abele 1997; Mercier et al. 2013). Indeed, exposure of glaciated rock-slopes may represent 'one of the most significant geomorphological consequences of deglaciation in mountain environments' (Ballantyne 2002b, p. 1938). Proposed explanations for this association reflect glacial erosion, retreat and thinning (downwastage) that may precondition, prepare or trigger RSF activity over timescales of years to millennia (McColl and Draebing 2019). These mechanisms are briefly summarised below; readers are directed to McColl (2012) and Pánek and Klimeš (2016) for more detailed accounts.

Rock-slope failure factors

Controls on paraglacial rock-slope (in)stability vary considerably with time, especially on glacial–interglacial timescales, and include the distribution and thermal regime of ice, glacial erosion, rockwall hydrological and mechanical conditions and seismicity. Factors that precondition rock-slopes for failure include geotechnical and topographic properties, and ultimately control the distributions, magnitudes and frequencies of failure. McColl and Draebing (2019) helpfully differentiate factors that prepare slopes, reducing slope stability to a critical state, from those that trigger final failure (Fig. 5.2). The former includes progressive growth of fractures and seismicity arising from the unloading of glacial and bedrock loads over century to millennial timescales. These reduce rock mass strength (resistance) and/or increase the magnitude of potential triggers. Triggering factors include the propagation of fractures (and consequent rock fatigue) caused by hydrological and thermal effects that vary on daily to seasonal timescales, but are superimposed on long-term climatic trends, as well as seismic shocks.

The role that glaciation and deglaciation play in priming slopes for failure or triggering failures can therefore span timescales of years to millennia, sometimes initiating failure long into subsequent interglacial periods (McColl 2012) and over multiple glacial cycles (Fig. 5.3; Grämiger et al. 2017, 2018; Hermanns et al. 2017b).

Prolonged glacial erosion can prime rock-slopes for postglacial failure by steepening and lengthening slopes (which may increase overburden shear stresses behind the rock face), and unloading them of rock overburden (Radbruch-Hall 1978; Bovis 1982; Augustinus 1995a, b). Subsequent glacier retreat removes lateral ice support, exposes slopes to a new weathering regime and enhances rock mass weakening initiated by unloading. Glacial debuttressing, often cited as a major driver of paraglacial rock-slope instability, may either trigger failure, or prepare a slope for failure at a later time (Holm et al. 2004; Cossart et al. 2008; McColl and Davies 2013; Kos et al. 2016; Grämiger et al. 2017). However, while it adds confinement to the slope stress field, over decadal and longer timescales glacier ice exhibits ductile behaviour, and so represents a weak buttress for glaciated valley walls. Stress redistribution

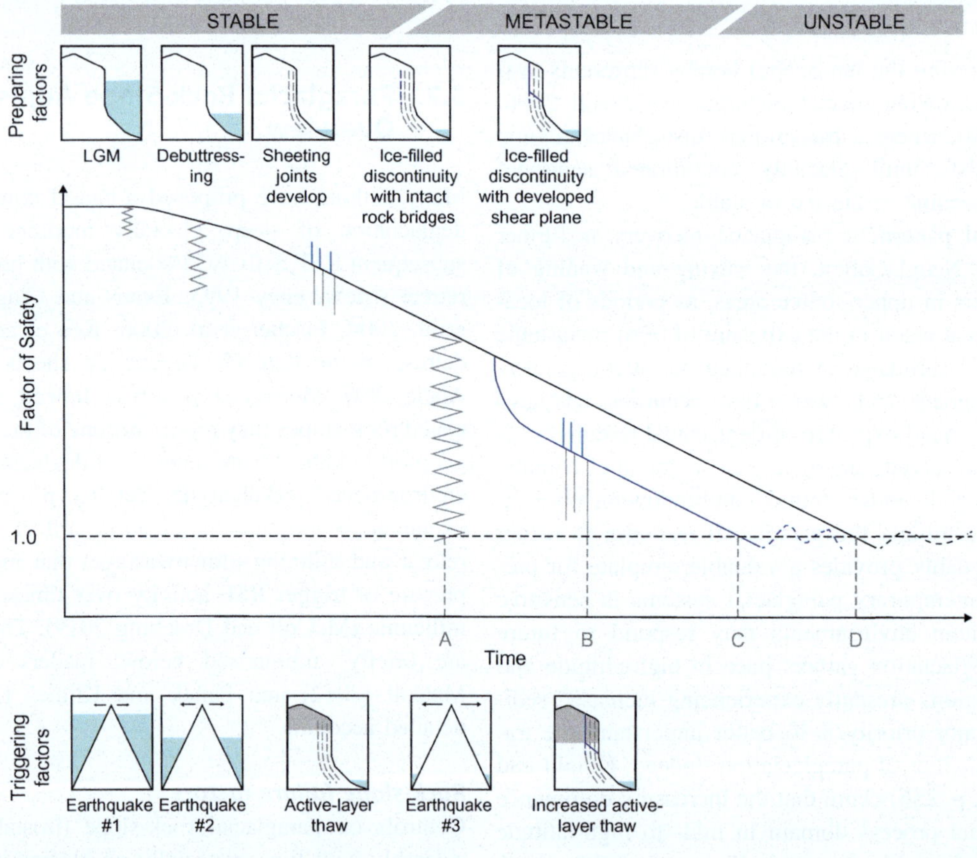

Fig. 5.2 A model for the reduction in stability of a rock-slope over time in response to changing internal and external factors that prepare slopes for failure and trigger final failure. Stability, expressed as the Factor of Safety, declines through time along different potential trajectories. Failure (a value below unity) occurs when resisting forces are exceeded by destabilising forces. New preparatory factors (depicted in the cartoons above) initiate more rapid reductions in stability as deglaciation and climate warming progress. Potential triggers are represented underneath the line and in the corresponding schematic cartoons below the main graph. A system-state change from rock-dominant state (black line) to an ice-dominant state (dark blue line) occurs when an ice-filled discontinuity with a continuous shear plane develops, and the system is more sensitive to failure. Letters A–D on the x-axis represent the breaching of external thresholds at A and B, and internal thresholds at C (sensitive rock state dominated by ice-filled fractures) and D (e.g. weathering triggered). (*Source* McColl and Draebing 2019, Fig. 2)

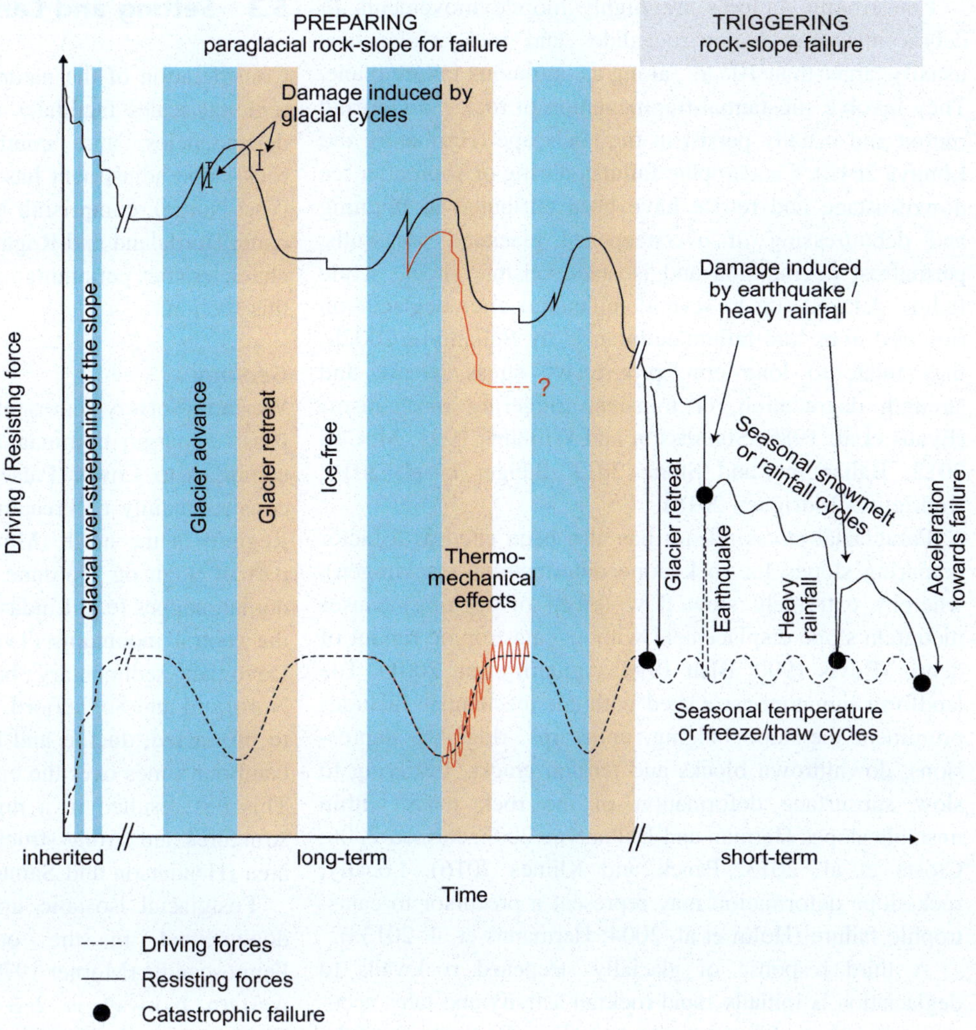

Fig. 5.3 Schematic representation of paraglacial preparation of a rock-slope until ultimate failure, with variation of driving and resisting forces during multiple glacial cycles. Incremental damage induced by glacier advance and retreat as purely mechanical loading and unloading (Grämiger et al. 2017) together with other preparatory factors during ice-free conditions reduces slope stability until a critical state is reached. A single small disturbance may ultimately trigger for catastrophic failure. Glacial cycles and other fatigue mechanisms (e.g. thermomechanical effects) may be more effective in preparing slopes for failure (red line). (*Source* Grämiger et al. 2018, Fig. 1a)

and thermomechanical stress effects (so-called 'paraglacial thermal shock') accompanying glacier fluctuations may be significantly more effective in weakening bedrock than the purely mechanical effects of glacier loading and unloading alone (Grämiger et al. 2018). The fact that the timing of RSF activity appears to often lag deglaciation by several millennia (Ballantyne et al. 2014a, b; Ballantyne and Stone 2013) indicates that glacial debuttressing (as often described) is of limited importance in explaining long histories of paraglacial rock-slope adjustment.

Ice loading of rock exerts high internal stresses, often causing elastic deformation of rock masses that is stored as residual strain energy (Wyrwoll 1977). During ice downwastage and unloading of glacially stressed rock, that energy is released, redistributing the orientation of principal stress fields within the rock, which may result in the development of a tensile stress zone behind the rock face (Ballantyne 2002b). As those surfaces are exposed, relaxation of tensile stresses causes lateral stress release (rebound), joint network propagation and reduced cohesion, which may either lead to immediate or delayed RSF activity, depending on the dissipation of residual stresses (Wyrwoll 1977), as well as rock mass properties, valley geometry and local environment.

Thus, rock fatigue is induced by both high, static overburden stresses enhanced by glacier erosion and downwastage, and a suite of cyclic stresses attributed to seismicity, thermal changes and fluctuations in water and ice loading in joints (McColl 2012; Krautblatter et al. 2013; Grämiger et al. 2017). Progressive strength degradation and the development of stress release fractures may in part explain the timing and distribution of some postglacial rock-slope failures (Ballantyne et al. 2014a).

Rock-slope failure responses

Steep rock-slopes can respond to the aforementioned stability changes in three ways (Ballantyne 2002b): catastrophic failure, deep-seated gravitational slope deformation and rockfall.

Catastrophic failures are highly mobile movements of debris including major rockslides and rock avalanches, usually larger than 10^6 m^3, along a continuous failure plane. They involve substantial fragmentation of rock mass during runout and usually persist in the landscape (Hermanns and Longva 2012). Catastrophic failures during or soon after ice downwastage and retreat have been attributed to thinning and debuttressing of oversteepened glaciated rockwalls, permafrost degradation and isostatic seismic shock, while failure delayed (up to several millennia) after deglaciation (referred to as 'pre-failure endurance' by Ballantyne 2002b) may relate to long-term, progressive stress release and strength degradation, or transient triggering mechanisms (Evans et al. 1989; Sigurdsson and Williams 1991; McColl 2012; Ballantyne and Stone 2013; Steiger et al. 2016; Rodríguez-Rodríguez 2018).

Paraglacial stress release has also been cited as a factor preparing slopes for rock-slope deformations (sackungen), whereby extremely slow flow (creep or sagging) causes mountain slope displacement without catastrophic runout of debris (Bovis 1990; Blair 1994; Agliardi et al. 2001). The landform signature associated with this mechanism includes prominent toe-slope bulging, antiscarps, ridge-top depressions, downthrown blocks and tension cracks, testifying to slow subsurface deformation of the rock mass within unstable slopes (Jarman and Ballantyne 2002; Jarman 2006; Crosta et al. 2013; Pánek and Klimeš 2016). Locally, rock-slope deformation may represent a precursor to catastrophic failure (Holm et al. 2004; Hermanns et al. 2013a).

A third response of glacially steepened rockwalls to deglaciation is initially rapid rockfall activity and talus accumulation below cliffs (Augustinus 1995a). On the basis of the large volumes of relict talus beneath rockwalls deglaciated in the Late Pleistocene, numerous authors have inferred that the rate of rockfall immediately after deglaciation greatly exceeded present (low) rates (Johnson 1984; Marion et al. 1995; Hinchliffe and Ballantyne 1999; Curry and Morris 2004). Disentangling paraglacial effects from periglacial forcing and rock mechanical properties presents a challenge in evaluating destabilising factors for rockfall activity in formerly glaciated areas (Wilson 2009, 2017; Cossart et al. 2014).

Paraglacial rock-slope adjustment: assessment

Until recently, though, evaluation of these factors responsible for glacial conditioning of rock-slope activity has been impeded by a lack of regional-scale datasets on the spatio-temporal distribution of postglacial RSFs from tectonically inactive regions (Ballantyne et al. 2014a). In this context, a rapidly growing inventory of exposure-dated RSFs in the mountains of Norway presents a rich opportunity to characterise the timing of this activity and assess competing explanations for long- and short-term causes.

5.3 Setting and Landscape Development

Consideration of the nature and effects of paraglaciation at regional scales facilitates understanding of spatio-temporal dependencies and controlling parameters. Paraglacial rock-slope adjustment has been studied in western Norway (vest-Norge), comprising Møre og Romsdal, Sogn og Fjordane, Hordaland and Rogaland counties (Fig. 5.4). Relevant characteristics pertaining to the study area are summarised in this section.

Geology

Western Norway represents the uplifted surface of a tilted passive margin dominated by metamorphic rocks of Precambrian to Lower Palaeozoic age (Fig. 5.4c). The area consists mainly of Precambrian basement (Western Gneiss Region) in the north (Møre og Romsdal) and the western part of Sogn og Fjordane, with the SW–NE aligned Caledonian nappes (e.g. Lindås, Finse, Jotun nappes) underlying the central mountains (Tveten et al. 1998). There are also Devonian sedimentary basins along the coast between Nordfjord and Sognefjord. Bedrock is highly tectonized due to protracted, ductile and brittle tectonics acting since Precambrian times over the entire region (Gee and Sturt 1985). This has resulted in a high density of brittle and ductile structures and strong structural control of RSFs in the study area (Henderson and Saintot 2011; Saintot et al. 2011).

Postglacial isostatic uplift rates at ~11 ka have been documented in the order of 50–500 mm yr^{-1} for Fennoscandia (Mørner 1979). Observed present-day uplift of western Norway is 2–3 mm yr^{-1} (Dehls et al. 2000a; Fjeldskaar et al. 2000), while debate surrounds the contribution of potential neotectonic processes. Generally, Norway has a low to intermediate seismic intensity (Fjeldskaar et al. 2000), though an area of earthquake activity is concentrated west of mid-Norway, related to a rifted passive continental margin (Bungum et al. 2005).

Quaternary glaciation

During Quaternary glacial cycles, the landscape was inundated beneath the Scandinavian ice sheet and mountain ice caps on multiple occasions, causing repeated bedrock loading, unloading and isostatic rebound. A mean bedrock lowering of ~520 m has been calculated for central Norway (Dowdeswell et al. 2010), though vertical linear erosion along fault-controlled valleys amounted to 1500–2000 m (Mangerud et al. 2011).

The last (Weichselian) glaciation started at the end of the Eemian (MIS 5e), with the Scandinavian ice sheet reaching its last glacial maximum (LGM) extent at different sectors between ~27 and 21 ka (Mangerud et al. 2011; Olsen et al. 2013; Hughes et al. 2016). Mangerud (2004) considered that at its LGM the ice sheet covered almost all mountains in

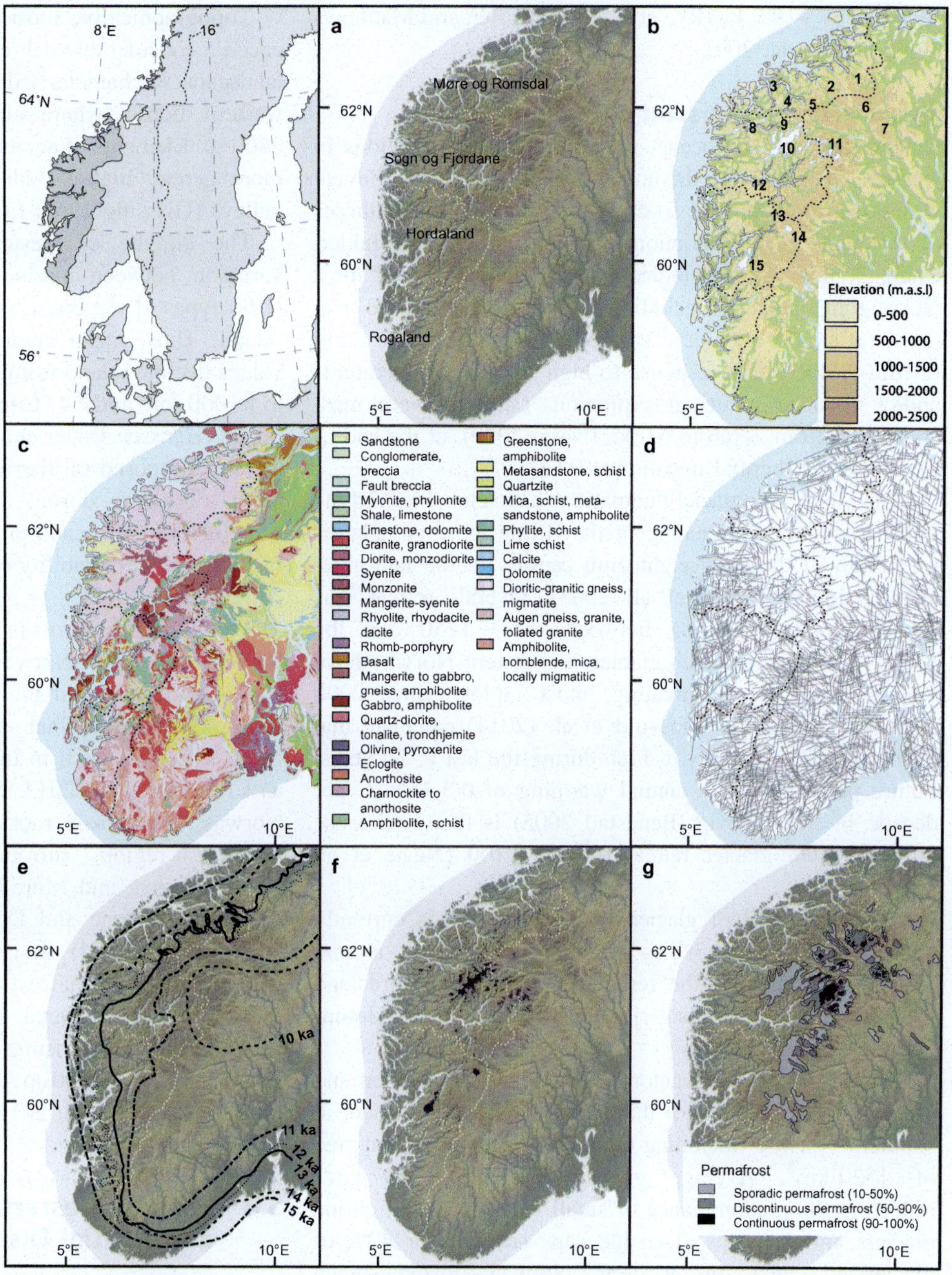

Fig. 5.4 Location and context of the study area in Norway: **a** the four counties of western Norway; **b** relief and selected localities referred to in the text: 1 Innerdalen, 2 Romsdalen, 3 Storfjord, 4 Årknes, 5 Tafjord, 6 Dovrefjell, 7 Rondane, 8 Nordfjord, 9 Loen, 10 Jostedalsbreen, 11 Jotunheimen, 12 Sognefjord, 13 Flåmsdalen, 14 Hardangerjøkulen and 15 Folgefonna; **c** geology (NGU 2015); **d** faults (NGU 2015); **e** pattern of Late Weichselian deglaciation of the Scandinavian ice sheet across western Norway since 15 ka (Hughes et al. 2016); **f** current glacier extent (Andreassen and Winsvold 2012) and **g** modelled distribution of permafrost (Gisnås et al. 2016). Norwegian Mapping authority base map: https://www.geonorge.no/

southern central Norway, though some mountains along the west coast fjords may have projected as nunataks.

Late Weichselian deglaciation commenced with the ice margin progressively withdrawing from its maximum limits, likely reaching the inner fjords of western Norway during the Bølling–Allerød interstadial, ~14.7–12.7 ka (Sollid and Sørbel 1979; Aarseth et al. 1997; Longva et al. 2009). However, ice sheet decay in western Norway appears to have been asynchronous from ~14–12 ka (Fig. 5.4e), with the ice front first retreating southwards from the coast in Møre og Romsdal (Hughes et al. 2016). Clearly, this ice retreat led to major thinning of the remaining ice sheet further inland, and progressive exposure of glaciated rockwalls. Deglaciation was interrupted during the Younger Dryas cold reversal (12.9–11.7 ka, Lohne et al. 2013), when significant ice thickening and readvances (several tens of kilometres) occurred in most, but not all western sectors of the Scandinavian ice sheet. Locally, ice reached thickness of 800–1200 m in fjords that had been ice-free during the Allerød (Mangerud 2004). After 11.5 ka, however, ice retreat was rapid. Deglaciation of the main valleys around the Jostedalsbre and Jotunheimen massifs is likely to have occurred by ~9.7 ka (Rye et al. 1987; Karlén and Matthews 1992; Dahl et al. 2002).

Neoglaciation and present-day glaciation
During the early Holocene, numerous glacier variations in western Norway were driven by abrupt, short-term climate variations, including the so-called '8.2 ka event'. Syntheses of Holocene glacier variations (Nesje et al. 2008; Winkler, Chap. 3, this volume) indicate the disappearance of most Norwegian glaciers, including the Jostedalsbre ice cap (Nesje et al. 1991) on at least one occasion during the Early or Mid-Holocene, in response to high summer temperatures and/or reduced winter precipitation. Glaciers were most contracted from ~6.6 to 6.0 ka (Nesje 2009), at the end of the Holocene thermal maximum (HTM).

Since then, Neoglacial fluctuations have characterised the Late Holocene, culminating in the 'Little Ice Age' glacier maximum of the early eighteenth century to the late nineteenth century (Nesje et al. 2008). Overall retreat was asynchronous, beginning between ~AD 1750 and the 1930s–1940s, since when glaciers in western Norway have predominantly been retreating, most rapidly since 2000 (Nesje et al. 2008). Winsvold et al. (2014) calculate that glacier area declined by *c.* 11% during the last *c.* 30 years. Future predicted mean annual warming of 0.3–0.4 °C per decade in Scandinavia (Benestad 2005) is likely to cause unprecedented glacier retreat by AD 2100 (Nesje et al. 2008).

More than half of glacier spatial coverage is currently concentrated in Sogn og Fjordane (on and around Jostedalsbreen), with the remainder mostly in Hordaland (Folgefonna and Hardangerjøkulen) and Oppland (Jotunheimen) (Fig. 5.4f).

The latest glacier inventory in Norway (Andreassen and Winsvold 2012) indicates that there were 1252 glaciers in southern Norway (including Oppland) covering a total area of 1520 km^2. Average glacier size was ~0.97 km^2, reflecting the predominance of small valley glaciers, cirque glaciers and ice caps. Two ice caps account for 42% of glacierized area in western Norway—Jostedalsbreen (474 km^2), the largest glacier in continental Europe, and Søndre Folgefonna (164 km^2).

Terrain and climate
Inland from the coast, the terrain of western Norway is dominated by the Scandes mountain chain, with 75% land area exceeding 300 m elevation and 35% above 1,000 m elevation (Fig. 5.4b). The highest summit is Galdhøpiggen (2469 m a.s.l.). Glaciations have formed numerous rockwalls, mostly located in the mountainous interior and along the western coast where steep overdeepened glacial troughs reach below sea level and form a network of fjords intruding inland for up to 200 km.

Topographically, most valley-side slopes have experienced a considerable degree of glacial erosion. Much of the landscape is characterised by high relief, especially in the western fjords, where alpine-type relief locally exceeds 2800 m. Elements of ancient paleic surfaces are preserved as more gentle, high-altitude plateaux above deeply incised valleys (Gjessing 1967; Etzelmüller et al. 2007).

The climate of western Norway exhibits significant variation between marine (Cfb) and subarctic or boreal (Dfc) types at the coast to tundra (ET) in high relief areas inland. Thus, mean annual precipitation and temperature values decline inland from >3500 mm and 6 °C on the coast to >1000 mm and −4 °C in more continental, montane areas inland (Hanssen-Bauer et al. 2017). Strong seasonal patterns are superimposed on these averages, with cyclonic precipitation occurring during autumn and winter, and intense snowmelt in spring and prolonged frost periods, which may increase the vulnerability of rock-slopes in western Norway (Blikra et al. 2006).

Mountain permafrost is recognised as an important factor for RSF activity in Norway (Blikra and Christiansen 2014) and is widespread in the high mountains. Its lower limit follows an altitudinal gradient in southern Norway from ~1600 m a.s.l. in the west to ~1300 m a.s.l. in the east (Gisnås et al. 2013; Steiger et al. 2016). In southern Norway, permafrost rockwalls are most common in the montane region surrounding inner Sognefjord and Jostedalsbreen, and Møre-Romsdal, as well as in the Hurrungane, Rondane and Dovre mountain areas (Fig. 5.4g). Hipp et al. (2014) described strong topographic aspect dependency of permafrost occurrence of around 500–600 m in the Jotunheimen area. Modelling of predicted warming suggests continued rising of the lower limit for mountain permafrost to ~1800 m a.s.l by the end of this century (Hanssen-Bauer et al. 2017).

5.4 RSFs in Western Norway: Processes and Spatial Distribution

Catastrophic failures of rock-slopes represent one of the most serious natural hazards in the glacially oversteepened mountain and fjord landscapes of western Norway. In addition to their direct impacts, possible secondary effects of valley impoundment (Hermanns et al. 2013b) and displacement waves in fjords and lakes (Hermanns et al. 2006a; Harbitz et al. 2014) represent especially high risk (Fig. 5.5). In the past century, 175 lives were claimed by three RSF activity and secondary effects (Blikra et al. 2006; Oppikofer et al. 2016). A firm appreciation of the mechanics, frequency and chronology of these events are of major importance for hazard evaluation and for understanding the behaviour of unstable rock-slopes under scenarios of future change.

Fig. 5.5 Slide scar and rock avalanche debris below Ramnefjellet (1,779 m) at Loen, Sogn og Fjordane. The top of the scarp is 900 m above the level of the lake Lovatnet, and displays outward dipping joint sets and foliation in generally massive granitic gneisses. Seven rock avalanches occurred here between AD 1905–1950, with four failures from September–November, 1936. Tragically, subsequent displacement (tsunami) waves of 40.5 and 74 m height killed 61 people in 1905 and 73 people in 1936, respectively. The total volume of displaced rock was estimated to be >3.2 × 10^6 m^3 (Grimstad and Nesdal 1990). Local clustering of RSF events at Loen highlights the potential for the same slope to undergo repeat failure over short timescales. *Photo* Paula Hilger

This section identifies the main processes, spatial and temporal distribution of paraglacial rock-slope responses in western Norway. Emphasis is placed on the prehistoric record of large-scale (>0.1 M m^3) failure events, for which an extensive data inventory is emerging. Establishing these patterns can assist in the identification of local and regional controlling parameters.

5.4.1 Mechanisms of Rock-Slope Failure

Forms of rock-slope adjustment in western Norway range from rock-slope deformations (sackungen), and translational slides of relatively intact bedrock to fully disintegrated rock avalanches and discrete rockfall activity. Assessing the significance of failure mechanisms is hampered, however, by inconsistent use of nomenclature. For example, terms such as megalandslide, rock avalanche and catastrophic failure are often used synonymously.

Nonetheless, it is widely considered that large rock avalanches dominate in western Norway's steep, high relief troughs and fjords (Blikra et al. 2006). Longva et al. (2009) distinguished rock avalanche from rockfall deposits in terms of a minimal volume of 0.1 M m^3 for the former. Rock avalanches are typically characterised by high velocities and significantly longer runout distances than for rockfalls (Hermanns and Longva 2012). Many rock avalanche deposits overlie deformed valley-fill sediments and are characterised by coarse debris cones and lobes supporting a chaotic surface topography of ridges, mounds and basins. Runout debris from valley wall failures often extends >1 km, unless constrained by the opposing valley slope. Nearly all historical rock avalanche events are deemed to have followed active rock-slope deformation, either shortly before or long in advance of failure (Hermanns et al. 2013a).

Failures may frequently be initiated, however, as translational slides. Of 72 unstable rock-slopes mapped by Saintot et al. (2011) in western Norway, the majority (48) were defined as (translational) rockslides. Yet failure mechanisms are often complex and their forms composite. A promising approach for better understanding the deformation and failure mechanisms of complex unstable rock-slopes appears to lie in the integration of detailed geological and monitoring data with structural and kinematic analysis and numerical modelling (e.g. Böhme et al. 2013; Booth et al. 2015; Oppikofer et al. 2017; Sandøy et al. 2017).

5.4.2 Slope-Scale Distribution of RSF Activity

At the local scale, the balance between shear stresses and shearing resistance of a rock mass is determined by the highly variable interplay between mechanical, thermal and hydrological bedrock characteristics, as well as rock weathering processes (Messenzehl et al. 2017). These authors assert that the relative contribution of geotechnical, topoclimatic, cryospheric and paraglacial properties is determined by a complex interaction with small-scale bedrock morphometry and overall valley topography. Intact rock strength, joint density and orientation relative to slope, bedrock roughness and morphometry constrain spatially variable stress fields, while bedrock moisture, the presence of permafrost and segregation ice strongly influence the effectiveness of weathering cycles.

The geological and topographic controls on rock-slope instability, including rock-slope morphometry, are clearly evident at the local scale in western Norway (Böhme et al. 2012, 2013). Gravitational slope deformation at Stampa (Sogn og Fjordane) is strongly controlled by inherited structures, such as pre-existing joint sets and the metamorphic foliation of the phyllites. These authors observed large open fractures or surface depressions developed along the main joint sets or a combination of two of them. Numerical modelling also supports structurally controlled failure, where discontinuities with a low strength dominate the rock mass behaviour.

Böhme et al. (2011) found that 79% of rock-slope instabilities in Sogn og Fjordane have developed either at convex breaks of slope (knickpoints) that are interpreted to be of glacial origin (Holm et al. 2004), or at the unstable edges of high plateaux (Fig. 5.6). The authors' findings agree with geometric models of rock-slope failure in Norway (Braathen et al. 2004) that predict rockfall and topple failure at plateau edges and rocksliding at convex profile knickpoints. These findings also echo those drawn from modelling the controls on rockfall in the Swiss Alps (Messenzehl et al. 2017), where the vulnerability of permafrost rockwalls to rockfall was linked to convex steep terrain (>40°) and north-facing valley flanks, promoting surface moisture supply and subsurface lateral heat fluxes.

Finally, at the slope scale, the distribution of postglacial RSFs can be driven by different mechanisms at different points on a slope profile, and different sectors of a slope can be activated at different times (Leith et al. 2010a, b). In this way, different parts of a rock mass can reach a state of critical conditional stability at different points in time, causing some unstable slopes to collapse repeatedly while others fail in a single event (Hermanns et al. 2013a). Generally, RSF activity can increase the probability of future failures in the vicinity because of accelerated unloading along the rock-slope (Hermanns, et al. 2006a).

5.4.3 Regional-Scale Distribution of RSF Activity

One way to evaluate the influence of glacial conditioning on RSF activity has been to relate the location of failures at the basin scale to former glacier limits, and other possible influences such as topogeometry, lithology, seismotectonics, permafrost degradation and freeze–thaw. The abundance of RSFs in glacially steepened and overdeepened valleys has been widely recognised. While spatial associations of long, steep, glacially modified terrain and clustering of RSFs make a strong case for (de)glacial conditioning of slope instability, they alone fail to explain failure mechanisms or triggers (McColl 2012). They also fail to explain sparsity of RSFs in areas of favourable geology and structure—many processes can produce long, steep slopes, but not all long, steep slopes fail. Moreover, generic factors (such as debuttressing, deglaciation meltwater and freeze–thaw) apply to entire montane areas and fail to explain spatial clustering of RSFs (Jarman and Harrison 2019).

Despite progress in understanding failure mechanics and controls at the process scale, assessment of the relative importance of different controls within a complex synergetic interplay is still lacking, especially at larger scales, where complex and emergent system behaviour (Phillips 2003) may mean that different causes trigger RSF activity at different scales (Messenzehl et al. 2017).

In a study of RSFs in the Scottish Highlands, Jarman (2006) noted that although glaciation affected the entire montane area in the last glacial cycle, slope failures are unevenly distributed (65% are found in seven main clusters, with the rest widely scattered) for which previous glaciological, lithological and seismotectonic explanations failed to adequately explain any pattern. Areas with similar relief and lithology could display either many or few failures. Jarman (2006, 2009) and Ballantyne (2008) identified RSFs in areas that experienced maximum glacial overburden, such as narrow troughs and breaches associated with particularly constrained glacier flow, and areas of flow convergence (such as confluent glacial valleys). Moreover, they recognised rock-slope instability focused in overdeepened basins, where steepening of the lower valley wall prolongs destabilisation of mountain slopes. In a review of more than 1,000 RSFs in the British mountains, Jarman and Harrison (2019) hypothesised that concentrated bedrock erosion in glacial breaches might have generated sufficient rebound stress differentials between lower and upper slopes to provoke failure, though this proposal is unverified.

Historical records and geological studies of western Norway show a high concentration of both postglacial gravitational slope failures and current rock-slope instabilities. As of 2015, systematic mapping in the three counties with most historic events (Sogn og Fjordane, Møre og

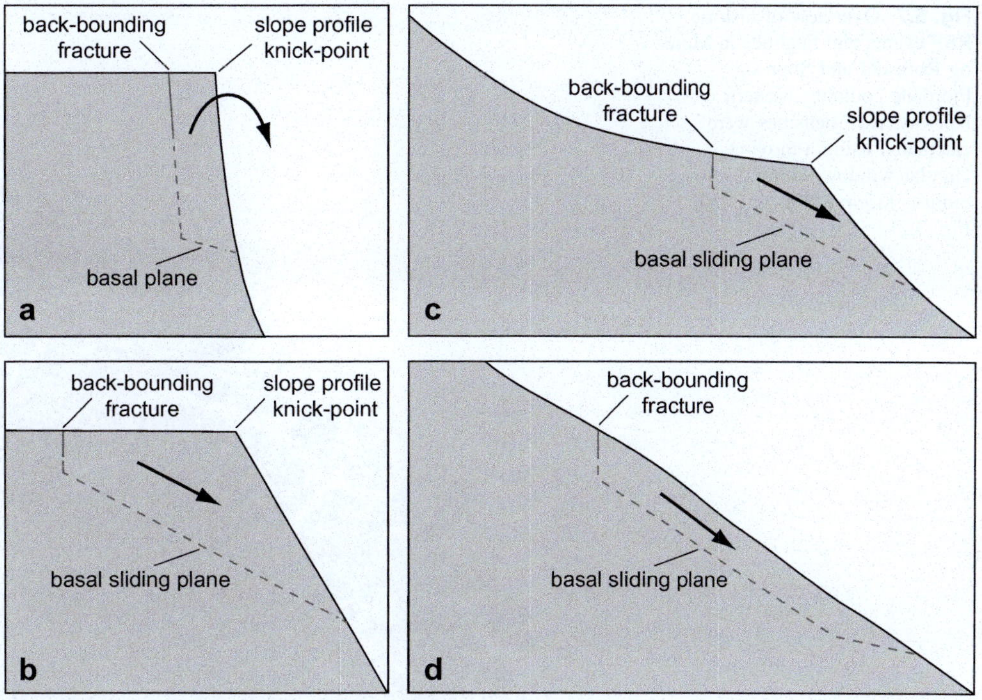

Fig. 5.6 Schematic representation of types of rock-slope instabilities in western Norway based on pre-failure geomorphology. Rock-slope instabilities in Sogn og Fjordane are located at unstable edges of plateau-like surfaces (**a** and **b**), and at knickpoints of slopes (**c**) but are not situated directly on a steep slope with constant slope angle (**d**). Typical modes of movement are fall or topple for (**a**) and slide for (**b**–**d**). After (Böhme et al. 2011, Fig. 9)

Romsdal and Troms) had revealed 253 unstable rock-slopes: 117 in Troms, 91 in Møre og Romsdal, 23 in Sogn og Fjordane and 13 in Rogaland (Devoli et al. 2011; Hermanns et al. 2013c; Oppikofer et al. 2013, 2015).

The highest frequency of failures was found in the high relief, inner fjord areas of Møre og Romsdal and Sogn og Fjordane (Fig. 5.7), while they are comparatively sparse in Rogaland, Hordaland and Rondane (Oppland) (Blikra et al. 2002, 2006; Böhme et al. 2011; Henderson and Saintot 2011). Almost 200 individual events have been mapped in Møre og Romsdal, with the greatest number in the Romsdalen and Tafjord valleys (Blikra et al. 2002, 2006). In Romsdalen (home to Trollveggen, Europe's tallest vertical rockwall), more than 15 large rock avalanches cover almost the entire valley floor over a distance of 25 km, while in Tafjord more than 10 rock avalanche deposits have been mapped in the fjord over a distance of less than 7 km (Blikra et al. 2006). Åknes on Sunnylvsfjord (Møre og Romsdal) is regarded the most hazardous rockslide area in Norway. In the same county, two smaller groups of failures occured in outer coastal locations around Otrøya and Syvdsfjord. Many RSFs are also found on weak schist (phyllite) in Aurlandsfjord and Flåmsdalen (Sogn og Fjordane).

Several studies have sought to quantify the spatial relations between RSF activity and geological and topographic parameters in western Norway (Böhme et al. 2011, 2014). These authors highlighted a strong spatial association between the occurrence of rockfalls in Sogn og Fjordane and its Quaternary geology, tectono–stratigraphic position and density of geological lineaments.

Certainly, given the region's complex and protracted ductile and brittle tectonic history, it may be expected that inherited bedrock geology and structures have a strong influence on the spatial distribution of RSFs on high steep (>35°) terrain. Lithology appears to be important in controlling the development of unstable rock masses, especially where failure planes coincide with the contact between hard basement gneissic rocks and soft weathered mafic and ultramafic rocks (Böhme et al. 2011). Saintot et al. (2011) demonstrated that parameters favouring instability and failure in western Norway include (1) relatively weak foliated metamorphic rocks, such as phyllites, schists and foliated gneisses; (2) fjord- or valley-dipping foliation or steep, slope-parallel foliation; (3) folds; (4) Caledonian thrusts cutting the slope and (5) regional brecciated (cataclastic) faults close to the slope. According to Henderson and Saintot (2011), large RSFs $>3 \times 10^6$ m^3 only tend to develop in western Norway when all critical structures are present (valley-dipping foliation and a weakened plane at the base of the potentially unstable block, as well as existing lateral boundaries of the unstable block). These findings highlight the importance of preconditioning geological parameters, coupled with glacially conditioned inheritance, for generating slope failure. Similarly, others have suggested that lithological and structural parameters are the most important long-term control on rock avalanche clusters in the European Alps (Hermanns et al. 2006a; Ostermann and Sanders 2017), with seismicity and climatic factors often triggering failure.

While present seismic intensity in Norway is low to intermediate (Fjeldskaar et al. 2000), an offshore

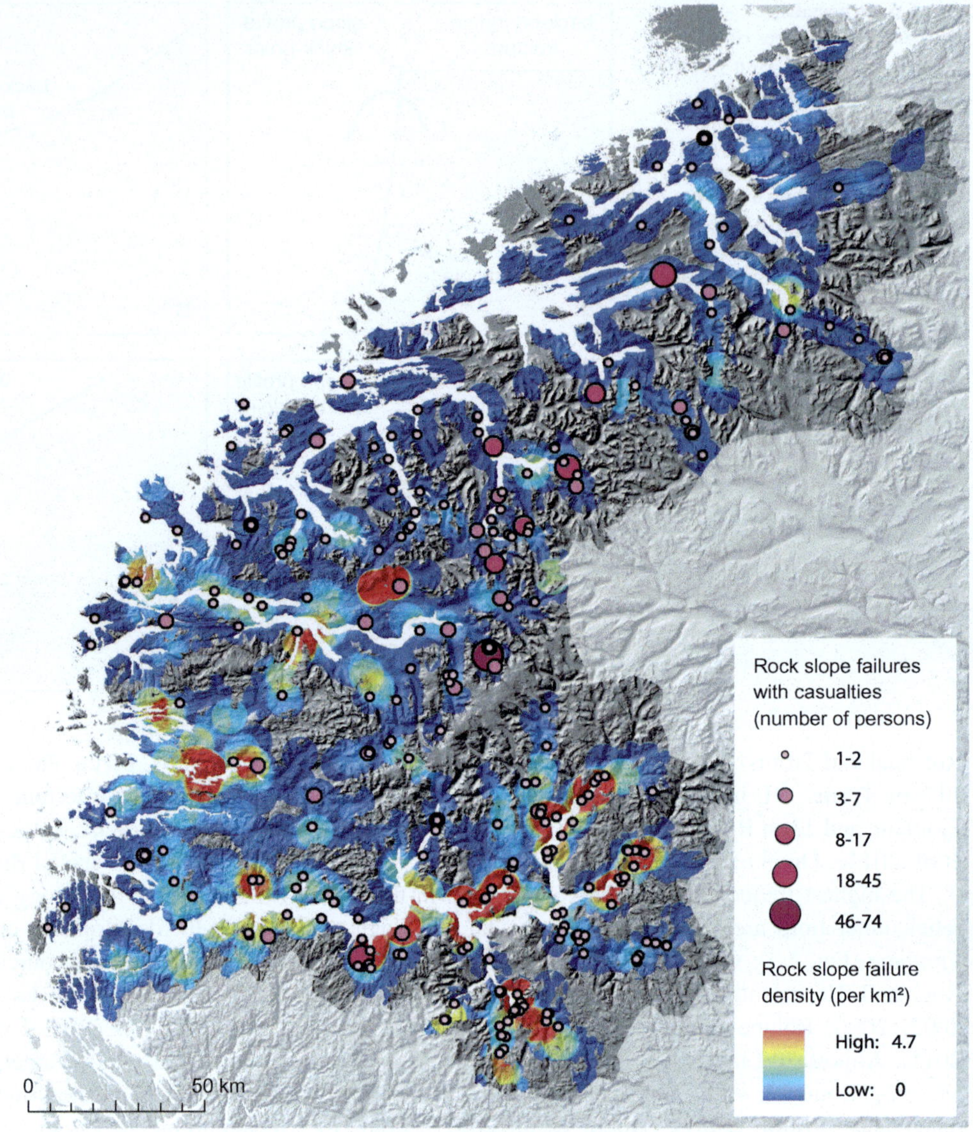

Fig. 5.7 Overview of historic RSF events and fatalities in Møre og Romsdal and Sogn og Fjordane counties, western Norway. RSF densities were calculated using a moving circular window with a 5 km radius. (*Source* Böhme 2014, Fig. 3.2)

concentration of earthquake activity is clustered west of central Norway (Bungum et al. 2005), and some authors (Böhme et al. 2011; Henderson and Saintot 2011) speculate whether the regional clustering of rockslides in Møre og Romsdal may reflect greater postglacial seismic activity as well as steep present-day uplift gradients (Fig. 5.8).

Permafrost thaw is widely recognised as an important factor for RSF activity due to melting of ice bonds in cracks and weakening of tensile and compressive strength in rock masses (Murton et al. 2006; Krautblatter et al. 2012, 2013). Preliminary modelling of permafrost distribution in Norwegian rockwalls (Steiger et al. 2016) also promises to shed light on the spatial clustering of RSFs. The results of this study imply that the highest spatial density of Norwegian rockwalls in permafrost follows an arc around the inner fjord areas of Møre og Romsdal, Sogn og Fjordane and the Rondane, associated with dissected high-altitude paleic surfaces. The authors proposed that unstable rock-slopes within the permafrost zone in southern Norway are mostly restricted to the large glacial valleys Romsdalen and Sunndalen in Møre og Romsdal, where valley wall elevations commonly exceed 1500 m a.s.l, and the area surrounding Jostedalsbreen (Sogn og Fjordane). By implication, thermal regime may be a further important factor conditioning rock-slope instability in this region, at least for high-elevation sites.

In their attempt to identify and quantify regional-scale controls (topoclimatic, cryospheric, paraglacial or/and rock mechanical properties) on rockfall activity in the Swiss Alps, Messenzehl et al. (2017) recognised permafrost distribution as the major control on the spatial distribution of rockfalls. The authors attributed clustering of rockfall source areas within a low-radiation altitudinal belt at 2900–3300 m a.s.l. to rock weathering via seasonal growth of segregation ice in shallow permafrost. Analysis of 90+ small RSFs in Jotunheimen lends

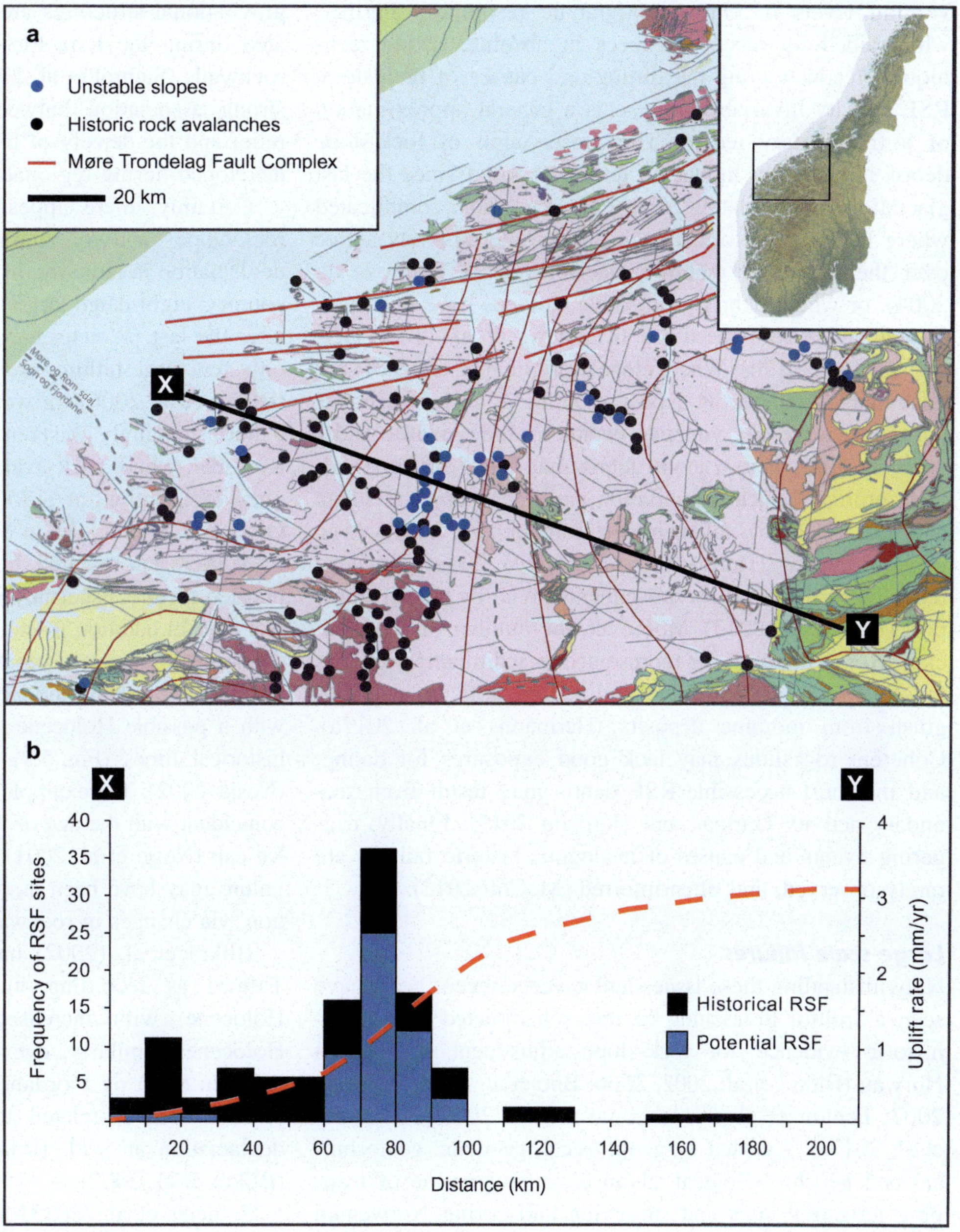

Fig. 5.8 Historical rock-slope failures and current rock-slope instabilities in Møre og Romsdal county, western Norway: **a** overview, showing current apparent uplift rates (mm/yr). Precambrian basement is shown in pink, Caledonian thrust sheets are in greens and yellows. Note clustering where uplift gradients are steep (Geological Survey of Norway database: http://geo.ngu.no/kart/berggrunn). **b** frequency histograms of RSF localities relative to uplift pattern (Vestøl 2006; Olesen et al. 2013). Profile X–Y includes landslides within 60 km of the profile. (*Source* Henderson and Saintot 2011, Fig. 1)

further support to this view. Using change detection and discreet Meyer wavelet analysis, combined with permafrost depth models, Matthews et al. (2018) identified a strong correlation between the spatial distribution of small ($<10^3$ m^3) RSFs and the aspect-dependent lower altitudinal limit of mountain permafrost in rockwalls.

5.5 Temporal Pattern of RSF Activity

Analysing the temporal pattern of slope failures is another means of assessing the influence of glaciation on rock-slope stability (McColl 2012; Ballantyne and Stone 2013; Ballantyne et al. 2014a). Timing datasets enable assessment of the random or clustered nature of landslide distributions, and comparison of these clusters with the timing of other events, for example, glacier retreat, climatic events or seismicity, has been crucial in understanding the causes of enhanced rock-slope adjustment (McColl 2012).

5.5.1 Timing of Prehistoric RSF Activity

Former RSF datasets represent important archives, though detailed comparisons should proceed with caution, given that they represent different morphological settings with varying levels of chronostratigraphic resolution. Further, while aided by recent advances in absolute dating techniques, reconstructing the timing and causes of prehistoric RSF activity invariably represents a general approximation of a fragmentary record. The preservation of rock-slope debris is generally limited to terrain exposed since the last glaciation, though dating control may be further complicated where failure or deformation has been intermittently active over the course of multiple glacial cycles (Tibaldi et al. 2004), or where debris or unstable surfaces have been preserved under cold-based ice (Matthews et al. 2013; Hermanns et al. 2017b). Where clear stratigraphic sequences are unavailable, reliance on exposure dating of surface deposits may over-represent younger events if they overlie older deposits. Elsewhere, runout debris may have dammed valleys, prompted local aggradation and been buried (Dadson and Church 2005), fallen into a fjord or lake and lack an onshore record (Hermanns and Longva 2012), or been recycled or removed by subsequent fluvial or glacial transport (Ballantyne 2013). Moreover, the number of supraglacial RSFs is likely under-represented in sediment records, as those deposits are reworked and can be difficult to distinguish from moraine deposits (Hermanns et al. 2017a). Coherent rockslides may lack good exposures for dating, and the most accessible RSF debris may result from secondary activity (Jarman and Harrison 2019). Finally, triggering events and causes of analogous historic failures are rarely observed, and often inferred (McColl 2012).

Large-scale failures

Notwithstanding these issues and caveats, recent years have seen a fruitful flourishing of research directed at the prehistoric evidence for rock-slope adjustment in western Norway (Blikra et al. 2002, 2006; Bøe et al. 2004; Aa et al. 2007; Fenton et al. 2011; Longva et al. 2009; Hermanns et al. 2017a). Assisted by geophysical, geodetic, geotechnical and geochronological advances, a programme of mapping terrestrial sites and surveying and coring Norwegian fjords has yielded a record of frequent large-scale RSFs throughout the prehistoric period.

The general opinion of early attempts to date postglacial rock-slope activity in western Norway (and beyond) was that most failures occurred shortly after deglaciation at the end of the Pleistocene, when glacio-isostatic uplift was most rapid, and pressure release likely promoted sudden unloading in the steepened and overdeepened alpine terrain, coupled with a high discharge of meltwater into bedrock fractures. Glacially conditioned rock mass fatigue in western Norway was recognised by Peulvast (1985), who attributed tensile stress conditions at the toe of glacial troughs in Sogn og Fjordane to increased overburden shear stresses associated with glacial erosion (*cf.* Augustinus 1995b; Jarman 2006, 2009). Further, now largely dormant or inactive widespread gravitational structures are considered to have been generated during the Late Pleistocene, in glacially conditioned rockwalls (Saintot et al. 2011). The latter study highlights a strong association between rock-slope weakening, steep relief and the density of brittle structures that most recently developed during regional deglaciation.

Certainly, there appears to be support for a peak in rock-slope activity during or shortly after regional deglaciation in Norway. In Troms county, in the north of the country, eight dated large RSFs occurred in a period shortly after the last (Weichselian) glaciation, ~10.5–11.5 ka, and only few large failures are recorded there in historical times (Blikra et al. 2006). In western Norway, a similar temporal pattern seemingly exists in the outer coastal area of Møre og Romsdal, where rock avalanches probably occurred shortly after deglaciation: at ~15–14 ka in Øtrefjellet, >11.5 ka in Ørsta and at <11.5 ka in Syvdsfjord (Blikra et al. 2002).

More recent research, however, has revealed a more complicated picture, especially in the inner fjord areas of western Norway. In addition to enhanced Late Pleistocene rock-slope adjustment, many catastrophic RSFs are now known to have occurred throughout the Holocene (Blikra et al. 2002, 2006), with a possible Holocene peak at ~3 ka, as well as during historical times (Fig. 5.9). A rock avalanche in Oldedalen (Nesje 2002), for example, occurred some 6000 years ago, coincident with the age of the formation of the Jostedalsbreen ice cap (Nesje et al. 2001). Nesje (2002) speculated whether failure may have been associated with renewed local glaciation, via changes in rockwall joint-water pressures.

Blikra et al. (2002) interpreted seismic stratigraphy in Tafjord as recording individual events throughout the Holocene, with increased frequencies during the Late Holocene. Similarly, data from Aurlandsfjord and Flåmsdalen in Sogn og Fjordane suggest that RSF activity in the phyllite area is related to two stages, one shortly after deglaciation at ~11–10 ka and a smaller event at ~3 ka (Blikra et al. 2002).

Schleier et al. (2015) also found evidence for two episodes of RSF activity in Innerdalen (Møre og Romsdal), in the form of a rock avalanche falling onto the surface of a downwasting glacier at ~14 ka, followed by a Holocene event at ~8 ka. In inner Nordfjord (Sogn og Fjordane), a multistage pattern is evident, where 11 large rock avalanches occurred during most of the Holocene until ~3.2 ka (Aa et al. 2007; Hermanns et al. 2017a). Four events were dated to within 2 ka of deglaciation at ~13.5 ka, while further six rock avalanches occurred between ~7.1 and 3.2 ka, at an average frequency of ~1 per 700 years. Finally, in Romsdal (Møre og Romsdal), Hilger et al. (2018) report exposure ages for three large RSFs between 12 and 10 ka, during a phase of paraglacial relaxation, as well as a period of between three and six catastrophic rock avalanche events that cluster at 4.5–5.5 ka.

Fig. 5.9 Vegetated debris cones below a triangular failure scarp on the south flank of Alstadfjellet (1,450 m), in Valldalen, Møre og Romsdal. Cosmogenic nuclide (^{10}Be) ages averaging 9.3 ± 1.0 ka were obtained for the slope-foot rock avalanche deposits (Hermanns et al. 2017a), approximately 5 ka after deglaciation at this locality. *Photo* Peter Wilson

A large dataset from inner Storfjord and its tributaries in Møre og Romsdal is especially intriguing in elucidating the temporal dynamics of rock-slope adjustment in this area. Based on a ^{14}C-dated sediment core and seismic stratigraphy, Longva et al. (2009) tentatively assigned ages to 108 rock avalanche deposits mapped on the floors of Storfjord, Nordalsfjord, Sunnylvsfjord, Geirangerfjord and Tafjord. Analysis of the Storfjord dataset in this relative chronostratigraphy suggested a rapid response and high frequency of rock avalanching following deglaciation, followed by a broadly constant frequency over the past 9 ka. 25 events (23% of the total number) deposited 89% of the total fjord rock avalanche volume (587 million m^3) during deglaciation (~14–11.5 ka), with five large events contributing 79% of the total debris volume. One-third of the Holocene events delivered half the total Holocene rock avalanche volume (65 million m^3) within 1,000 years of final deglaciation. The authors suggested this high frequency may reflect frost activity, permafrost thaw, exfoliation and strong earthquakes associated with rapid isostatic rebound. After a period of high rock-slope adjustment during Younger Dryas and Preboreal times (~13–9 ka), the remainder of the Holocene at Storfjord appears to have been characterised generally by an even distribution of smaller rock avalanches, averaging 5–8 events per 1,000 years for the whole Storfjord system. Given the likely under-representation of prehistoric events preserved in the geological record, these figures should probably be considered minimal.

Exposure ages of further 22 rock avalanches in western Norway were compiled by Hermanns et al. (2017a), who interpret their asynchronous temporal distribution of failures as primarily tracking the progressive decay of the Weichselian ice sheet. Half of the dated events in their inventory occurred within the first millennium of ice retreat. A further five events (22%) occurred during the Holocene Thermal Maximum (~8–5 ka), a period characterised by the warmest Holocene temperatures in Norway (Nesje 2009; Lilleøren et al. 2012). Whether or not this signifies an association with climate cannot be concluded on the basis of this small dataset. Four remaining events (20%) are relatively evenly distributed throughout the rest of the Holocene. These findings are in accordance with detailed analyses of rock avalanche deposits in Storfjord that show that 51 out of 109 rock avalanches occurred within three millennia following deglaciation, and that more than 80% of rock avalanche debris was deposited in the fjord in the first millennia after deglaciation (Böhme et al. 2015).

RSF activity is frequently preceded by a period of accelerated slope deformation, especially under non-seismic conditions (Eisbacher and Clague 1984). Dating of six sliding planes was undertaken for unstable rock-slopes in Møre og Romsdal and Sogn og Fjordane by Hermanns et al. (2012, 2013b, 2017b) to determine the timing of initiation of deformation and long-term slip rates. At Oppstadhornet (Otrøya) and Skjeringahaugane (Lustrafjord), sliding has taken place at a constant rate since deformation initiated during deglaciation at ~14.2 and 10 ka, respectively. Preliminary interpretation of Storehornet (Homindal) suggests a period of deformation shortly after deglaciation was followed much later by a rock avalanche event during the Late Holocene. Delayed deformation is also evident at Ivasnasen

(Sunndalen), where simultaneous sliding and failure (rock avalanches) were dated to ~3.5 ka (Oppikofer et al. 2017).

Finally, it is relevant to briefly mention the fjord and lake record of colluvial mass movements in western Norway, which suggests increased frequencies of turbidites, floods, debris flow and snow avalanche activity immediately after deglaciation, with episodic enhanced activity in the Late Holocene (Blikra and Nemec 1993; Bøe et al. 2002, 2003, 2004; Lepland et al. 2002; Bellwald et al. 2016). These latter pulses of soft sediment reworking have been primarily attributed to in washing during regional climatic irregularities, for example, at 5.6–5.3 ka and especially at 3.2–2.8 ka, though seismic activity is considered an important regional trigger for increased mass movements at ~11.7–11.0 and 2.2–2.0 ka.

In a recent compilation of sediment cores and seismic profiles from 22 fjords and lakes in Sogn og Fjordane and Mør og Romsdal, Bellwald et al. (2019) dated 125 turbidites and mass movement deposits, and propose temporal clustering of increased mass movement activity for two periods in the Early Holocene (11.0–9.7 and 8.3–7.8 ka), followed by quiescence, and renewed mass movement activity since ~4.2 ka. Notwithstanding limited chronological control, the authors suggested broad, regional correspondence in the timing of episodic enhanced mass movement, offshore slides and rock avalanche activity, and proposed that seismic activity was the most plausible regional trigger, driven at least during the Early Holocene by glacio-isostatic rebound.

Small-scale failures and rockfalls

While much prior research has focused on catastrophic RSFs in Norway, less attention has been directed to small ($<10^3$ m^3, Matthews et al. 2018) features in the context of assessing the significance of paraglacial RSF activity. This is important, given that the occurrence, magnitude and frequency of RSFs are influenced by a host of geological, topographic and climatic factors. Small RSFs are generally considered to respond to deglaciation over a shorter period of time than large RSFs (Ballantyne 2002a). Improved understanding of these controls on RSF occurrence requires a consideration of RSF magnitude, and is essential for accurate modelling of future, climate-driven RSF activity (Gariano and Guzzetti 2016).

In the largest study of its kind, Matthews et al. (2018) applied Schmidt hammer exposure-age dating (SHD) to construct a detailed regional chronology of the frequency and magnitude of 92 small RSFs in Jotunheimen. The authors failed to detect evidence for a strong activity peak immediately after deglaciation at ~10 ka, and attributed this to considerable downwastage of the Late Weichselian ice sheet by this time (Mangerud et al. 2011; Hughes et al. 2016), which may have reduced the scale of paraglacial effects on subsequent RSF activity. Using a weighted age–frequency distribution and probability density function analysis, they indicated four 10^2–10^3 year periods of enhanced, small RSF frequency (at ~8.9, 7.3, 5.9 and 4.5 ka). Maximum activity was reached at ~4.5 ka. Since then, the incidence of small RSFs in this region appears to have declined significantly. Matthews et al. (2018) interpret the timing of peak RSF frequency (and its lagging the Holocene Thermal Maximum by ~2 ka) as primarily a response to long-term changes in permafrost depth during episodes of relatively warm climate and transition to a seasonal-freezing climatic regime.

Similarly, Marr et al. (2019) applied SHD to five small RSFs in Opplendskedalen (Møre og Romsdal), which was deglaciated at ~11.5 ka. Their results show that three features stabilised during the Holocene Thermal Maximum and do not support the hypothesis that RSF activity predominately occurs shortly after local deglaciation. While assuming that these small RSFs likely reflect glacially conditioned, long-term stress release (in part), the authors identified climatic factors related to warming (e.g. permafrost degradation, snowmelt and increased joint-water pressure) as triggering mechanisms. While based on a very small sample size, the wider representativeness of their conclusions appears to be supported by Matthews et al. (2018).

Finally, and notwithstanding the difficulty of distentangling the effects of periglacial conditions and intrinsic paraglacial rockwall instability, the spatio-temporal distribution of rockfall talus-derived landforms (protalus rock glaciers and protalus/pronival ramparts) may also shed light on the occurrence, magnitude and frequency of small RSFs. SHD of three relict rock glaciers in Ottadalen (Oppland), north of the Jotunheim massif yielded SHD ages of ~11–9 ka (Matthews et al. 2013). These may represent paraglacial landforms that formed in response to enhanced rockfall during early Holocene deglaciation, though alternatively, they may be much older features that were preserved beneath cold-based ice.

Similarly, dating of five relict protalus ramparts and rock glaciers in the Romsdalsalpane mountains (Møre og Romsdal) and at Nystølsnovi (Sogn og Fjordane) yielded SHD ages of ~15–8 ka (Matthews and Wilson 2015; Matthews et al. 2017). These data were interpreted as suggesting these rockfall-derived landforms accumulated over a period of ~6 ka between regional deglaciation and the end of the Younger Dryas, when glacial debuttressing and permafrost degradation enhanced the release of rock debris from cliffs. Matthews et al. (2017) considered at least two of the rock glaciers to be primarily paraglacial landforms, at least in terms of debris supply. A protalus rampart in Opplendskedalen that yielded an SHD age of 5.6 ka was regarded by Marr et al. (2019) to have initially formed after deglaciation and added to during the Early Holocene by rock debris supplied from a rockwall weakened by permafrost degradation.

5.5.2 Timing of Historic RSF Activity

The historical record of failure of rock-slopes, including rock avalanches and large rockfalls in western Norway has been collected from historical archives (Furseth 2006), and modern activity (Fig. 5.10) is registered in databases maintained by the Geological Survey of Norway (NVE 2018).

Large rock avalanches are known to have caused a series of catastrophes throughout the historical period in western Norway, including at Loen in 1905 and 1936 (Fig. 5.5) and Tafjord in 1934 (3 M m^3) in which altogether 175 people were killed by displacement waves (Furseth 2006). The frequency and chronology of such events are of major importance for hazard evaluation and for understanding the behaviour of unstable rock-slopes (Aa et al. 2007).

Rockfall activity in Norway reportedly increased during the 'Little Ice Age' (notably between the years AD 1650 and 1760), a phenomenon attributed to the increased frequency of extreme rainstorm events at that time (Grove 1972). However, there is little evidence for an increase in larger RSFs associated with post-'Little Ice Age' glacier retreat in mainland Norway (Laute and Beylich 2013). The absence of a strong post-'Little Ice Age' rock-slope response may simply reflect significant lagging of bedrock permafrost thaw behind air temperature warming (Hermanns 2018, pers. comm.). Alternative explanations include the extremely limited and brief loading and unloading of rock-slopes by 'Little Ice Age' glaciers, combined with low rates of seismotectonic activity. The relative paucity of RSFs associated with 'Little Ice Age' ice retreat has been recognised elsewhere (e.g. Evans and Clague 1994; Jaboyedoff et al. 2012; Grämiger et al. 2018).

Both frequency and magnitude of historical RSF activity should be considered in temporal context. Extensive historical records of deadly or damaging RSF events in Norway during the past few centuries can give an impression of high levels of activity, though few accounts detail the volumes of failed rock. Based on historical records over the past 400 years, a frequency of 2–6 rock avalanches per century has been cited as an historical average for all Norway (Furseth 2006; Blikra et al. 2006). These historic rates appear to approximate Holocene averages at least for several specific active sites.

The Mannen area in lower Romsdalen, for example, is one such locality, where more than 30 historical RSF events were documented in the valley (NVE 2018), though these are primarily small (<100 m^3) rockfall events. Blikra et al. (2016) suggested a present-day recurrence interval in Romsdalen of <100 year for a 0.15 M m^3 failure, while the recurrence of a 2–4 M m^3 volume was estimated to be 100–1000 years. These present-day averages accord with dated records of 3–6 large rock avalanches within 1,000 years at Mannen, 4.5–5.5 ka (Hilger et al. 2018). A similar average rock avalanche frequency of 6–10 events per 1,000 years was estimated for the last 9,000 years in Innerdalen and Storfjord (Aa et al. 2007; Longva et al. 2009). Low recurrence intervals for RSF activity imply post-failure rock-slope relaxation on decadal timescales (Schleier et al. 2017; *cf.* Hermanns et al. 2006a; Crosta et al. 2017). Sudden stress release due to failure may cause weakening of the rock-slope via reorganisation of the rockwall stress field (Hermanns et al. 2006b).

Several rockwalls in western Norway are known to have experienced multiple failures in recent years (Fig. 5.5). The

Fig. 5.10 Examples of small- to moderate-sized modern RSFs in western Norway: **a** a rockfall that occurred about AD 2005 on the south-facing slope of lower Langfjelldalen (Valldalen, Møre og Romsdal), at an altitude of 900 m; **b** the Nystølsnovi RSF (Langedalen/Austerdalen area of Jostedalsbreen) occurred in AD 1994. *Photos* John Matthews

Loen site (inland Sogn og Fjordane), for instance, failed repeatedly within a few decades in the early twentieth century (Hermanns et al. 2006a). This observation raises questions for the interpretation of overlapping age ranges for prehistoric failures at the same site, and whether such data represent one event or multiple failures from the same slope.

5.5.3 Causes and Triggers of Rock-Slope Failure: Synthesis

The occurrence of RSFs of varying magnitude in western Norway's mountains reflects preconditioning and triggering by a wide range of environmental factors whose relative importance are likely scale-dependent. These factors include paraglacial stress release and fracture propagation due to unloading, debuttressing, palaeoseismicity, permafrost degradation and other climatic triggers (Ballantyne et al. 2014a).

This section seeks to characterise the timing of postglacial RSFs in western Norway and evaluates the significance of these factors. A preliminary analysis of the prehistoric RSF age inventories is presented, and timing data for 49 prehistoric RSFs are summarised in Table 5.1. Figure 5.11 identifies the location of the sites. This work focuses on moderate to large (>10^3 m^3) RSF events with absolute dating control. Accordingly, records of colluvial mass movements are omitted from this preliminary investigation, and two important large databases of Norwegian RSFs were also excluded—the Storfjord relative chronostratigraphic archive (Longva et al. 2009) and Matthews et al.'s (2018) record of small RSFs in Jotunheimen.

Chronological control for 34 sites was established by terrestrial cosmogenic nuclide (^{10}Be) exposure dating, described in the source references. ^{10}Be cosmogenic exposure ages were calculated by the cited authors using the CRONUS (Balco et al. 2008) and are reported here as mean ages and ±1 sigma internal uncertainties. Unless otherwise indicated, ages were calibrated to global production rates and Sa (LSDn) scaling models according to Lifton et al. (2014) and Borchers et al. (2016). Eight further samples were dated through ^{14}C analyses of organic material sampled from terrestrial deposits or cores of fjord sediments. For consistency, all radiometric ^{14}C ages were recalibrated to calendar (cal.) years BP using the IntCal13 dataset (Reimer et al. 2013), at the ±1 sigma uncertainty level. Where multiple dates had been secured at a given site, a 'best-estimate' age for the dated RSFs was obtained by calculating the uncertainty-weighted mean age for each RSF. The ages of four landforms were estimated by Schmidt hammer exposure dating (SHD), and three deposits were dated by morphostratigraphic correlation (Table 5.1).

Aggregation and more detailed interrogation of these datasets is ongoing, and forms part of a broader project. The distribution of ages (A) of the 49 dated RSFs are plotted with the cumulative number of RSFs (N) in Fig. 5.12, and are described by the regression equation:

$$N = 3.7583A - 5.2945 \ (\mathrm{n} = 49,\ \mathrm{r}^2 = 0.993) \quad (5.1)$$

The data demonstrate that RSF activity has occurred throughout almost the entire postglacial period, from 15.4 ± 1.7 ka to 1.0 ± 0.4 ka. The frequency of activity, however, was non-linear through time, with the cumulated number of RSFs increasing slightly more steeply with age at ~12.1, 10.1, 7.2 and 5.3 ka, than during intervening periods. The periodicity for RSFs during these episodes is extremely high, for example, between ~12–11, 10–9 and 5.3–4.9 ka, the sampled RSFs occur on average every ~130 years. This is compared with an overall average periodicity of ~310 years for all 49 RSFs since ~15 ka. Figure 5.13a plots the frequency of RSFs per millennia.

The timing of deglaciation, however, varies from ~9.9 ka (for inland, high-elevation sites that were reoccupied by glacier ice during the Younger Dryas) to ~14.2 ka (for coastal localities in Møre og Romsdal that were first to emerge from the retreating Scandinavian Ice Sheet). Figure 5.13b takes account of these differences, depicting the distribution of RSF ages since ice retreat for each site, aided by recent time-slice reconstructions of ice retreat (Hughes et al. 2016).

Figure 5.13b reveals a fascinating temporal distribution of RSFs. The immediate response of recently deglaciated rock-slopes to nonglacial conditions appears comparatively muted. Only 12% of total postglacial failures took place in the first millennium after ice retreat, though subsequently there is an intensification of RSF activity, with nearly one-third of recorded failures having occurred 1–3 ka following deglaciation. RSF frequency then experienced an overall decrease over a duration of 4 ka. Some 7–6 ka after ice retreat, however, Holocene RSF activity accelerated rapidly, and 20% of postglacial failure occurred at these times (~4–5 ka and 7–8 ka BP). Since ~4 ka BP, RSF frequency declined markedly. This pattern resonates with McColl's (2012) global survey of major RSFs in which some major failures were shown to have initiated prior to or synchronous with deglaciation, while most documented cases typically occurred some millennia after ice retreat.

It is instructive to further evaluate the RSF timing data by considering the volume of material reworked, as well as event frequency, as estimated for many of the sampled RSF deposits. Figure 5.13c and d illustrate the findings, which resonate with those of Böhme et al. (2015). While both magnitude and frequency of RSFs during the initial stages of

Table 5.1 Collated timing data for 49 moderate to large (>10^3 m^3) prehistoric RSFs in western Norway, showing their main characteristics and estimated lag time following glacier retreat.

RSF#	Location		Lithology	Type[1]	Volume (10^6 m^3)[2]	^{14}C age		Calibrated age		Method[3]	Lag ka[5]	References
						ka BP	Error 1σ	ka BP	Error 1σ			
	Møre og Romsdal											
1	Innerdalen (A–E)	Innerdalen	gneiss	RA	12			14.1	0.4	^{10}Be	<1	Schleier et al. (2015)
2	Innerdalen (F)	Innerdalen	gneiss	RA	22.5			7.97	0.94	^{10}Be	3.5	Schleier et al. (2015)
3	Ivasnasen (IF1)	Sunndalen	gneiss	RS/RA	5.4			13.0		STR	0	Oppikofer et al. 2017
4	Ivasnasen (IF2)	Sunndalen	gneiss	RS/RA	1.6			3.3	0.1	^{10}Be[(4)]	9.0	Oppikofer et al. 2017
5	Oppstadhornet	Otrøya	gneiss	DSGSD	10			15.4	1.2	^{10}Be	<1	Hermanns et al. (2013b)
6	Gråura	Romsdalsfjord	gneiss	RA				14.1	1.9	^{10}Be	1.0	Hermanns et al. (2017a)
7	Innfjorddalen (A1)	Innfjorddalen	gneiss	RA	20.6			14.3	1.4	^{10}Be	1.0	Schleier et al. (2017)
8	Innfjorddalen (C)	Innfjorddalen	gneiss	RA	6.3			8.79	0.94	^{10}Be	6.5	Schleier et al. (2017)
9	Innfjorddalen (D)	Innfjorddalen	gneiss	RA	0.4			1.03	0.38	^{10}Be	14.0	Schleier et al. (2017)
10	Mannen 1	Romsdalen	gneiss	RA				11.25	0.75	STR	0.4	Hilger et al. (2018)
11	Mannen 2	Romsdalen	gneiss	RA				9.39	0.3	^{10}Be	2.3	Hilger et al. (2018)
12	Mannen 3	Romsdalen	gneiss	RA				10.0	0.5	STR	1.7	Hilger et al. (2018)
13	Mannen 4	Romsdalen	gneiss	RA				4.93	0.09	^{10}Be	6.8	Hilger et al. (2018)
14	Mannen 5	Romsdalen	gneiss	RA				4.96	0.15	^{10}Be	6.7	Hilger et al. (2018)
15	Mannen 6	Romsdalen	gneiss	RA				4.95	0.1	^{10}Be	6.8	Hilger et al. (2018)
16	Skiri (1)	Romsdalen	gneiss	RA				11.7	1.3	^{10}Be	1.0	Hermanns et al. (2017a)
17	Skiri (2)	Romsdalen	gneiss	RA				11.0	1.3	^{10}Be	1.0	Hermanns et al. (2017a)
18	Svarttinden	Romsdalen	gneiss	RA				8.7	1.1	^{10}Be	3.0	Hermanns et al. (2017a)
19	Alstadfjellet	Valldalen	gneiss	RA				9.3	1.0	^{10}Be	5.0	Hermanns et al. (2017a)
20	Kallen	Tafjord	gneiss	RA				11.4	2.1	^{10}Be	1.0	Hermanns et al. (2017a)
21	Onilsavatnet	Tafjord	gneiss	RA	200			11.4	0.21	^{10}Be	1.1	Hermanns et al. (2009)
22	Blåhornet	Sunnylvsfjord	gneiss	RA	0.009			2.11	0.31	^{10}Be	9.9	Böhme et al. (2015)
23	Norangsdalen (Skylstad)	Sunnmøre	gneiss	RA				6.0	0.7	SHD	7.0	Nesje et al. (1994)
24	Norangsdalen (Uraseter)	Sunnmøre	gneiss	RA				4.2	0.8	SHD	9.0	Nesje et al. (1994)
25	Nakkaneset	Geirangerfjord	gneiss	RA				7.25	1.14	^{10}Be[(4)]	4.8	Böhme et al. (2015)
26	Nokkenibba	Geirangerfjord	gneiss	RA				7.06	1.25	^{10}Be[(4)]	4.9	Böhme et al. (2015)
	Sogn og Fjordane											
27	Erdalen	Erdalen	gneiss	RS		8.81	0.13	9.862	0.18	^{14}C	0.0	Hansen et al. 2009
28	Vora (a)	Myklebustdalen	gneiss	RA	80			12.9	1.9	^{10}Be	1.0	Hermanns et al. (2017a)
29	Vora (b)	Myklebustdalen	gneiss	RA				12.1	1.3	^{10}Be	1.0	Hermanns et al. (2017a)

(continued)

Table 5.1 (continued)

RSF#	Location		Lithology	Type[1]	Volume (10^6 m^3)[2]	^{14}C age		Calibrated age		Method[3]	Lag ka[5]	References
						ka BP	Error 1σ	ka BP	Error 1σ			
30	Vora (c)	Myklebustdalen	gneiss	RA				9.49	0.5	SHD	2.0	Aa et al. (2007)
31	Vora (d)	Myklebustdalen	gneiss	RA				10.8	1.4	^{10}Be	1.0	Hermanns et al. (2017a)
32	Vora (Grisemyra B)	Myklebustdalen	gneiss	RA		6.2	0.04	7.09	0.08	^{14}C	5.0	Aa et al. (2007)
33	Vora (Grisemyra D)	Myklebustdalen	gneiss	RA		5.51	0.04	6.39	0.06	^{14}C	5.5	Aa et al. (2007)
34	Vora (Grisemyra F)	Myklebustdalen	gneiss	RA		5.065	0.035	5.82	0.07	^{14}C	6.0	Aa et al. (2007)
35	Vora (Grisemyra J)	Myklebustdalen	gneiss	RA		4.585	0.035	5.27	0.18	^{14}C	6.5	Aa et al. (2007)
36	Vora (Grisemyra O)	Myklebustdalen	gneiss	RA		4.47	0.05	5.13	0.15	^{14}C	8.5	Aa et al. (2007)
37	Oldedalen	Nordfjord	monzonite	RA	0.4	5.22	0.08	6.04	0.13	^{14}C	4.0	Nesje (2002)
38	Bøyadalen	Fjærlandsfjord	monzonite	RA				3.5	0.6	^{10}Be	8.1	Hermanns et al. (2011)
39	Øyrahagestolen	Fjærlandsfjord	gneiss	RA				10.0	1.4	^{10}Be	1.6	Hermanns et al. (2011)
40	Uraneset	Fjærlandsfjord	gneiss	RA				10.1	1.4	^{10}Be	1.5	Hermanns et al. (2011)
41	Skjeringahaugane P1	Lusterfjord	phyllite	RS				7.0	0.8	^{10}Be	3.0	Hermanns et al. (2012)
42	Skjeringahaugane P2	Lusterfjord	phyllite	RS				6.8	0.7	^{10}Be	3.2	Hermanns et al. (2012)
43	Skjeringahaugane P3	Lusterfjord	phyllite	RS				4.0	0.6	^{10}Be	6.0	Hermanns et al. (2012)
44	Storurdi	Urdadalen	pyroxene-granulite	RA	10			1.83	0.76	SHD	7.5	Wilson (2009)
45	Aurlandsfjorden	Aurlandsfjorden	phyllite, granodiorite	RA		2.64	0.07	2.77	0.06	^{14}C	8.2	Bøe et al. (2004)
46	Stampa (Joasete)	Flåm	phyllite	RA	4.27			3.754	0.38	^{10}Be (4)	7.0	Böhme et al. (2013)
47	Stampa (Ramnanosi 1)	Flåm	phyllite	RA	0.3			2.579	0.22	^{10}Be (4)	8.2	Böhme et al. (2013)
48	Stampa (Ramnanosi 2)	Flåm	phyllite	RA	2.71			12.87	1.12	^{10}Be (4)	0	Böhme et al. (2013)
	Telemark											
49	Urdbøuri	Rauland	granite, granodiorite	RA				11.3	1.3	^{10}Be	1.0	Hermanns et al. (2017a)

Notes

1. Type: *RA* rock avalanche; *RF* rockfall; *RS* rockslide and *DSGSD* deep-seated gravitational slope deformation
2. Volume calculations are often approximations based on surficial area and average thickness of the deposits (or extracted from swath bathymetry and seismic surveys for fjord sediments)
3. Method

(i) ^{10}Be cosmogenic exposure ages were calculated by cited authors using the CRONUS (Balco et al. 2008) and are reported here as mean ages and ±1 sigma uncertainties. Unless otherwise stated, ages were calibrated to global production rates and Sa (LSDn) scaling models according to Lifton et al. (2014) and Borchers et al. (2016). Derivation of ^{10}Be ages from local production rates (Fenton et al. 2011; Goehring et al. 2012) would make the ages approximately 10% or 5% older, respectively (Hermanns et al. 2013b)

(ii) STR refers to age estimation based on morpho-stratigraphic relations

(iii) For consistency, radiometric ^{14}C ages were recalibrated to calendar (cal.) years B.P. using the IntCal13 dataset (Reimer et al. 2013), at the ±1 sigma uncertainty level

(iv) SHD (Schmidt hammer exposure dating) was undertaken using Type N hammers and local calibration curves

4. These authors derived ^{10}Be ages using St scaling (Lal 1991; Stone 2000)
5. Lag time represents the maximum difference between the calibrated ages of local deglaciation and RSF at each locality

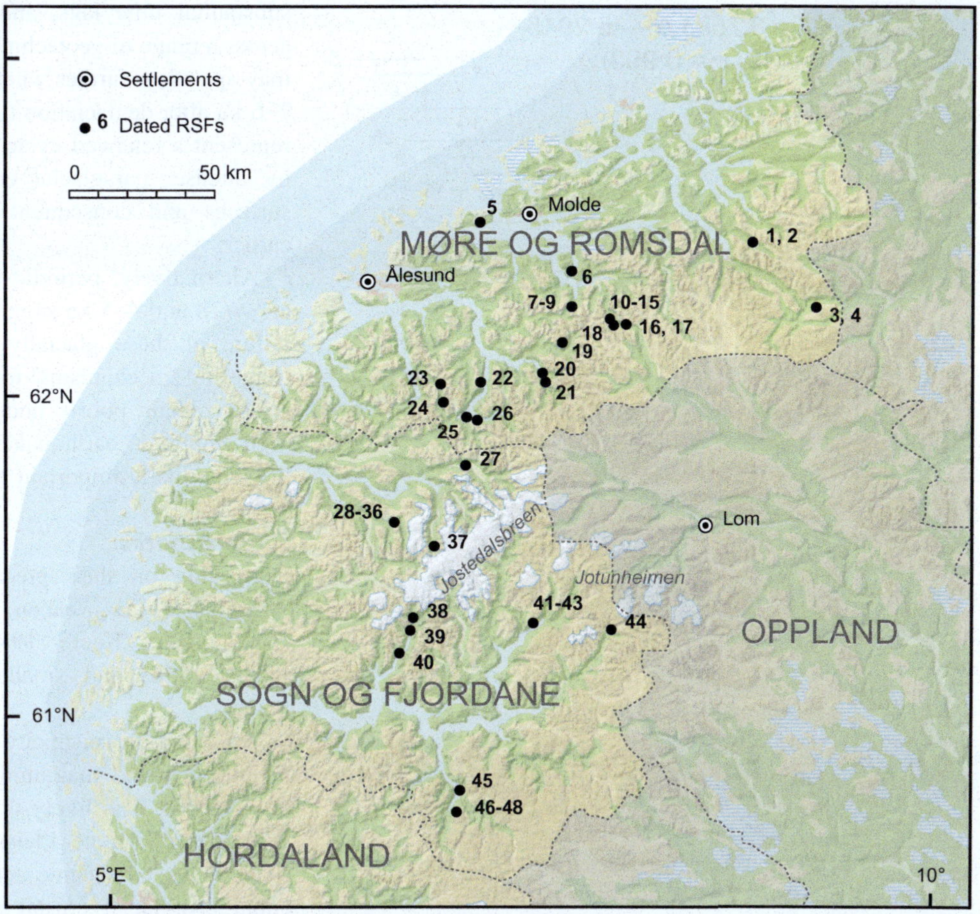

Fig. 5.11 Location of 49 moderate to large (>10^3 m^3) prehistoric RSFs in western Norway (numbering as in Table 5.1). Norwegian Mapping authority base map: https://www.geonorge.no/

deglaciation appear to have been somewhat subdued, some 80% of total postglacial rock-slope sediment flux appears to have been released from unstable and metastable stores during the second millennia after ice retreat. Of course, volume calculations are rough estimations. Smaller events may be under-represented for older time periods, with their signature more likely to have been removed by erosion or hidden by larger deposits (Böhme et al. 2015). Notwithstanding some occasional exceptions, though, the prehistoric record after 10 ka is dominated by RSF events whose individual and cumulative volume were several orders of magnitude lower than during the early paraglacial period.

Proposed long-term and short-term causes of RSF activity are considered below, in the light of these observations.

Glacial debuttressing and thermal shock

In contrast to a rapid (century timescale) response to deglaciation, many major RSF events in western Norway evidently lagged deglaciation by ~1–2 ka (Blikra et al. 2006; Böhme et al. 2015). Furthermore, as suspected by workers in Norway for some time, the frequency (not magnitude) of RSF activity appears to have reached levels comparable with the Late Weichselian–Holocene transition as recently as 4 ka (Blikra et al. 2006). These characteristics of the Norwegian dataset contrast with models that envisage paraglacial rock-slope adjustment in terms of an immediate ‘rapid response’ to deglaciation, succeeded by approximately constant event frequency averaged over the timescale of interglacial recovery. In particular, apart from six (12%) cases of dated RSFs that were synchronous with deglaciation, the patterns summarised in Table 5.1 do not fit well with ‘deglacial debuttressing’ or ‘paraglacial thermal shock’, given the close association of these mechanisms with rapid collapse of unstable rockwalls during or immediately after glacier downwastage.

Paraglacial stress release

Conversely, the majority of large failures observed in this inventory have post-dated ice retreat by at least one thousand years. Paraglacial stress release is the development of stresses perpendicular to and parallel to a rock-slope that result from glacial erosion and deglacial unloading (McColl 2012). Such time-dependent stress changes may initiate gradual rock mass fatigue over century to millennial timescales through progressive shear plane development within fractured rock masses. As noted, for example, at Ivasnasen

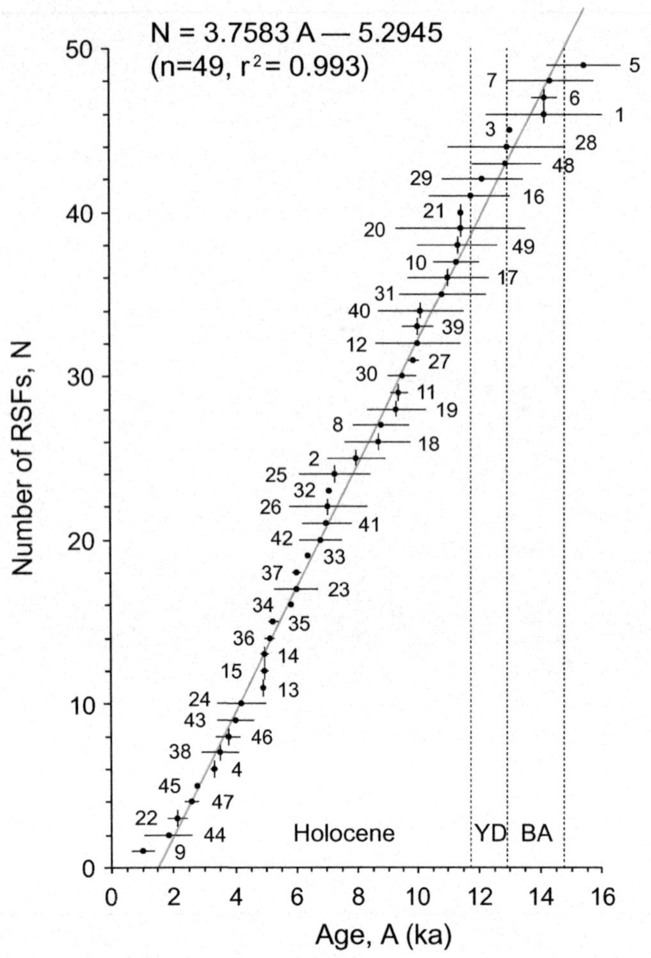

Fig. 5.12 Calibrated ages for 49 RSFs, arranged cumulatively by ascending RSF age. Numbering follows Table 5.1. Length of the horizontal bars represents $\pm 1\sigma$ total uncertainties, and vertical bars indicate weighted mean ages. The number of RSFs (N) increases non-linearly with increasing age (A). Vertical dashed lines separate the Bølling–Allerød Interstadial (BA: 14.7–12.9 ka), Younger Dryas Stade (YD: 12.9–11.7 ka) and Holocene (<11.7 ka)

(Oppikofer et al. 2017), slope instabilities may 'endure' slow deformation over such timescales. Modelling the response of glacially oversteepened slopes to Pleistocene deglaciation in Canada suggests that average pre-failure endurance may locally exceed 5 ka (Cruden and Hu 1993).

An implication of this interpretation is that the observed pattern of timing of RSFs in Fig. 5.13 may reflect prolonged stress release initiated by Late Weichselian unloading. Such time-dependent degradation of rock strength may result in spontaneous kinematic release of rock centuries or millennia after deglaciation, with or without involvement of an extrinsic trigger mechanism.

This explanation for the delayed or lagged response of rock-slope systems is consistent with the behaviour of at least 19 (39%) of the sampled RSFs that failed during the first 2 ka after deglaciation. Further research is needed to assess 'pre-failure endurance' as an explanation for these substantial time lags, and to identify its temporal limits across a range of geotechnical and lithological contexts. One may speculate further whether the clustering of dated RSFs 7–6 ka after deglaciation (i.e. at ~4–5 and 7–8 ka BP) may represent a renewed cycle of stress release failures, caused by intense 'paraglacial erosional unloading' of rockwall surfaces and consequent rock mass degradation 7–5 ka earlier.

Alternatively, periodic renewal of recorded RSF events at ~7–8 and 4–5 ka suggests that other factors may trigger failure of these glacially conditioned rock-slopes. While triggering mechanisms for large rock avalanches in the region remain poorly understood, Blikra et al. (2002) cite high-amplitude earthquakes and extreme rainfall events as the most likely important factors.

Seismic shocks

Depending on their pre-existing state, powerful (M >6) earthquakes can weaken rock masses and trigger catastrophic RSFs (Keefer 1984; Eisbacher and Clague 1984; Jibson 1996) and smaller shocks may trigger smaller failures.

Although the seismicity of Norway is low to intermediate in intensity, and magnitudes rarely exceed 5.5 (Fjeldskaar et al. 2000), it is likely that earthquake activity was much greater during Late Pleistocene deglaciation. A pulse of shallow crustal seismic activity can be attributed to differential postglacial crustal rebound and associated shallow faulting, in conjunction with elastically stored tectonic stresses (Cave and Ballantyne 2016). The prolonged response time of resultant crustal stress changes mean that associated seismic activity may lag deglaciation by centuries or millennia (Stewart et al. 2000).

Notwithstanding Norway's poorly controlled paleoseismic records (Redfield and Hermanns 2016), paleoseismological studies suggest that large-magnitude (M >6) intraplate earthquakes occurred in Scandinavia at ~13–9 ka (Mörner 2005, 2013), which correlates with the peak rate of postglacial uplift in western Norway (Svendsen and Mangerud 1987; Fjeldskaar 1997). Of 53 M > 6 palaeoearthquakes in Sweden (Mörner 2013, and references therein), 28 occurred between 11–9 ka (averaging 14 ka^{-1}), compared with 21 earthquakes for the last 9 ka (at an average rate of 2.3 ka^{-1}). Six out of seven M >7 earthquakes occurred during regional deglaciation, 13–9 ka.

This pattern was also observed in west and mid-Norwegian fjords, with frequent earthquakes inferred during the Early Holocene (~11.2–8.3 ka) and triggered by glaco-isostatic rebound, followed by low frequency in the Mid-Holocene, and a slight reactivation in the Late Holocene with clusters from 4.2–1.8 ka (Bellwald et al. 2019). Late Holocene seismic reactivation has been observed in all regions previously covered by the Scandinavian Ice Sheet.

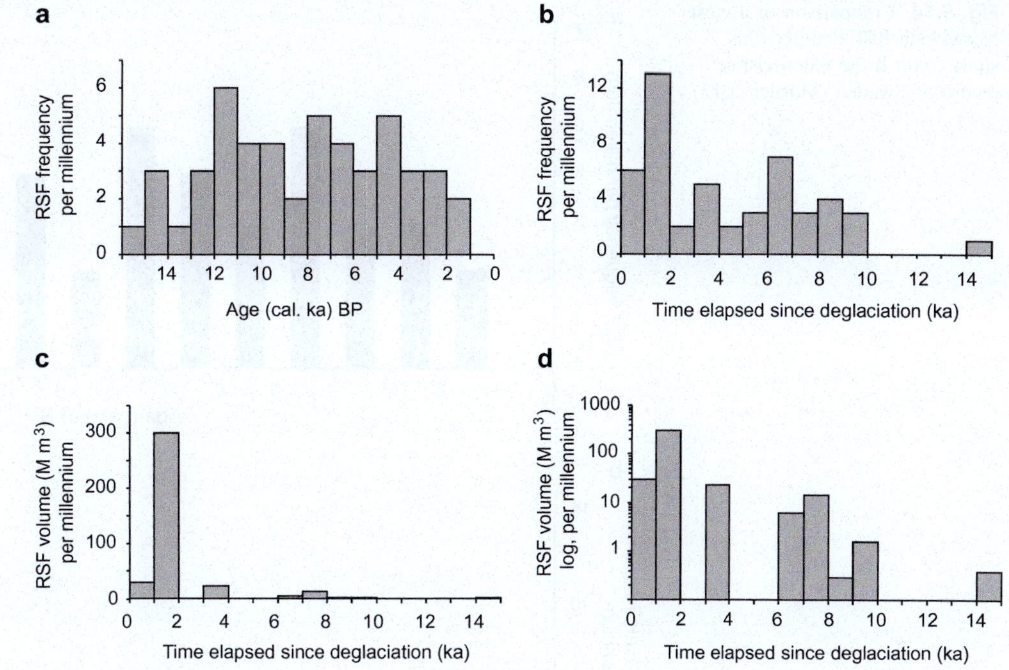

Fig. 5.13 Temporal distribution of the ages of 49 dated RSFs in western Norway: **a** frequency of the number of RSFs per millennium; **b** frequency of the number of RSFs per millennia since local deglaciation; **c** total RSF volume failed per millennium since local deglaciation and **d** total log RSF volume failed per millennium since local deglaciation.

Analysis by Bellwald et al. (2019) of 125 postglacial mass movement deposits (subaquatic mass flows, debris flows, slumps, slides and turbidites) in western Norwegian fjords led to the identification of clusters of enhanced activity in the Early Holocene (11–9.7 ka), at around 8.0 ka, and in the Late Holocene (~4.2 ka to present). The authors interpreted most of these events as reflecting regional seismic activity triggering failure of climatically preconditioned slopes. Interpreting their record of mass movement clusters as a proxy for palaeoseismic activity, they suggest that at least 36 individual regional earthquakes (M >6) occurred in west and mid-Norway through the Holocene.

Coastal areas of western Norway represent the region of greatest historic earthquake activity in the south of the country (Dehls et al. 2000b). Moreover, the concentration of gravitational faults and slope failures in Sogn og Fjordane and Hordaland and in parts of Møre og Romsdal may point to powerful prehistoric earthquakes in western Norway (Olesen et al. 2004, 2013; Olsen et al. 2013). However, triggering mechanisms for a cluster of large RSFs in Innfjord (Møre og Romsdal) formerly attributed to Holocene neotectonic seismicity and earthquake shaking (Anda et al. 2002; Blikra et al. 2006) require fresh evaluation following reinterpretation of a supposed Holocene reverse fault (the so-called Berill Fault) by Schleier et al. (2016).

If palaeoseismicity played a significant role in triggering postglacial RSFs in western Norway, a close correspondence between the observed RSF ages (Table 5.1) and the timing of rapid, regional postglacial uplift and palaeoearthquakes might be expected. Many RSFs within a given radius would occur on or shortly after the same day. At $\pm 1\sigma$ uncertainties, the calibrated age ranges for 28 (57%) of events in the RSF dataset are bracketed by or overlap one of the aforementioned periods of enhanced postglacial seismicity in western Norway (11.2–8.3 ka and 4.2–1.8 ka). However, many of these ages also overlap with other likely triggers and causal factors, notably the period of assumed paraglacial stress release. Initial comparison of the RSF database and the regional plaeoseismic record suggests only weak association (Fig. 5.14).

Climatic triggers

As indicated above, changes in rockwall hydrology and thermal regime may have a critical bearing on rock-slope instability (Fig. 5.15), via postglacial permafrost melt, as well as seasonal snowmelt, extreme rainstorms and weathering of shear surfaces (Krautblatter et al. 2013; Blikra and Christiansen 2014; Schleier et al. 2015; Hilger et al. 2018). At present, permafrost in western Norway may exist to considerable depths in rockwalls above 1,300–1,700 m elevation, depending on aspect (Hipp et al. 2014). During the Younger Dryas, permafrost probably extended to sea level, so rock mass stability during and shortly after deglaciation would likely have been compromised via excess joint-water pressures induced by permafrost degradation. Schleier et al. (2015) explored this idea further, with regard to the dating evidence from Innerdalen (Møre og Romsdal). The rock avalanche events they documented occurred during the Late Pleistocene Bølling–Allerød Interstadial (~14.7–12.9 ka) under paraglacial conditions, and during the Atlantic chronozone, or Holocene Thermal Maximum (~8–5 ka). The authors observed that both

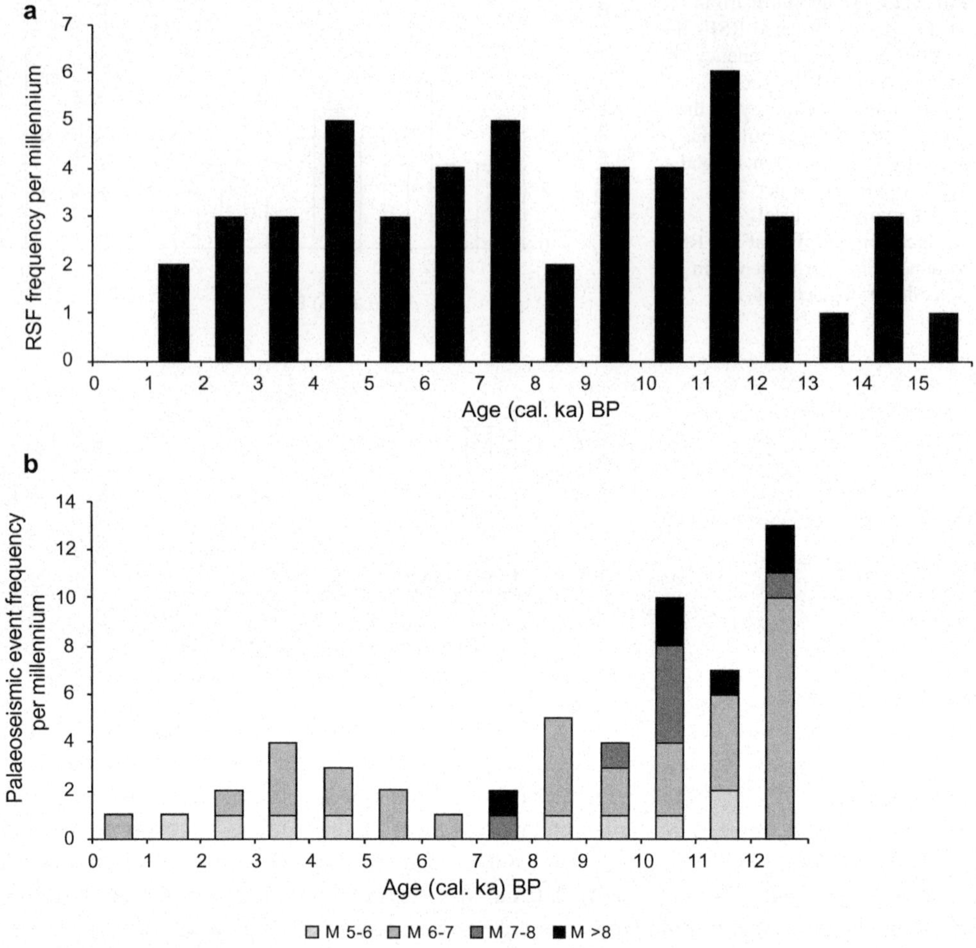

Fig. 5.14 Comparison of **a** west Norwegian RSF activity (this study) with **b** the paleoseismic record of Sweden (Mörner 2013)

events immediately followed significant cooling periods in the northern hemisphere (Older Dryas stadial 14.1–13.8 ka and the so-called 8.2 ka event), when the permafrost limit was lowered, and likely contributed to greater rockwall stability. However, during the subsequent warming periods, rising permafrost limits are considered to increase the probability of failure. As outlined above, Matthews et al. (2018) also highlighted Holocene permafrost degradation as a conditioning factor for small RSFs; a finding that carries important implications for slope instability under future warming scenarios.

Permafrost degradation was tested as an explanation for periodic increases in rock-slope activity by interrogating the timing data in Table 5.1. Very rapid warming occurred twice during the postglacial period, terminating stadial conditions at the Older Dryas Stade–Bølling-Allerød Interstade transition (~14.7–14.2 ka) and at the Younger Dryas Stade–Holocene transition (~12.2–11.7 ka). Rapid warming also characterised the transition towards the Holocene Thermal Maximum (~8.5–8.0 ka). Each of these periods likely caused permafrost degradation, increased joint-water pressure and enhanced freeze–thaw activity (Ballantyne et al. 2013; McColl 2012). However, the number of RSF age ranges ($\pm 1\sigma$) that overlap these three periods of rapid warming (19) is actually lower than those that occur within the subsequent 500 years after these warming transitions (23). Accordingly, the present database gives no clear evidence for widespread RSF activity having been triggered by warming and thaw of permafrost and/or increased water pressure within rock masses. The timing of the majority of dated RSFs appears to be independent of periods of rapid warming.

However, mechanical modelling of permafrost degradation by Krautblatter et al. (2013) suggests that permafrost degradation may slowly condition RSF activity on long-term (millennial) timescales, by both external controls (via gradual and cyclical thermal changes during the glacial–interglacial transition), and internal responses (via progressive rock fatigue and joint propagation). In consequence, thermal effects should not be dismissed as triggering or preparing a higher number of RSF events than those bracketed by these narrow windows of rapid warming.

Nesje et al. (1994) and Nesje (2002) speculated whether climatic deterioration (cooler and wetter climate with

Fig. 5.15 The 20 m high backscarp crowning the Mannen rock-slope instability, Møre og Romsdal. The site is classified high risk by the Geological Survey of Norway (NGU), owing to deformation rates of 2–5 cm per year of ~5 × 10^6 m^3 rock mass. Seasonal acceleration at the sliding surface occurs at the end of the summer, when local ground temperatures exceed 0 °C. Extensive, prehistoric rock avalanche deposits litter the Rauma (Romsdalen) valley floor some 1,200 m below and represent 6–9 catastrophic RSF events dated to ~12–4.5 ka (Hilger et al. 2018). With the summit (1,294 m) corresponding with the lower limit of permafrost, near-future warming is expected to further lower rock mass stability at this and similar sites. *Photo* Paula Hilger

enhanced seasonality) following the post-Holocene Thermal Maximum may have initiated three Mid-Holocene rock avalanches dated to ~4.2 and 6.0 ka in Oldedalen (Sogn og Fjordane) and at ~6.0 ka in Norangsdalen (Møre og Romsdal), via elevated joint-water pressures. Precipitation in southern Norway was 170% greater at ~6–5 ka than during the reference period 1961–1990 (Bøe et al. 2006).

The significance of high, fluctuating groundwater within bedrock is powerfully exemplified in the case of the 1756 Tjellefonna (Møre og Romsdal) rockslide, which followed two weeks of heavy rain (Furseth 2006). Numerical slope stability modelling and historical accounts suggest that heavy, long-lasting rainfall was the triggering factor for this slide, the largest (9.3–15 M m^3) RSF historically recorded in Norway (Sandøy et al. 2017). Climatic deterioration (higher precipitation connected to a strong seasonality) at the end of the Holocene Thermal Maximum (when permafrost distribution was at a minimum) has also been proposed as a possible explanation for temporal clustering of multiple failures at Mannen (Møre og Romsdal) at ~5.5–4.5 ka (Hilger et al. 2018).

However, the millennial-scale delay between deglaciation and peak failure frequency at the sampled RSFs implies that high joint-water pressures associated with deglaciation played a limited part in triggering failure at these sites. Nonetheless, increased precipitation during periodic climatic deterioration may account for some of the RSF ages during the Mid-Late Holocene, such as the frequency peaks at ~5–4 ka. As described above, regional climatic irregularities causing increased rainfall have been tentatively linked to enhanced colluvial reworking of sediments in western Norway at 5.6–5.3 ka and 3.2–2.8 ka. However, none of the calibrated ages of observed RSFs in Table 5.1 fall within these periods, and only five (10%) of the $\pm 1\sigma$ age ranges exhibit any overlap with these periods of apparent enhanced rainfall.

Nonetheless, a lack of temporal connection between RSF events and phases of enhanced precipitation does not necessarily exclude specific, short-term meteorological events that may have prepared or even triggered failure (*cf.* Ostermann and Sanders 2017). Further analysis of the timing data and regional climatic records may facilitate more robust evaluation of this and other climatic triggers.

Temporal patterns: summary

In general, the results of this preliminary analysis support those from previous research in the European Alps (Prager et al. 2008; McColl 2012) and Scotland (Ballantyne et al. 2014a, b) indicating that the wastage of Late Pleistocene ice sheets in tectonically stable montane regions was followed by a period of greatly enhanced RSF activity that extended over several millennia. The most important factors for preconditioning and triggering failure appear to have been paraglacial stress release, progressive rock mass weakening and possibly seismic activity driven by rapid rates of glacio-isostatic rebound during the period of maximum uplift rates. Glacial debuttressing and climatic factors appear to have been of more limited, local importance in triggering postglacial RSFs in northern and central Europe; though the relationship between rockwall thermal changes, increased water supply and Late Holocene RSF activity in Norway merit further attention.

5.6 Modelling RSF Frequency Through Time

Four general models have been proposed (Ballantyne and Stone 2013) for changing RSF frequency with time elapsed since deglaciation (Fig. 5.16), namely (1) constant frequency (steady state) model (the frequency of RSFs does not change with time; Cruden and Hu 1993); (2) steady-state decline model (the frequency of RSF activity declines linearly over time; (3) exhaustion model (the frequency of RSF activity declines exponentially as sediment sources ('unfailed' metastable rockwalls) are depleted; Cruden and Hu 1993) and (4) rapid response model (the frequency of RSF activity is greatest during or immediately after deglaciation; Church and Ryder 1972).

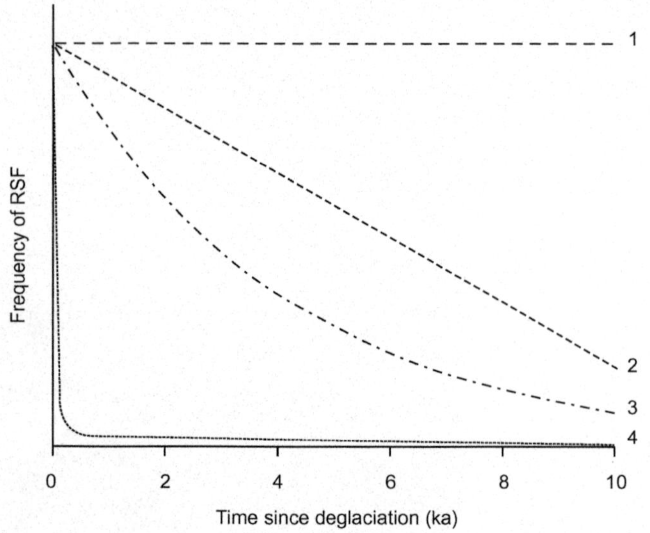

Fig. 5.16 Four possible models for the changing frequency of RSF activity after deglaciation: 1. constant frequency model; 2. steady-state decline; 3. exhaustion model; 4. rapid response model. For explanation, see text. (*Source* Ballantyne and Stone 2013, Fig. 1)

The value of these general models of paraglacial landscape response lies in their simplification and generalisation of complex realities. By definition, therefore, they hide much of the local variability related to specific environments and process domains (Cossart et al. 2013), which becomes increasingly relevant as the scale of inquiry is reduced (Kirkbride and Deline 2018). In identifying a number of unresolved paraglacial research priorities, Olivia et al. (2019) highlight modelling of complex paraglacial landscape response and sediment cascades to allow landscape trajectories to be anticipated and their effects managed.

Collection of a large inventory of published RSF ages spanning the Late Weichselian–Holocene has facilitated evaluation of these models (Fig. 5.17). The timing data presented fail to conform closely to any of the proposed models but suggest that the temporal distribution of postglacial RSFs is best described by a modification of the exhaustion model that incorporates episodic, external perturbations and renewed, stochastic, paraglacial responses. Figure 5.17b illustrates a proposed 'episodic response model'. While Ballantyne (2002a) outlined a visually similar scheme for the episodic release of glacially conditioned sediment, exemplified in the case of a marine transgression through primary glacigenic deposits, there are important distinctives.

In recent decades, since exhaustion curves were first applied to paraglacial rock-slope systems, the temporal complexity of the latter has become increasingly evident. In particular, greater attention has been given to pre-failure endurance and the post-deglaciation time lag that appear to characterise many (but not all) rock-slope systems (McColl 2012; Ballantyne et al. 2014a, b). Modelling rock-slope responses to deglaciation must accommodate these different trajectories.

The episodic response model presented here also rejects an assumption of the exhaustion model that each site fails only once. As outlined above, multiple, closely spaced, large RSFs have occurred on the same slope at several localities in Norway. The work outlined in this chapter also tentatively proposes that the history (memory) of failure at a given site may increase the likelihood of repeat failures by exposing fresh, unloaded, metastable bedrock surfaces and joint networks and reorganising stress fields (*cf.* Crosta et al. 2017; Hilger et al. 2018).

Further, the probability of failure at any given site is not constant, and supply of metastable and available material changes spatially and over time, and is to a limited extent, replenishable. Valley impoundment by RSF debris may lead to localised fluvial aggradation (upstream) and enhanced incision adjacent to and downstream of RSFs, thereby increasing slope instability and metastable sediment availability with time (Rodríguez-Rodríguez et al. 2018; Olivia et al. 2019).

The index for illustrating the magnitude and direction of paraglacial response is therefore measured in terms of volume of failed RSF debris rather than sediment availability, which can vary spatially and temporally over an interglacial timescale. While sediment cascades represent unifying conceptual frameworks for postglacial landscape evolution, often in terms of an exponential decay in paraglacial

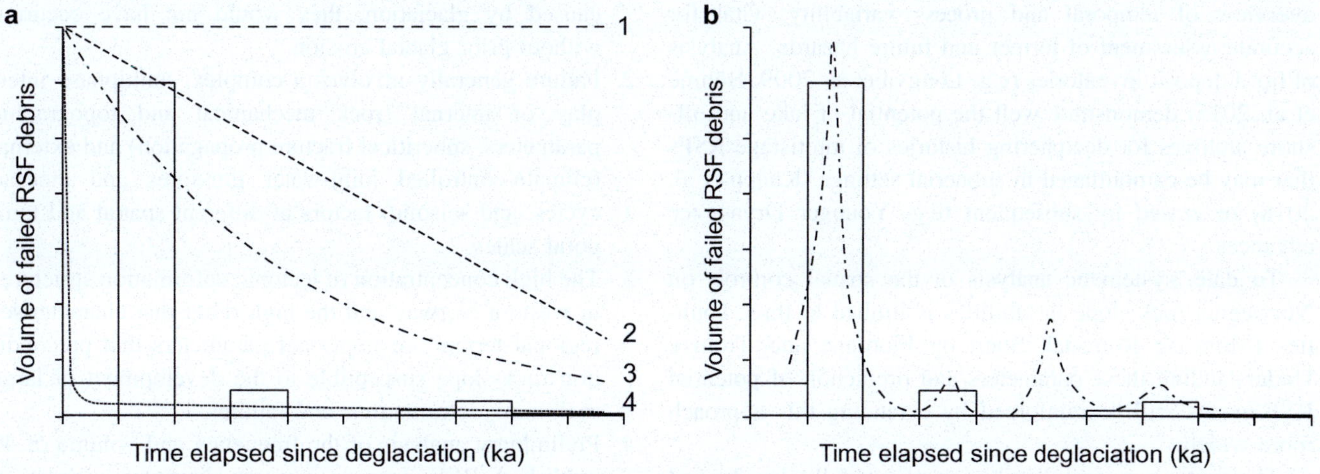

Fig. 5.17 **a** Different models of time-dependent behaviour of post-glacial RSF activity have been fitted to the volume distribution of failed RSF debris, described in this study, but failed to account sufficiently for the observed post-deglaciation time lag. **b** In contrast, a model following an episodic response to deglaciation shows the best fit for the dataset, and presents multiple possible pathways for subsequent delayed or renewed RSF activity, as widely observed in deglaciated mountain environments

sediment supply through time, the episodic response model gives attention to these non-uniform and complex elements of paraglacial rock-slope adjustment.

5.7 Future Research

Although the Norwegian montane landscape is dominated by the distinctive signature of recurrent Pleistocene glaciation, a range of postglacial geomorphological responses has led to extensive modification of the glacial landscape. Explicit recognition of the nature and effects of this glacially conditioned landscape evolution is relatively new, and significant gaps in our understanding remain. This section presents a number of exciting avenues for future study of rock-slope activity in glaciated mountain regions, particularly in Norway.

The research outlined in this chapter provides clear empirical evidence for glacially influenced or conditioned RSF activity in western Norway. The spatio-temporal distribution of prehistoric failures imply that multiple, temporary increases in RSF activity occurred at different times following deglaciation, characterised by varying amplitude, duration and geomorphic work. While the initiation of paraglacial rock-slope adjustment to deglaciation can be very rapid (Kos et al. 2016), a widely observed (but only partly understood) characteristic of many deglaciated rock-slope systems is the significant time lag between glacier retreat and maximum RSF activity. To better explain these lags, McColl (2012) recommends that attention is focused on (i) assessing the timescale of stress redistribution and development of rock mass degradation, (ii) evaluating the links between RSF event distribution and climate data (periodic warming and increased rainfall) and (iii) with enhanced postglacial seismicity. Cosmogenic dating of fault scarps will shed further light on the temporal connection between palaeoseismicity and enhanced RSF activity.

The heterogeneous spatial distribution, magnitude and mechanics of RSF events within mountain landscapes reflect a complex interplay of multiple causal factors that collectively define the sensitivity of rock-slopes to fail at different scales, and yet vary in their relative importance between different scales. Multivariate modelling approaches (e.g. multiple logistic regressions) are being successfully applied to explaining the complex system behaviour of multiple, scale-dependent controls on rockfall (e.g. Messenzehl et al. 2017), and may represent a fruitful approach to improving our understanding of the relative significance of RSF controls, and effective hazard management.

Integrating advances in geochronology, geodetics and remote survey and visualisation tools have facilitated the development of regional-scale RSF inventories that allow research to extend beyond reductionist, landform-based enquiries to better explain (and predict) the mechanics and evolution of mountain landsystems. Monitoring of unstable rockwalls has yielded clearer insights into the drivers of slope failure and deformation, and allows calibration of slope stability models for back analyses of former failure. Combining analytical or numerical modelling with field monitoring of contemporary mechanisms therefore represents an important approach for improved simulation and understanding of complex, multiple causes of failure, as well as assessment of ‘pre-failure endurance’ as an explanation for millennial-scale time lags.

Several studies illustrate the rich potential of coupling onshore and offshore datasets to gain more representative overview of temporal and process variability, vital for accurate assessment of former and future hazards. Analysis of fjord deposit inventories (e.g. Longva et al. 2009; Böhme et al. 2015) demonstrate well the potential of lake and offshore archives for deciphering histories of multistage RSFs that may be camouflaged in subaerial settings (Knapp et al. 2018) or erased by subsequent (e.g. Younger Dryas) ice advances.

To date, systematic analysis of the spatial controls on Norwegian rock-slope instabilities is limited to three counties (Møre og Romsdal, Sogn og Fjordane and Troms). Understanding these parameters and prediction of potential RSF events will be reinforced by extending this approach more widely.

The spatio-temporal distribution of rockfall talus-derived landforms (protalus rock glaciers and pronival (protalus) ramparts) also promises to shed light on the occurrence, magnitude and frequency of small RSFs in western Norway (e.g. Lilleøren and Etzelmüller 2011). Establishing the age and extent of these features may clarify understanding of rockfall and small-scale RSF activity in western and southern Norway, though their history may represent protracted, multistage activity conditioned by multiple factors of varying importance.

Finally, the research summarised in this chapter provides compelling evidence of the efficiency of paraglacial rock-slope activity for dismantling mountain slopes after deglaciation. Vast quantities of coarse debris are frequently released within the first few millennia of ice retreat, and accumulate in valley floor or fjord stores prior to further glaciation. The implications for interpreting postglacial sediment budgets, and evaluating the role of paraglacial RSF activity for glacial trough widening, cirque enlargement and subsequent glacial erosion, merit further consideration.

5.8 Summary

The findings of this synthesis are broadly congruent with wider research that emphasises the persistence and transient nature of the paraglacial period in rock-slope systems. The following conclusions are highlighted:

1. Deglaciation marks a complex transition of boundary conditions for rock-slope stability in which steep, unstable bedrock stores respond non-linearly to glacial inheritance. Most prehistoric RSFs in the heavily glaciated terrain of western Norway can be considered 'paraglacial' in the sense that they have occurred in a transitional period of environmental adjustments following the disturbance of deglaciation and involve changes in the intensity of landscape-forming processes (Oliva et al. 2019). Although they were not directly caused by glaciation, they would not have occurred without prior glacial erosion.
2. Failure generally involves a complex, multiphase interplay of internal (rock mechanical and topographic parameters, subcritical fracture propagation) and external (climate-controlled joint-water pressures and thermal cycles, and seismic) factors at different spatial and temporal scales.
3. The high concentration of tectonic deformation structures in western Norway and the high relief that characterises regional terrain are important parameters that precondition rock-slope susceptible to the development of large rock mass deformations and failures.
4. Preliminary analysis of the frequency and volume of 49 published RSF ages in western Norway suggests a modest increase in major RSF activity occurred during or immediately after Late Weichselian deglaciation.
5. Time-dependent paraglacial stress release and progressive rock mass degradation, in combination with glacio-isostatic crustal adjustments and possibly large-magnitude earthquakes, appear to have played an essential role in preconditioning, preparing and triggering a much larger number of RSFs and gravitational deformation structures up to several thousand years after ice sheet decay. 80% of total RSF volume surveyed in western Norway appears to have been delivered to fjords and valley floors during a period ~1–2 ka after deglaciation.
6. Catastrophic RSF activity in western Norway has continued through the Holocene, suggesting that rock-slope response to deglaciation is not solely attributable to glacial debuttressing, postglacial seismicity and associated rock mass weakening. High event frequencies were identified at 8–7 and 5–4 ka BP, though the volume of rock material released during the Mid- and Late Holocene was at least one order of magnitude lower than during the second millennium after deglaciation. These periods of enhanced, delayed or time-lagged RSF activity are ascribed to Holocene climatic irregularities, including seismic activity, and possibly increased precipitation rates, high air temperatures and associated degradation of permafrost in rock-slopes.
7. Glacial and paraglacial inheritance, alpine relief, abundant brittle and ductile bedrock structures, seasonal heavy precipitation, snow melt and freeze–thaw cycles continue to render the fjords and valleys of western Norway vulnerable to RSF events. These hazards may increase under global change scenarios, especially in relation to permafrost degradation.
8. The timing data presented fails to conform to existing general models of changing paraglacial RSF frequency (rapid response, constant frequency (steady state), exhaustion and steady-state decline), but are best

described by a modification of the exhaustion model that incorporates episodic, external perturbations and renewed, stochastic, paraglacial responses. This 'episodic response model' recognises this complexity, as well as the post-deglaciation time lag and pre-failure endurance that appear to characterise many rock-slope systems.

9. Several future research priorities are identified, including (a) understanding the significant time lag between glacier retreat and maximum RSF activity, especially the need to assess the timescale of stress redistribution and development of rock mass degradation; (b) the application of multivariate modelling approaches to explain complex behaviour of RSF controls; (c) the use of numerical modelling and field monitoring of failure mechanisms to simulate multiple causes of failure; (d) harnessing the potential of fjord deposit inventories for reconstructing RSF histories; (e) extending systematic surveying of spatial controls on rock-slope instabilities to other counties of Norway and (f) evaluating the use of talus-foot landforms as potential inventories of rockfall and small RSF events. Finally, several wider implications of the research presented in this chapter are recognised for the evolution of glaciated, montane landscapes.

Acknowledgements The author is especially grateful to Paula Hilger, John Matthews and Peter Wilson for kindly supplying photographs, and to Reginald Hermanns for giving valuable feedback on an early draft. The author, however, is solely responsible for the final outcome. Technical support from Mr. Aiden Bygrave is gratefully acknowledged.

References

Aa AR, Sjåstad J, Sønstegaard E, Blikra LH (2007) Chronology of Holocene rock-avalanche deposits based on Schmidt-hammer relative dating and dust stratigraphy in nearby bog deposits, Vora, inner Nordfjord, Norway. The Holocene 17:955–964

Aarseth I, Austbø P, Risnes H (1997) Seismic stratigraphy of Younger Dryas ice-marginal deposits in western Norwegian fjords. Norw J Geol 77:65–86

Abele G (1997) Influence of glacier and climatic variation on rockslide activity in the Alps. Paläoklimaforschung 19:1–6

Agliardi F, Crosta GB, Zanchi A (2001) Structural constrains on deep-seated slope deformations kinematics. Eng Geol 59:83–102

Anda E, Blikra LH, Braathen A (2002) The berill fault—first evidence of neotectonic faulting in southern Norway. Norw J Geol 82:175–182

André M-F (2009) From climatic to global change geomorphology: contemporary shifts in periglacial geomorphology. In: Knight J, Harrison S (eds) Periglacial and paraglacial processes and environments. Geological Society, London, Special Publications 320, pp 5–28

Andreassen LM, Winsvold SH (eds) (2012) Inventory of Norwegian Glaciers. Norwegian Water Resources and Energy Directorate (NVE), Oslo, Rapport 38–2012

Augustinus PC (1995a) Rock mass strength and the stability of some glacial valley slopes. Zeitschrift für Geomorphol 39:55–68

Augustinus PC (1995b) Glacial valley cross-profile development: the influence of in situ rock stress and rock mass strength, with examples from the Southern Alps, New Zealand. Geomorphology 14:87–97

Balco G, Stone JO, Lifton NA, Dunai TJ (2008) A complete and easily accessible means of calculating surface exposure ages or erosion rates from ^{10}Be and ^{26}Al measurements. Quat Geochronol 3:174–195

Ballantyne CK (2002a) A general model of paraglacial landscape response. The Holocene 12:371–376

Ballantyne CK (2002b) Paraglacial geomorphology. Quatern Sci Rev 21:1935–2017

Ballantyne CK (2003) Paraglacial landsystems. In: Evans DJA (ed) glacial landsystems. Edwin Arnold, London, pp 432–461

Ballantyne CK (2008) After the ice: holocene geomorphic activity in the Scottish highlands. Scott Geograph J 124:8–52

Ballantyne CK, Stone JO (2013) Timing and periodicity of paraglacial rock-slope failures in the Scottish Highlands. Geomorphology 186:150–161

Ballantyne CK, Sandeman GF, Stone JO, Wilson P (2014a) Rock-slope failure following Late Pleistocene deglaciation on tectonically stable mountainous terrain. Quatern Sci Rev 86:144–157

Ballantyne CK, Wilson P, Gheorghiu D, Rodés À (2014b) Enhanced rock-slope failure following ice-sheet deglaciation: timing and causes. Earth Surf Proc Land 39:900–913

Bellwald B, Hjelstuen BO, Sejrup HP, Haflidason H (2016) Postglacial mass movements and depositional environments in a high-latitude fjord system—Hardangerfjorden, western Norway. Mar Geol 379:157–175

Bellwald B, Hjelstuen BO, Sejrup HP, Stokowy T, Kuvås J (2019) Holocene mass movements in west and mid-Norwegian fjords and lakes. Mar Geol 407:192–212

Benestad RE (2005) Climate change scenarios for Northern Europe from multi-model IPCC AR4 climate simulations. Geophys Res Lett 32(L17704):1–3

Blair RW (1994) Mountain and valley wall collapse due to rapid deglaciation in Mount Cook National Park, New Zealand. Mt Res Dev 14:347–358

Blikra LH, Nemec W (1993) Postglacial avalanche activity in western Norway: depositional facies sequences, chronostratigraphy and palaeoclimatic implications. Paläoklimaforschung 11:143–162

Blikra LH, Braathen A, Anda E, Stalsberg K, Longva O (2002) Rock avalanches, gravitational bedrock fractures and neotectonic faults onshore northern west Norway: examples, regional distribution and triggering mechanisms. NGU Rapport 2002.016. Geological Survey of Norway, Trondheim, Norway

Blikra LH, Longva O, Braathen A, Anda E, Dehls JF, Stalsberg K (2006) Rock slope failures in Norwegian fjord areas: examples, spatial distribution and temporal pattern. In: Evans SG, Scarawcia Mugnozza G, Strom AL, Hermanns RL (eds) Landslides from massive rock slope failure: proceedings of the NATO advanced research workshop on massive rock slope failure: new models for hazard assessment, Celano, Italy, 16–21 June 2002 (NATO science series: IV Earth and environmental sciences; 49) Springer, Dordrecht, pp 475–496

Blikra LH, Christiansen HH (2014) A field-based model of permafrost-controlled rockslide deformation in northern Norway. Geomorphology 208:34–49

Blikra LH, Majala G, Anda E, Berg H, Eikenæs O, Helgås G, Oppikofer T, Hermanns R, Böhme M (2016) Fare- og risikoklassifisering av ustabile fjellparti—Faresoner, arealhåndtering og tiltak. Rapport no. 77–2016. Norwegian Water Resources and Energy Directorate (NVE), Oslo (in Norwegian)

Bøe A-G, Dahl SO, Lie Ø, Nesje A (2006) Holocene river floods in the upper Glomma catchment, southern Norway: a high-resolution multiproxy record from lacustrine sediments. The Holocene 16:445–455

Bøe R, Lepland A, Blikra LH, Longva O, Sønstegaard E (2002) Postglacial mass movements in western Norway with special emphasis on the 2000–2200 BP and 2800–3200 BP periods—final report. NGU Rapport 2002.020. Geological Survey of Norway, Trondheim, Norway

Bøe R, Rise L, Blikra LH, Longva O, Eide A (2003) Holocene mass movements in Trondheimsfjorden, Central Norway. Norw J Geol 83:3–22

Bøe R, Longva O, Lepland A, Blikra LH, Sønstegaard E, Haflidason H, Bryn P, Lien R (2004) Postglacial mass movements and their causes in fjords and lakes in western Norway. Norw J Geol 84:35–55

Böhme M, Saintot A, Henderson IHC, Henriksen H, Hermanns RL (2011) Rock slope instabilities in Sogn and Fjordane County, Norway: a detailed structural and geomorphological analysis. In: Jaboyedoff M (ed) Slope tectonics. Geological Society, London, Special Publications 351, pp 97–111

Böhme M, Hermanns RL, Fischer L, Oppikofer T, Bunkholt HSS, Derron M-H, Carrea D, Jaboyedoff M, Eiken T (2012) Detailed assessment of the deep-seated gravitational deformation at Stampa above Flåm, Norway. In: Eberhardt E, Froese C, Turner K, Leroueil S (eds) Landslides and engineered slopes. Protecting society through improved understanding: proceedings of the 11th international and 2nd North American symposium on Landslides, Banff, Canada, 3–8 June 2012. Balkema, Rotterdam, pp 647–652

Böhme M, Hermanns RL, Oppikofer T, Fischer L, Bunkholt HSS, Eiken T, Pedrazzini A, Derron M-H, Jaboyedoff M, Blikra LH, Nilsen B (2013) Analyzing complex rock slope deformation at Stampa, western Norway, by integrating geomorphology, kinematics and numerical modelling. Eng Geol 154:116–130

Böhme M, Derron M-H, Jaboyedoff M (2014) Quantitative spatial analysis of rockfalls from road inventories: a combined statistical and physical susceptibility model. Nat Hazards Earth Syst Sci Discuss 2:81–121

Böhme M, Oppikofer T, Longva O, Jaboyedoff M, Hermanns RL, Derron MH (2015) Analyses of past and present rock slope instabilities in a fjord valley: implications for hazard estimations. Geomorphology 248:464–474

Booth AM, Dehls J, Eiken T, Fischer L, Hermanns RL, Oppikofer T (2015) Integrating diverse geologic and geodetic observations to determine failure mechanisms and deformation rates across a large bedrock landslide complex: the Osmundneset landslide, Sogn og Fjordane, Norway. Landslides 12:745–756

Borchers B, Marrero S, Balco G, Caffee M, Goehring B, Lifton N, Nishiizumi K, Phillips F, Schaefer J, Stone J (2016) Geological calibration of spallation production rates in the CRONUS-Earth project. Quat Geochronol 31:188–198

Bovis MJ (1982) Uphill-facing (antislope) scarps in the Coast Mountains, southwest British Columbia. Geol Soc Am Bull 93:804–812

Bovis MJ (1990) Rock-slope deformation at Affliction Creek, southern Coast Mountains, British Columbia. Can J Earth Sci 27:243–254

Braathen A, Blikra LH, Berg SS, Karlsen F (2004) Rock-slope failures in Norway: type, geometry, deformation mechanisms and stability. Norw J Geol 84:67–88

Brierley GJ (2010) Landscape memory: the imprint of the past on contemporary landscape forms and processes. Area 42:76–85

Bungum H, Lindholm C, Faleide J-I (2005) Postglacial seismicity offshore mid-Norway with emphasis on spatio-temporal-magnitudal variations. Mar Pet Geol 22:137–148

Cave JAS, Ballantyne CK (2016) Catastrophic rock-slope failures in NW Scotland: quantitative analysis and implications. Scott Geogr J 132:185–209

Church M (2002) Fluvial sediment transfer in cold regions. In: Hewitt K, Byrne M-L, English M, Young G (eds) Landscapes of transition: landform assemblages and transformations in cold regions. Springer, Dordrecht, pp 93–118

Church M, Ryder JM (1972) Paraglacial sedimentation: a consideration of fluvial processes conditioned by glaciation. Geol Soc Am Bull 83:3059–3071

Church M, Slaymaker O (1989) Disequilibrium of Holocene sediment yield in glaciated British Columbia. Nature 337:452–454

Cossart E, Fort M (2008) Sediment release and storage in early deglaciated areas: towards an application of the exhaustion model from the case of Massif des Écrins (French Alps) since the Little Ice Age. Norsk Geografisk Tidsskrift—Norw J Geogr 62:115–131

Cossart E, Braucher R, Fort M, Bourles D, Carcaillet J (2008) The consequences of glacial debuttressing in deglaciated areas: evidence from field data and cosmogenic datings. Geomorphology 95:3–26

Cossart E, Mercier D, Decaulne A, Feuillet T (2013) An overview of the consequences of paraglacial landsliding on deglaciated mountain slopes: typology, timing and contribution to cascading fluxes. Quaternaire 24:13–24

Cossart E, Mercier D, Decaulne A, Feuillet T, Jónsson HP, Sæmundsson Þ (2014) Impacts of post-glacial rebound on landslide spatial distribution at a regional scale in Northern Iceland (Skagafjörður). Earth Surf Proc Land 39:336–350

Crosta GB, Frattini P, Agliardi F (2013) Deep seated gravitational slope deformations in the European Alps. Tectonophysics 605:13–33

Crosta GB, Hermanns RL, Dehls J, Lari S, Sepulveda S (2017) Rock avalanches clusters along the northern Chile coastal scarp. Geomorphology 289:27–43

Cruden DM, Hu XQ (1993) Exhaustion and steady state models for predicting landslide hazards in the Canadian rocky mountain. Geomorphology 8:279–285

Curry AM, Morris CJ (2004) Lateglacial and Holocene talus slope development and rockwall retreat on Mynydd Du, UK. Geomorphology 58:85–106

Dadson SJ, Church M (2005) Postglacial topographic evolution of glaciated valleys: a stochastic landscape evolution model. Earth Surf Proc Land 30:1387–1403

Dahl S-O, Nesje A, Lie Ø, Fjordheim K, Matthews JA (2002) Timing, equilibrium-line altitudes and climatic implications of two early-Holocene glacial re-advances during the Erdalen Event at Jostedalsbreen, western Norway. The Holocene 12:17–25

Dehls JF, Olesen O, Olsen L, Blikra LH (2000a) Neotectonic faulting in northern Norway; the Stuoragurra and Nordmannvikdalen postglacial faults. Quatern Sci Rev 19:1447–1460

Dehls JF, Olesen O, Bungum H, Hicks EC, Lindholm CD, Riis F (2000b) Neotectonic map: Norway and adjacent areas. Scale 1:3 million. Geological Survey of Norway, Trondheim

Devoli G, Eikenæs O, Taurisano A, Hermanns RL, Fischer L, Oppikofer T, Bunkholt H (2011) Plan for skredfarekartlegging—delrapport steinsprang steinskred og fjellskred. NVE rapport 15/2011. Norges Vassdrags-og Energidirektorat, Oslo, Norway (in Norwegian 120 pp.)

Dowdeswell JA, Ottesen D, Rise L (2010) Rates of sediment delivery from the Fennoscandian Ice Sheet through an ice age. Geology 38:3–6

Eisbacher GH, Clague JJ (1984) Destructive mass movements in high mountains: Hazard and management. Geolog Surv Can Pap 84–16

Etzelmüller B, Romstad B, Fjellanger J (2007) Automatic regional classification of topography in Norway. Norw J Geol 87:167–180

Evans SG, Clague JJ (1994) Recent climatic change and catastrophic geomorphic processes in mountain environments. Geomorphology 10:107–128

Evans SG, Clague JJ, Woodsworth GJ, Hungr O (1989) The Pandemonium Creek rock avalanche, British Columbia. Can Geotech J 26:427–446

Fenton CR, Hermanns RL, Blikra LH, Kubik PW, Bryant C, Niedermann S, Meixner A, Goethals MM (2011) Regional ^{10}Be production rate calibration for the past 12 ka deduced from the radiocarbon-dated Grøtlandsura and Russenes rock avalanches at 69° N, Norway. Quat Geochronol 6:437–452

Fischer L, Kääb A, Huggel C, Noetzli J (2006) Geology, glacier retreat and permafrost degradation as controlling factors of slope instabilities in a high-mountain rock wall: the Monte Rosa east face. Nat Hazards Earth Syst Sci 6:761–772

Fjeldskaar W (1997) Flexural rigidity of Fennoscandia inferred from the postglacial uplift. Tectonics 16:596–608

Fjeldskaar W, Lindholm C, Dehls JF, Fjeldskaar I (2000) Post-glacial uplift, neotectonics and seismicity in Fennoscandia. Quatern Sci Rev 19:1413–1422

Furseth A (2006) Skredulykker i Norge. Tun Forlag, Oslo, Norway (in Norwegian 207 pp.)

Gardner JS (1980) Frequency magnitude and spatial distribution of mountain rockfalls and rockslides in the Highwood Pass area, Alberta, Canada. In: Coates DR, Vitek JD (eds) Thresholds in geomorphology. Allen and Unwin, London, pp 267–295

Gariano SL, Guzzetti F (2016) Landslides in a changing climate. Earth Sci Rev 162:227–252

Gee DG, Sturt BA (eds) (1985) The Caledonide orogen–Scandinavia and related areas, vol 1. Wiley, Chichester

Gisnås K, Etzelmüller B, Farbrot H, Schuler TV, Westermann S (2013) Cryo-GRID 1.0: permafrost distribution in Norway estimated by a spatial numerical model. Permafrost Periglac Process 24:2–19

Gisnås K, Westermann S, Schuler TV, Melvold K, Etzelmüller B (2016) Small-scale variation of snow in a regional permafrost model. The Cryosphere 10:1201–1215

Goering BM, Lohne ØL, Mangerud J, Svendsen JI, Gyllencreutz R, Schaefer J, Finkel R (2012) Late Glacial and Holocene ^{10}Be production rates for western Norway. J Quat Sci 27:89–96

Gjessing J (1967) Norway's paleic surface. Nor Geogr Tidsskr 21:69–132

Grämiger LM, Moore JR, Gischig VS, Ivy-Ochs S, Loew S (2017) Beyond debuttressing: mechanics of paraglacial rock slope damage during repeat glacial cycles. J Geophys Res Earth Surf 122:1004–1036

Grämiger LM, Moore JR, Gischig VS, Loew S (2018) Thermomechanical stresses drive damage of Alpine Valley Rock Walls during repeat glacial cycles. J Geophys Res Earth Surf 123:2620–2646

Grimstad E, Nesdal S (1990) The Loen rockslides—a historical review. In: Barton M, Stephansson W (eds) Rock joints: proceedings of the international symposium on rock joints, Loen, Norway, 4–6 June 1990. Balkema, Rotterdam, pp 3–8

Grove JM (1972) The incidence of landslides, avalanches, and floods in western Norway during the Little Ice Age. Arct Alp Res 4:131–138

Hansen L, Beylich A, Burki V, Eilertsen RS, Fredin O, Larsen E, Lyså E, Nesje A, Stalsberg K, Tønnesen JF (2009) Stratigraphic architecture and infill history of a deglaciated bedrock valley based on georadar, seismic profiling and drilling. Sedimentology 56:1751–1773

Hanssen-Bauer I, Førland EJ, Haddeland I, Hisdal H, Lawrence D, Mayer S, Nesje A, Nilsen JEØ, Sandven S, Sandø AB, Sorteberg A, Ådlandsvik B (2017) Climate in Norway 2100—a knowledge base for climate adaptation. Rapport No. 1/2017. The Norwegian Centre for Climate Services, Oslo

Harbitz CB, Glimsdal S, Løvholt F, Kveldsvik V, Pedersen GK, Jensen A (2014) Rockslide tsunamis in complex fjords: from an unstable rock slope at Åkerneset to tsunami risk in western Norway. Coast Eng 88:101–122

Henderson IHC, Saintot A (2011) Regional spatial variations in rockslide distribution from structural geology ranking: an example from Storfjorden, western Norway. In: Jaboyedoff M (ed) Slope Tectonics. Geological Society, London, Special Publications 351, pp 79–95

Hermanns RL, Longva O (2012) Rapid rock slope failures. In: Clague JJ, Stead D (eds) Landslides: types, mechanisms, and modelling. Cambridge University Press, Cambridge, pp 59–70

Hermanns RL, Blikra LH, Naumann M, Nilsen B, Panthi KK, Stromeyer D, Longva O (2006a) Examples of multiple rock-slope collapses from Köfels (Ötz valley, Austria) and western Norway. Eng Geol 83:94–108

Hermanns RL, Niedermann S, Garcia AV, Schellenberger A (2006b) Rock avalanching in the NW Argentine Andes as a result of complex interactions of lithologic, structural and topographic boundary conditions, climate change and active tectonics. In: Evans SG, Scarawcia Mugnozza G, Strom AL, Hermanns RL (eds) Landslides from Massive Rock Slope Failure: Proceedings of the NATO advanced research workshop on massive rock slope failure: new models for hazard assessment, Celano, Italy, 16–21 June 2002, (NATO Science Series: IV, Earth and Environmental Sciences; 49). Springer, Dordrecht, pp 497–520

Hermanns RL, Blikra LH, Ivy-Ochs S, Kubik P, Naumann R (2009) Geomorphic control on the size of outburst floods: examples from Tafjord, Norway and Patagonia. In: Nakrem HA (ed) Proceedings of Vinterkonferansen, Norsk Geologisk Forening, Bergen, pp 50–51

Hermanns RL, Fischer L, Oppikofer T, Bøhme M, Dehls JF, Henriksen H, Booth A, Eilertsen R, Longva O, Eiken T (2011) Mapping of unstable and potentially unstable rock slopes in Sogn og Fjordane (work report 2008–2010). NGU Rapport 2011.055. Geological Survey of Norway, Trondheim, Norway

Hermanns RL, Redfield T, Bunkholt H, Fischer L, Oppikofer T, Gosse J, Eiken T (2012) Cosmogenic nuclide dating of slow moving rockslides in Norway in order to assess long-term slide velocities. In: Eberhardt E, Froese C, Turner K, Leroueil S (eds) Landslides and engineered slopes. Protecting society through improved understanding: Proceedings of the 11th international and 2nd North American symposium on landslides, Banff, Canada, 3–8 June 2012. Balkema, Rotterdam, pp 3–8

Hermanns RL, Oppikofer T, Anda E, Blikra LH, Böhme M, Bunkholt H, Crosta GB, Dahle H, Devoli G, Fischer L, Jaboyedoff M, Loew S, Sætre S, Yugsi Molina FX (2013a) Hazard and risk classification for large unstable rock slopes in Norway. In: Proceedings of the international conference on Vaiont 1963–2013—thoughts and analyses after 50 years since the Catastrophic Landslide. Italian journal of engineering geology and environment—Book Series 6. Sapienza University, Rome, pp 245–254

Hermanns RL, Oppikofer T, Dahle H, Eiken T, Ivy-Ochs S, Blikra LH (2013b) Understanding long-term slope deformation for stability assessment of rock slopes: the case of the Oppstadhornet rockslide, Norway. In: Proceedings of the international conference on Vaiont, 1963–2013—thoughts and analyses after 50 years since the catastrophic landslide. Italian journal of engineering geology and environment—Book Series 6. Sapienza University, Rome, pp 255–264

Hermanns R, Blikra L, Anda E, Saintot A, Dahle H, Oppikofer T, Fischer L, Bunkholt H, Böhme M, Dehls J, Lauknes T, Redfield T, Osmundsen P, Eiken T (2013c) Systematic mapping of large unstable rock slopes in Norway. In: Margottini C, Canuti P, Sassa K (eds) Landslide science and practice, vol 1. Landslide inventory and susceptibility and hazard zoning. Springer, Heidelberg, pp 29–34

Hermanns RL, Schleier M, Böhme M, Blikra LH, Gosse K, Ivy-Ochs S, Hilger P (2017a) Rock-avalanche activity in W and S Norway peaks after the retreat of the Scandinavian ice sheet. In: Mikoš M, Vilímek V, Yin Y, Sassa K (eds) Advancing Culture of Living with Landslides, vol 5. Landslides in Different Environments. Springer-Verlag, Heidelberg, pp 331–338

Hermanns RL, Böhme M, Oppikofer T, Gosse J, Penna IM (2017b) When does progressive rock-slope failure start and how does it develop in time? an analysis from Norway: a landscape dominated by steep alpine relief excavated by multiple glacial cycles. In: Leith K, Ziegler M, Perras M, Loew S (eds) Progressive rock failure conference, Monte Verità, 5–9 June 2017, extended abstracts. ETH Zurich, Zurich, pp 40–43

Hewitt K (2002) Introduction: landform assemblages and transitions in cold regions. In: Hewitt K, Byrne M-L, English M, Young G (eds) Landscapes of transition: landform assemblages and transformations in cold regions. Springer, Dordrecht, pp 1–8

Hewitt K (2006) Disturbance regime landscapes: mountain drainage systems interrupted by large rockslides. Prog Phys Geogr 30:365–393

Hilger P, Hermanns RL, Gosse JC, Jacobs B, Etzelmüller B, Krautblatter M (2018) Multiple rock-slope failures from Mannen in Romsdal Valley, western Norway, revealed from Quaternary geological mapping and ^{10}Be exposure dating. The Holocene 28:1841–1854

Hinchliffe S, Ballantyne CK (1999) Talus accumulation and rockwall retreat, Trotternish, Isle of Skye Scotland. Scott Geograph J 115:53–70

Hipp T, Etzelmüller B, Westermann S (2014) Permafrost in Alpine Rock faces from Jotunheimen and Hurrungane, Southern Norway. Permafrost Periglac Process 25:1–13

Holm K, Bovis M, Jakob M (2004) The landslide response of alpine basins to post-Little Ice Age glacial thinning and retreat in southwestern British Columbia. Geomorphology 57:201–216

Hughes ALC, Gyllencreutz R, Lohne ØS, Mangerud J, Svendsen JI (2016) The last Eurasian ice sheets—a chronological database and time-slice reconstruction, DATED-1. Boreas 45:1–45

Hungr O, Leroueil S, Picarelli L (2014) The Varnes classification of landslide types, an update. Landslides 11:167–194

Jaboyedoff M, Derron M, Jakubowski J, Oppikofer T, Pedrazzini A (2012) The 2006 Eiger rockslide, European Alps. In: Clague JJ, Stead D (eds) Landslides: types, mechanisms and modeling. Cambridge University Press, Cambridge, pp 282–296

Jarman D (2006) Large rock slope failures in the Highlands of Scotland: characterisation, causes and spatial distribution. Eng Geol 83:161–182

Jarman D (2009) Paraglacial rock slope failure as an agent of glacial trough widening. In: Knight J, Harrison S (eds) Periglacial and paraglacial processes and environments. Geological Society, London, Special Publications 320, pp 102–131

Jarman D, Ballantyne CK (2002) Beinn Fhada, Kintail: a classic example of paraglacial rock-slope deformation. Scott Geograph J 118:59–68

Jarman D, Harrison S (2019) Rock slope failure in the British mountains. Geomorphology 340:202–233

Jibson RW (1996) Using Landslides for Paleoseismic Analysis. In: McCalpin JP (ed) Paleoseismology. Academic Press, San Diego, International Geophysics Series 62, pp 397–438

Johnson PG (1984) Paraglacial conditions of instability and mass movement: a discussion. Zeitschrift für Geomorphologie 28:235–250

Karlen W, Matthews JA (1992) Reconstructing Holocene glacier variations from glacial lake sediments: studies from Nordvestlandet and Jostedalsbreen-Jotunheimen, southern Norway. Geogr Ann 74A:327–348

Keefer DK (1984) Landslides caused by earthquakes. Geol Soc Am Bull 95:406–421

Kirkbride MP, Deline P (2018) Spatial heterogeneity in the paraglacial response to post-Little Ice Age deglaciation of four headwater cirques in the Western Alps. Land Degrad Dev 29:3127–3140

Knapp S, Gilli A, Anselmetti FS, Krautblatter M, Hajdas I (2018) Multistage rock-slope failures revealed in lake sediments in a seismically active Alpine region (Lake Oeschinen, Switzerland). J Geophy Res Earth Surf 123:658–677

Knight J, Harrison S (2014) Mountain glacial and paraglacial environments under global climate change: lessons from the past, future directions and policy implications. Geogr Ann 96A:245–264

Knight J, Harrison S (2018) Transience in cascading paraglacial systems. Land Degrad Dev 29:1991–2001

Kos A, Amann F, Strozzi T, Delaloye R, von Ruette J, Springman S (2016) Contemporary glacier retreat triggers a rapid landslide response, Great Aletsch Glacier, Switzerland. Geophys Res Lett 43 (24):12466–12474

Krautblatter M, Huggel C, Deline P, Hasler A (2012) Research perspectives on unstable high-alpine bedrock permafrost: measurement, modelling and process understanding. Permafrost Periglac Process 23:80–88

Krautblatter M, Funk D, Günzel FK (2013) Why permafrost rocks become unstable: a rock-ice-mechanical model in time and space. Earth Surf Proc Land 38:876–887

Lal D (1991) Cosmic ray labeling of erosion surfaces: in situ nuclide production rates and erosion models. Earth Planet Sci Lett 104:424–439

Laute K, Beylich AA (2013) Holocene hillslope development in glacially formed valley systems in Nordfjord, western Norway. Geomorphology 188:12–30

Leith K, Amann F, Moore JR, Kos A, Loew S (2010a) Conceptual modelling of near-surface extensional fracture in the Matter and Saas Valleys, Switzerland. In: Williams AL, Pinches GM, Chin CY, McMorran TJ, Massey CI (eds) Geologically active: proceedings of the 11th IAEG congress, Auckland, New Zealand. CRC Press, Boca Raton, pp 363e371

Leith K, Moore J, Amann F, Loew S (2010b) Slope failure induced by post-glacial extensional fracturing in the Matter and Saas Valleys, Switzerland. Geophysical Research Abstracts 12 (EGU2010–4599)

Lepland A, Bøe R, Sønstegaard E, Haflidason H, Hovland C, Olsen H, Sandnes R (2002) Sedimentological descriptions and results of analytical tests of sediment cores from fjords and lakes in northwest Western Norway. NGU Rapport 2002.014. Geological Survey of Norway, Trondheim, Norway

Lifton N, Sato T, Dunai TJ (2014) Scaling in situ cosmogenic nuclide production rates using analytical approximations to atmospheric cosmic-ray fluxes. Earth and Planetary Science Letters 386:149–160

Lilleøren K, Etzelmüller B (2011) A regional inventory of rock glaciers and ice-cored moraines in Norway. Geogr Ann 93A:175–191

Lilleøren KS, Etzelmüller B, Schuler TV, Gisnås K, Humlum O (2012) The relative age of permafrost—estimation of Holocene permafrost limits in Norway. Global Planet Change 92–93:209–223

Lohne ØS, Mangerund J, Birks HH (2013) Precise ^{14}C ages of the Vedde and Saksunarvatn ashes and the Younger Dryas boundaries from western Norway and their comparison with the Greenland Ice Core (GICC05) chronology. J Quat Sci 28:490–500

Longva O, Blikra LH, Dehls JF (2009) Rock avalanches—distribution and frequencies in the inner part of Storfjorden, Møre og Romsdal County, Norway. NGU Rapport 2009.002. Geological Survey of Norway, Trondheim, Norway

Mangerud J (2004) Ice sheet limits on Norway and the Norwegian continental shelf. In: Ehlers J, Gibbard PL (eds) Quaternary glaciations—extent and chronology, vol 1. Europe. Elsevier, Amsterdam, pp 271–294

Mangerud J, Gyllencreutz R, Lohne Ø, Svendsen JI (2011) Glacial history of Norway. In: Ehlers J, Gibbard PL, Hughes PD (eds) Developments in quaternary science, vol 15. Elsevier Amsterdam, pp 279–298

Marion J, Filion L, Hétu B (1995) The Holocene development of a debris slope in subarctic Québec, Canada. The Holocene 5:409–419

Marr P, Winkler S, Löffler J (2019) Schmidt-hammer exposure-age dating (SHD) performed on periglacial and related landforms in Opplendskedalen, Geirangerfjellet, Norway: implications for mid- and late-Holocene climate variability. The Holocene 29:97–109

Matthews JA, Wilson P (2015) Improved Schmidt-hammer exposure ages for active and relict pronival ramparts in southern Norway, and their palaeoenvironmental implications. Geomorphology 246:7–21

Matthews JA, Nesje A, Linge H (2013) Relict Talus-Foot Rock Glaciers at Øyberget, Upper Ottadalen, Southern Norway: schmidt hammer exposure ages and palaeoenvironmental implications. Permafrost Periglac Process 24:336–346

Matthews JA, Wilson P, Mourne RW (2017) Landform transitions from pronival ramparts to moraines and rock glaciers: a case study from the Smørbotn cirque, Romsdalsalpane, southern Norway. Geogr Ann 99A:15–37

Matthews JA, Winkler S, Wilson P, Tomkins MD, Dortch JM, Mourne RW, Hill JL, Owen G, Vater AE (2018) Small rock-slope failures conditioned by Holocene permafrost degradation: a new approach and conceptual model based on Schmidt-hammer exposure-age dating, Jotunheimen, southern Norway. Boreas 47:1144–1169

McColl ST (2012) Paraglacial rock-slope stability. Geomorphology 153–154:1–16

McColl ST, Davies TRH (2013) Large ice-contact slope movements: glacial buttressing, deformation and erosion. Earth Surf Proc Land 38:1102–1115

McColl ST, Draebing D (2019) Rock slope instability in the proglacial zone: state of the art. In: Heckman T, Morche D (eds) geomorphology of proglacial systems: landform and sediment. Springer, Cham, pp 119–141

McSaveney MJ (1993) Rock avalanches of 2 May and 6 September 1992, Mount Fletcher, New Zealand. Landslide News 7:2–4

Mercier D, Cossart E, Decaulne A, Feuillet T, Jónsson HP, Sæmundsson Þ (2013) The Höfðahólar rock avalanche (sturzström): chronological constraint of paraglacial landsliding on an Icelandic hillslope. The Holocene 23:432–446

Messenzehl K, Meyer H, Otto J-C, Hoffmann T, Dikau R (2017) Regional-scale controls on the spatial activity of rockfalls (Turtmann Valley, Swiss Alps)—a multivariate modeling approach. Geomorphology 287:29–45

Mörner N-A (2005) An interpretation and catalogue of palaeoseismicity in Sweden. Tectonophysics 408:265–307

Mörner N-A (2013) Patterns in seismology and palaeoseismology and their application in long-term hazard assessments—the Swedish case in view of nuclear waste management. Pattern Recogn Phys 1:75–89

Murton JB, Peterson R, Ozouf J-C (2006) Bedrock fracture by ice segregation in cold regions. Science 314:1127–1129

Nesje A (2002) A large rockfall avalanche in Oldedalen, inner Nordfjord, western Norway, dated by means of a sub-avalanche Salix sp. tree trunk. Norw J Geol 82:59–62

Nesje A (2009) Latest Pleistocene and Holocene alpine glacier fluctuations in Scandinavia. Quatern Sci Rev 28:2119–2136

Nesje A, Kvamme M, Rye N, Løvlie R (1991) Holocene glacial and climate history of the Jostedalsbreen region, western Norway: evidence from lake sediments and terrestrial deposits. Quatern Sci Rev 10:87–114

Nesje A, Blikra LH, Anda E (1994) Dating rockfall-avalanche deposits from degree of rock-surface weathering by Schmidt-hammer tests: a study from Norangsdalen, Sunnmøre, Norway. Norw J Geol 74:108–113

Nesje A, Matthews JA, Dahl SO, Berrisford M, Andersson C (2001) Holocene glacier fluctuations of Flatebreen and winter-preci-pitation changes in the Jostedalsbreen region, western Norway, based on glaciolacustrine sediment records. The Holocene 11:267–280

Nesje A, Bakke J, Dahl SO, Lie Ø, Matthews JA (2008) Norwegian mountain glaciers in the past, present and future. Global Planet Change 60:10–27

NGU (2015) Bedrock—national bedrock database. Geological survey of Norway. http://geo.ngu.no/kart/berggrunn_mobil/ . Accessed 31 July 2020 (in Norwegian)

NVE (2018) NVE Atlas. https://atlas.nve.no/ . Accessed 31 July 2020 (in Norwegian)

Olesen O, Blikra LH, Braathen A, Dehls JF, Olsen L, Rise L, Roberts D, Riis F, Faleide JI, Anda E (2004) Neotectonic deformation in Norway and its implications: a review. Norw J Geol 84:3–34

Olesen O, Bungum H, Dehls J, Lindholm C, Pascal C, Roberts D (2013) Neotectonics seismicity and contemporary stress field in Norway—mechanisms and implications. In: Olsen L, Fredin O, Olesen O (eds) Quaternary geology of Norway. Geological Survey of Norway Special Publication 13, Trondheim, pp 145–174

Oliva M, Mercier D, Ruiz-Fernández J, McColl S (2019) Paraglacial processes in recently deglaciated environments. Land Degrad Dev (in press)

Olsen L, Sveian H, Bergstrøm B, Ottesen D, Rise L (2013) Quaternary glaciations and their variations in Norway and on the Norwegian continental shelf. In: Olsen L, Fredin O, Olesen O (eds) Quaternary geology of Norway. Geological Survey of Norway Special Publication 13, Trondheim, pp 27–78

Oppikofer T, Saintot A, Otterå S, Hermanns RL, Anda E, Dahle H, Eiken T (2013) Investigations on unstable rock slopes in Møre og Romsdal—status and plans after field surveys in 2012. NGU Rapport 2013.014. Geological Survey of Norway, Trondheim, Norway

Oppikofer T, Nordahl B, Bunkholt H, Nicolaisen M, Jarna A, Iversen S, Hermanns RL, Böhme M, Yugsi Molina FX (2015) Database and online map service on unstable rock slopes in Norway—from data perpetuation to public information. Geomorphology 249:69–81

Oppikofer T, Hermanns RL, Sandøy G, Böhme M, Jaboyedoff M, Horton P, Roberts NJ, Fuchs H (2016) Quantification of casualties from potential rock-slope failures in Norway. In: Aversa S, Cascini L, Picarelli L, Scavia C (eds) Landslides and engineered slopes. Experience, theory and practice: proceedings of the 12th international symposium on Landslides (Napoli, Italy, 12–19 June 2016). CRC Press/Balkema, Leiden, pp 1537–1544

Oppikofer T, Saintot A, Hermanns RL, Böhme M, Scheiber T, Gosse J, Dreiås GM (2017) From incipient slope instability through slope deformation to catastrophic failure—different stages of failure development on the Ivasnasen and Vollan rock slopes (western Norway). Geomorphology 289:96–116

Ostermann M, Sanders D (2017) The Benner pass rock avalanche cluster suggests a close relation between long-term slope deformation (DSGSDs and translational rock slides) and catastrophic failure. Geomorphology 289:44–59

Pánek T, Klimeš J (2016) Temporal behavior of deep-seated gravitational slope deformations: a review. Earth-Sci Rev 156:14–38

Peulvast J-P (1985) Relief, érosion différentielle et morphogenèse dans un bourrelet montagneux de haute latitude: Lofoten-vesteralen et Sogn-Jotun (Norvège). Unpublished PhD thesis, University of Paris 1 Panthéon-Sorbonne, Paris, 1642 pp

Phillips JD (2003) Sources of nonlinearity and complexity in geomorphic systems. Prog Phys Geogr 27:1–23
Prager C, Zangerl C, Patzelt G, Brandner R (2008) Age distribution of fossil landslides in the Tyrol (Austria) and its surrounding areas. Nat Hazards Earth Syst Sci 8:377–407
Radbruch-Hall DH (1978) Gravitational creep of rock masses on slopes. In: Voight B (ed) Rockslides and avalanches, vol 1. Natural phenomena. Elsevier, Amsterdam, pp 607–657
Redfield TF, Hermanns RL (2016) Gravitational slope deformation, not neotectonics: revisiting the Nordmannvikdalen feature of northern Norway. Norw J Geol 96:245–273
Reimer PJ, Bard E, Bayliss A et al (2013) IntCal13 and Marine13 radiocarbon age calibration curves 0–50,000 years cal BP. Radiocarbon 55:1869–1887
Rodríguez-Rodríguez L, González-Lemos S, Ballesteros D, Valenzuela P, Domínguez-Cuesta MJ, Llana-Fúnez S, Jiménez-Sánchez M (2018) Timing of paraglacial rock-slope failures and denudation signatures in the Cantabrian Mountains (North Iberian Peninsula). Land Degrad Dev 29:3159–3173
Ryder JM (1971) The stratigraphy and morphology of paraglacial alluvial fans in south-central British Columbia. Can J Earth Sci 8:279–298
Rye N, Nesje A, Lien R, Anda E (1987) The late Weichselian ice sheet in the Nordfjord-Sunnmøre area and deglaciation chronology for Nordfjord, western Norway. Nor Geogr Tidsskr 41:23–43
Saintot A, Henderson IHC, Derron M-H (2011) Inheritance of ductile and brittle structures in the development of large rock slope instabilities: examples from western Norway. In: Jaboyedoff M (ed) Slope tectonics. Geological Society, London, Special Publications 351, pp 27–78
Sandøy G, Oppikofer T, Nilsen B (2017) Why did the 1756 Tjellefonna rockslide occur? A back-analysis of the largest historic rockslide in Norway. Geomorphology 289:78–95
Schleier M, Hermanns RL, Rohn J, Gosse J (2015) Diagnostic characteristics and paleodynamics of supraglacial rock avalanches, Innerdalen, Western Norway. Geomorphology 245:23–39
Schleier M, Hermanns RL, Krieger I, Oppikofer T, Eiken T, Rønning JS, Rohn J (2016) Gravitational reactivation of a pre-existing post-Caledonian fault system: the deep-seated gravitational slope deformation at Middagstinden, western Norway. Norw J Geol 96:201–222
Schleier M, Hermanns RL, Gosse JC, Oppikofer T, Rohn J, Tønnesen AF (2017) Subaqueous rock-avalanche deposits exposed by post-glacial isostatic rebound, Innfjorddalen, Western Norway. Geomorphology 289:117–133
Sigurdsson O, Williams RS (1991) Rockslides on the terminus of 'Jökulsárgilsjökull', southern Iceland. Geogr Ann 73A:129–140
Slaymaker O (2009) Proglacial, periglacial or paraglacial? In: Knight J, Harrison S (eds) Periglacial and paraglacial processes and environments. Geological Society, London, Special Publications 320, pp 71–84
Slaymaker O (2011) Criteria to distinguish between periglacial, proglacial and paraglacial environments. Quaest Geogr 30:85–94
Sollid JL, Sørbel L (1979) Deglaciation of western Central Norway. Boreas 8:233–239
Steiger C, Etzelmüller B, Westermann S, Myhra KS (2016) Modelling the permafrost distribution in steep rock walls in Norway. Norw J Geol 96:329–341
Stewart IS, Sauber J, Rose J (2000) Glacio-seismotectonics: ice sheets, crustal deformation and seismicity. Quatern Sci Rev 19:1367–1389
Stone J (2000) Air pressure and cosmogenic isotope production. J Geophys Res 105:23753–23760
Svendsen JI, Mangerud J (1987) Late Weichselian and Holocene sea-level history for a cross-section of western Norway. J Quat Sci 2:113–132
Tibaldi A, Rovida A, Corazzato C (2004) A giant deep-seated slope deformation in the Italian Alps studied by paleoseismological and morphometric techniques. Geomorphology 58:27–47
Tveten E, Lutro O, Thorsnes T (1998) Geologisk kart over Norge, 1:250,000. Geological Survey of Norway, Trondheim, Norway
Vestøl O (2006) Determination of postglacial land uplift in Fennoscandia from levelling, tide-gauges and continuous GPS stations using least squares collocation. J Geodesy 80:248–258
Wilson P (2009) Rockfall talus slopes and associated talus-foot features in the glaciated uplands of Great Britain and Ireland: periglacial, paraglacial or composite landforms? In: Knight J, Harrison S (eds) Periglacial and paraglacial processes and environments. Geological Society, London, Special Publications 320, pp 133–144
Wilson P (2017) Periglacial and paraglacial processes, landforms and sediments. In: Coxon P, McCarrron S, Mitchell F (eds) Advances in Irish quaternary studies, vol 1. Atlantis Press, Paris, pp 217–254
Winkler S (2021) Terminal moraine formation processes and geomorphology of glacier forelands at selected outlet glaciers of Jostedalsbreen, South Norway. In: Beylich AA (ed) Landscapes and Landforms of Norway. Springer, Dordrecht, pp XXX–XXX
Winsvold SH, Andreassen LM, Kienholz C (2014) Glacier area and length changes in Norway from repeat inventories. The Cryosphere 8:1885–1903
Wyrwoll K-H (1977) Causes of rock-slope failure in a cold area: Labrador-Ungava. Geol Soc Am Rev Eng Geol 3:59–67

6 The Snow-Avalanche Impact Landforms of Vestlandet, Southern Norway

John A. Matthews and Geraint Owen

Abstract

Abundant valley-floor craters, lacustrine craters, stream-bank ramparts and related snow-avalanche impact landforms characterise the landscape of Vestlandet, where topography and snow climate appear to be optimal for their formation. In this chapter, the characteristics of these enigmatic landforms are described and current knowledge and understanding of their formative processes, age and development are reviewed. Near-circular craters, up to 185 m in diameter with rims defined by prominent erosional scars have much in common with meteorite impact craters of similar size, except that they have formed incrementally as a result of frequent avalanches during the Holocene, rather than during single-impact events. Erosional and depositional features can be explained by proximal and distal jets produced by the impact of large volumes of snow at the valley-floor break of slope carrying sedimentary material upwards and outwards from the craters. Air launch of avalanches produces a steeper impact angle, enhancing the energy available for erosion of craters and the uplift of sedimentary material onto depositional mounds and ramparts. Most craters and ramparts in Vestlandet involve snow-avalanche impact on water surfaces, which generates impulse waves that may enhance or modify these landforms.

Keywords

Snow-avalanche impact landforms • Valley-floor craters • Lacustrine craters • Stream-bank ramparts • Meteorite craters • Impulse waves

J. A. Matthews (✉) · G. Owen
Department of Geography, College of Science, Swansea University, Singleton Park, Swansea, SA2 8PP, UK
e-mail: J.A.Matthews@Swansea.ac.uk

6.1 Introduction

Snow avalanches have been widely investigated as important natural hazards in mountain landscapes (Haeberli and Whiteman 2015) where it has been demonstrated that they can be one of the dominant mass movement processes in areas of high relief (Laute and Beylich 2014a, b, c). Nevertheless, the landforms produced by high-velocity snow-avalanche impact constitute a neglected area of geomorphology. The geomorphological role of snow avalanches has been viewed mainly as one of deposition and redeposition of sedimentary material rather than erosional activity (Gardner 1970; Luckman 1977; Hewitt 1989). Recognition of erosion by snow avalanches has been limited mainly to the abrasion of bedrock in chutes close to avalanche source areas (Allix 1924; Rapp 1959) and to relatively minor erosional features on depositional fans (Jomelli and Bertran 2001; Jomelli and Francou 2000; Sanders 2013).

The first descriptions of erosional and depositional landforms produced by high-velocity snow-avalanche impact appear to have been made in North America and Scandinavia (Davis 1962; Liestøl 1974; Corner 1975). Since then, further investigations of similar landforms have been made in these regions (Corner 1980; Hole 1981; Blikra et al. 1989; Matthews and McCarroll 1994; Smith et al. 1994; Owen et al. 2006; Johnson and Smith 2010; Matthews et al. 2015, 2017) and a few examples have been investigated in the Southern Alps of New Zealand (Fitzharris and Owens 1984), the Scottish Highlands (Ballantyne 1989) and the English Lake District (Brown et al. 2011; Evans et al. 2015). With the exception of southern Norway, however, understanding of these enigmatic landforms has been based almost entirely on case studies. This makes the exceptional number and diversity of snow-avalanche impact landforms in Vestlandet, southern Norway, of unique international importance.

In this chapter, we review several aspects of snow-avalanche impact landforms in Vestlandet that are of general significance, including their classification,

A. A. Beylich (ed.), *Landscapes and Landforms of Norway*, World Geomorphological Landscapes,
https://doi.org/10.1007/978-3-030-52563-7_6

morphological and sedimentary characteristics, erosional and depositional processes, and environmental controls. In addition, the dating of snow-avalanche deposits and comparisons with small meteorite impact craters shed light on the magnitude and frequency of the snow-avalanche impact events that result in the development of these unusual landforms. Particular snow-avalanche impact landforms mentioned in the text are located in Fig. 6.1.

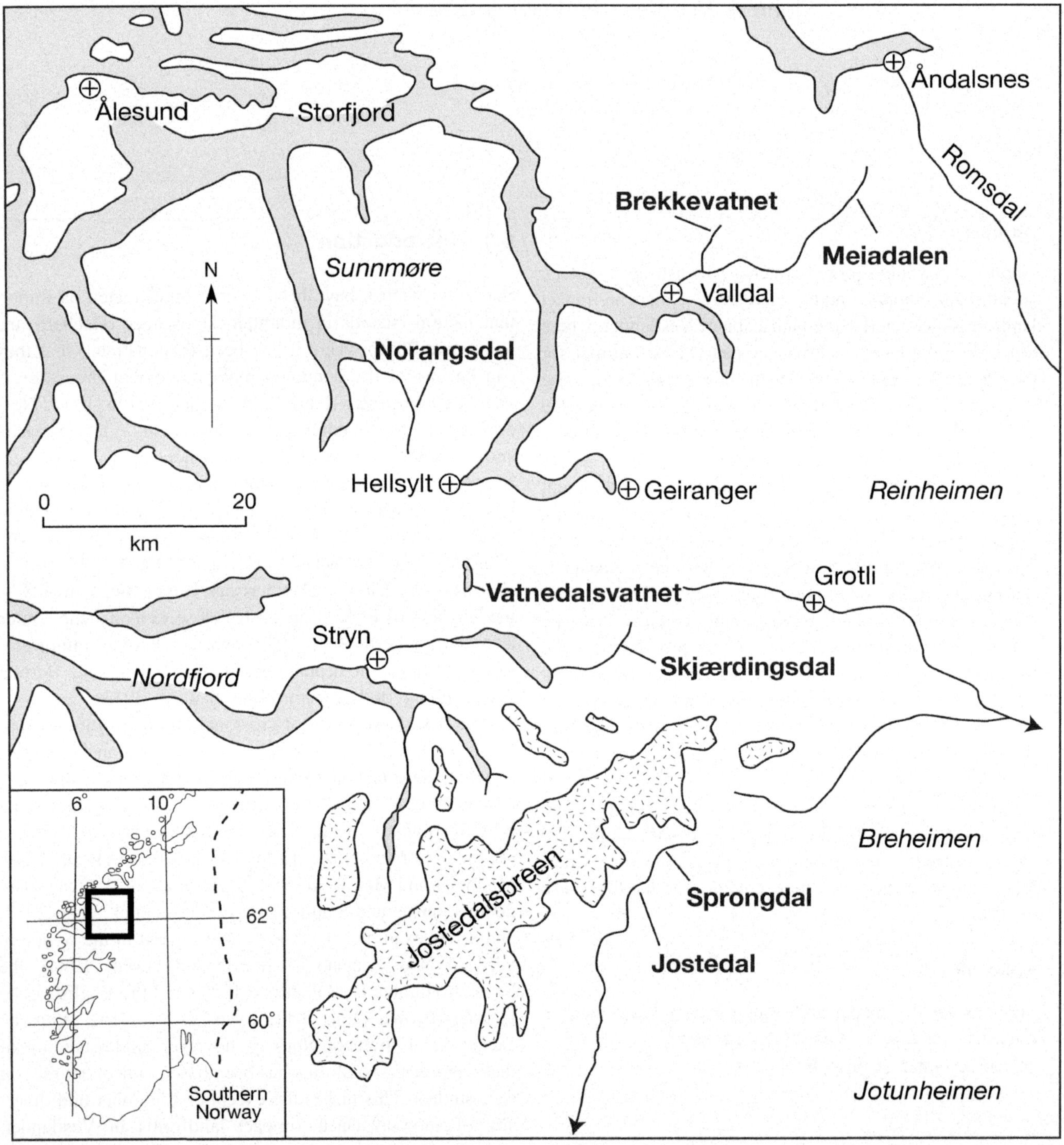

Fig. 6.1 Locations mentioned in the text, Vestlandet, southern Norway

6.2 Classification and Terminology of Snow-Avalanche Impact Landforms

Although transitional forms exist, three distinctive types of snow-avalanche impact landforms can be recognised (Fig. 6.2). Here we use the terms valley-floor craters, lacustrine craters and riverbank ramparts for essentially similar erosional and depositional landforms described elsewhere as snow-avalanche pits, pools and tongues, respectively (cf. Corner 1980; Luckman et al. 1994).

Valley-floor snow-avalanche impact craters are more-or-less circular depressions eroded in sedimentary material often with associated semi-circular distal depositional ridges or mounds forming part of the crater rim. An impressive example from Meiadalen is shown in Fig. 6.3. The craters tend to be water filled due to the prevailing climate.

Lacustrine snow-avalanche impact craters are similar but eroded in lake-floor sediments close to lake shorelines (e.g. Fig. 6.4a from Vatnedalsvatnet). The semi-circular distal depositional mounds or ridges are largely, if not wholly, submerged below lake level and enclose 'pools' of relatively deep water.

Riverbank snow-avalanche impact ramparts are elongated mounds or ridges of sedimentary material deposited on and largely parallel to the distal bank of rivers (e.g. Fig. 6.4b from Jostedalen). The mounds include material eroded from riverbeds as well as material derived directly from the avalanches. Avalanche impact also results in crater-like erosional widening of river channels and banks where they are composed of sedimentary material rather than bedrock.

Additional terms used to describe and measure these landforms, and their associated avalanche paths, are defined in this chapter as follows (see also Fig. 6.5):

- Crater diameter (D) = the diameter of the crater rim (minimum diameter for non-circular craters).
- Crater wall height (w) = maximum height of the crater wall above water level; normally the height of the proximal erosional scar (w_p), rather than the height of the distal scar (w_d).

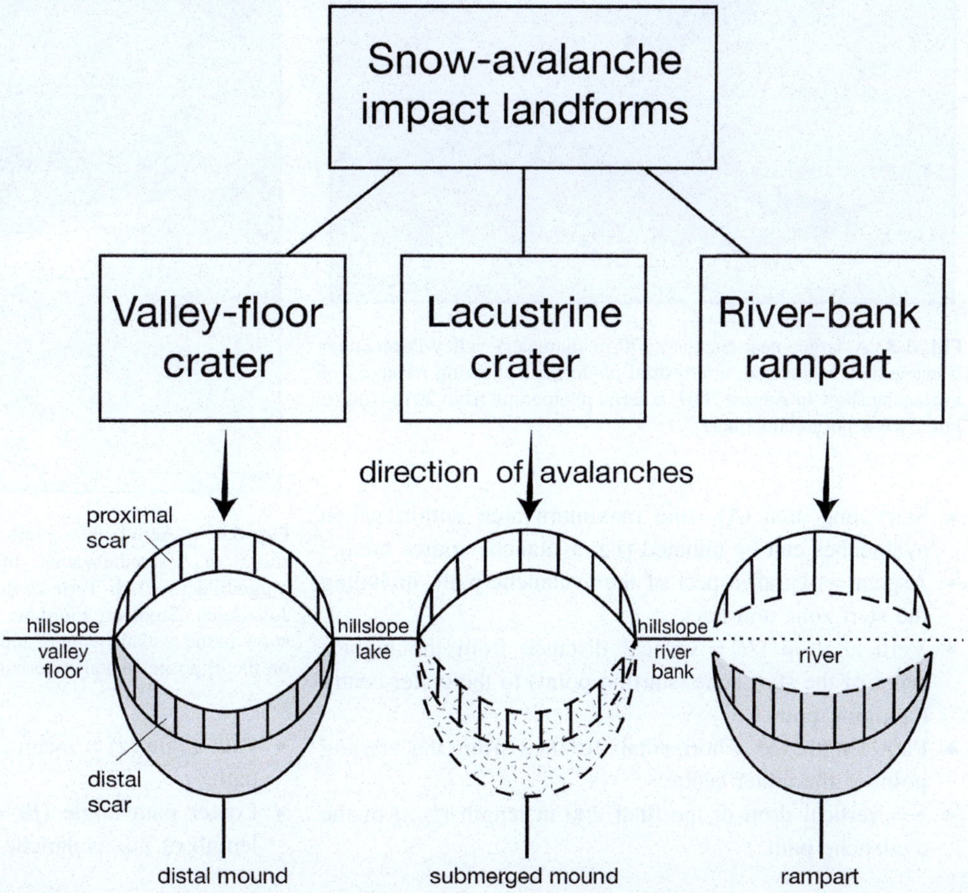

Fig. 6.2 Three main types of snow-avalanche impact landforms: their classification and interrelationships

Fig. 6.3 A large, near-circular (170-m diameter) valley-floor crater (Ørstevatnet, Meiadalen: **a** terrestrial photograph showing remnants of avalanche snow in August 2007; **b** aerial photograph from 2014 (source http://www.norgeibilder.no/)

Fig. 6.4 **a** Aerial photograph of lacustrine craters close to the SW shore of Vatnedalsvatnet in August 2010 (source http://www.norgeibilder.no/). **b** Terrestrial photograph of a riverbank rampart in Jostedalen, Sogn og Fjordane in 1991. Note remnants of avalanche snow in the avalanche track and the boulder composition of the rampart on the opposite (distal) riverbank

- Start-zone area (A) = the maximum area within which avalanches can be initiated (i.e. avalanche source area).
- Aspect = overall aspect of the avalanche path, including the start zone and track.
- Vertical drop (H) = vertical distance from the highest point of the start zone (starting point) to the crater centre (stopping point).
- Path length (L) = horizontal distance from the starting point to the crater centre.
- h = vertical drop of the final 200 m length (l_{200}) of the avalanche path.
- Path angle (α) = mean slope angle of the entire avalanche path.
- Lower path angle (β) = mean slope of the final 200 m length of the avalanche path.

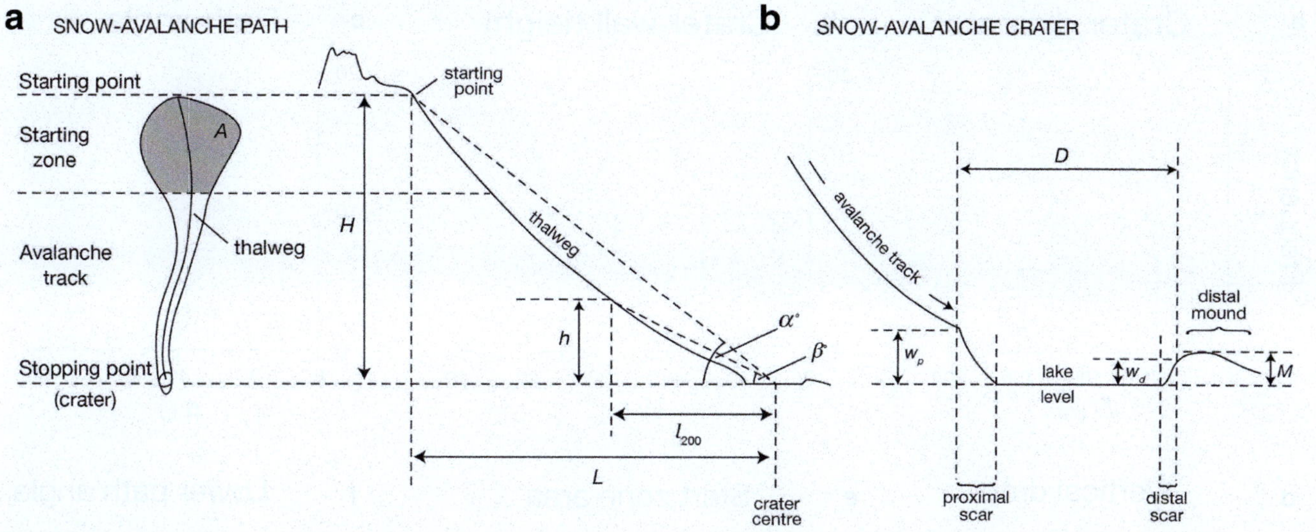

Fig. 6.5 **a** Plan and cross-section of a typical snow-avalanche path and crater. **b** Enlarged cross-section of the crater. See text for explanation of terms and symbols

6.3 Morphology and Sedimentology of Snow-Avalanche Impact Landforms

There are fundamental similarities in the morphology of the three types of snow-avalanche impact landforms but there is considerable variability in their size and shape. The diameters (*D*) of 52 valley-floor and lacustrine craters in Vestlandet (Matthews et al. 2017) range from 10 to 185 m with an average of 85 m (Fig. 6.6a). Crater wall height (*w*) ranges from 1 to 40 m with an average of 13 m (Fig. 6.6b) and is normally represented by the height of the proximal scar (w_p), which is eroded into the hillslope. Proximal scars are often impressive features due to their scale and instability where craters are highly active, as in Norangsdalen (Fig. 6.7a, b). However, crater wall height is only a minimum estimate of crater depth as it does not take account of the depth of the crater floor below water level, which has been measured for two of the largest craters at 8.2 m (the Meiadalen crater shown in Fig. 6.3; Matthews et al. 2011) and 11.4 m (the crater at Urdvatnet in Norangsdalen shown in Fig. 6.7b; Owen et al. 2006).

Distal mounds and ridges associated with craters are often relatively low (Figs. 6.3 and 6.7) and, in the case of lacustrine craters, commonly submerged (e.g. Fig. 6.4a), but some are more impressive, attaining heights of 12 m and 8 m at valley-floor craters in Skjærdingsdalen (Fig. 6.8a) and Brekkevatnet (Fig. 6.8b), respectively. Distal crater scars commonly occur on the proximal side of the larger distal mounds but they are usually smaller than the proximal crater scars, which were termed 'blast zones' by Matthews et al. (2017).

Mounds and riverbank ramparts tend to be asymmetrical in cross profile with relatively steep proximal slopes that reflect impact processes (see below). Matthews and McCarroll (1994) showed that 12 ramparts in Sprongdalen and Jostedalen (Luster, Sogn og Fjordane) had proximal slope angles of 18–42°, which were consistently steeper than the distal slope angles, on average by 17°. Several minor features of distal mounds are exemplified by the valley-floor crater near Urdvatnet, (Figs. 6.7b and 6.9). These include gravel ridges, boulder lines, beaded ridges, fine sediment banked against and covering large boulders, and gravel clumps (Owen et al. 2006), all associated with a debris spread blanketing and extending beyond the distal mound.

The sedimentary characteristics of snow-avalanche impact deposits, like other colluvial deposits, depend largely on the source material. These deposits consist of a component derived from material transported by the snow avalanche prior to impact and a component representing material excavated on impact. Both components of snow-avalanche impact deposits can therefore vary considerably in their sedimentary characteristics, though coarse, angular material originating from frost weathering in the avalanche source areas would be expected to dominate the former component. Matthews and McCarroll (1994) carried out a detailed investigation of clast roundness of the boulder ramparts in Sprongdalen and Jostedalen, and included comparisons with clasts in the adjacent avalanche tracks and river channels. Avalanche-track clasts were characterised by very angular-to-angular modal classes according to the roundness chart of Powers (1953), while the modal class of the river clasts ranged from sub-angular to rounded, and the clasts of the ramparts were intermediate in character. From

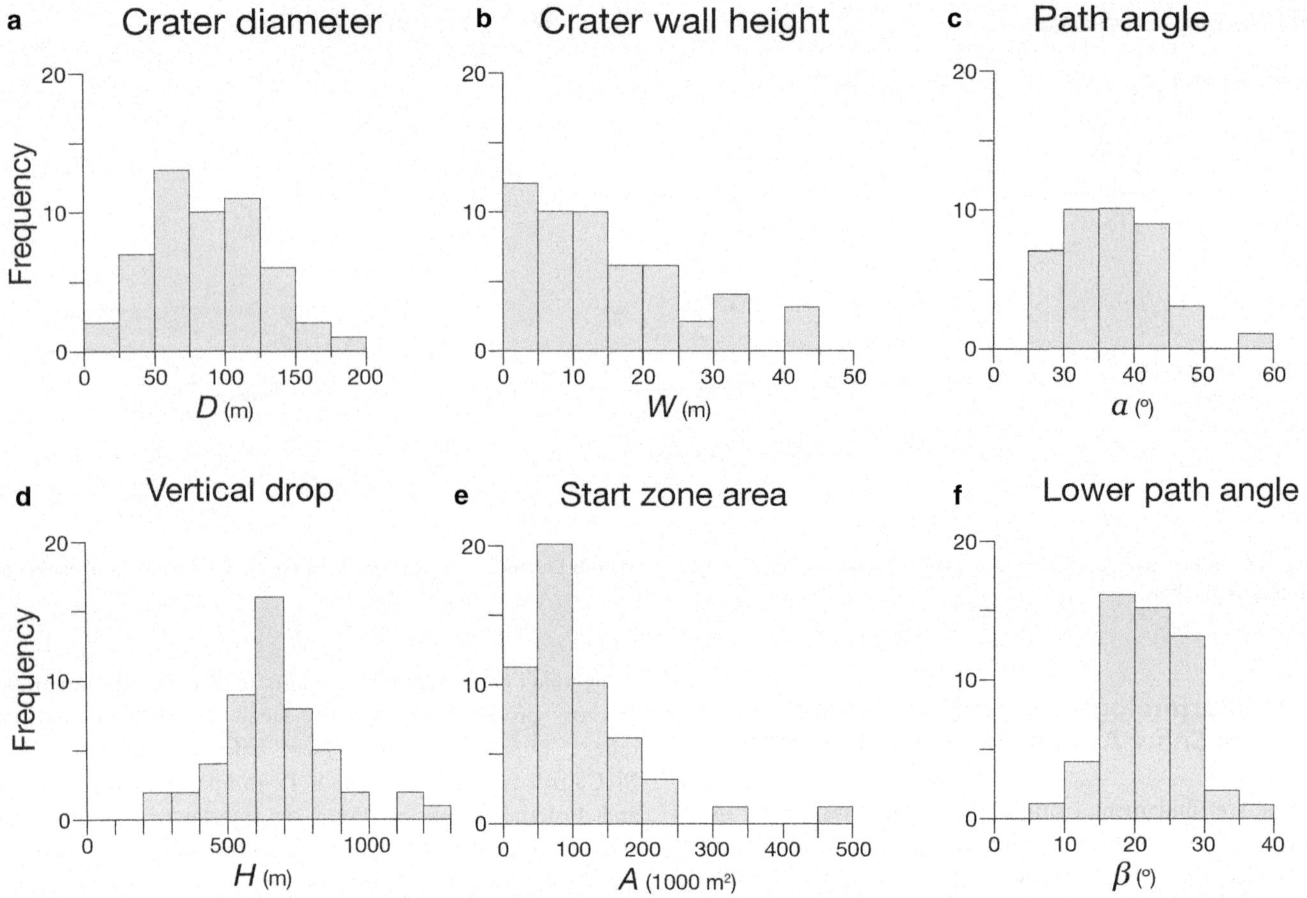

Fig. 6.6 Characteristics of 52 valley-floor and lacustrine craters in Vestlandet: **a** crater diameter; **b** crater wall height; **c** path angle; **d** vertical drop; **e** start-zone area and **f** lower path angle

the different proportions of clasts in the various roundness categories, Matthews and McCarroll (1994) showed that some ramparts were dominated by clasts originating from the river bedload, whereas others were composed largely of clasts transported from avalanche source areas and tracks. On average, 50–60% of the clasts in the ramparts originated from excavation of the river bedload.

6.4 Characteristics of the Snow-Avalanche Paths

In Vestlandet, craters invariably occur at the foot of a steep avalanche path (including the avalanche source area and track) with a path angle (α) of 28–59° (average 35.6°), a vertical drop (H) of 200–1300 m (average 670 m) and a source area (A) of 20,000–470,000 m^2 (average 110,000 m^2) (Fig. 6.6c–e). However, Matthews et al. (2017) found no relationship between crater diameter or crater wall height and path angle or vertical drop. In contrast, crater diameter and wall height (variables that are strongly inter-correlated at $r > 0.80$, $p < 0.01$, and reflect crater size) are significantly correlated ($r = 0.25$–0.31; $p < 0.05$) with avalanche source area. These data indicate that the source area of the avalanche (and hence the volume of snow available for avalanching) is more important than the morphology of the avalanche track as a determinant of crater size.

In addition, crater wall height (but not crater diameter) is significantly correlated ($r = 0.34$; $p < 0.02$) with the lower path angle (β) defined by Matthews et al. (2017) as the slope angle of the final 200 m length of the avalanche path. Average lower path angle for Vestlandet is 22° with a range of 9–37° (Fig. 6.6f). It can therefore be concluded that, for craters of similar size, the size of proximal scars is related to the lower path angle, and that large proximal scars are produced where a snow avalanche erodes the foot of a sufficiently steep slope.

Topographic focusing of snow avalanches accounts for the location and near-circular shape of craters. Such focusing begins in the avalanche source areas with concave cross-slope profiles (Gleason 1995; McClung 2001). Further focusing then occurs in the avalanche-track zone, where a distinct channel may be eroded. Thus, large volumes of snow are repeatedly transported towards approximately the

Fig. 6.7 Prominent proximal scars defining crater rims in Norangsdalen: **a** the 15-m high scar of a crater in Gailskredvatnet in July 1999; **b** the 40-m high scar of a crater at Urdvatnet in July 1999

Fig. 6.8 Relatively large distal mounds associated with **a** a valley-floor crater in Skjærdingsdalen in August 2011 and **b** a lacustrine crater in Brekkevatnet in July 2010. Note people standing on the crest of both mounds, remnant avalanche snow banked against both mounds and the fresh debris deposited on mound (**b**)

same impact point at the foot of the avalanche path. The maximum diameter of craters in Vestlandet with an elongated shape is invariably aligned at right angles to the avalanche path, which can be accounted for by variations in the precise route taken by successive avalanches. The central position of the highest points of snow-avalanche impact deposits, which taper away to either side, can be similarly attributed to the most frequented avalanche route.

The path angles (α) of snow avalanches that form craters are similar to those of non-crater-forming avalanches. However, the lower path angles (β) of the former are significantly greater than those of the latter. Slope angles in the source areas of all types of snow avalanches are normally 30–50°, occasionally 60° (Perla 1977; Schweizer and Jamieson 2001 Schweizer et al. 2003; Pudasaini and Hutter 2007). In non-crater-forming snow avalanches, the lowermost part of the avalanche path generally consists of a run-out zone, which extends for distances of up to ~500 m on a slope as low as 5° (McClung et al. 1989; Perla and Martinelli 2004). Typical slope angles of run-out zones are 15° or less and an angle of <10° tends to be used to define the beginning of such zones (McClung and Schaerer 2006), the associated landforms of which are typically low-angle depositional fans. In contrast, β at 52 craters investigated by Matthews et al. (2017) was <15° in five cases only, of which the lowest value was 9°. Thus, typical craters form in the absence of a run-out zone, where snow avalanches impact valley- (or lake-) floor sediments at a relatively steep angle. The crucial difference between crater-forming and non-crater-forming snow avalanches therefore appears to be the angle of impact of the avalanche in relation to the break of slope at the slope foot, as reflected in the lower path angle.

6.5 Snow-Avalanche Impact Processes: The Dynamics of Erosion and Deposition

Processes associated with snow-avalanche impact craters and other snow-avalanche impact landforms have much in common with craters formed by meteorite impact (Melosh 1996; Collins et al. 2012; Osinski and Pierazzo 2013) and

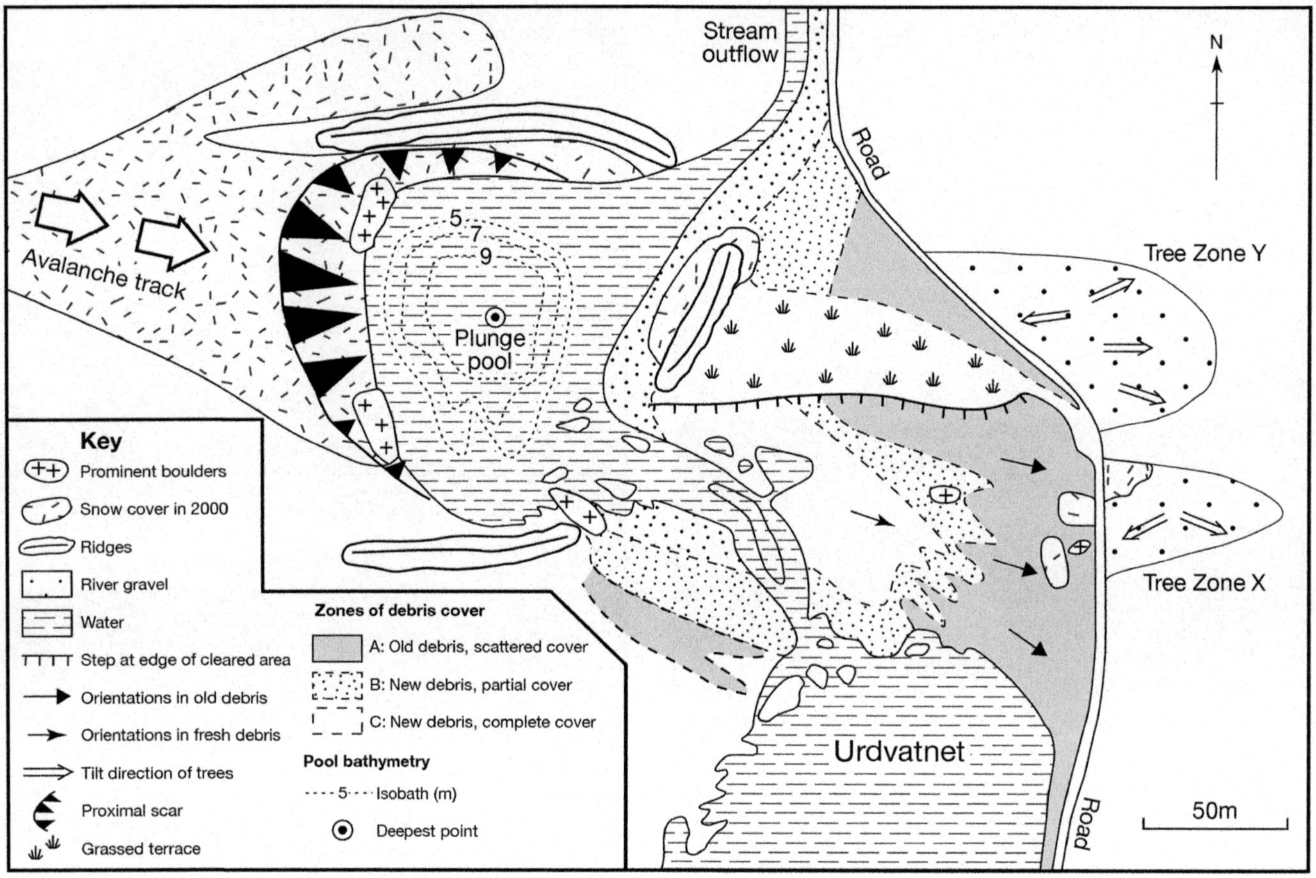

Fig. 6.9 Sketch map of a valley-floor crater and associated features at Urdvatnet, Norangsdaklen (based on Owen et al. 2006)

explosions (Roddy et al. 1977) and may be discussed in terms of erosion, transport and deposition. The erosion or excavation process involves the sudden contact of moving material (projectile or impactor) with the ground surface (target), whereby a shock wave initiated by impact pressure penetrates the ground surface. This sets material in motion leading to an upward and outward transport of eroded material (the excavation flow and impact plume). In all but the most oblique meteorite impacts, the excavated material is directed radially away from the impact point leaving a crater that is hemispherical, rather than ellipsoidal, because of the explosive behaviour of the impactor. Most meteorite craters are circular because 90% of meteorites impact with the Earth's surface at >15° (Melosh 1996; Pierazzo and Melosh 2000; Davison et al. 2011). This fits well as an analogue for near-circular snow-avalanche impact craters and related landforms since the lower path angle exceeds 15° for about 90% of sampled snow-avalanche impact craters. However, the erosion of slope-foot sedimentary deposits is not the only potential source of the deposited material. Owen et al. (2006) recognised additional sources at Urdvatnet, Norangsdalen, including direct transport by avalanches from the upper valley sides and the redistribution of debris from the lower valley-side slopes.

For meteorite impact, ejecta tend to be evenly concentrated in downrange and uprange jets (with most material directed downrange) if impact angles are in the range 60–45° but at impact angles of 45–30° the uprange jet disappears and at 20–10° the downrange jet disappears (Osinski et al. 2011, 2013). These features of meteorite impact are consistent with impact angles inferred from the lower path angles of crater-forming snow avalanches in Vestlandet. They also provide at least a partial explanation for the downrange deposition of material in distal mounds and ramparts and the uprange erosion of proximal scars favoured by relatively steep impact angles. Impact angle may be enhanced by the air launch of snow avalanches caused by downslope convexities in the avalanche path (Schweizer et al. 2003). Where air launch affects the lower avalanche path, the impact angle can be steepened considerably, which would be consistent with the formation of uprange jets and the erosion of proximal scars.

Matthews et al. (2017) also recognise the phenomenon of secondary craters which, in the context of snow-avalanche

Fig. 6.10 Imbricated boulder line indicative of basal shear during deposition by snow avalanche at Urdvatnet, Norangsdalen. Direction of flow from right to left is indicated by the scale bar (length 15 cm)

cratering, are attributed to individual boulders and/or snow packets thrown out from the main body of flowing snow. These may be only a few metres in diameter and are analogous to the 'impact holes', 'dug craters', 'penetration craters' or 'penetration funnels' described from some meteorite crater fields (Krinov 1960, 1963; Elston and Scott 1971; Hodge 1994; Wright et al. 2007; Osinski and Pierazzo 2013).

In relation to deposition by snow avalanches, Owen et al. (2006) recognised three possible emplacement mechanisms: (1) basal shear, (2) fall-out and (3) melt-out. High shear stresses at the base of an avalanche may transport debris as bedload (Jomelli and Bertran 2001) producing strongly oriented longitudinal bedforms including gravel ridges, imbricated boulder lines (Fig. 6.10), gravel clusters and sediment tails in the lee of large boulders. Coarser debris carried in suspension falls through the moving avalanche producing less prominently oriented forms, which may reflect lack of grain-to-grain interaction (Jomelli and Bertran 2001). Large clasts supported in a snow matrix melt-out to produce forms with no particular orientation (randomly distributed conical gravel clumps, precariously perched blocks and the absence of fine sediment piled against boulder flanks).

In Vestlandet, most snow-avalanche impact craters and ramparts involve avalanching into water-filled craters or rivers, which are likely to be ice-covered when most avalanches occur. This is due to the combination of topographic factors, prevailing climate and the spring avalanche season (see below). Although water appears not to be necessary for the formation of snow-impact landforms, it should be regarded as an enhancing or moderating factor (Matthews et al. 2017). Snow-avalanche impact onto a water body generates an initial hydrodynamic crater, the collapse of which produces impulse waves with possible seiche effects within the confined space defined by the crater walls (Fritz et al. 2003; Heller et al. 2009; Zitti et al. 2016). In particular, collapse of the hydrodynamic crater causes a backward-moving (uprange) impulse wave to entrain an air cavity against the proximal crater wall. The collapse of this air cavity and its rebound produces an explosive, near-vertical jet, which is likely to contribute to the erosion of proximal scars (aptly termed blast zones in this context). Meanwhile, a less explosive, forward-moving (downrange) impulse wave is likely to overtop and/or deposit sediment on distal mounds and/or erode distal scars.

6.6 Distribution and Environmental Controls on Snow-Avalanche Impact Landforms

Snow-avalanche activity involves the interaction of topography, snowpack and meteorological conditions (Schweizer et al. 2003; Eckerstorfer and Christiansen 2011) and most avalanches take place after heavy snowfalls (Armstrong and Williams 1992; Pudasaini and Hutter 2007). The oversteepened valley-side slopes and flat valley floors of the glacial troughs in Vestlandet, combined with a mean annual snow accumulation of >4 m throughout the region in the avalanche source areas (http://www.senorge.no/?p=klima), are particularly conducive to snow-avalanche impact landforms. Remarkable concentrations of these landforms can be found in particular valleys or associated with particular lakes in many parts of Møre og Romsdal and Sogn og Fjordane fylker

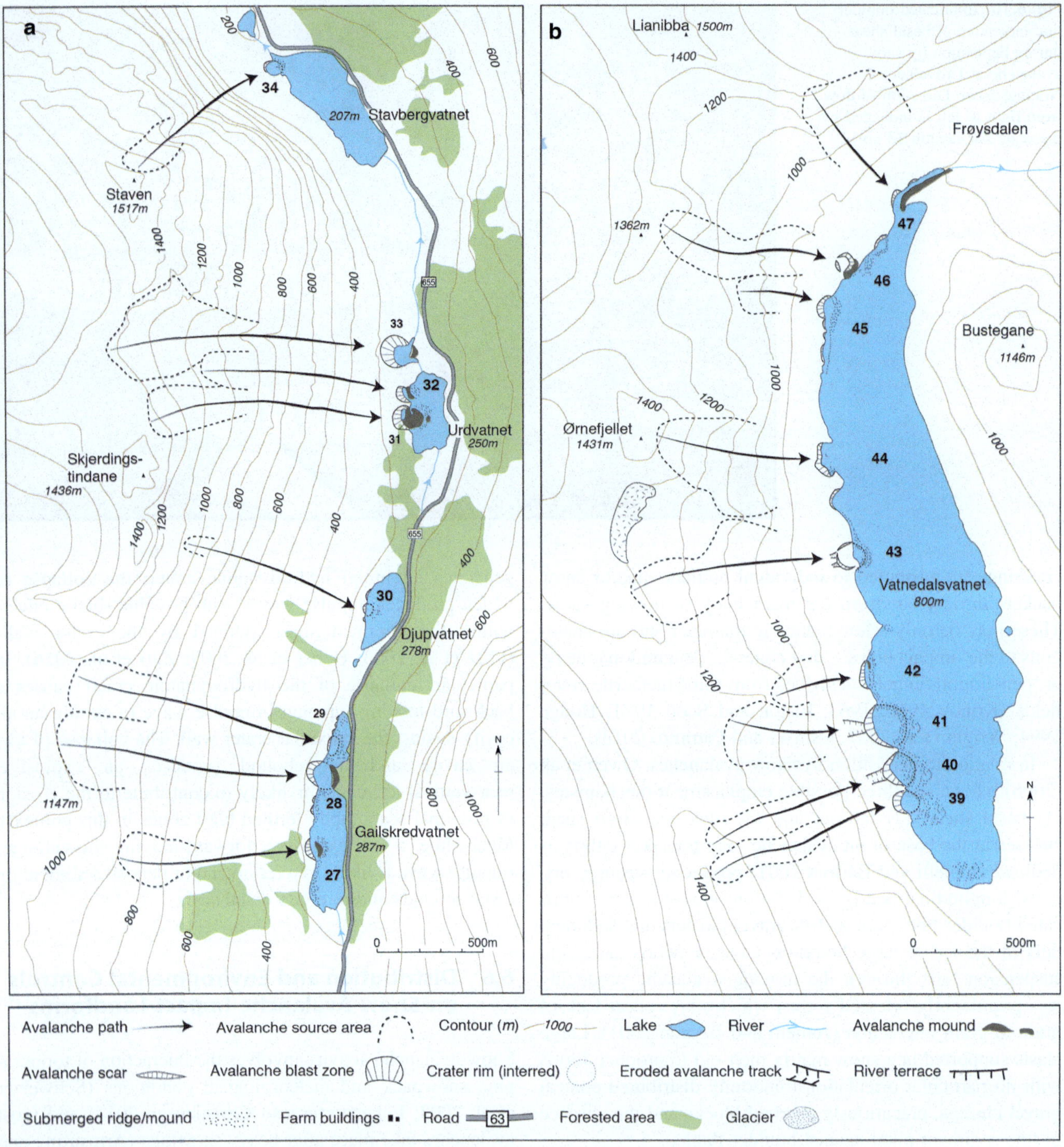

Fig. 6.11 High concentrations of snow-avalanche impact craters in **a** Norangsdalen and **b** Vatnedaldsvatnet. Note the preferred easterly aspect of the avalanche paths

(counties), such as Norangsdalen and Vatnedalsvatnet (Fig. 6.11; see also other examples in Matthews et al. 2017). Farther east, where snowfall is greatly reduced and slopes are generally less steep, snow-avalanche impact landforms are rarely encountered, though individual examples occur where local conditions are favourable. For example, in Jotunheimen, only three sites with such landforms are known to the authors.

There is no simple relationship between temperature and the triggering of snow avalanches: complex relationships between meteorological elements over a period of days to weeks are involved, and these relationships may alter as climate changes (e.g. Baggi and Schweizer 2009; Lazar and Williams 2008; Castebrunet et al. 2012; Fouinat et al. 2018). The main meteorological controls on snow-avalanche timing and frequency within Vestlandet were shown by Laute and Beylich (2014a, b) to be snowfall intensity, intervals with strong winds leading to snow drifting and rapid increases in air temperature, all within the peak avalanche season of March to May. It can be assumed that similar meteorological conditions pertain in the valleys where snow-avalanche craters and ramparts have been investigated.

Although trees may hinder the triggering of avalanches, decelerate avalanches and limit avalanche run-out distances of small- to medium-sized snow avalanches (Teich 2013), all of the avalanche source areas and many of the avalanche paths associated with the craters and ramparts investigated in Vestlandet are located in the alpine zone, and hence are little affected by trees. Furthermore, tree growth in the avalanche path at lower altitudes is likely to have little or no effect on the high-magnitude snow-avalanche events responsible for the snow-impact landforms discussed in this chapter.

6.7 Dating of Snow-Avalanche Impact Landforms: Age and Activity

Snow-avalanche impact ramparts in Sprongdalen and Jostedalen have been dated using lichenometry (Matthews and McCarroll 1994) and Schmidt-hammer exposure-age dating (SHD, Matthews et al. 2015). It was possible to date these particular ramparts because they are composed largely of boulders which are conducive to the use of both of these exposure-age techniques. In many other cases, ramparts and crater mounds are composed of sand and gravel with insufficient boulders for dating. Although radiocarbon dating and dendrochronology have been used to date particular snow-avalanche impact landforms in the Rocky Mountains (Smith et al. 1994), suitable material for dating using these alternative techniques has yet to be found in southern Norway.

Conventional lichenometry using the mean size of the five largest thalli of *Rhizocarpon* subgenus *Rhizocarpon* and the lichenometric dating curves of Erikstad and Sollid (1986) and Bickerton and Matthews (1992) yielded lichenometric ages of 250–1990 years for 12 ramparts. Schmidt-hammer exposure-age dating of five of the same ramparts using mean R-values derived from 800 boulders per rampart (one impact per boulder) yielded SHD_{mean} ages of 1030 ± 260 to 3225 ± 320 years (Matthews et al. 2015). In general, SHD has the potential to date older landforms than lichenometric dating, the more limited temporal range of which can nevertheless be effective as an indicator of the extent and timing of relatively recent snow-avalanche activity.

As these results are estimates of age derived from boulders on the surface of extensive landforms that have developed over a considerable period of time, they require careful interpretation. Both sets of results reflect boulder surface age, rather than landform age. Lichenometric age may be interpreted as an estimate of the age of the oldest surface boulders, whereas SHD_{mean} age estimates the average age of the surface boulders. Despite the age discrepancy between the two techniques, the above results are consistent with rampart formation over at least the last 2000 and 3000 years, respectively, and continuing to the present day.

Matthews et al. (2015) extended their SHD dating to differences in age between the crests and distal fringes of ramparts, and also attempted to obtain more realistic estimates of the length of time over which rampart formation occurred (Fig. 6.12). The SHD_{mean} age of rampart crests tends to be younger than the distal fringes, which yielded ages of up to 5375 ± 965 years. Furthermore, use of SHD_{max}, which estimates the maximum age of surface boulders (defined as the age exceeded by the oldest 5% of surface boulders) yielded ages of up to 9065 years for the distal fringes. This result strongly suggests not only that relatively old evidence can be preserved, unburied by later snow-avalanche activity, in the distal fringes of ramparts, but also that the ramparts have been forming as a result of snow-avalanche activity throughout the Holocene.

Two further approaches involving lichenometry were developed by McCarroll (1993, 1994) based on lichen-size frequency distributions obtained from the 100 largest lichens (each measured lichen from a separate boulder). These were used by Matthews and McCarroll (1994) to explore further the extent of relatively recent activity on the diachronous surfaces of ramparts. First, lichen-size frequency distributions indicated minor modes at 20–105 mm that were calibrated to c. 45, 95, 165, 205, 275 and 360 years ago (Matthews and McCarroll 1994) indicating variations in snow-avalanche frequency over the last few centuries. Five other modes in the distributions were identified at 110–180 mm but could not be calibrated. They nevertheless suggest older phases of relatively high avalanche activity when boulders were added to the surface of ramparts.

The second lichenometric approach involved comparing the lichen-size cumulative percentage frequency curves from the ramparts with the results of two simulation models reflecting contrasting temporal patterns of snow-avalanche activity (Fig. 6.13). One simulated distribution was based on a record of historically dated avalanche damage to farms in valleys to the west of the Jostedalsbreen ice cap, where maximum snow-avalanche activity in the eighteenth century correlates with glaciers advancing towards their maximum extent during the coldest phase of the Little Ice Age (Grove 1972, 1988; cf. Matthews and Briffa 2005). The other

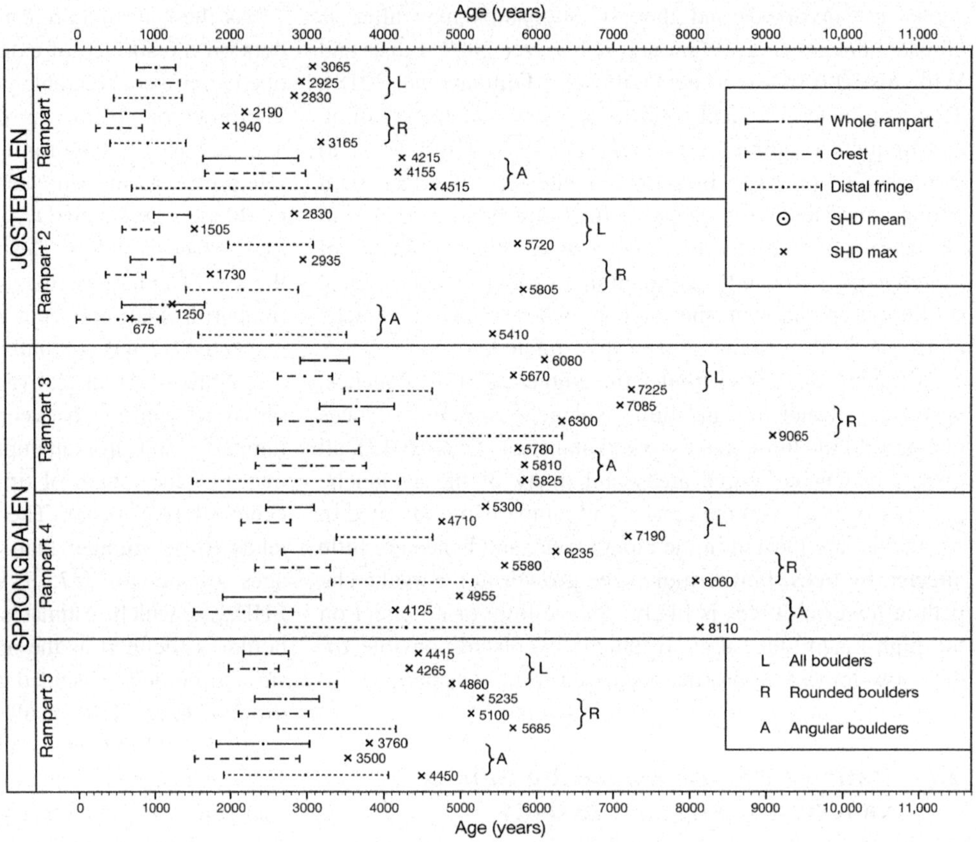

Fig. 6.12 Riverbank ramparts in Jostedalen and Sprongdalen, dated by Schmidt-hammer exposure-age dating (SHD) (based on Matthews et al. 2015)

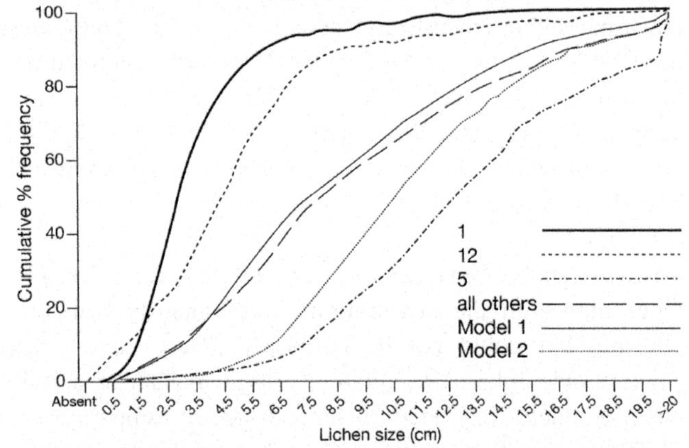

Fig. 6.13 Cumulative lichen-size frequency distributions for 12 riverbank ramparts (numbered 1, 5, 12 and all others) and two models (labelled 1 and 2) (based on Matthews and McCarroll 1994). See text for further explanation

simulation was based on a record of lichenometrically dated debris-flow activity in Austerdalen, to the southeast of Jostedalsbreen (Innes 1985), where a nineteenth century maximum in activity is more closely related to precipitation variations than to temperature. Further details of the simulation modelling are described in McCarroll (1993). Matthews and McCarroll (1994) concluded that snow-avalanche activity associated with nine of the twelve ramparts can be explained by Model 1 (Fig. 6.13). In other words, the temporal pattern of rampart-forming avalanches appears more closely related to the precipitation-dependent debris-flow record than the more temperature-dependent glacier record (cf. Matthews et al. 2009).

6.8 Magnitude and Frequency of Snow-Avalanche Impacts in Landform Development

Relatively little is known about the magnitude and frequency of the snow-avalanche events that produce snow-avalanche impact landforms. They have rarely been observed, although Liestøl (1974) reported seeing a boulder of about 40 cm in diameter, together with snow and water being thrown fan-like onto a riverbank rampart from the spot where an avalanche hit the river. They are formed by relatively large snow avalanches with sufficient energy to erode craters and transport large boulders over considerable heights and distances. Even the largest events appear to result only in scattered, fresh lichen-free boulders on otherwise lichen-covered surfaces, which is consistent with the development of snow-avalanche impact landforms from infrequent large impact events of sufficient magnitude over a relatively long period of time. Luckman (2004) has commented that there may be some contribution of debris to depositional ramparts or mounds from frequent small 'dirty' avalanches once excavated pits or pools are full of previously avalanched snow. However, such small and perhaps frequent events are more likely to lead to superficial modification of these landforms rather than produce them.

On the basis of morphological evidence combined with the approximate ages of trees damaged and felled by air displacement ahead of larger avalanches, Owen et al. (2006) recognised three scales of avalanche magnitude and frequency involved in the formation of the remarkable snow-avalanche impact landforms at Urdvatnet (Fig. 6.7). Air-launched and topographically focused avalanches of intermediate scale drop steeply onto the valley floor, where they erode the plunge pool and proximal scar and eject debris to form the distal mound. The largest avalanches, which may occur with decadal to centennial frequency, tend to erode the distal mound and distribute sedimentary material across the valley floor. Smaller avalanches, which may occur many times per year, scatter smaller amounts of debris across the lower slopes and tend to infill the crater and plunge pool.

Two lines of empirical evidence are available with which to estimate the magnitude and frequency of the snow-avalanche events that are capable of producing snow-avalanche impact craters in Vestlandet. First, dendrochronological estimates of the frequency of extreme-magnitude snow-avalanche events in Bødalen over the last 100 years (Decaulne et al. 2014) indicate a recurrence interval of 15–20 years may be applicable. The second line of evidence is based on lacustrine sedimentological evidence of 47 snow-avalanche event layers deposited during the last 7300 years in Oldenvatnet (Vasskog et al. 2011), which suggests a recurrence interval of ~ 155 years. Thus, the two approaches together suggest lower and upper limits of 60 and 600 avalanches, respectively, to the number of avalanches required to excavate the snow-avalanche craters in Vestlandet since deglaciation around 10,000 years ago (Matthews et al. 2017).

These estimates are consistent with the conclusions of comparative modelling of the kinetic energy required to form meteorite and snow-avalanche impact craters of similar size. Matthews et al. (2017) showed that for snow-avalanche craters of average size (85 m diameter in Vestlandet), the kinetic energy at impact is likely to be of the order of 1.3×10^{11} J, which is two orders of magnitude less than the 3.1×10^{9} J involved in excavating an 85-m diameter meteorite crater. It would be expected, therefore, that hundreds of snow-avalanche impacts are necessary for similar excavation potential to an equivalent single-impact high-energy meteorite impact event.

6.9 Summary

Southern Norway, in general, and Vestlandet, in particular, appear to provide the only location on Earth where there are sufficient examples to enable a full investigation of the nature of snow-avalanche impact landforms. Three main types of snow-avalanche impact landforms can be identified: (1) valley-floor craters, (2) lacustrine craters and (3) riverbank ramparts, essentially equivalent terms for the pits, pools and tongues, respectively, of Corner (1980). Fifty-two craters and 12 riverbank ramparts have been investigated in some detail from this region. Many of these landforms are active today and are worthy of conservation.

The average diameter of snow-avalanche impact craters in Vestlandet is 85 m and the largest is 185 m. The occurrence of craters depends on the topographic focusing of avalanches at the foot of steep valley-side slopes (α = 28–59°) where the gradient of the final 200 m of the avalanche path (β) exceeds $\sim 15°$ and the avalanche impacts onto sedimentary deposits. Their size is most closely related to avalanche snow volume, as reflected in the start-zone area (A), which varies from 18,000 to 467,000 m^2. Crater-forming avalanches are distinguished from non-crater-forming snow avalanches by the steep lower path angle combined with a sharp break of slope at the slope foot. In Vestlandet, topography and snow climate (>4.0 m mean annual snow accumulation, high snowfall intensity, strong winds and rapid temperature changes during the March–May peak avalanche season) combine to provide optimal environmental conditions for the development of snow-avalanche impact landforms.

Erosional and depositional processes can be compared with those involved in the formation of small meteorite craters. Crater rims are defined by proximal scars eroded into hillsides and smaller distal scars eroded into distal depositional mounds. Erosion of these scars is carried out by uprange and downrange jets, respectively. The jets eject sedimentary material upwards and outwards from the craters. A substantial proportion of the sedimentary material in riverbank ramparts is scooped out of river channels by downrange jets and deposited on the distal bank. Air launch, which steepens the impact angle, may also be necessary to produce snow-avalanche impact craters and ramparts. Impact onto a water body may be an enhancing or moderating factor due to the generation of impulse waves, which follow the collapse of an initial hydrodynamic crater.

Depositional processes including basal shear, fall-out and melt-out produce distinctive minor landforms ranging from strongly oriented longitudinal bedforms, including gravel ridges, imbricated gravel clusters and sediment tails in the lee of large boulders, to randomly distributed conical gravel clumps and precariously perched blocks. Small secondary craters attributed to erosion by individual boulders and/or snow packets thrown out from the main body of flowing snow also occur.

Where they are composed of boulders, the age and activity of riverbank ramparts can be assessed using exposure-age dating techniques (lichenometry and Schmidt-hammer dating). Combined with the results of comparative modelling of snow avalanche and meteorite craters of similar size, and existing knowledge of the recurrence interval of major avalanches, the dating evidence suggests that the currently active snow-avalanche impact landforms of Vestlandet have developed incrementally over at least the last 10,000 years of the Holocene. They are likely to have been most active during phases of enhanced winter snowfall, such as during the Little Ice Age.

Acknowledgements We have investigated the snow-avalanche impact landforms of Vestlandet as part of the research programme of the Jotunheimen Research Expeditions 1991, 1999, 2000, 2010, 2011 and 2014. This chapter constitutes Jotunheimen Research Expeditions, Contribution No. 209 (see http://jotunheimenresearch.wixsite.com/home). We are grateful to Anna Ratcliffe for preparation of figures.

References

Allix A (1924) Avalanches. Geogr Rev 14:519–560

Armstrong BR, Williams K (1992) The avalanche book. Fulcrum Publishing, Golden CO

Baggi S, Schweizer J (2009) Characteristics of wet snow avalanche activity: 20 years of observations from a high alpine valley (Dischma, Switzerland). Nat Hazards 50:97–108

Ballantyne CK (1989) Avalanche impact landforms on Ben Nevis, Scotland. Scott Geogr Mag 105:38–42

Bickerton RW, Matthews JA (1992) On the accuracy of lichenometric dates: an assessment based on the 'Little Ice Age' moraine sequence of Nigardsbreen, southern Norway. The Holocene 2:227–237

Blikra LH, Hole PA, Rye N (1989) Skred i Norge. Hurtige massebevegelser og avsetningstyper i alpine områder, Indre Nordfjord. Nor Geol Unders Skr 92:1–17

Brown VH, Evans DJA, Evans IS (2011) The glacial geomorphology and surficial geology of the south-west English Lake District. J Maps 2011:221–243

Castebrunet H, Eckert N, Giraud G (2012) Snow and weather climate control on snow avalanche occurrence fluctuations over 50 yr in the French Alps. Clim Past 8:855–875

Collins GS, Melosh HJ, Osinski GR (2012) The impact-cratering process. Elements 8:25–30

Corner GD (1975) Rundvatnet—avalanche plunge-pool or meteorite impact crater? Nor Geogr Tidsskr 29:75–76

Corner GD (1980) Avalanche impact landforms in Troms, North Norway. Geogr Ann Ser A (Phys Geogr) 62:1–4

Davis GH (1962) Erosional features of snow avalanches, Middle Fork Kings River, California. US Geol Surv Prof Pap 450D:122–125

Davison TM, Collins GS, Elbeshausen D, Wünnemann K, Kearsley AT (2011) Numerical modelling of oblique hypervelocity impacts in strong ductile targets. Meteorit Planet Sci 46:1510–1524

Decaulne A, Eggertsson Ó, Laute K, Beylich AA (2014) A 100-year extreme snow-avalanche record based on tree-ring research in upper Bødalen, Inner Nordfjord, western Norway. Geomorphology 218:3–15

Eckerstorfer M, Christiansen HH (2011) Topographical and meteorological control on snow avalanching in the Longyearbyen area, central Svalbard 2006–2009. Geomorphology 134:186–196

Elston DP, Scott GR (1971) Pueblito de Allende penetration craters and experimental craters formed by free fall. J Geophys Res 76:5756–5764

Erikstad L, Sollid JL (1986) Neoglaciation in South Norway using lichenometric methods. Nor Geogr Tidsskr 40:85–105

Evans DJA, Brown VH, Roberts DH, Innes JB, Bickerdike HL, Vieli A, Wilson P (2015) Wasdale Head. In: McDougall DA, Evans DJA (eds) The Quaternary of the lake district: field guide. Quaternary Research Association, London, pp 213–238

Fitzharris BB, Owens IF (1984) Avalanche tarns. J Glaciol 30:308–312

Fouinat L, Sabatier P, David F, Montet X, Schoeneich P, Chaumillon E, Poulenard J, Arnaud F (2018) Wet avalanches: long-term evolution in the Western Alps under climate and human forcing. Clim Past Discuss 14:1299–1313

Fritz HM, Hager WH, Minor H-E (2003) Landslide generated impulse waves. 2. Hydrodynamic impact craters. Exp Fluids 35:520–532

Gardner JS (1970) Geomorphic significance of avalanches in the Lake Louise area, Alberta, Canada. Arct Alp Res 2:135–144

Gleason JA (1995) Terrain parameters of avalanche starting zones and their effect on avalanche frequency. In: Proceedings of the international snow science workshop, Snowbird, Utah, USA, 30 October–3 November 1994, pp 393–404

Grove JM (1972) The incidence of landslides, avalanches, and floods and floods in western Norway during the Little Ice Age. Arct Alp Res 4:131–138

Grove JM (1988) The Little Ice Age. Methuen, London

Haeberli W, Whiteman C (eds) (2015) Snow and ice-related hazards, risks, and disasters. Elsevier, Amsterdam

Heller V, Hager WH, Minor H-E (2009) Landslide generated impulse waves in reservoirs: basics and computation. Versuchsanstalt für Wasserbau, Hydrologie und Glaziologie, ETH Zürich, Zürich

Hewitt K (1989) The altitudinal organisation of Karakoram geomorphic processes and depositional environments. Zeitschrift für Geomorphologie, NF, Supplementband 76:9–32

Hodge PW (1994) Meteorite craters and impact structures of the earth. Cambridge University Press, Cambridge

Hole J (1981) Groper danna av snøskred i Sunnylven og tilgrensande områder på Sunnmøre. Førbels resultat. Norsk Geografisk Tidsskrift 35:167–172

Innes JL (1985) Lichenometric dating of debris-flow deposits on alpine colluvial fans in southwest Norway. Earth Surf Proc Land 10:519–524

Johnson AL, Smith DJ (2010) Geomorphology of snow avalanche impact landforms in the southern Canadian Cordillera. Can Geogr 54:87–103

Jomelli V, Bertran P (2001) Wet snow avalanche deposits in the French Alps: structure and sedimentology. Geogr Ann Ser A (Phys Geogr) 83:15–28

Jomelli V, Francou B (2000) Comparing the characteristics of rockfall talus and snow-avalanche landforms in an alpine environment using a new methodological approach: Massif des Ecrins, French Alps. Geomorphology 35:181–192

Krinov EL (1960) Principles of meteoritics. Pergamon Press, London

Krinov EL (1963) The Tunguska and Sikhote-Alin meteorites. In: Middlehurst B, Kuiper G (eds) The solar system. Moon, meteorites and craters, vol 4, pp 208–234. University of Chicago Press, Chicago

Laute K, Beylich AA (2014a) Morphometric and meteorological controls on recent snow avalanche distribution and activity on hillslopes in steep mountain valleys in western Norway. Geomorphology 218:16–34

Laute K, Beylich AA (2014b) Environmental controls and geomorphic importance of a high-magnitude/low frequency snow avalanche event in Bødalen, Nordfjord, western Norway. Geogr Ann Ser A (Phys Geogr) 96:465–484

Laute K, Beylich AA (2014c) Environmental controls, rates and mass transfers of contemporary hillslope processes in the headwaters of two glacier-connected drainage basins in western Norway. Geomorphology 216:93–113

Lazar B, Williams M (2008) Climate change in western ski areas. Potential changes in the timing of wet avalanches and snow quality for the Aspen ski area in the years 2030 and 2100. Cold Reg Sci Technol 51:219–228

Liestøl O (1974) Avalanche plunge-pool effect. Norsk Polarinstitutt Arbok 1972:179–181

Luckman BH (1977) The geomorphic activity of snow avalanches. Geogr Ann Ser A (Phys Geogr) 59:31–48

Luckman BH (2004) Avalanche, snow. In: Goudie AS (ed) Encyclopedia of geommorphology. Routledge, London, pp 41–44

Luckman BH, Matthews JA, Smith DJ, McCarroll D, McCarthy DP (1994) Snow-avalanche impact landforms: a brief discussion of terminology. Arct Alp Res 26:128–129

Matthews JA, Briffa KR (2005) The 'Little Ice Age': re-evaluation of an evolving concept. Geogr Ann Ser A (Phys Geogr) 87:17–36

Matthews JA, McCarroll D (1994) Snow-avalanche impact landforms in Breheimen, southern Norway: origin, age and paleoclimatic implications. Arct Alp Res 26:103–115

Matthews JA, Dahl S-O, Dresser PQ, Berrisford MS, Lie Ø, Nesje A, Owen G (2009) Radiocarbon chronology of Holocene colluvial (debris-flow) events at Sletthamn, Jotunheimen, southern Norway: a window on the changing frequency of extreme climatic events and their landscape impact. The Holocene 19:1107–1129

Matthews JA, Shakesby RA, Owen G, Vater AE (2011) Pronival rampart formation in relation to snow-avalanche activity and Schmidt-hammer exposure-age dating (SHD): three case studies from southern Norway. Geomorphology 130:280–288

Matthews JA, McEwen LJ, Owen G (2015) Schmidt-hammer exposure-age dating (SHD) of snow-avalanche impact ramparts in southern Norway: approaches, results and implications for landform age, dynamics and development. Earth Surf Proc Land 40:1705–1718

Matthews JA, Owen G, McEwen LJ, Shakesby RA, Hill JL, Vater AE, Ratcliffe AC (2017) Snow-avalanche impact craters in southern Norway: their morphology and dynamics compared with small terrestrial meteorite craters. Geomorphology 296:11–30

McCarroll D (1993) Modelling late-Holocene snow-avalanche activity: incorporating a new approach to lichenometry. Earth Surf Proc Land 18:527–539

McCarroll D (1994) A new approach to lichenometry: dating single-age and diachronous surfaces. The Holocene 4:383–396

McClung DM (2001) Characteristics of terrain, snow supply and forest cover for avalanche initiation by logging. Ann Glaciol 32:223–229

McClung DM, Schaerer P (2006) The avalanche handbook. The Mountaineers Books, Seattle, WA

McClung DM, Mears AI, Schaerer P (1989) Extreme avalanche run-out: data from four mountain ranges. J Glaciol 13:180–184

Melosh HJ (1996) Impact cratering: a geological process. Oxford University Press, Oxford

Osinski GR, Pierazzo E (2013) Impact cratering: processes and products. In: Osinski GR, Pierazzo E (eds) Impact cratering: processes and products. Wiley-Blackwell, Chichester, pp 1–20

Osinski GR, Tornabene LL, Grieve RAF (2011) Impact ejecta emplacement on terrestrial planets. Earth Planet Sci Lett 310:167–181

Osinski GR, Grieve RAF, Tornabene LL (2013) Excavation and impact ejecta emplacement. In: Osinski GR, Pierazzo E (eds) Impact cratering: processes and products. Wiley-Blackwell, Chichester, pp 43–59

Owen G, Matthews JA, Shakesby RA, He X (2006) Snow-avalanche impact landforms, deposits and effects at Urdvatnet, southern Norway: implications for avalanche style and process. Geogr Ann Ser A (Phys Geogr) 88:295–307

Perla R (1977) Slab avalanche measurements. Can Geotech J 14:206–213

Perla RI, Martinelli M Jr (2004) Avalanche Handbook. Hawaii, University Press of the Pacific, Honolulu

Pierazzo E, Melosh HJ (2000) Understanding oblique impacts from experiments, observations and modelling. Annu Rev Earth Planet Sci 28:141–167

Powers MC (1953) A new roundness scale for sedimentary particles. J Sediment Petrol 23:117–119

Pudasaini SP, Hutter K (2007) Avalanche dynamics: dynamics of rapid flows of dense granular avalanches. Springer, Berlin

Rapp A (1959) Avalanche boulder tongues in Lappland: a description of little-known landforms of periglacial debris accumulation. Geogr Ann 41:34–48

Roddy DJ, Pepin RO, Merrill RB (eds) (1977) Impact and explosion cratering. Oxford, Pergamon. In: Proceedings of the symposium on planetary cratering mechanics, Flagstaff, Arizona, 13–17 Sept 1976

Sanders D (2013) Features related to snow avalanches and snow glides, Nordkette range (Northern Calcareous Alps). Geo Alp (Jahreszeitschrift zur Alpengeologie) 10:71–92

Schweizer J, Jamieson JB (2001) Snow cover properties for skier triggering of avalanches. Cold Reg Sci Technol 33:207–221

Schweizer J, Jamieson JB, Schneebeli M (2003) Snow avalanche formation. Rev Geophys 41:4/1016/2003

Smith DJ, McCarthy DP, Luckman BH (1994) Snow-avalanche impact pools in the Canadian Rocky Mountains. Arct Alp Res 16:116–127

Teich M (2013) Snow avalanches in forested terrain. Dr. Sc. degree. ETH, Zurich

Vasskog K, Nesje A, Støren EN, Waldmann N, Chapron E, Aritzegu D (2011) A Holocene record of snow avalanche and flood activity reconstructed from a lacustrine sedimentary sequence at Oldenvatnet, western Norway. The Holocene 21:597–614

Wright SP, Vesconi MA, Spagnuolo MG, Cerutti C, Jacob RW, Cassidy WA (2007) Explosion craters and penetration funnels in the Campo del Cielo, Argentina crater field. In: 38th lunar and planetary science conference, abstracts #2017

Zitti G, Ancey C, Postacchini M, Brocchini M (2016) Impulse waves generated by snow avalanches: momentum and energy transfer to a water body. J Geophys Res Earth Surf 121:2399–2423

7 Fluvial Processes and Contemporary Fluvial Denudation in Different Mountain Landscapes in Western and Central Norway

Achim A. Beylich and Katja Laute

Abstract

In the mountainous landscapes of mainland Norway, fluvial processes and contemporary fluvial denudation occur in both non-glacierized and glacierized (glacier-connected) drainage basin systems. Fluvial processes in non-glacierized cold climate drainage basins are driven by snowmelt or rainfall, or by combinations of snowmelt and rainfall, whereas fluvial processes in glacier-connected drainage basin systems are generated by snowmelt, glacier melt or rainfall, or by combinations of snowmelt/glacier melt and rainfall. Contemporary chemical and mechanical fluvial denudation rates in drainage basins of the partly glacierized and very steep mountain landscape of the inner Nordfjord in western Norway and of the non-glacierized and less steep mountain landscape located south of the Trondheim fjord in central Norway are found to be rather low, with mechanical fluvial denudation being significantly higher in the steeper and partly glacierized mountain landscape of the inner Nordfjord as compared to the studied mountain landscape in central Norway. In the partly glacierized mountain landscape of the inner Nordfjord mechanical fluvial denudation clearly dominates over chemical denudation, whereas the non-glacierized mountain landscape in central Norway is characterized by a clear dominance of chemical denudation over mechanical fluvial denudation. In the inner Nordfjord pluvially triggered denudational process events and pluvially activated sediment sources in ice-free surface areas as well as pluvially induced fluvial suspended sediment transport are, with respect to their overall contribution to drainage basin-wide annual suspended sediment yields, more important than the mostly thermally induced suspended sediment transport during the summer (glacier ablation) or in spring (snowmelt). However, the highest quantitative share of fluvial bedload transport occurs here during thermally generated snowmelt peak runoff in spring which is due to a significantly higher availability of bedload material within the mainstream channels as compared to time periods later in the hydrological year. Compared to that, in the studied non-glacierized mountain landscape in central Norway, the highest quantitative share of the total fluvial transport occurs during thermally induced snowmelt peak runoff in spring. With respect to the total annual sediment mobilization and total annual fluvial sediment transport, the sum of snowmelt-generated peak runoff events occurring over the hydrological year is here altogether more important than the annual sum of purely rainfall-generated peak runoff events. While the larger topographic relief and the partly glacial coverage in the mountain landscape of the inner Nordfjord are important factors for the higher mechanical fluvial denudation rates, the clearly higher importance of extreme pluvial events with respect to mobilization and transport of sediments within the drainage basins of this mountain landscape as compared to drainage basins in the non-glacierized mountain landscape south of the Trondheim fjord in central Norway must be highlighted. Fluvial transport in the studied fluvial systems in western and central Norway is clearly supply limited. It is expected that postulated changes of the current wind, air temperature and precipitation regimes will lead to increased chemical and mechanical fluvial denudation rates in both partly glacierized drainage basin systems in western Norway and non-glacierized drainage basin systems in central Norway.

A. A. Beylich (✉) · K. Laute
Geomorphological Field Laboratory (GFL), Sandviksgjerde, Strandvegen 484, 7584 Selbustrand, Norway
e-mail: achim.beylich@geofieldlab.com

K. Laute
e-mail: katja.laute@geofieldlab.com

A. A. Beylich (ed.), *Landscapes and Landforms of Norway*, World Geomorphological Landscapes,
https://doi.org/10.1007/978-3-030-52563-7_7

Keywords

Fluvial processes • Denudation • Glacierized and non-glacierized drainage basin • Mountain landscape • Environmental drivers • Climate change

7.1 Introduction

Fluvial denudation, including both chemical and mechanical processes, is of high relevance for landscape development and for the mobilization and transfer of solutes and sediments from headwater systems through main stem of drainage basin systems to the ocean. Fluvial denudation is controlled by a range of environmental drivers and can be significantly affected by anthropogenic activities (Beylich and DENUCHANGE Team 2018). In the mountainous landscapes of mainland Norway, fluvial processes and contemporary fluvial denudation occur in both non-glacierized and glacierized (glacier-connected) drainage basin systems. Due to the given high latitude and/or high elevation, most drainage basin systems in mainland Norway are partly or entirely situated in cold climate environments (Beylich 2021; Ketzler et al. 2021). Fluvial processes in non-glacierized cold climate drainage basins are driven by snowmelt or rainfall, or by combinations of snowmelt and rainfall, whereas fluvial processes in glacier-connected drainage basin systems are generated by snowmelt, glacier melt and rainfall, or by combinations of snowmelt/glacier melt and rainfall (e.g. Clark 1988; Barsch et al. 1994; Beylich and Gintz 2004; Beylich and Sandberg 2005; Beylich et al. 2010, 2017; Orwin et al. 2010; Beylich 2011, 2016; Beylich and Laute 2012, 2015, 2016, 2018). Detailed quantitative studies on the characteristics and the overall absolute and relative denudational importance of thermally and pluvially induced fluvial events are still comparably rare (e.g. Orwin et al. 2010; Beylich 2011, 2016; Beylich and Laute 2015, 2016, 2018; Beylich et al. 2017). Despite the fact that there is a good number of studies from glacierized and non-glacierized cold climate drainage basin systems that present data on fluvial transport and fluvial yields as well as on intra- and inter-annual variations of fluvial transport and fluvial yields, there is, however, general agreement that a better knowledge of environmental controls and an improved understanding of the absolute and relative importance of thermally and pluvially generated runoff events and the connected fluvial yields are still required to better understand the relative importance of glacial versus non-glacial processes and the complex fluvial transport dynamics within the respective drainage basin systems (e.g. Orwin et al. 2010; Beylich 2011, 2016; Beylich and Laute 2012, 2015, 2016, 2018; Beylich et al. 2017).

In addition, only few really detailed and quantitative process geomorphological studies on environmental drivers and rates of fluvial processes and fluvial denudation have been carried out in glacierized and non-glacierized drainage basin systems in mainland Norway (e.g. Bogen 1989, 1996; Beylich et al. 2010, 2017; Bogen et al. 2011, 2013, 2015; Beylich and Laute 2012, 2014, 2015, 2016, 2018; Liermann et al. 2012; Kennie and Bogen 2013, 2014; Xu et al. 2013, 2015a, b).

In this book chapter, we present, compare and discuss data on fluvial processes and contemporary chemical and mechanical fluvial denudation rates in selected and representative cold climate drainage basin systems in different mountain landscapes in western Norway and central Norway. We compare fluvial processes and contemporary fluvial denudation rates in partly glacierized drainage basin systems of the oceanic and mountainous fjord landscape of the inner Nordfjord in western Norway with fluvial processes and contemporary fluvial denudation rates in non-glacierized drainage basin systems in the boreal-oceanic and mountainous landscape south of the Trondheim fjord in central Norway. Special focus is on environmental drivers and the absolute and relative importance of chemical and mechanical fluvial denudation, and on the quantitative role of thermally and pluvially generated runoff events for the total contemporary fluvial denudation rates in the different mountain landscapes. Possible effects of postulated climatic changes on the contemporary denudation rates are presented.

7.2 The Mountainous Fjord Landscape in Western Norway

The cool-oceanic and mountainous fjord landscape in western Norway is characterized by its high local relief with elevation typically ranging from sea level up to 1500–2000 m a.s.l. With an elevation of 2083 m a.s.l., the Lodalskåpa is the highest peak of the Jostedalsbreen region (Fig. 7.1). The Jostedalsbreen ice cap is the largest glacier in continental Europe. It covers an area of 487 km^2 and encompasses around 50 outlet glaciers which descend within the adjacent valleys. All of these glacially carved and currently still glacier-connected and partly glacierized tributary valleys exhibit a steep and parabolic-shaped valley cross profile in combination with a plateau-like relief situated above the deeply incised valley systems. The longitudinal valley profiles are usually stepped and characterized by convex knickpoints separating defined valley infill areas (Beylich et al. 2009; Hansen et al. 2009). The predominant bedrock is rather homogenous and belongs to the Jostedal Complex as part of the Western Gneiss Region. The Jostedal Complex consists mostly of Precambrian migmatites, orthogneisses and granites (Ramberg et al. 2013). The thicknesses of glaciogenic deposits are generally very low and the highest share of the ice-free surface areas is characterized by bare bedrock (Beylich and Laute 2012, 2015, 2016; Laute and Beylich 2014, 2016). Along the fjords, in the valley bottoms and along the lower hillslopes within the mountain valleys, a widespread vegetation cover exists. The vegetation is mainly composed of the two boreal tree species grey alder (*Alnus incana* (L) Moench) and the European white birch (*Betula pubescens* Ehrh.) in combination with dwarf shrubs and grasses. Above the tree line, only sparsely spread grasses, mosses and lichens exist (Beylich et al. 2009; Laute and Beylich 2018). Within the very steep mountain valleys of this region (Fig. 7.1), relevant denudational surface processes include rock and boulder falls, snow avalanches, slush flows, debris flows and slides, creep processes, wash and chemical denudation, as well as fluvial transport of solutes, suspended sediments and bedload (Beylich and Laute 2012, 2015, 2016; Laute and Beylich 2014, 2016, 2018; Beylich et al. 2017). Due to the ongoing retreat of the various outlet glaciers of the Jostedalsbreen ice cap (e.g. Nesje et al. 2008; Burki et al. 2009; Laumann and Nesje 2009; Winkler et al. 2009; Ketzler et al. 2021; Laute and Beylich 2021; Winkler 2021), a significant enlarging of proglacial areas occurs (Laute and Beylich 2021).

It is existing knowledge that the comparably mild winters along the Norwegian west coast are primarily caused by two factors: (i) Low-pressure cells which cross the North Atlantic Ocean and transport warm and humid air to the north, and (ii) the North Atlantic Current (NAC) as the northern extension of the Gulf Stream which transports warm seawater from the Mexican Gulf to the Norwegian coast (e.g. Glässer 1993). West from the Jostedalsbreen ice cap, the climate is determined by the location between the maritime west coast and the Jostedalsbreen ice cap which is situated within the Scandinavian mountain range (the Scandinavian Caledonides). Along the western side of the Jostedalsbreen, frontal and orographic precipitation prevails (e.g. Glässer 1993). The orographic effect forces the warm and moist air to rise above the mountain range causing locally very high amounts of precipitation in western Norway. Maxima of precipitation occur usually during autumn (mainly rainfall) and winter (mainly snowfall) due to cyclonic activity (Beylich and Laute 2012; Laute and Beylich 2018). Due to the change of relief with many summits reaching higher than 1800 m a.s.l. within the innermost parts of the fjords (e.g. Nordfjord), a recognizable gradient from the outlets to the highest peaks within the valleys and drainage basins is detected for both air temperature and precipitation. Mean annual areal precipitation sums in the inner Nordfjord are around 1500 mm, and the mean annual air temperature is 5.7 °C (at 50 m a.s.l.) (Beylich et al. 2017).

7.3 The Mountain Landscape South of the Trondheim Fjord in Central Norway

This boreal-oceanic and mountainous landscape (Fig. 7.1) belongs to the central part of the Scandinavian mountain chain with peaks reaching lower elevations than in the more northern and southern parts of the Caledonides (see above and e.g. Beylich 2003; Fredin et al. 2013; Ramberg et al. 2013; Beylich et al. 2017; Beylich and Laute 2018; Laute and Beylich 2018). The lithology of the area is dominated by greenstone and amphibolites, grey-green phyllite and greywacke with locally found rhyolite-tuff (Lutro 2005). The Quaternary glaciations (e.g. Fredin et al. 2013; Olsen et al. 2013; Vorren and Mangerud 2013) created here a morphology that is characterized by rather flat to hilly areas between approximately 200 and 400 m a.s.l. with single peaks reaching maximum elevations above 600 m a.s.l. and well-defined and partly deeply incised main valleys (Fig. 7.1). The longitudinal profiles of most mainstream channels are characterized by several convex knickpoints. Quaternary deposits and/or covers (e.g. Rise et al. 2006; Fredin et al. 2013; Olsen et al. 2013) include larger surface areas with till and ice-marginal moraines, peat and humus in combination with surface areas of bare bedrock in elevations above 200 m a.s.l. and mostly glaciomarine deposits along the main rivers in lower elevations. Glaciofluvial and fluvial deposits are generally found along the mainstreams. The largely closed and continuous vegetation cover is dominated by large areas with boreal forests and bogs in combination with surface areas of bare bedrock in elevations above 200 m a.s.l., and grassland areas with agricultural use along the main rivers in lower elevations. In addition to buildings, roads and trails, human activities include forestry, agriculture and water power with several dammed and artificially regulated lakes. Relevant geomorphological processes include chemical weathering and fluvial solute transport resulting in chemical denudation, and mechanical weathering, local rock falls, slides and mud/debris flows, surface wash, fluvial erosion and down-cutting, stream bank erosion, and fluvial suspended sediment and bedload transport resulting in mechanical denudation (Beylich and Laute 2018). In the areas located along the main rivers in lower elevations and being characterized by comparably thick glaciomarine deposits, grassland covers and agricultural use, ravines and gullies are found frequently. In nearly flat surface areas in higher elevations, meanders are found along several mainstream stretches (Beylich and Laute 2018).

Also, the mountains of central Norway are located close to the North Atlantic Ocean (Norwegian Sea) in a prevailing westerly wind regime which gives rise to a pronounced orographic effect and strong west–east gradient in cloudiness

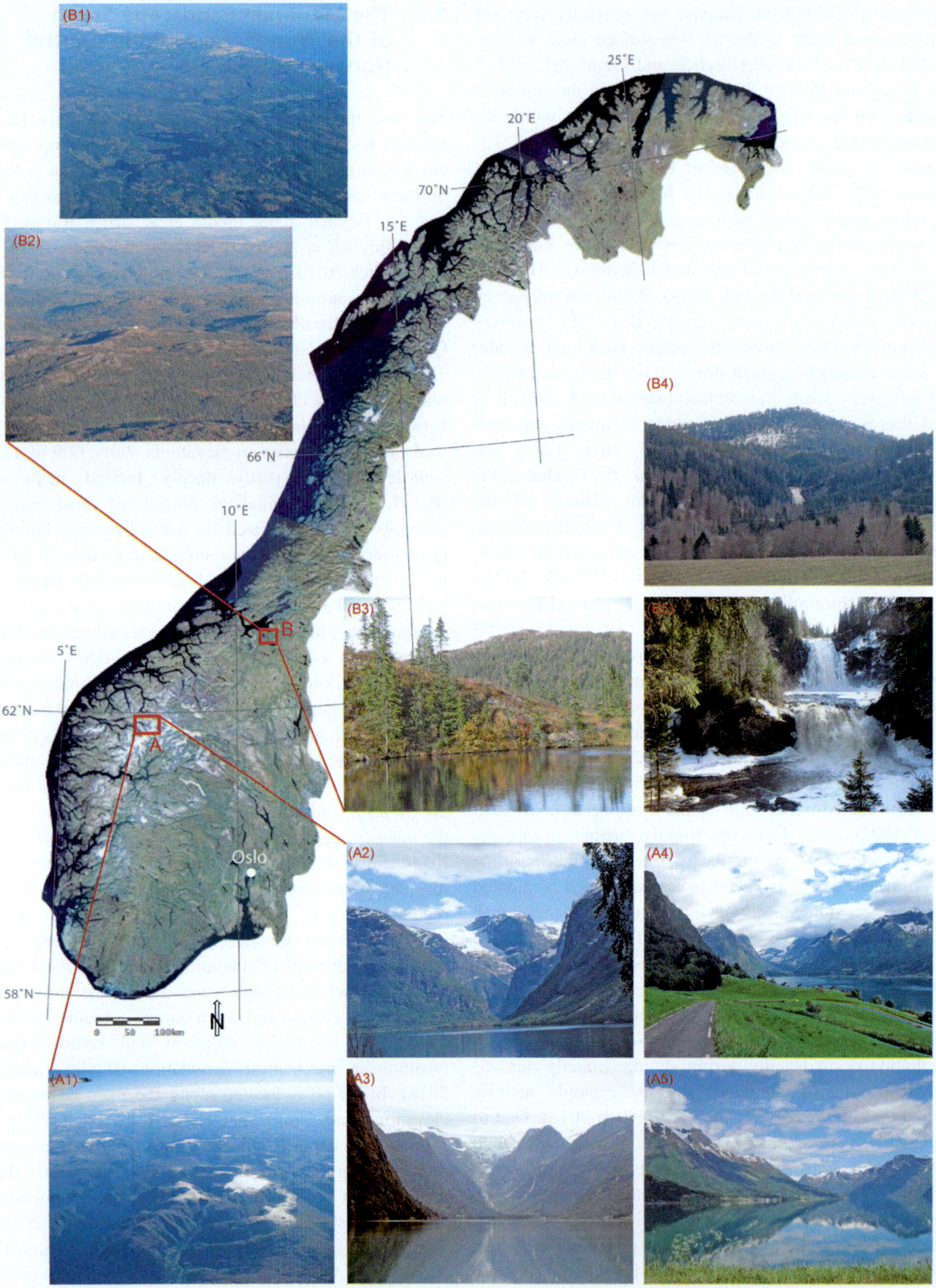

Fig. 7.1 Location of the study regions in western and central Norway and views of the partly glacierized mountain landscape of the inner Nordfjord in western Norway (A1–A5) and of the non-glacierized mountain landscape south of the Trondheim fjord in central Norway (B1–B5). *Source* of Norway map ©NVE Atlas 3.0 (Kartverket, Geovekst og kommuner-Geodata AS, NVE), *Photographs* © K. Laute

and precipitation. Accordingly, precipitation is again largely connected with cyclonic activity and the northerly position of the area is partly counteracted by the influence of the North Atlantic Current (NAC). Mean annual precipitation sums south of the Trondheim fjord are around 770 mm, and the mean annual air temperature is 6.6 °C (at 12 m a.s.l.) (Beylich and Laute 2018).

7.4 Fluvial Processes and Denudation in Partly Glacierized Drainage Basins in Western Norway

The fluvial processes and the chemical and mechanical fluvial denudation rates reported in this book chapter were studied and measured in the two neighbouring and partly glacierized Erdalen and Bødalen drainage basins being part of the inner Nordfjord situated on the western side of the Jostedalsbreen ice cap (Figs. 7.2 and 7.3). Erdalen is located at 61° 50′ N, 07° 10′ E (location of the drainage basin outlet), has a total surface area of 79.5 km^2 and ranges in elevation from 29.0 m a.s.l. to 1888.0 m a.s.l. The Bødalen drainage basin is located at 61° 48′ N, 07° 05′ E (location of the drainage basin outlet) with a total surface area of 60.1 km^2 and elevation ranging from 52.0 m a.s.l. to 2083.0 m a.s.l. (Beylich and Laute 2015, 2016; Beylich 2016; Beylich et al. 2017). Both drainage basins exhibit principal geological, geomorphological, hydrological and vegetation cover characteristics of the mountainous fjord landscape west from the Jostedalsbreen ice cap and were thus selected as representative test areas for the field studies and measurements.

7.4.1 Fieldwork and Methods

Fieldwork has been conducted in both drainage basins since 2004 and has included detailed field observations, geomorphological mapping, GIS and DEM computing for the generation of slope maps and surface area proportions, and the continuous and year-round monitoring of meteorological parameters, ground temperatures, runoff, and fluvial solute, suspended sediment and bedload transport (2004–2013 in Erdalen and 2008–2013 in Bødalen) (Fig. 7.2). Two automatic weather stations (Campbell CR10X-based weather stations) installed at carefully selected points of the main valley longitudinal profiles at 360 m a.s.l. in Erdalen and at 577 m a.s.l. in Bødalen for continuous (hourly) and year-round measurement of precipitation, snow depth, air temperature, radiation, wind speed and direction and ground temperature (at 20 cm ground depth) were in operation in the Erdalen and Bødalen drainage basins over the entire investigation period (Beylich and Laute 2012, 2015, 2016; Beylich et al. 2017). In addition, ground temperature sensors (Hobo Pendant Loggers, Long life for art) have been installed at 20 cm depth in different aspects and elevations in the different sampling areas for collecting hourly data on ground temperatures. Snow cover in both drainage basins was monitored on a daily basis using three automatic time-lapse cameras (Moultrie I-60 Cameras) in each valley operating year-round and covering most of the sampling areas as well as by daily manual photo documentation during field campaigns (Laute and Beylich 2014, 2016; Beylich et al. 2017). Snow cores covering the complete vertical depth of snowpack were taken during winters along defined profiles using a plastic tube device (10 cm diameter) and, in combination with rainwater samples collected across the drainage basin areas with rainfall collectors during field campaigns, used for the quantitative analysis of atmospheric solute inputs to the drainage basin systems (Beylich and Laute 2012, 2016). Channel discharge at in total ten selected sites (five sites in the Erdalen drainage basin and five sites in the Bødalen drainage basin) was measured with stationary hydrometric stations providing a continuous (hourly) and year-round monitoring of water levels using pressure sensors (Global Water). The continuous automatic monitoring of water levels was combined with frequently repeated direct channel discharge measurements in the turbulent stream reaches using the salt dilution method (Hongve 2006) at different defined water levels. Based on the water-level recordings and the calculated stage-discharge relations, hydrographs were created. Daily specific runoff (mm d^{-1}) was then calculated by dividing the computed daily discharges by the corresponding drainage basin area (e.g. Beylich and Laute 2012; Beylich et al. 2017). Daily discharge-weighted solute concentrations were computed from hourly readings of electric conductivity values at the hydrometric stations, using electric conductivity sensors (Global Water), in combination with frequent direct stream water samplings. The estimations of daily, monthly and annual solute yields were based on the relationships between electric conductivity ($\mu S\ cm^{-1}$) measured in field and concentration of total dissolved solids (mg l^{-1}) analysed in the laboratory (e.g. Beylich and Laute 2012). Daily discharge-weighted suspended sediment concentrations were calculated from hourly readings of optical turbidity at the hydrometric stations using turbidity sensors (Global Water), in combination with frequent direct stream water samplings. Water samplings carried out parallel to the automatic turbidity measurements were accomplished both manually (vertically integrating 1 l samples at 12 h intervals during field campaigns with wide-necked polyethylene bottles) and with two ISCO automatic water samplers collecting 1 l samples at (i) different defined time intervals (60, 45, 30, 15, 10, 5 min) and (ii) water-level-dependent time intervals for filtration of samples in field (portable pressure filter) and laboratory analysis of concentrations of inorganic suspended

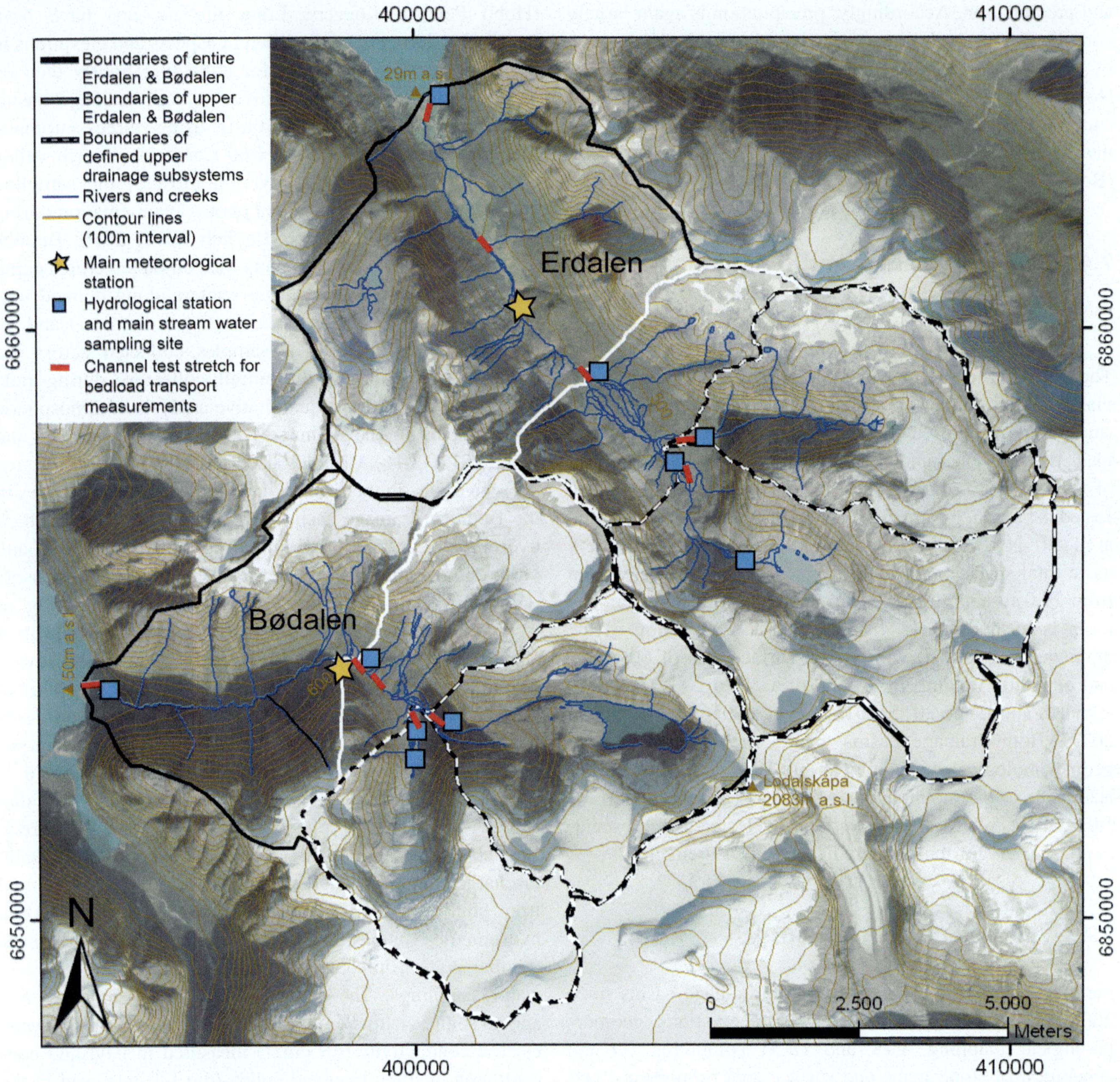

Fig. 7.2 Location, instrumentation, and main sampling and measurements sites of the partly glacierized Erdalen and Bødalen drainage basins in the inner Nordfjord in western Norway *Source* of digital topographic map and orthophoto (2010) © Kartverket (www.kartverket.no)

sediments (mg l^{-1}). Hourly turbidity values were correlated with suspended sediment concentrations as based on the conducted stream water samplings and sample field and laboratory analyses (e.g. Beylich et al. 2017). The high-resolution samplings carried out with the two ISCO automatic water samplers revealed that the hourly time interval applied for automatic turbidity measurements was sufficient to quantify daily discharge-weighted suspended sediment concentrations in the streams of the Erdalen and Bødalen drainage basins which are characterized by generally rather low suspended sediment concentrations. Bedload transport measurements were conducted with designed impact sensors installed at selected stream channel stretches in combination with tracer measurements and measurements carried out with a 3 × 3 in. Helley-Smith hand-held sampler. All impact-sensor, tracer and Helley-Smith measurements were carried out across the entire active bedload transport channel width, and each single Helley-Smith measurement was done over a defined time span of 10 min (e.g. Beylich and Laute 2014, 2015). The opening of the Helley-Smith sampler was wide enough for the grain sizes transported in the stream channels (e.g. Bunte et al. 2008).

Fig. 7.3 Views of the partly glacierized Erdalen (**a–d**) and Bødalen (**e–h**) drainage basins. *Photographs* **a–c**, **e–h** © K. Laute, **d** © A.A. Beylich

7.4.2 Contemporary Fluvial Processes and Denudation Rates

Based on the conducted field programme in Erdalen and Bødalen, annual atmospheric solute inputs, annual solute gross yields and the associated drainage basin-wide mean annual chemical denudation rates were calculated for the Erdalen and the Bødalen drainage basin systems. With a mean annual areal precipitation sum of 1500 mm and a mean weighted TDS value of 4.0 mg l^{-1} for precipitation (based on the samplings of snowpack and rainwater), the mean annual solute input for the inner Nordfjord is calculated to be 6.0 t km^{-2} yr^{-1}. Runoff in the inner Nordfjord occurs year-round, and the computed mean annual solute gross yield is 12.0 t km^{-2} yr^{-1} for the entire Erdalen drainage basin and 8.8 t km^{-2} yr^{-1} for the entire Bødalen drainage basin, which results in a mean annual chemical denudation rate of 6.0 t km^{-2} yr^{-1} in Erdalen and 2.8 t km^{-2} yr^{-1} in Bødalen. As a result, chemical denudation accounts for 24% of the total contemporary fluvial denudation in Erdalen but only for 6% of the total fluvial denudation in Bødalen (Fig. 7.4) (Beylich 2016; Beylich and Laute 2016). The detected difference in the absolute and relative importance of chemical denudation in the Erdalen and Bødalen drainage basins is largely explained by the varying drainage basin characteristics given in Erdalen and Bødalen. The Bødalen drainage basin system is characterized by (i) slightly higher elevation, (ii) larger topographic relief, (iii) greater mean slope angles, (iv) a clearly higher share of surface areas being occupied by glacier ice and/or bedrock, (v) a clearly smaller share of surface areas with sedimentary covers and (vi) a smaller mean thickness of sedimentary covers as compared to the Erdalen drainage basin system, with all these factors favouring a lower intensity of chemical weathering and chemical denudation and a higher intensity of mechanical fluvial denudation in Bødalen compared to the Erdalen drainage basin (Beylich and Laute 2012, 2015, 2016; Beylich et al. 2017). Solute gross concentrations are clearly highest in the winter period December–March, thus reflecting the dominance of comparably ion-rich base flow from the drainage basins during this period of low air temperatures, dominance of snowfall over rainfall and low to very low runoff. Compared to that, solute gross concentrations are lowest in the summer period July–August, reflecting the clear dominance of ion-poor glacier meltwater in this period with high air temperatures and high runoff. The spring period April–June displays lower solute concentrations than the winter period, thus reflecting the influence of ion-poor snowmelt water, which was mostly accumulated and stored as snow over the previous winter months, in this period with increasing air temperatures and high runoff. In autumn (September–November), which is the period with clearly most rainfall over the year, the solute concentrations are clearly lower than in winter. Thus, the detected intra-annual variations of solute gross concentrations reflect obvious diluting effects by (i) thermally controlled snowmelt in spring, (ii) thermally controlled glacier melt in summer and (iii) more frequent, longer and more intense rainfall events in autumn (Fig. 7.5). As a result, in spring and summer, the measured increase of solute gross yields with increasing runoff is weaker than in autumn, with the decrease in solute concentrations due to diluting effects being, however, more than compensated for by the increasing runoff (Fig. 7.5) (Beylich and Laute 2012, 2016).

Looking in detail at the four defined periods winter (December–March), spring (April–June), summer (July–August) and autumn (September–November), it becomes obvious that over the entire time of the field programme only 3.2% (entire Erdalen drainage basin) or 3.3% (entire Bødalen drainage basin) or the total annual suspended sediment transport occurs during the four winter months December–March, whereas 22.9% (Erdalen) or 21.0% (Bødalen) occur in spring (3 months April–June), 33.1% (Erdalen) or 32.9% (Bødalen) in summer (2 months July–August) and 40.8% (Erdalen) or 42.8% (Bødalen) during the 3 autumn months September–November (Figs. 7.5 and 7.6). Accordingly, the intra-annual temporal pattern of suspended sediment transport appears to be very similar for the entire Erdalen and the entire Bødalen drainage basin systems and fluvial suspended sediment transport in the inner Nordfjord is largely restricted to the 8-month period from April to November. Seasonal snowmelt in the inner Nordfjord usually starts in March/April and can last in the higher parts of the drainage basins until July. While air temperatures significantly increase, precipitation is usually comparatively low during the spring months April to June. Measured increases in runoff are accordingly most thermally determined and are associated with moderate increases of suspended sediment concentrations. Fluvially mobilized and transported sediments are mostly provided by snow Avalanche and slush flow deposits, and by local stream channel bank erosion. Larger rainfall events are not frequent in spring but can, when they occur, significantly increase runoff and can lead to increased suspended sediment concentrations due to increased delivery of sediments through small tributaries and increased mainstream channel bank erosion (Beylich et al. 2017). Runoff in summer (July–August) is most thermally determined by glacier ablation and the detectable increase in suspended sediment concentrations and yields is mostly due to clearly increased sediment delivery from the glacierized areas of the drainage basin systems. An additional source for suspended sediments during summer is fluvial bank erosion in the mainstream channels, whereas most small tributaries without connection

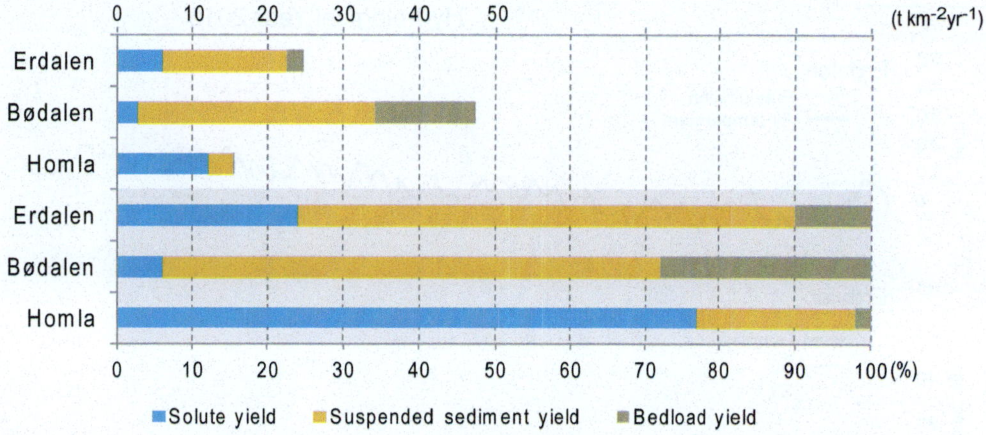

Fig. 7.4 Mean annual and drainage basin-wide solute, suspended sediment, bedload and total fluvial yields in the Erdalen, Bødalen and Homla drainage basins

to glaciers or larger and longer lasting snow patches or fields are usually dried out for most of the time. Larger rainfall events are not frequent in summer but can, when they occur, cause detectable increases in runoff and slight increases of suspended sediment transport due to further intensified mainstream channel bank erosion. However, in the ice-free drainage basin surface areas with sedimentary covers, such summer rainfall events have, due to at this time generally low water content of slope deposits and regolith, no significant effects on runoff and on the delivery of suspended sediments into the mainstream channels (Beylich et al. 2017). In autumn (September–November), runoff and suspended sediment transport are mainly pluvially determined by longer lasting and/or heavy rainfall events, and suspended sediments are mainly delivered from the ice-free surface areas within the drainage basins (Beylich et al. 2017). On days with occurring saturation overland flow and slope wash processes significant increases of discharge in small tributaries draining slope areas can be observed. In slope areas where these slope tributaries are incised into sedimentary covers, the increased discharge leads to a mobilization of the tributary channel beds and locally also to significant tributary bank erosion which, in addition to slope wash, further increases the delivery of suspended sediments to mainstream channels (Laute and Beylich 2014, 2016; Beylich and Laute 2015; Beylich et al. 2017). The clearly highest suspended sediment concentrations are measured on days when debris flows are triggered by extreme rainfall events. Debris flows can cause significant transfers of sediments from ice-free surface areas with sedimentary covers into mainstream channels (Beylich et al. 2010, 2017; Laute and Beylich 2014). Channel bank erosion in the mainstreams forms an additional sediment source for suspended sediment transport in autumn. As a result, the intensity of fluvial suspended sediment transport in autumn strongly depends on the number of specific rainfall events causing saturation overland flow and connected slope wash, increased delivery of suspended sediments through small tributaries and/or the triggering of debris flows. In years with relevant debris flow activity, the total annual suspended sediment transport rate in Erdalen and Bødalen is strongly determined by these single extreme events (Laute and Beylich 2014; Beylich et al. 2010, 2017). When comparing the described periods spring, summer and autumn, most fluvial suspended sediment transport is, as already mentioned, occurring during the 3-month period of autumn, followed by the 2-month period of summer and by the 3-month period of spring (Fig. 7.6). This interesting finding points out that pluvially triggered denudational process events and pluvially activated sediment sources in ice-free surface areas of the drainage basins as well as pluvially induced fluvial suspended sediment transport are, with respect of their overall contribution to drainage basin-wide annual suspended sediment yields, actually more important than the most thermally induced suspended sediment transport during the 2 summer months July and August (glacier ablation) or in spring (snowmelt) (Beylich et al. 2017).

While the intra-annual temporal patterns of the activation of various sediment sources and of fluvial suspended sediment transport are mostly controlled by meteorological events and are very similar across the inner Nordfjord, a significant spatial variability of fluvial suspended sediment transport and suspended sediment yields is detected. It is found that the mean annual suspended sediment yield for entire Bødalen is with 31.3 t km^{-2} yr^{-1} almost twice as high as the mean annual suspended sediment yield for entire Erdalen which accounts for 16.4 t km^{-2} yr^{-1} (Fig. 7.4). This significant spatial variability of suspended sediment yields is due to the already mentioned varying drainage basin characteristics and can largely be explained by the clearly higher sediment supply from the glacierized surface areas given in Bødalen as compared to the glacierized surface areas found in Erdalen (Beylich et al. 2017). In addition to the share of glacier coverage which is the most important drainage basin characteristic for explaining the spatial variability of suspended sediment yields between different drainage basins,

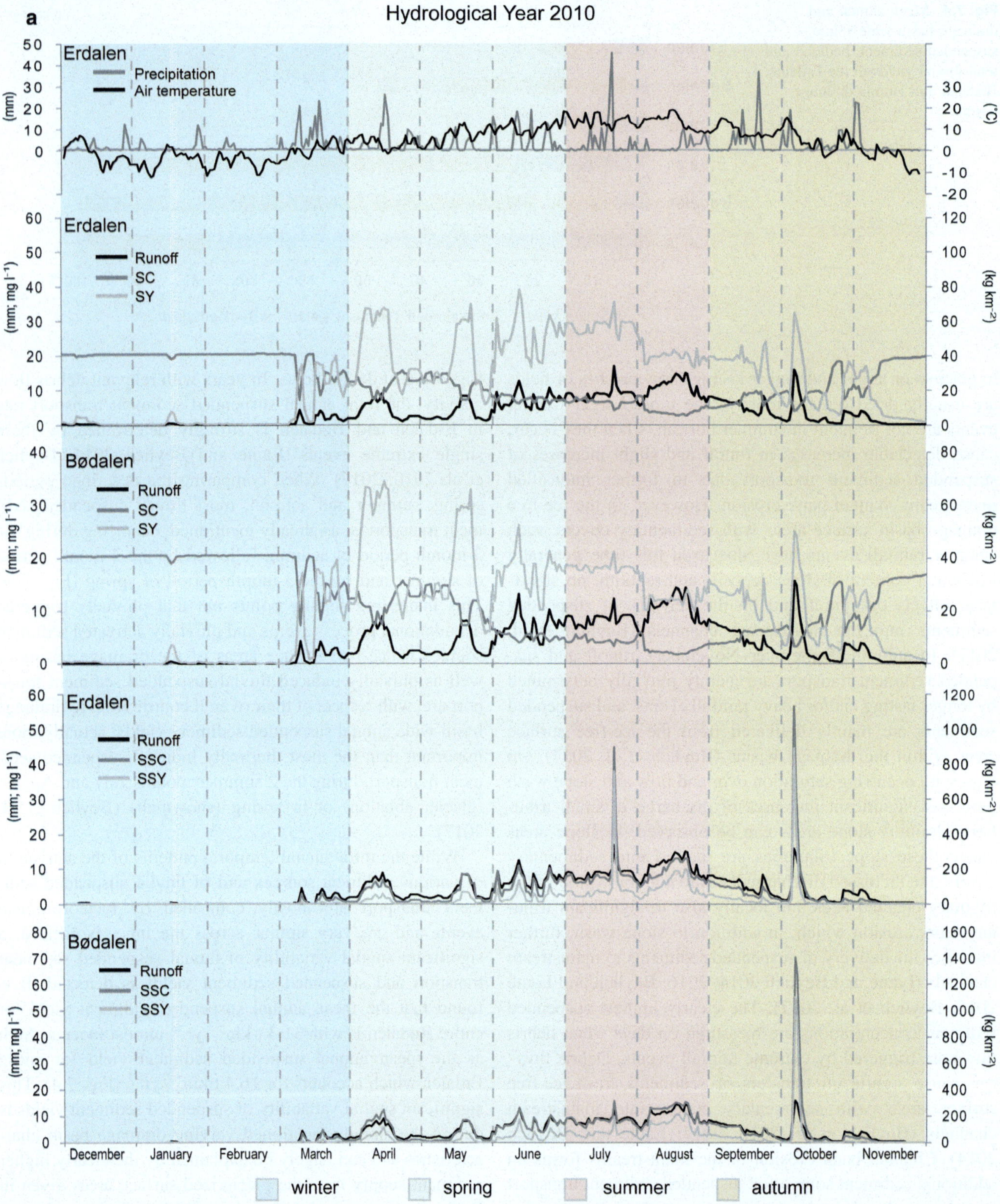

Fig. 7.5 Time series of daily precipitation sums, daily mean air temperatures, daily runoff, mean daily discharge-weighted solute gross concentrations (SC), daily solute gross yields (SY), mean daily discharge-weighted suspended sediment concentrations (SSC) and daily suspended sediment yields (SSY) in the selected hydrological years 2010 (**a**), 2011 (**b**) and 2012 (**c**) for the entire Erdalen and Bødalen drainage basins

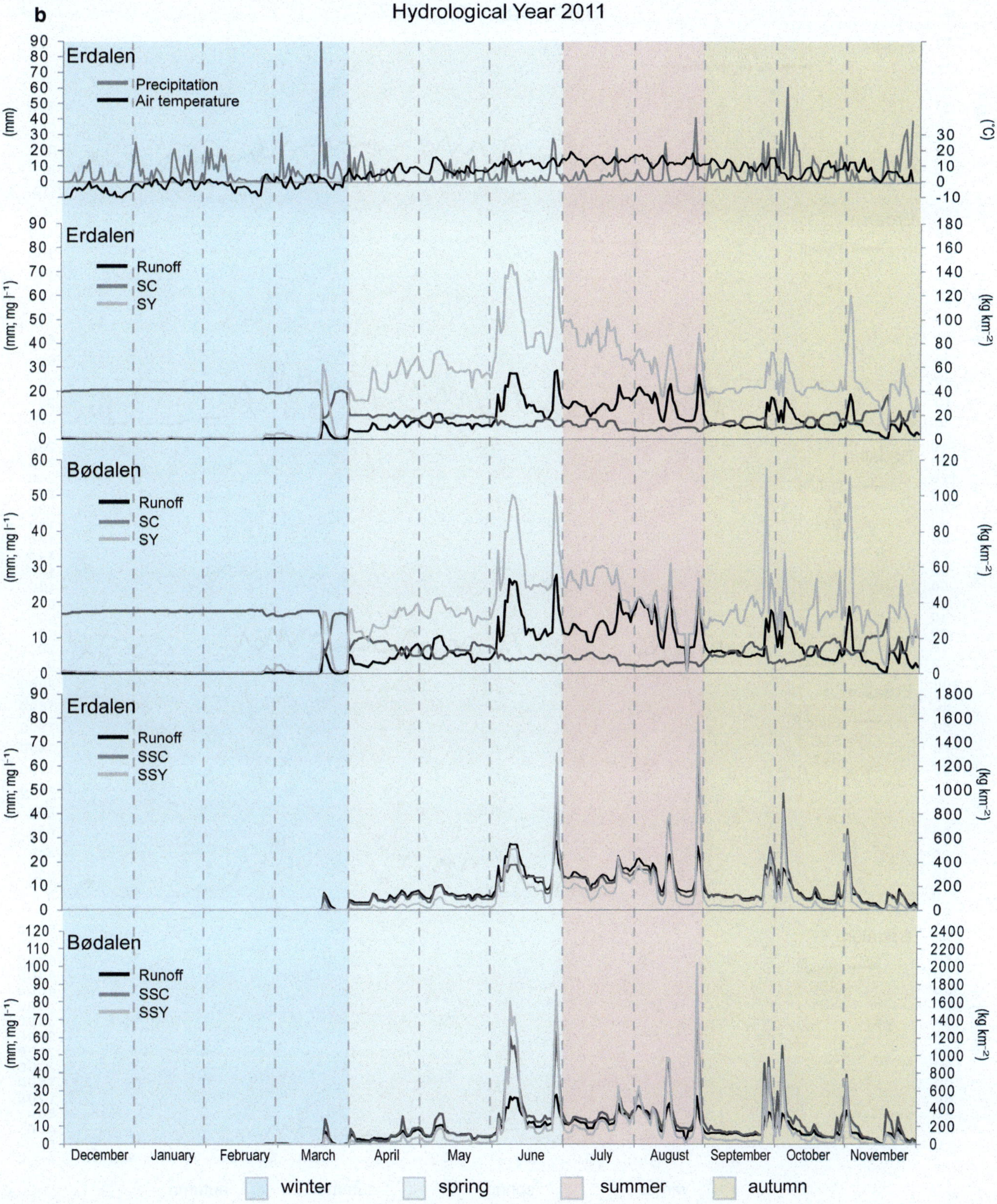

Fig. 7.5 (continued)

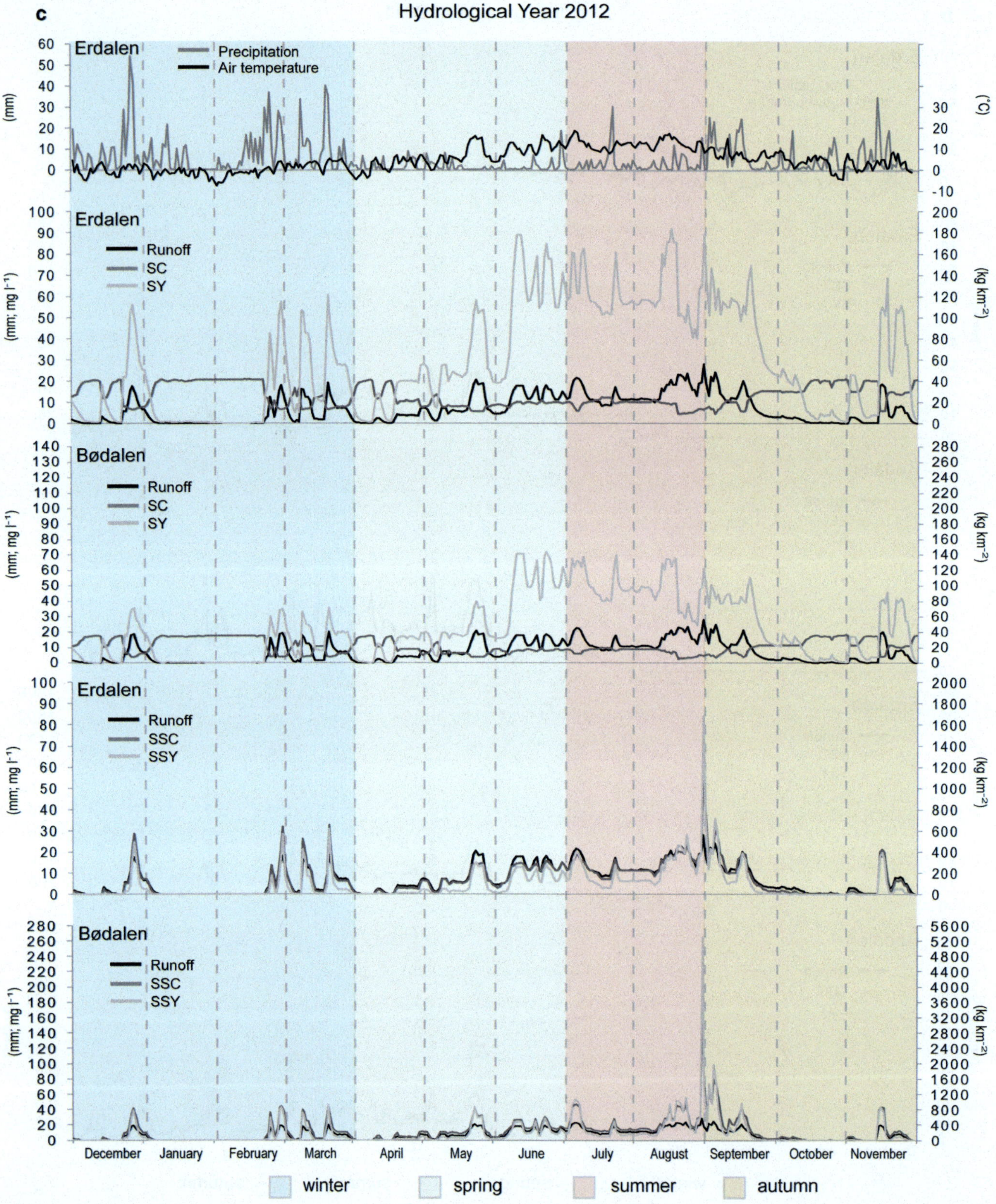

Fig. 7.5 (continued)

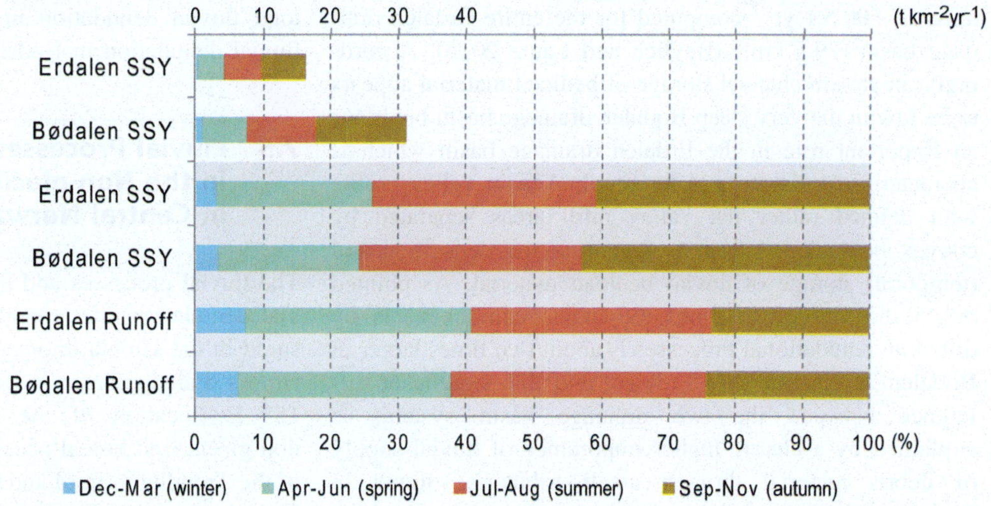

Fig. 7.6 Runoff and suspended sediment yields (SSY) for different periods (winter, spring, summer, autumn) of the year in the entire Erdalen and Bødalen drainage basins

the steepness of slopes and the degree of vegetation cover of ice-free drainage basin surface areas with sedimentary covers are important factors controlling suspended sediment yields and their spatial variability. The suspended sediment concentrations and suspended sediment yields measured in the Erdalen and Bødalen drainage basin systems are altogether rather low. It should be pointed out that, despite the detected high spatial variability, sediment delivery from glacierized surface areas is altogether very small within the area of the inner Nordfjord on the western side of the Jostedalsbreen ice cap, especially when compared to the sediment delivery that is postulated for the Nigardsbreen outlet glacier on the eastern side of the Jostedalsbreen ice cap, and to other ice caps in comparable climates and topographies (e.g. Bogen 1989, 1996; Harbor and Warburton 1993; Hallet et al. 1996; Beylich et al. 2010, 2017; Orwin et al. 2010; Liermann et al. 2012; Kennie and Bogen 2013; Bogen et al. 2011, 2013, 2015; Xu et al. 2013, 2015a, 2015b; Beylich and Laute 2015). Major reasons for the altogether low sediment supply from both glacierized and non-glacierized surface areas in the inner Nordfjord are certainly the high resistance of the predominant bedrock towards glacier erosion and weathering, and—even more important—the only small amounts of sediments being generally available within the entire drainage basin systems of the inner Nordfjord. Only 32% of the total surface area in Erdalen and only 16% of the surface area in Bødalen are occupied by slope deposits or regolith, and again 71% of these surface areas in Erdalen and 70% of these surface areas in Bødalen are covered by (in most places) stable vegetation cover developed below 1000 m a.s.l. As a result, fluvial sediment transport in the inner Nordfjord is clearly supply limited (Beylich et al. 2010, 2017; Laute and Beylich 2014, 2016; Beylich and Laute 2015). In addition, the significant storage of suspended sediments, being supplied by present outlet glaciers, in proglacial lakes situated close to the termini of these outlet glaciers needs to be considered in the interpretation of drainage basin-wide suspended sediment yields. For the proglacial lake systems located in Erdalen and Bødalen, the computed trap efficiencies range from 23% to 52% in Erdalen and reach 68% in Bødalen (Liermann et al. 2012; Beylich et al. 2017). Altogether, fluvial suspended sediment transport accounts for two-thirds (66%) of the total fluvial transport and total fluvial yields in the entire Erdalen and Bødalen drainage basins (Fig. 7.4).

Fluvial bedload transport in the inner Nordfjord shows its highest intensity and an intra-annual peak during snowmelt-generated high runoff in June (Fig. 7.7), which is explained by the fact that spring (April–June) is the period when the largest share of sediments is delivered from ice-free slope systems into the mainstream channels by rockfalls, snow avalanches, stream channel bank erosion, and fluvial transfers through small tributaries (Laute and Beylich 2014; Beylich and Laute 2015). Over summer (July–August) significant amounts of material can be supplied by stream channel bank erosion and especially by fluvial transfers from glacier-connected tributaries, and are fluvially transported within the mainstream channels during glacier–melt-induced high runoff. Especially in the narrow lower part of the Bødalen drainage basin, pluvially induced high runoff leads also in fall (especially in October) to high debris supplies from steep tributaries incised into sedimentary covers (Beylich and Laute 2015). The mean annual bedload yield of the entire Bødalen drainage basin is 13.3 t km^{-2} yr^{-1} which is 5.5 times higher than the mean annual bedload yield for the entire Erdalen drainage basin system (2.4 t km^{-2} yr^{-1}) (Fig. 7.4). At the same time, the total annual mass of material being transferred into the mainstream channels by rockfalls, snow avalanches, fluvial stream channel bank erosion and fluvial transfers through small tributaries is 878.4 t yr^{-1} in the entire Bødalen (60.1 km^2) which is 1.8 times larger than the total annual mass of 500.5 t yr^{-1} computed for the entire Erdalen drainage basin (79.5 km^2) (Beylich and Laute 2015). Accordingly, in-stream channel storage of bedload material appears to be low in the very steep Bødalen drainage basin but plays an important role in the Erdalen drainage basin which is characterized by a stepped longitudinal main valley profile with defined rather flat valley infill areas separated by convex bedrock knickpoints favouring the deposition and (temporal) storage of fluvial bedload material. As pointed out, sediment delivery into the mainstream channels by different denudational processes is about two times larger in Bødalen as compared to Erdalen, and this significant difference between the two drainage basin systems is explained by a clearly higher importance of fluvial supply of debris material through small tributaries found in Bødalen (Beylich and Laute 2015). The detected high importance of small tributaries serving as sediment sources for bedload material in Bødalen is largely a consequence of the given drainage basin morphometry of Bødalen, which is characterized by a narrow main valley with steep and till-covered slopes in lower Bødalen. A second important factor is that several tributaries in upper Bødalen are, in addition to draining subareas with a comparably high availability of erodible till and glaciofluvial sediments, glacier-connected and accordingly characterized by high discharges over the entire summer (Beylich and Laute 2015). Altogether, fluvial bedload transport in the glacier-connected Erdalen and Bødalen drainage basins is relatively low and is much more related to the availability of sediments than to channel discharge (Beylich and Laute 2014, 2015). The comparably little supply of sediments from hillslope systems into the mainstream channels is a result of the highly resistant bedrock and the nearly closed and stable vegetation cover found in the inner Nordfjord (Laute and Beylich 2014, 2016; Beylich and Laute 2015; Beylich 2016; Beylich et al. 2017). The main outlet glaciers in Erdalen and Bødalen are providing significant amounts of sediments but are not of importance as current sediment source for fluvial bedload transport in the mainstream channels as the proglacial lakes situated in front of these outlet glaciers trap all bedload material being delivered directly from these outlet glaciers (Liermann et al. 2012; Beylich and Laute 2015; Beylich et al. 2017). Compared to suspended sediment yields and chemical denudation rates, the fluvial bedload yields computed for Bødalen and Erdalen are clearly of quantitative importance, ranging from 10% of the total fluvial transport and total fluvial yield in the entire Erdalen drainage basin to 28% of the total fluvial transport and total fluvial yield of the entire Bødalen drainage basin (Fig. 7.4) (Beylich and Laute 2015, 2016; Beylich 2016). Mechanical fluvial denudation clearly dominates over chemical denudation in the inner Nordfjord, with mechanical fluvial denudation reaching 76% of the total fluvial denudation in Erdalen and 94% of the total fluvial denudation in Bødalen (Fig. 7.4).

7.5 Fluvial Processes and Denudation in the Non-glacierized Drainage Basins in Central Norway

The fluvial processes and the chemical and mechanical fluvial denudation rates reported here were studied and measured in the Homla drainage basin (Figs. 7.8 and 7.9). The Homla drainage basin is located at 63° 24′ 50″ N, 10° 48′ 15″ E (location of the drainage basin outlet) in the non-glacierized boreal-oceanic mountain environment south of the Trondheim fjord and drains to the north into the fjord. The drainage basin system has a total surface area of 156.3 km^2 and ranges in elevation from 0.0 m a.s.l. to 697.1 m a.s.l. It exhibits principal geological, geomorphological, hydrological and vegetation cover characteristics of the mountain landscape south of the Trondheim fjord in central Norway and thus was selected as representative test area for the field studies and measurements.

7.5.1 Fieldwork and Methods

The fieldwork and measurements that have been carried out in the Homla drainage basin since 2011 are methodologically largely identical with field measurements carried out in the Erdalen and Bødalen drainage basins (see above). The continuous and year-round monitoring of runoff and fluvial solute, suspended sediment and bedload transport in the Homla drainage basin was carried out over a 5-year period (2011–2016) using a stationary hydrometric station close to the outlet of the Homla drainage basin in combination with frequent and year-round manual samplings and measurements carried out close to the drainage basin outlet and at other selected stream channel stretches within the Homla drainage basin system and selected sub-basins (Fig. 7.8) (Beylich and Laute 2018).

7.5.2 Contemporary Fluvial Processes and Denudation Rates

Mean monthly shares of annual runoff together with intra-annual variations of the mean monthly shares of annual net solute, suspended sediment and bedload yields for the 5-year investigation period from November 2011 until October 2016 are shown in Fig. 7.10. As a result of the given boreal-oceanic morphoclimate in the area, runoff is occurring year-round and the runoff regime is clearly nival, with a minimum of monthly runoff during the coldest month

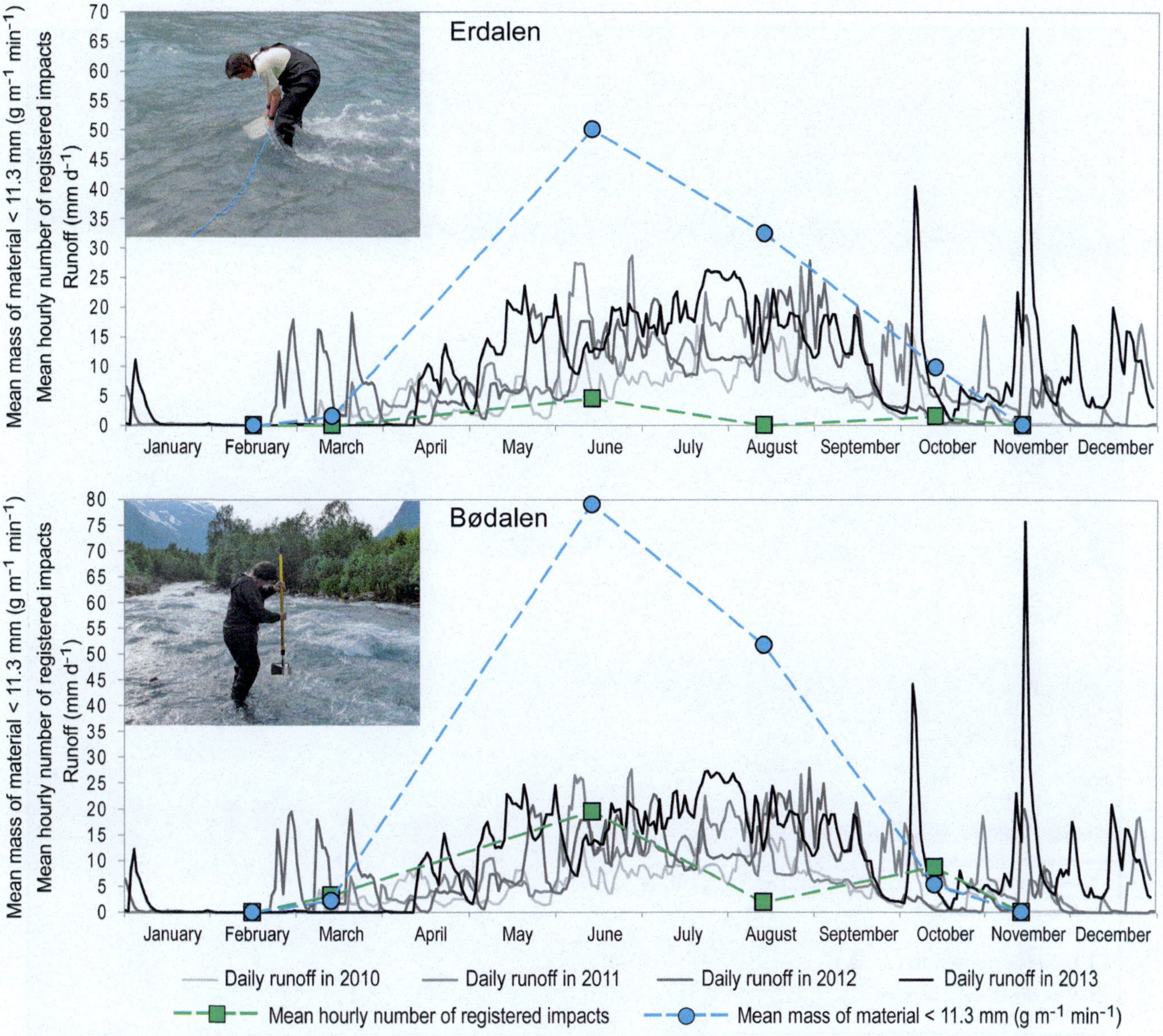

Fig. 7.7 Fluvial bedload transport measured with impact sensors (bedload particles >11.3 mm) and Helley-Smith sampler (bedload particles <11.3 mm) in the Erdalen and Bødalen drainage basins (modified after Figs. 4 and 5 in: Beylich and Laute (2015). *Photographs* © K. Laute

January and a maximum of monthly runoff in May. Between November and March, a high share of the occurring precipitation is accumulating as snow, and distinct peak runoff during the spring snowmelt in April and May is most thermally determined. Compared to that, after spring snowmelt and until late fall runoff is mostly pluvially controlled. Figure 7.10 displays that the intra-annual variability of fluvial transport is largely following the intra-annual variation of runoff with lowest fluvial transport being measured in January and highest fluvial transport being recorded in May. However, it is detected that the three main components of fluvial transport, solute transport, suspended sediment transport and bedload transport, show some distinct differences in their temporal variation over the year. Fluvial solute transport is characterized by an altogether clearly smaller intra-annual variation than the fluvial transport of suspended sediments and bedload material. In the winter months, discharge in the Homla mainstream channel is, apart from short-term peak runoff events caused by thermally determined thaw and snowmelt events, to a high share groundwater-fed, leading to comparably high mean daily discharge-weighted solute concentrations of up to 50–55 mg l^{-1} and still relevant chemical denudation during these months being characterized by low runoff. In January, 3.0% of the annual chemical denudation occurs, whereas the values are 3.1% for February and 5.5% for December. The

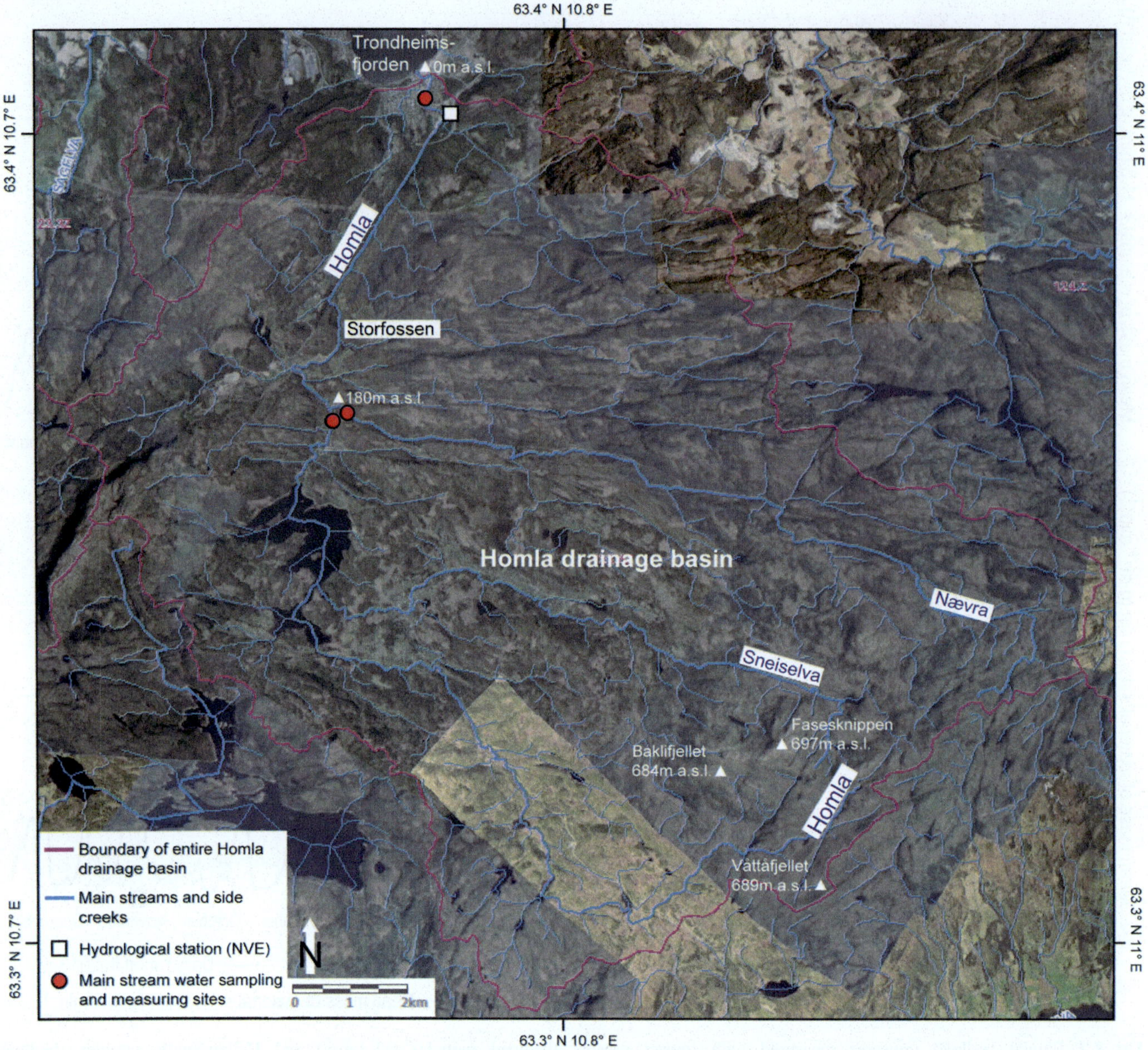

Fig. 7.8 Location, instrumentation, and main sampling and measurements sites of the non-glacierized Homla drainage basin south of the Trondheim fjord in central Norway. *Source* of orthophoto, layer drainage basin boundary and layer mainstreams and side creeks © NVE Atlas 3.0 (Kartverket, Geovekst og kommuner-Geodata AS, NVE)

detectable diluting effect during spring snowmelt, with mean daily discharge-weighted solute concentrations down to 15–17 mg l^{-1}, is more than compensated for by the clearly increased runoff, and 13.3% of the annual chemical denudation is recorded in May. Solute concentrations increase during the warm summer months and stay with mean daily discharge-weighted solute concentrations around 30–40 mg l^{-1} comparably high during fall resulting in rather constant monthly shares of annual chemical denudation from July (9.8%) over August (11.0%) and September (10.4%) until October (10.9%) despite the slightly declining monthly shares of annual runoff (Fig. 7.10). Mean annual atmospheric solute inputs to the Homla drainage basin are 6.0 t km^{-2} yr^{-1} which is based on a mean annual solute concentration of 7.8 mg l^{-1} computed from collected snow-core and rainwater samples. The atmospherically corrected chemical denudation rate for the entire Homla drainage basin system is 12.1 t km^{-2} yr^{-1} (Beylich and Laute 2018) (Fig. 7.4). Altogether, chemical weathering and chemical denudation are of comparably low intensities which is

Fig. 7.9 Views of the non-glacierized Homla drainage basin. *Photographs* © K. Laute

explained by the rather cool climate of the boreal-oceanic study area in combination with only small thicknesses of sedimentary covers found in large shares of the drainage basin system.

The computed mean annual suspended sediment yield (3.3 t km^{-2} yr^{-1}) and the measured mean annual bedload yield (0.3 t km^{-2} yr^{-1}) of the Homla drainage basin are clearly smaller than the mean annual chemical denudation

Fig. 7.10 Mean monthly shares (%) of mean annual runoff (**A**), mean annual solute yield (**B**), mean annual suspended sediment yield (**C**) and mean annual bedload yield (**D**) in the Homla drainage basin (modified after Fig. 14 in: Beylich and Laute (2018)

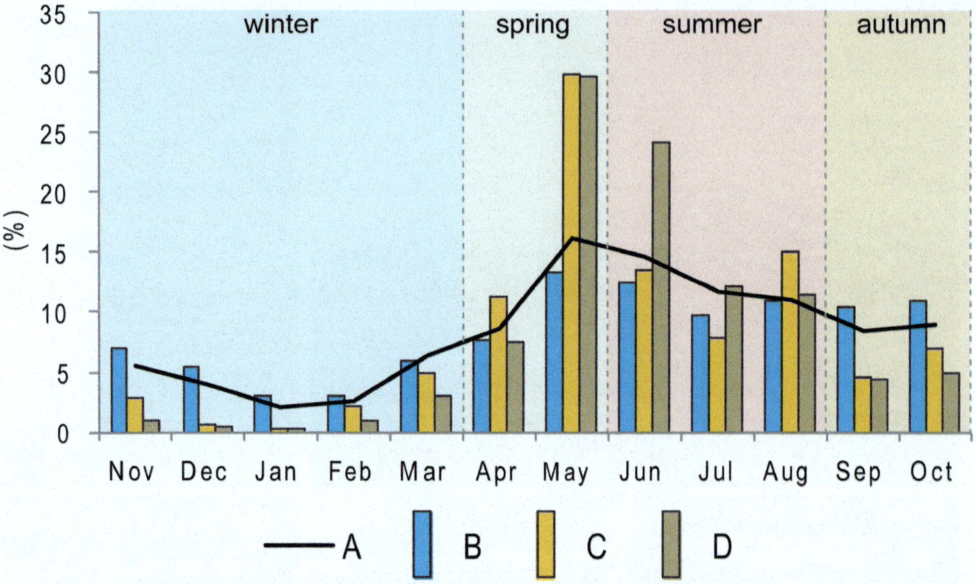

rate of this drainage basin system (Fig. 7.4). The intra-annual variability of fluvial suspended sediment and fluvial bedload transport as displayed in Fig. 7.10 is largely dependent on the temporal variation of activation of various sediment sources found within the Homla drainage basin system. The clearly highest monthly share of suspended sediment (29.9%) and bedload material (29.7%) is transported in May, whereas hardly any fluvial solid transport is detected during the winter months. Accordingly, over the year most intensive activation of sediment sources through wash processes, fluvial activity and triggering of slides and mud/debris flows is recorded during snowmelt when sedimentary covers within the drainage basin are water saturated. Fluvial suspended sediment transport rapidly declines after May and only 13.6% of the annual suspended sediment transport is recorded in June. In comparison to that, bedload transport remains at a clearly higher level in June (24.1% of the total annual bedload transport) which is explained by the delayed transport and the delayed (partly) export out of the Homla drainage basin system of bedload material that was transferred into the Homla mainstream channels during spring snowmelt. During the following months (July–October), both suspended sediment and bedload transport are mostly related to relevant pluvial events. The highest monthly share (15.1%) of the annual suspended sediment transport occurring between July and October is recorded in the month (August) that displays the highest frequency of extreme rainfall events which can trigger significant transfers of sediments into mainstream channels (Beylich and Laute 2018). The comparably high monthly share of annual bedload transport detected in July (12.2%) is explained by the still continuing transport and export of bedload material that was transferred into the mainstream channels during spring snowmelt, whereas the bedload transport in August (11.4% of the total annual bedload transport) also includes bedload material that is newly delivered into mainstream channels through new rainfall-triggered denudational events in August. Both suspended sediment and bedload transport clearly decline after August. With respect to total annual sediment mobilization and total annual fluvial sediment transport, the sum of snowmelt-generated peak runoff events occurring over the hydrological year is altogether more important than the annual sum of purely rainfall-generated peak runoff events. The detected clear dominance of annual suspended sediment transport over annual fluvial bedload transport, with suspended sediment transport being on average more than 10 times higher than bedload transport, can largely be explained by the fact that the highest share of the total fluvial sediment load is supplied as fine-grained sediments from the fluvially dissected glaciomarine deposits found in the lower Homla drainage basin. Mechanical fluvial denudation in the Homla drainage basin is altogether of low intensity which is mostly explained by the nearly closed and continuous vegetation cover, the dominance of surface areas with only small topographic relief in the upper Homla drainage basin and the high weathering resistance of the bedrock found in the study area. Human activities, particularly agriculture and forestry, increase mechanical denudation locally through significant disturbances of the vegetation cover (Beylich and Laute 2018). Altogether, chemical denudation clearly dominates over mechanical fluvial denudation with chemical denudation reaching 77% of the total fluvial denudation (Fig. 7.4).

7.6 Chemical and Mechanical Fluvial Denudation in Mainland Norway

The Norwegian Water Resources and Energy Directorate (Norges Vassdrags-og Energidirektorat NVE) is running a national monitoring network with measurements of discharge, erosion and fluvial transport in numerous drainage basins across Norway including both glacierized and non-glacierized drainage basin systems (e.g. Bogen et al. 2011; Norwegian Water Resources and Energy Directorate 2019).

As already described, the mean annual atmospheric solute input for the inner Nordfjord in western Norway was computed to be 6.0 t km^{-2} yr^{-1}. The mean annual solute gross yield is 12.0 t km^{-2} yr^{-1} for the entire Erdalen drainage basin and 8.8 t km^{-2} yr^{-1} for the entire Bødalen drainage basin, which results in a mean annual chemical denudation rate of 6.0 t km^{-2} yr^{-1} in Erdalen and 2.8 t km^{-2} yr^{-1} in Bødalen (Fig. 7.4). Compared with other drainage basin systems across northern Europe, the annual chemical denudation rates calculated for the Erdalen and Bødalen drainage basins are in a similar order of magnitude as annual chemical denudation rates found for other drainage basins situated close or rather close to the west coast of Norway (Beylich et al. 2004, 2006; Beylich 2011, 2016; Beylich and Laute 2012, 2016). The fluvial suspended sediment and fluvial bedload transport recorded for the partly glacierized Erdalen and Bødalen drainage basins are with total annual sediment yields of 18.8 t km^{-2} yr^{-1} in Erdalen and 44.6 t km^{-2} yr^{-1} in Bødalen rather low (Fig. 7.4). Despite the detected high spatial variability, sediment delivery from glacierized surface areas is altogether very small within the area of the inner Nordfjord, especially when compared to the sediment delivery that is postulated for the eastern side of the Jostedalsbreen ice cap, and for other ice caps in comparable climates and topographies in Norway and worldwide (Bogen 1989, 1996; Harbor and Warburton 1993; Hallet et al. 1996; Beylich et al. 2010, 2017; Orwin et al. 2010; Liermann et al. 2012; Kennie and Bogen 2013; Bogen et al. 2011, 2013, 2015; Xu et al. 2013, 2015a, 2015b; Beylich and Laute 2015). Major reasons for the low sediment supply from both glacierized and non-glacierized surface areas in the inner Nordfjord are the high resistance of the predominant bedrock towards glacier erosion and weathering, the significant trap efficiency of proglacial lakes, the only small amounts of sediments being generally available within the drainage basin systems and the nearly closed and (in most places) stable vegetation cover developed below 1000 m a.s. l. in surface areas with sedimentary covers (Laute and Beylich 2014; Beylich and Laute 2015; Beylich et al. 2017). Altogether, fluvial sediment transport in the inner Nordfjord is clearly supply limited. However, mechanical denudation dominates with 76% (Erdalen) and 94% (Bødalen) clearly over chemical denudation (24% for Erdalen, 6% for Bødalen) in the partly glacierized drainage basin systems of the inner Nordfjord in western Norway (Fig. 7.4).

Annual atmospheric solute inputs to the non-glacierized boreal-oceanic Homla drainage basin in central Norway are 6.0 t km^{-2} yr^{-1}, which is the same annual rate that is found for the inner Nordfjord, with comparable rates of annual atmospheric solute inputs also having been found for other oceanic drainage basin systems situated close or rather close to the Norwegian west coast (Beylich et al. 2004, 2006; Beylich and Laute 2012, 2016). Also, the atmospherically corrected chemical denudation rate for the entire Homla drainage basin system is with 12.1 t km^{-2} yr^{-1} (Fig. 7.4) in a similar range of magnitude as for a number of other northern European drainage basins (e.g. Beylich et al. 2004, 2006; Beylich and Laute 2012, 2016, 2018; Beylich 2011, 2016; Dixon 2016). The mean annual suspended sediment yield (3.3 t km^{-2} yr^{-1}) and the mean annual bedload yield (0.3 t km^{-2} yr^{-1}) in the Homla drainage basin are clearly smaller than the mean annual chemical denudation rate (Fig. 7.4). As a result, the mean total annual sediment yield computed for the Homla drainage basin system (3.6 t km^{-2} yr^{-1}) is similar to the published mean values of the annual sediment yield supplied from "sediment-source areas" classified as "forest" (2.88 t km^{-2} yr^{-1}) and "mountain areas outside the glacier areas" (5.0 t km^{-2} yr^{-1}) in entire Norway (e.g. Bogen et al. 2011). Comparably low mechanical fluvial denudation rates as in the Homla drainage basin were also found in non-glacierized sub-arctic oceanic environments in Swedish and Finnish Lapland being, as the Homla drainage basin, characterized by nearly closed and continuous vegetation covers and moderate to low topographic relief (Beylich 2011, 2016; Dixon 2016).

Comparing the partly glacierized drainage basin systems in western Norway with the non-glacierized drainage basin system in central Norway, a clear dominance of mechanical denudation over chemical denudation is found for the partly glacierized drainage basin systems whereas clear dominance of chemical denudation over mechanical denudation is detected for the non-glacierized drainage basin systems (Fig. 7.4). Both chemical and mechanical denudations are generally low across the studied drainage basin systems in mainland Norway which is explained by highly resistant bedrock, an only low availability of sediments within the drainage basin systems, nearly closed and continuous vegetation covers in the surface areas with sedimentary covers, and accordingly clearly supply-limited conditions.

7.7 Summary and Prospect

Contemporary chemical and mechanical fluvial denudation rates in the partly glacierized drainage basin systems in the inner Nordfjord in western Norway are rather low with mechanical fluvial denudation being 3.1 times higher than chemical denudation in Erdalen and mechanical fluvial denudation being 15.9 times higher than chemical denudation in Bødalen. The total annual denudation in Bødalen is 1.9 times higher than the total annual denudation with mechanical fluvial denudation being 2.4 times higher in Bødalen compared to Erdalen and chemical denudation being 2.1 times higher in Erdalen as compared to Bødalen (Fig. 7.4). The low chemical denudation rates in the inner Nordfjord are explained by the cool climate and the small shares of surface areas with sedimentary covers combined with only small thicknesses of sedimentary covers, whereas the low mechanical fluvial denudation rates are largely explained by the high resistance of the predominant bedrock towards glacier erosion and weathering, the significant trap efficiency of the existing proglacial lakes, the generally only low availability of sediments within the drainage basin systems and the nearly closed and largely stable vegetation cover developed in ice-free surface areas with sedimentary covers. As a result, mechanical fluvial denudation clearly dominates over chemical denudation in the inner Nordfjord, with mechanical denudation reaching 76% of the total fluvial denudation in Erdalen and 94% of the total fluvial denudation in Bødalen. Pluvially triggered denudational process events and pluvially activated sediment sources in ice-free surface areas of the drainage basins as well as pluvially induced fluvial suspended sediment transport are, with respect to their overall contribution to drainage basin-wide annual suspended sediment yields, more important than the most thermally induced suspended sediment transport during the summer (glacier ablation) or in spring (snowmelt). At the same time, the highest quantitative share of fluvial bedload transport occurs during thermally generated snowmelt peak runoff in spring which is due to the significantly higher availability of bedload material within the mainstream channels compared to time periods later in the hydrological year. Altogether, fluvial sediment transport in the inner Nordfjord is clearly supply limited.

Also in the non-glacierized boreal-oceanic Homla drainage basin in central Norway, both chemical and mechanical fluvial denudations are of low intensity with chemical denudation being 3.4 times higher than mechanical denudation and accordingly clearly dominating over mechanical denudation (Fig. 7.4). Low chemical denudation in central Norway is mostly explained by the cool climate in combination with the small thicknesses of sedimentary covers in larger areas of the drainage basin systems, whereas low mechanical denudation is mostly explained by the nearly closed and continuous vegetation cover in combination with the high weathering resistance of the predominant bedrock and the high share of surface areas with only low topographic relief especially in the upper parts of the drainage basin systems. The clearly highest quantitative share of the total fluvial sediment load is supplied as fine-grained sediments from the fluvially dissected glaciomarine deposits found in the lower parts of the drainage basins. Altogether, chemical denudation clearly dominates over mechanical fluvial denudation with chemical denudation accounting for 77% of the total fluvial denudation. The highest quantitative share of fluvial transport occurs during thermally generated snowmelt peak runoff in spring. With respect to the total annual sediment mobilization and total annual fluvial sediment transport, the sum of snowmelt-generated peak runoff events occurring over the hydrological year is altogether more important than the annual sum of purely rainfall-generated peak runoff events. Human activities, especially agriculture and forestry, increase mechanical denudation locally through significant disturbances of the existing vegetation cover.

Total fluvial denudation in the partly glacierized mountain landscape of the inner Nordfjord in western Norway is 1.6 times (Erdalen drainage basin) or 3.0 times (Bødalen drainage basin) higher than in the non-glacierized mountain landscape south of the Trondheim fjord (Homla drainage basin) which is due to clearly higher mechanical fluvial denudation rates in the steeper and partly glacierized mountain landscape of the inner Nordfjord. While the larger topographic relief and the partly glacial coverage are important factors for the higher mechanical fluvial denudation rates, the clearly higher importance of pluvial events with respect to the mobilization and transport of sediments in the drainage basins of the inner Nordfjord as compared to the Homla drainage basin in the non-glacierized mountain landscape south of the Trondheim fjord in central Norway must be highlighted. The altogether low contemporary fluvial denudation rates found for the supply-limited fluvial systems in western and central Norway are in a similar range of magnitude than fluvial denudation rates found for other partly glacierized and non-glacierized drainage basin systems in Norway and across northern Europe. However, the mechanical fluvial denudation rates measured in the glacier-connected drainage basin systems of the inner Nordfjord on the western side of the Jostedalsbreen ice cap are clearly smaller than mechanical fluvial denudation rates reported for the eastern side of this ice cap.

A possible higher frequency of strong storm events, as postulated for Norway, may cause more frequent damages to the existing tree vegetation cover resulting in a growing potential for the erosion of sedimentary covers in the affected surface areas. At the same time, reduced precipitation sums in late fall and winter and increasing air temperatures with the related shifts of the relative shares of rain versus snow precipitation may lead to less-distinct runoff peaks and, connected to this, to reduced mechanical denudation during spring snowmelt. However, increasing air temperatures will clearly cause accelerated glacier ablation rates and, accordingly, increase mechanical denudation rates in partly glacierized drainage basin systems. A future increase of the frequency and intensity of heavy rainfall events, as postulated for Norway, will certainly increase mechanical denudation rates, particularly in the very steep drainage basin systems in western Norway, and higher annual precipitation sums in combination with increasing annual air temperatures as generally expected and postulated for the future (IPCC 2013; Hanssen-Bauer et al. 2017; Laute and Beylich 2018) will lead to higher chemical denudation rates in western and central Norway. In addition, it can be expected that the expected changes of the wind, air temperature and precipitation regimes will be of particularly high impact in surface areas that are affected by disturbances of the existing vegetation cover by human activities.

As a result, it is expected that the postulated changes of the current wind, air temperature and precipitation regimes will lead to increased annual chemical and mechanical fluvial denudation rates in both partly glacierized drainage basin systems in western Norway and non-glacierized drainage basin systems in central Norway.

References

Barsch D, Gude M, Mäusbacher R, Schukraft G, Schulte A (1994) Recent fluvial sediment budgets in glacial and periglacial environments, NW Spitsbergen. Zeitschrift für Geomorphologie Suppl 97:111–122

Beylich AA (2003) Present morphoclimates and morphodynamics of Lanjavagge, the Northern Swedish Lapland and Austdalur, East Iceland. Jökull 52:33–54

Beylich AA (2011) Mass transfers, sediment budgets and relief development in cold environments: results of long-term geomorphologic drainage basin studies in Iceland, Swedish Lapland and Finnish Lapland. Zeitschrift für Geomorphologie 55:145–174

Beylich AA (2016) Environmental drivers, spatial variability and rates of chemical and mechanical fluvial denudation in selected glacierized and nonglacierized cold climate catchment geosystems: from coordinated field data generation to integration and modeling. In: Beylich AA, Dixon JC, Zwolinski Z (eds) Source-to-sink fluxes in undisturbed cold environments. Cambridge University Press, Cambridge, pp 385–397

Beylich AA (2021) Geomorphological landscapes, earth surface processes and landforms in Norway. In: Beylich AA (ed) Landscapes and Landforms of Norway. Springer

Beylich AA, Gintz D (2004) Effects of high-magnitude/low-frequency fluvial events generated by intense snowmelt or heavy rainfall in Arctic periglacial environments in northern Swedish Lapland and northern Siberia. Geogr Ann 86A:11–29

Beylich AA, Laute K (2012) Spatial variations of surface water chemistry and chemical denudation in the Erdalen drainage basin, Nordfjord, western Norway. Geomorphology 167–168:77–90

Beylich AA, Laute K (2014) Combining impact sensor field and laboratory flume measurements with other techniques for studying fluvial bedload transport in steep mountain streams. Geomorphology 218:72–87

Beylich AA, Laute K (2015) Sediment sources, spatiotemporal variability and rates of fluvial bedload transport in glacier-connected steep mountain valleys in western Norway (Erdalen and Bødalen drainage basins). Geomorphology 228:552–567

Beylich AA, Laute K (2016) Chemical denudation in partly glacierized mountain catchments of the fjord landscape in western Norway: contemporary rates, environmental controls, and possible effects of climate change. In: Beylich AA, Dixon JC, Zwolinski Z (eds) Source-to-sink fluxes in undisturbed cold environments. Cambridge University Press, Cambridge, pp 275–292

Beylich AA, Laute K (2018) Morphoclimatic controls of contemporary chemical and mechanical denudation in a boreal-oceanic drainage basin system in central Norway (Homla drainage basin, Trøndelag). Geogr Ann 100A:116–139

Beylich AA, Sandberg O (2005) Geomorphic effects of the extreme rainfall event of 20-21 July 2004 in the Latnjavagge catchment, northern Swedish Lapland. Geogr Ann 87A:409–419

Beylich AA, Kolstrup E, Thyrsted T, Linde N, Pedersen LB, Dynesius L (2004) Chemical denudation in Arctic-alpine Latnjavagge (Swedish Lapland) in relation to regolith as assessed by radio magnetotelluric-geophysical profiles. Geomorphology 57:303–319

Beylich AA, Schmidt KH, Neuvonen S, Forbrich I, Schildt A (2006) Solute fluxes in the Kidisjoki catchment, subarctic Finnish Lapland. Géomorphol: Relief, Processus, Environ 3:205–2012

Beylich AA, Laute K, Liermann S, Hansen L, Burki V, Vatne G, Fredin O, Gintz D, Berthling I (2009) Subrecent sediment dynamics and sediment budget of the braided sandur system at Sandane, Erdalen (Nordfjord, western Norway). Norw J Geogr 63:123–131

Beylich AA, Liermann S, Laute K (2010) Fluvial transport during thermally and pluvially induced peak runoff events in a glacier-fed mountain catchment in western Norway. Geogr Ann 92A:237–246

Beylich AA, Laute K, Storms JEA (2017) Contemporary suspended sediment dynamics within two partly glacierized mountain drainage basins in western Norway (Erdalen and Bødalen, inner Nordfjord). Geomorphology 287:126–143

Beylich AA, DENUCHANGE Team (2018) The new I.A.G./A.I.G. Working Group on Denudation and Environmental Changes in Different Morphoclimatic Zones (DENUCHANGE): scientific need, key research questions and planned activities. In: 1st Workshop of the IAG/AIG Working Group DENUCHANGE: Denudation and Environmental Changes in Different Morphoclimatic Zones. Storkowo-Szczecinek (Poland), 25–27 September 2018. Book of Abstracts: 3–5

Bogen J (1989) Glacial sediment production and development of hydro-electric power in glacierized area. Ann Glaciol 13:6–11

Bogen J (1996) Erosion rates and sediment yields of glaciers. Ann Glaciol 22:48–52

Bogen J, Bønsnes TE, Kennie P, Ottesen RT, Eggen OA (2011) The sediment budget of Norwegian rivers and the flux of particle bound chemical elements. Geophys Res Abs 13. EGU2011-8694-1

Bogen J, Xu M, Kennie P (2013) The impact of proglacial lakes on downstream sediment delivery. Geophys Res Abs 15. EGU2013-11005

Bogen J, Xu M, Melvold K (2015) The impact of climate change on sediment delivery from the Jostedalsbreen ice cap, Southern Norway. Geophys Res Abs 17. EGU2015-1203

Bunte K, Abt SR, Potyondy JP, Swingle KW (2008) A comparison of coarse bedload measured with bedload traps and Helley-Smith samplers. Geodin Acta 21:53–66

Burki V, Larsen E, Fredin O, Nesje A (2009) Glacial remobilization cycles as revealed by lateral moraine sediment, Bødalen glacier foreland, western Norway. Holocene 19:415–426

Clark MJ (1988) Periglacial hydrology. In: Clark MJ (ed) Advances in periglacial geomorphology. Wiley, Chichester, pp 415–462

Dixon JC (2016) Contemporary solute and sedimentary fluxes in Arctic and subarctic environments: current knowledge. In: Beylich AA, Dixon JC, Zwolinski Z (eds) Source-to-sink fluxes in undisturbed

cold environments. Cambridge University Press, Cambridge, pp 39–51
Fredin O, Bergstrøm B, Eilertsen R, Hansen L, Longva O, Nesje A, Sveian H (2013) Glacial landforms and quaternary landscape development in Norway. In: Olsen L, Fredin O, Olesen O (eds) Quat Geol Norway. 13:5–25
Glässer E (1993) Norwegen. Wissenschaftliche Länderkunden 14. Darmstadt
Hallet B, Hunter L, Bogen J (1996) Rales of erosion and sediment evacuation by glaciers: a review of field data and their implications. Global Planet Change 12:213–235
Hansen L, Beylich AA, Burki V, Eilertsen R, Fredin O, Larsen E, Lyså A, Nesje A, Stalsberg K, Tønnesen JF (2009) Stratigraphic architecture and infill history of a deglaciated bedrock valley based on georadar, seismic profiling and drilling. Sedimentology 56:1751–1773
Hanssen-Bauer I, Førland EJ, Haddeland I, Hisdal H, Mayer S, Nesje A, Nilsen JEØ, Sandven S, Sandø AB, Sorteberg A, Ådlandsvik B (2017) Climate in Norway 2100—a knowledge base for climate adaptation. NCCS report 1/2017
Harbor J, Warburton J (1993) Relative rates of glacial and non-glacial erosion processes in Alpine environments. Arct Alp Res 25:1–7
Hongve D (2006) A revised procedure for discharge measurement by means of the salt diluting method. Hydrol Process 1:267–270
IPCC (2013) Summary for policymakers. In: Stocker TF, Qin D, Plattner GK; Tignor M, Allen SK, Boschung J, Nauels A, Xia Y, Bex V, Midgley PM (eds) Climate change 2013: the physical science basis. Contribution of working group I to the fifth assessment report of the intergovernmental panel on climate change. Cambridge University Press, Cambridge
Kennie P, Bogen J (2013) Varved lake sediment sequences and their relation to climate-controlled hydrological processes in proglacial Lake Nigardsvatn, western Norway. Geophys Res Abs 15. EGU2013-13706
Kennie P, Bogen J (2014) Changes in sedimentation patterns in proglacial lake Engabrevatn as a consequence of Svartisen hydropower operations, Nordland, Norway. Geophys Res Abs 16. EGU2014-11786
Ketzler G, Römer W, Beylich AA (2021) The Climate of Norway. In: Beylich AA (ed) (2021) Landscapes and Landforms of Norway. Springer
Laumann T, Nesje A (2009) A simple method of simulating the future frontal position of Briksdalsbreen, western Norway. Holocene 19:221–228
Laute K, Beylich AA (2014) Environmental controls, rates and mass transfers of contemporary hillslope processes in the headwaters of two glacier-connected drainage basins in western Norway. Geomorphology 216:93–113
Laute K, Beylich AA (2016) Sediment delivery from headwater slope systems and relief development in steep mountain valleys in western Norway. In: Beylich AA, Dixon JC, Zwolinski Z (eds) Source-to sink fluxes in undisturbed cold environments. Cambridge University Press, Cambridge, pp 293–312
Laute K, Beylich AA (2018) Potential effects of climate change on future snow avalanche activity in western Norway deduced from meteorological data. Geogr Ann 100A:163–184
Laute K, Beylich AA (2021) Recent glacier changes and formation of new proglacial lakes at the Jostedalsbreen ice cap in southwest Norway. In: Beylich AA (ed) Landscapes and landforms of Norway. Springer
Liermann S, Beylich AA, van Welden A (2012) Contemporary suspended sediment transfer and accumulation processes in the small proglacial Sætrevatnet sub-catchment, Bødalen, western Norway. Geomorphology 167–168:91–101
Lutro O (2005) Berggrunnskart Stjørdal 16211. 1:50.000. Norges geologiske undersøkelse, Trondheim
Nesje A, Bakke J, Dahl SO, Lie Ø, Matthews JA (2008) Norwegian mountain glaciers in the past, present and future. Global Planetary Change 60:10–27
Norwegian Water Resources and Energy Directorate (2019) https://www.nve.no (September 2019)
Olsen L, Sveian H, Bergstrøm B, Ottesen D, Rise L (2013) Quaternary glaciations and their variations in Norway and on the Norwegian continental shelf. In: Olsen L, Fredin O, Olesen O (eds) Quat Geol Norway. 13:27–78
Orwin JF, Lamoureux SF, Warburton J, Beylich AA (2010) A framework for characterizing fluvial sediment fluxes from source to sink in cold environments. Geogr Ann 92A:155–176
Ramberg IB, Bryhni I, Nøttvedt A, Rangnes K (eds) (2013) Landet blir til. Norges geologi. 2. utgave. Norsk Geologisk Forening, Trondheim, 655p
Rise L, Bøe R, Sveian H, Lyså A, Olsen L (2006) The deglaciation history of Trondheimsfjorden and Trondheimsleia, central Norway. Norw J Geol 86:419–438
Vorren TO, Mangerud J (2013) Istider kommer og går. In: Ramberg IB, Bryhni I, Nøttvedt A, Rangnes K (eds) Landet blir til. Norges geologi. 2. utgave. Norsk Geologisk Forening, Trondheim, pp 494–547
Winkler S (2021) Terminal moraine formation processes and geomorphology of glacier forelands at the selected outlet glaciers of Jostedalsbreen, South Norway. In: Beylich AA (ed) Landscapes and Landforms of Norway. Springer
Winkler S, Elvehøy H, Nesje A (2009) Glacier fluctuations of Jostedalsbreen, western Norway, during the past 20 years: the sensitive response of maritime mountain glaciers. Holocene 19:395–414
Xu M, Bogen J, Bønsnes TE, Gytri S (2013) Proglacial lake sedimentation from jökulhlaups, Blåmannsisen, northern Norway. Geophys Res Abs 15. EGU2013-11062-1
Xu M, Bogen J, Ragulina G, Read A (2015a) Long-term sediment yield of the Austdalsbreen glacier, southwest Norway. Geophys Res Abs 17. EGU2015-1204
Xu M, Bogen J, Ragulina G, Read A (2015b) Early and late Holocene sediment yield of Austdalsbreen glacier, southwest Norway. Geomorphology 246:277–289

Periglacial Landforms in Jotunheimen, Central Southern Norway, and Their Altitudinal Distribution

8

Stefan Winkler, Anika Donner, and Angela Tintrup gen. Suntrup

Abstract

Jotunheimen in central South Norway is home of the highest peaks in Norway. It is characterised by omnipresent high-altitudinal plateaus creating a specific gross morphology and providing diverse natural environments for the periglacial process-system. A variety of periglacial landforms including different types of patterned ground, rock glaciers and ice-cored moraines, and solifluction features occur in Jotunheimen. Their altitudinal distribution is strongly controlled by the regional distribution of permafrost, but may also be influenced by other factors like topography, bedrock properties, sedimentology, or vegetation. The regional distribution of permafrost modelled on basis of meteorological parameters and ground temperature measurements accompanied by local air temperature records provide a good basis for the investigation of causal links between the altitudinal zonation of certain periglacial landforms and the lower limit of permafrost. Whereas the fluctuations of this lower limit of permafrost since Deglaciation are reasonably well constraint and its uppermost position during the Holocene Thermal Maximum followed by a subsequent late Holocene lowering seems confirmed, there is hardly any available chronological data on the periglacial landforms themselves. Exceptions are small-scale features on recently deglaciated glacier forelands, but other well-developed landforms seem to have formed shortly after Deglaciation and appear quite inactive today.

S. Winkler (✉) · A. Tintrup gen. Suntrup
Department of Geography and Geology, University of Würzburg, Am Hubland, 97074 Würzburg, Germany
e-mail: stefan.winkler@uni-wuerzburg.de

A. Tintrup gen. Suntrup
e-mail: angela.tintrup-suntrup@uni-wuerzburg.de

A. Donner
Institute of Geology, University of Innsbruck, Innrain 52f, 6020 Innsbruck, Austria
e-mail: Anika.Donner@uibk.ac.at

Keywords

Periglacial landforms • Periglacial process-system • Altitudinal zonation • Permafrost • Jotunheimen

8.1 Introduction

The Jotunheimen Mountains in central South Norway are part of the Scandinavian Caledonides, the 1800 km long and up to 300 km wide geologically old mountain range striking SSW–NNE that constitutes the central structure of the Scandinavian Peninsula (Sigmond et al. 1984; Roberts and Gee 1985; see Sect. 8.2.1). They cover an area of approximately 3500 km^2 and are home of the highest peaks in Norway (Galdhøpiggen, 2469 m a.s.l.; Glittertinden 2464 m a.s.l.) and about 100 other peaks higher than 2000 m a.s.l. (Fig. 8.1). Unlike other high mountain ranges that can be subdivided into individual ranges following the overall tectonic structure or major faults, Jotunheimen can be divided into 14 separate mountain massifs (e.g. Hurrugane, Smørstabtindan, or Memurutindan; Fig. 8.2a, b, f). These massifs are surrounded by wide plateaus with flat surfaces that clearly dominate among Jotunheimen's different geomorphological units and are characteristic for the whole mountain region. Most of these plateaus lie at altitudes above 1500 m a.s.l. and a few major valleys like Leirdalen, Visdalen, or Veodalen topographically subdivide Jotunheimen into different sectors, albeit they mostly do not represent structural geology- or tectonic-related boundaries (Fig. 8.2c). An exception is the comparatively deeply incised Bøverdalen that coincidentally constitutes Jotunheimen's northern boundary towards the Breheimen region (Fig. 8.3c). The large lakes Bygdin and Tyin are Jotunheimens's southern boundary and Sjodalen represents its eastern one together with the somewhat diffuse transition to Valdresflye (Fig. 8.3d). Finally, the Hurrungane Massif constitutes the westernmost massif in Jotunheimen and

A. A. Beylich (ed.), *Landscapes and Landforms of Norway*, World Geomorphological Landscapes,
https://doi.org/10.1007/978-3-030-52563-7_8

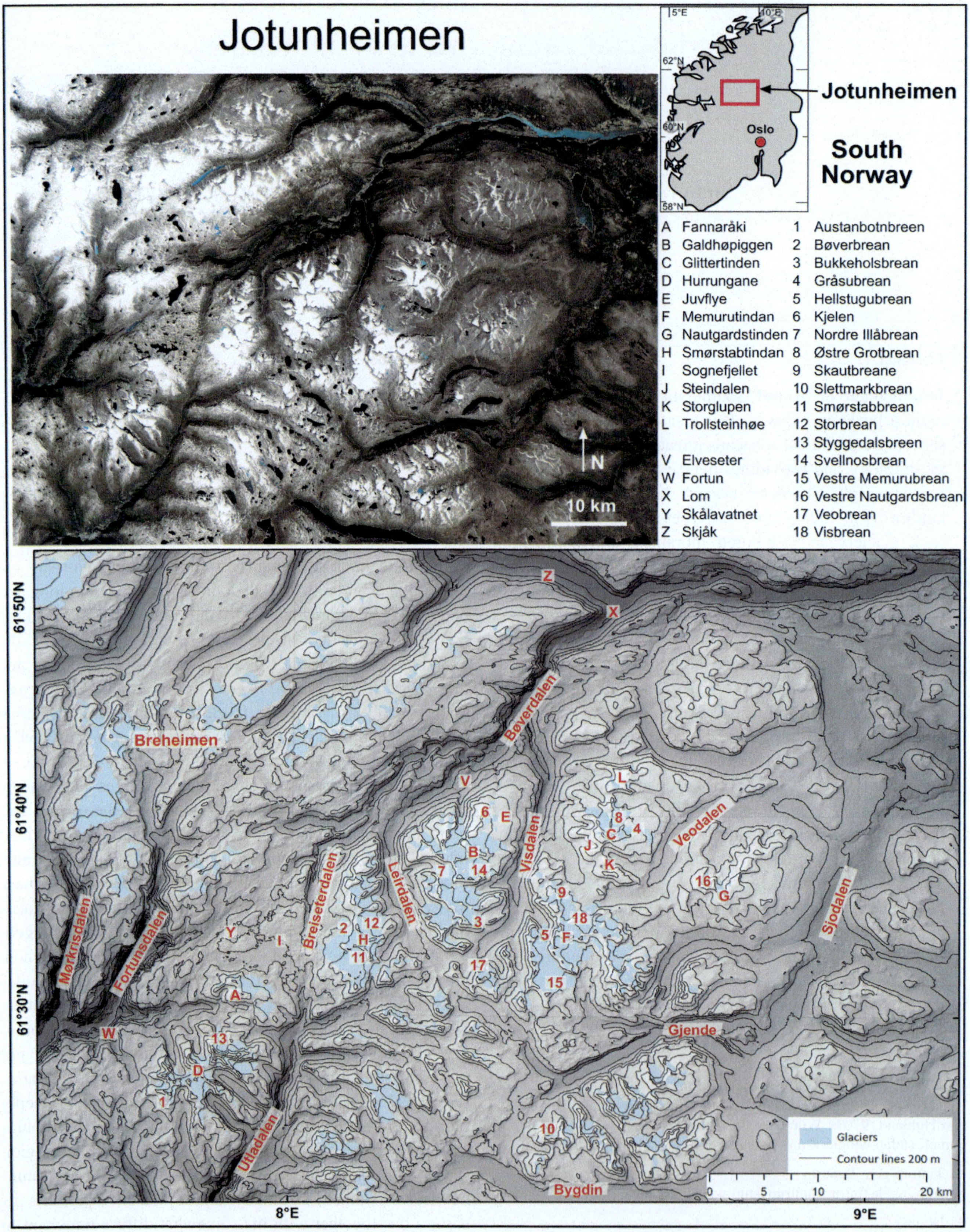

Fig. 8.1 Satellite image (2018; adapted from www.norgeibilder.no) and map/digital elevation model of Jotunheimen (adapted/modified from www.høydedata.no, both © Kartverket/www.kartverket.no). The small insert depicts Jotunheimen's location within South Norway. Locations referred to in the text are indicated on the map and include few meteorological stations (cf. Table 8.1)

Fig. 8.2 Gross morphology of Jotunheimen: **a** Hurrungane Massif in western Jotunheimen, one of the most 'alpine'/rugged-shaped parts of Jotunheimen, Styggedalsbreen in the centre of the image. **b** View from Keilhaus Top on the ascent to Galdhøpiggen towards East. The upper part of Svellnosbrean (foreground) and the Memurutindane Massif with Hellstugubrean (background) are visible. The deeply incised valleys contrast to the gently/moderately undulating summit levels. **c** Upper Visdalen with its wide valley cross-profile, typical for some mayor valleys in Jotunheimen. **d** Gravdalen in western-central Jotunheimen. **e** View from Trollsteinkvelven (tributary valley of Veodalen) in southeasterly direction towards Nautgardstinden (far distance) depicting the smooth surfaces for summit levels at high altitudes. **f** View from the summit of Fannaråki towards Smørstabtindan in western-central Jotunheimen, note the blockfield in the foreground (Photos: Stefan Winkler, a/f 07.08.1994, b 08.08.1992, c 09.08.2000, d 24.07.2016, e 15.07.2017)

combined with Sognefjellet (Fig. 8.3a) is usually referred to as West Jotunheimen.

Jotunheimen is characterised by a current co-existence of both glacial and periglacial geomorphological process-systems. According to the latest glacier inventory representing the status in 2003 the total glacier area in Jotunheimen is c. 190 km^2 (Baumann et al. 2009). Most glaciers are small individual valley-type or cirque-type glaciers, several simultaneously connected to others forming glacier complexes of multiple glacier units that are not entirely separated by steep rock walls

Fig. 8.3 Gross morphology of Jotunheimen: **a** Sognefjellet in western Jotunheimen, Smørstabbtindan in the background. **b** Part of Juvflye with Juvvasshøe (far distance) and Juvvasshytta (centre), note the patterned ground cover. **c** View from the lower slope of Juvflye towards Breheimen showing a similar gross morphology with isolated peaks (some with cirques) 'sitting' on the top of high-altitude plateaus. The deeply incised valley in the centre of the photo is Bøverdalen. **d** View from Valdresflye towards the southeastern part of Jotunheimen depicting the diffuse transition without clear morphological boundary. **e** Skautflye in eastern-central Jotunheimen, note the ice-cored moraines at glaciers in background (Photos: Stefan Winkler, a 03.09.2004, b 26.07.2018, c 14.07.2018, d 24.06.1998, e 17.08.2008)

due to the specific morphology of the summit areas/peaks of some massifs. Although the largest glacier complex Smørstabbrean had an area of about 15.8 km^2 in 2003 (Andreassen and Winsvold 2012), the concurrent mean area of Jotunheimen's 262 individual glaciers was 0.87 km^2 and their mean length 1.02 km (Baumann et al. 2009). Current glaciers roughly range from 1300 to 2300 m a.s.l. but the glaciation limit/climatic equilibrium line shows a significant rise from western Jotunheimen eastwards reflecting the transitional zone from maritime to continental climate within South Norway. Eastern

Jotunheimen is the most continental glacier area in South Norway (Østrem et al. 1988) and the general centre of glaciation in Jotunheimen clearly is located in its western and central parts.

The restricted current (and Holocene: see Sect. 8.3.1) glaciation of Jotunheimen combined with considerable proportions of its land surface located at high altitudes promotes the existence of permafrost and occurrence of related periglacial processes. Jotunheimen constitutes the largest coherent area of mountain permafrost in South Norway and a number of relevant parameters like permafrost depth, ground temperature regimes, or altitudinal lower limits of permafrost occurrence and its temporal variability through the Holocene have been studied using a number of different methodological attempts including empirical ground measurements and sophisticated modelling (e.g. Ødegård et al. 1992, 1996; Isaksen et al. 2002, 2011; Etzelmüller et al. 2003; Farbrot et al. 2011; Hipp et al. 2012; Lilleøren et al. 2012; Gisnås et al. 2013, 2017; see Sects. 8.2.2 and 8.3.2). Periglacial processes and landforms have accordingly been addressed in a considerable number of studies during the past decades. These studies do, however, mostly focus on individual sites and/or individual features. A region-wide comprehensive overview has still not been undertaken. This selective summary of studies conducted on periglacial processes and landforms in Jotunheimen cannot substitute such an attempt but aims to highlight relevant characteristics of the periglacial process-system in Jotunheimen. In particular, it should outline which periglacial landforms do occur and dominate the region potentially allowing subsequent comparison to periglacial process-systems in other mountain regions worldwide. To enable such comparative attempt, the relevant natural framework besides the regional distribution of permafrost will briefly be outlined, for example, the Pleistocene/Holocene glaciation history or Jotunheimen's specific gross morphology. Finally, the altitudinal zonation of periglacial landforms will be discussed to provide a good basis for investigations of its causal links to the lower limit of permafrost and other controlling factors.

8.2 Natural Environment

8.2.1 Geology and Gross Morphology

Jotunheimen is geologically part of the Scandinavian Caledonides (Scandes) formed by orogenetic collision processes between Laurentia and Baltica related to the closure of the Iapetus Ocean from Mid Silurian to Early Devonian c. 425 to 400 Ma ago (Roberts and Gee 1985; Fossen et al. 2008). The bedrock of the Jotun Nappe Complex exposed in Jotunheimen itself is of early-Proterozoic age and underwent significant metamorphosis and folding approximate 900 Ma ago. During the Caledonian orogeny these Precambrian high-grade metamorphic rocks of the Jotun Nappe Complex were pushed as a thrust nappe stack from the Northwest over Baltica basement (the Western Gneiss Complex today exposed, e.g. in the Jostedalsbreen and Breheimen regions west/north of Jotunheimen) into its current position and subsequently exposed by tectonic uplift and erosion. Pyroxene-granulite gneiss is the dominating lithology of the Jotun Nappe Complex and central Jotunheimen (Battey and Bryhni 1981), but locally in comparatively small areas few other bedrock types like gabbroic gneiss, mylonitic gneiss, amphibolite, or ultramafic bedrock like peridotite crop out (Lutro and Tveten 1996). In the context of geomorphological processes it is, however, most relevant that the dominating pyroxene-granulite gneiss is comparatively homogeneous (Battey and McRitchie 1973, 1975), especially in its resistance to weathering and erosion (cf. Matthews et al. 2016). This allows not only application of Schmidt-hammer exposure-age dating (SHD) for surface-exposure dating of boulders and bedrock on a regional scale (Matthews and Owen 2010; Matthews and Winkler 2011; Matthews et al. 2014, 2018; Winkler and Matthews 2014; Winkler et al. 2016), it also contributes to generally low to moderate regional talus production rates if compared to geologically younger mountain ranges with less resistant bedrock. The latter is in combination with Jotunheimen's specific topography of significant relevance for the occurrence of specific periglacial landforms like rock glaciers that are, among other factors, depending on available debris sources.

Following the abovementioned orogenetic processes, the Jotunheimen region was exposed to subaerial weathering and erosion during most of late Palaeozoic, Mesozoic, and Cenozoic in combination with periods of tectonic uplift, the latest commencing in late Tertiary (Lidmar-Bergström et al. 2000). These periods of subaerial erosion have been causally linked to a characteristic of Jotunheimen's modern topography: the wide plateaus as gently undulating flat surface features at high altitudes above the modern valley levels (Figs. 8.2 and 8.3). Although some parts of Jotunheimen like the Hurrungane Massif in western Jotunheimen exhibit a rugged 'Alpine' orography with sharp peaks, cirques, arêtes, and deeply incised valleys or gorges (Fig. 8.2a), these specific features of widespread flat plateaus leave beyond controversy their great imprint on the gross morphology of Jotunheimen. Simultaneously they represent one of the most important geomorphological differences to other high mountain ranges worldwide.

These plateaus have since the early twentieth century been interpreted as relict landforms and evidence of preglacial landform development summarised by the term 'paleic surface' (Reusch 1901; Ahlmann 1919, 1922; Gjessing 1967). Lidmar-Bergström et al. (2000) attempt to reconstruct different levels of preglacial surfaces in South Norway and relate them to different episodes of tectonic uplift and surface erosion during the late Mesozoic and Tertiary. The highest summits of Jotunheimen are, furthermore, interpreted as 'residual massifs' that exhibit a tentative summit level ('gipfelflur') likely caused by more resistant bedrock (Gjessing 1978). But irrespective of their influence on the spatial distribution and diversity/selective formation of periglacial landforms in Jotunheimen, detailed formation mechanisms of these high-altitude plateaus and the processes involved remain mostly unknown alongside their age. Deep chemical weathering combined with specific processes of denudation at a passive continental margin is widely considered, albeit few preglacial weathering residua that survived multiple Pleistocene glaciations in Scandinavia (e.g. Lidmar-Bergström 1988) hardly allow any definite judgement (Ebert 2009). Recent research has, furthermore, revealed that unlike previous assumptions glacial erosion should also have affected the 'paleic surface' (Egholm et al. 2017, Andersen et al. 2018). Even if Jotunheimen was presumably covered by erosional ineffective cold-based ice during most of the latest Pleistocene glaciations (see below) it is well possible that both flat surfaces and gentle shaped summit regions in Jotunheimen constitute an inherent morphology rather than actual former Mesozoic or Tertiary surfaces. Like the fjord region of western South Norway (e.g. Nesje et al. 1992; Nesje and Whillans 1994), Jotunheimen had developed fluvial valleys during the late Tertiary uplift (presumably incised to a lesser extent) prior to the first Pleistocene glaciations providing linear drainage patterns subsequently utilised by the Scandinavian Ice Sheet (Hall et al. 2013). Since their occurrence rather than their detailed mode of formation or age is relevant in the context of periglacial landforms and processes described here, it is referred to the existing literature for more insights into the discussion on preglacial landform development of Norway and Scandinavia (e.g. Klemsdal and Sjulsen 1988; Rudberg 1988; Lidmar-Bergström et al. 2000; Goodfellow et al. 2008; Kleman et al. 2013). The automatic regional classification of topography in Norway (Etzelmüller et al. 2007) describes the majority of Jotunheimen just as similar regions (covering 12% of entire South Norway) as 'high paleic mountains with glacial incision, mostly moderate slopes'.

8.2.2 Climate and Current Permafrost

Jotunheimen has a position in the transitional climatic zone between the strong maritime climate of western South Norway and the more continental climate of eastern South Norway. Although it owns a mountain climate with its specific influencing factors like altitude or aspect, the general climate gradient of decreasing precipitation within Jotunheimen from the West/Southwest towards the East/Northeast is visible in the climate data (Table 8.1). Annual precipitation decreases from about 2000–1500 mm in Hurrugane to about 1500–1000 mm at Galdhøpiggen and 1000–750 mm in upper Veodalen in eastern Jotunheimen (www.senorge.no). Mean annual air temperatures in their dependency on altitude do not depict a comparable West-East gradient in Jotunheimen, but with increasing continentality annual air temperature amplitudes increase and are basis for its 'weak oceanic' and accordingly 'transitional' eco-climatological classification (Moen 1998). Unfortunately, most climate data available for the Jotunheimen region are from its outer fringe and hardly representative for its central high-altitude plateaus. The data generally reflect, however, Jotunheimen's transitional climatic character and the abovementioned general trends of air temperature and precipitation including more local factors like a significant rain shadow affecting meteorological stations at its north-eastern boundary (Lom, Gjeilo i Skjåk; Table 8.1b).

Because the (former) summit station Fannaråki needs to be treated with caution due to its exposure to westerly air flow and specific local conditions (Winkler 2001), only the high-altitude meteorological stations Sognefjellet and Juvvasshøe can be regarded as representative for the zones dominated by periglacial processes and permafrost in central Jotunheimen. Consequently, most studies referring to meteorological data utilised their data series for both interpretation and modelling. Detailed data reveals, however, that climatic conditions are quite variable on a year-to-year basis (Table 8.2). The meteorological station Sognefjellet (1413 m a.s.l.) has a mean annual air temperature (MAT) of −3.1 °C for the normal period 1961–90 whereas the station Juvvasshøe on Juvflye (1894 m a.s.l.; operating since 1999) reports an average mean annual air temperature of −3.4 °C for the operational drift period 1999–2018 and a calculated −4.6 °C for the same normal period 1961–90 (eKlima data base by www.met.no). Previously, Ødegård et al. (1992) calculated MATs ranging from −2.6 °C at 1500 m a.s.l. to −6.4 °C at 2200 m a.s.l. on Juvflye based on data from 11 meteorological stations around Jotunheimen. These values correspond well to the 1 km-gridded MAT normals (1971–2000) between −2.0 and −4.0 °C (adapted from www.met.no) given for the five study sites of Winkler et al. (2016) in a transect from 1500 to 1925 m a.s.l. at the same location. Estimates for annual precipitation at Juvflye/Juvvasshøe range from 800 to 1000 mm (Farbrot et al. 2011), but detailed data are not available and the 1 km-gridded mean annual precipitation normals (www.met.no) show slightly higher values. Ødegård et al. (1992) mention, however, that strong winds typical for Juvflye result in comparatively little snow cover and a (late) maximum snow depth of 0.5 m in May. Winkler et al. (2016) note that Juvflye was largely snow-free during their fieldwork in July 2015 when in most other parts of central Jotunheimen the terrain above about 1200 m a.s.l. retained snow cover after a snowy winter and

Table 8.1 Climate normals (1961–90) of air temperature (a) and precipitation (b) for selected meteorological stations in and around Jotunheimen (data: www.met.no). The station list is arranged according to the geographical location of the stations: West (top), North, East, South (bottom; cf. Fig. 8.1 but note that few stations are located outside the area covered on the map)

Station	Altitude (m a.s.l.)	Jan (° C)	Feb	Mar	Apr	May	Jun	Jul	Aug	Sep	Oct	Nov	Dec	Year ∅
(a)														
Fannaråki	2062	−9.5	−9.7	−8.6	−7.3	−2.7	1.2	2.7	2.4	−1.3	−3.5	−7.2	−9.0	−4.4
Fortun	27	−5.1	−4.9	−1.0	3.8	9.6	13.0	14.2	13.2	8.9	4.8	−0.4	−3.4	4.4
Sognefjell	1413	−10.7	−10.2	−9.4	−5.8	−0.2	4.2	5.7	5.4	1.2	−2.1	−6.9	−8.8	−3.1
Bøverdal–Sletten	594	−9.7	−9.1	−4.3	0.6	6.3	10.5	12.1	11.0	6.8	2.5	−4.3	−7.3	1.3
Elveseter	674	−9.3	−8.6	−4.5	−0.3	6.1	10.2	11.5	10.3	6.3	2.2	−4.4	−7.4	1.0
Juvvasshøe	1894	−11.4	−11.1	−10.1	−7.4	−2.4	2.3	3.9	3.3	−1.2	−3.1	−8.4	−9.7	−4.6
Gjeilo i Skjåk	378	−9.4	−8.2	−2.7	2.7	8.5	12.7	13.9	12.8	8.4	3.8	−2.9	−6.6	2.8
Lom	382	−9.6	−8.4	−3.0	2.4	8.4	12.6	13.9	12.8	8.3	3.7	−3.0	−7.0	2.6
Beitostølen	822	−9.8	−9.0	−6.2	−1.5	4.5	9.8	11.1	9.9	5.0	1.3	−4.8	−8.0	0.2
Hindseter	896	−10.7	−9.8	−6.3	−1.9	4.7	9.4	10.8	9.6	5.4	0.8	−5.2	−9.2	−0.2
Tyinkrysset	864	−8.7	−8.0	−6.7	−1.6	3.8	9.1	10.3	9.4	5.4	1.7	−3.7	−6.5	0.4

Station	Altitude (m a.s.l.)	Jan (mm)	Feb	Mar	Apr	May	Jun	Jul	Aug	Sep	Oct	Nov	Dec	Year Σ
(b)														
Fannaråki	2062	119	85	85	74	59	72	104	113	115	119	133	122	1200
Fortun	27	72	43	51	22	34	46	57	65	90	99	80	80	739
Skålavatn	1014	83	53	66	34	48	69	72	87	122	114	90	92	930
Sognefjell	1413	76	48	63	30	40	60	78	85	109	108	82	81	860
Bøverdal	701	37	21	26	13	22	33	45	44	54	55	49	40	439
Bøverdal–Sletten	594	35	20	24	12	23	36	48	45	52	53	45	37	430
Elveseter	674	46	24	32	16	28	41	57	55	64	66	52	49	530
Gjeilo i Skjåk	378	24	13	14	8	16	28	44	34	30	32	25	27	295
Lom	382	22	14	14	10	17	34	48	36	37	35	27	27	321
Beitostølen	822	51	32	40	29	57	74	84	84	79	80	64	46	720
Bygdin	1055	73	50	63	47	77	89	104	116	117	122	93	70	1022
Grov	808	31	20	23	16	30	53	68	56	47	49	36	34	463
Hindseter	896	35	22	25	18	34	54	97	55	48	52	42	38	490
Tyinkrysset	864	90	58	68	40	73	90	103	103	116	124	103	87	1055

Table 8.2 Maximum, minimum, and average monthly air temperatures (period September 1999 to September 2018) for the meteorological station Juvvasshøe (1894 m a.s.l.) on Juvflye (data: www.met.no)

Air temperature	Jan (°C)	Feb	Mar	Apr	May	Jun	Jul	Aug	Sep	Oct	Nov	Dec
Maximum	−7.2	−5.6	−5.4	−2.1	4.1	4.8	9.6	8.5	4.3	0.2	−1.4	−5.1
Minimum	−12.3	−14.2	−13.3	−9.0	−4.8	−0.3	3.1	2.8	−1.7	−7.3	−12.1	−13.9
Mean	−9.84	−9.94	−9.08	−5.53	−1.50	2.14	5.54	4.18	1.08	−3.04	−6.82	−8.91

an unusual cold spring season. Little snow cover can, therefore, be expected for such exposed high-altitude surfaces that are heavily affected by wind drift of snow, an important factor for the occurrence and development of permafrost.

Jotunheimen is one of the major permafrost regions in whole Scandinavia (Gisnås et al. 2013, 2017). According to Etzelmüller et al. (2003), it represents about 35% of the total permafrost area in South Norway and has the unique characteristic of being simultaneously considerably glacierised. A number of studies on the properties of permafrost (e.g. ground temperatures, permafrost depths, lower limits of continuous/discontinuous/sporadic permafrost) have been conducted in Jotunheimen, the vast majority of these at Juvflye/Juvasshøe and its surroundings. Based on their abovementioned MAT calculations, Ødegård et al. (1992, 1996) report an altitudinal lower limit of permafrost in central Jotunheimen (i.e. around Juvflye) of 1450 m a.s.l. confirming previous calculations of King (1984). They additionally measured shallow mean annual ground temperature (MGT) data for Galdehøe (2195 m a.s.l.) between −4.2 and −4.4 °C, a site near Juvvatnet (1855 m a.s.l.) between −1.7 and −1.9 °C, and for Dugurdsmålkampen (1547 m a.s.l.) between −0.6 and −0.7 °C in order to relate MGT to MAT data. For their lower limit of permafrost with 0 °C MGT they give an equivalent MAT of about −2.5 °C. Subsequent research applying different techniques like borehole-temperature measurements, bottom temperature of snow (BTS) measurements, and geophysical exploration utilising ground-penetrating RADAR, DC-resistivity, or seismic refraction has meanwhile largely confirmed and refined these first results. The relationships between ground, ground surface, and air temperatures have now been well established (Isaksen et al. 2003) and can, for example, be applied for remote sensing approaches or detailed modelling approaches to delineate the spatial distribution of permafrost over wider regions (Ødegård et al. 1999; Gisnås et al. 2013). Farbrot et al. (2011) describe that the 'surface offset', i.e. the difference between MAT and mean annual ground surface temperature (MGST), can vary between <1 °C for exposed, bare sites and up to 4.5 °C for sites that experience a prolonged snow cover. Their findings highlight that local snow conditions may be decisive for the spatial and altitudinal distribution of permafrost in Jotunheimen, especially given its specific gross morphology (see Sect. 8.2.1). Additionally, they conclude that the West-East gradient of lower mountain permafrost limits towards more continental regions is less (or not) caused by lower MAT but decreased snow fall and lower limits of blockfields, the latter promoting negative temperature anomalies with their coarse boulder accumulations.

Isaksen et al. (2002) place the lower limit of permafrost based on BTS and DC-resistivity measurements at 1460 m a.s.l. on Juvflye, Hauck et al. (2004) by applying three different geophysical techniques into a zone between 1470 and 1410 m a.s.l. Farbrot et al. (2011) measure MGTs at 10 m depths for a borehole-transect at Juvflye between 1894 and 1307 m a.s.l. and calculate an altitudinal gradient of −0.005 °C/m^{-1}, less than at more eastern and continental study sites. Five boreholes between 1894 and 1559 m a.s.l. reveal MGTs ranging from −2.5 °C and −0.3 °C at 10 m depth indicating permafrost. Two lower boreholes at 1458 m a.s.l. and 1307 m a.s.l. do not show negative MGTs but only seasonal frost of several metres thickness, i.e. indicate no surface-near permafrost. Based on a 129-m deep borehole at Juvvasshøe (1894 m a.s.l.), Isaksen et al. (2001) calculate a total thickness of permafrost of ~380 m at the site. Ødegård et al. (1999) highlight that in contrast to this site permafrost between 1600 and 1450 m a.s.l. at Juvflye is in a degrading stage and at 1600 m a.s.l. restricted to a depth below 20 m. They conclude that the permafrost likely formed in the more recent past during the 'Little Ice Age' (LIA), but new results from Winkler et al. (2016) open for the possibility that it is in fact much older (see Sect. 8.3.2). At a different site near Glittertinden in eastern Jotunheimen, Harris and Cook (1986, 1988) place the lower limit of continuous permafrost at 1810 m a.s.l. and for discontinuous one at 1640 m a.s.l. Sporadic (mountain) permafrost may locally occur at lower altitudes in eastern Jotunheimen, but accurate field data is lacking and current maps/models (Gisnås et al. 2013, 2017) naturally have restrictions regarding their resolution. Current and future climate change will also affect the properties of permafrost in Jotunheimen, in particular its near-surface temperature, the depth of the active layer, and its altitudinal lower limits (Hipp et al. 2012). It will likely lead to significant permafrost degradation and related change within the periglacial process-system and the entire geo-ecosystems (Isaksen et al. 2007, 2011).

8.3 Palaeoclimate

8.3.1 Deglaciation and Holocene Glacier Chronology

The Pleistocene glaciation history of Jotunheimen provides a necessary framework for the development of permafrost and related periglacial landforms. Climatic conditions and their variability through time are decisive for the regional intensity of periglacial processes and the spatial distribution of permafrost. In turn, periglacial landforms may be utilised as palaeoclimatic archives, especially medium- and long-term climate changes.

Central Jotunheimen has been at or near the culmination centres of Late Pleistocene Scandinavian Ice Sheets including the Last Glacial Maximum (Mangerud et al. 2011). The preservation of relict/paleic landforms (Kleman et al. 2008), cosmogenic nuclide dating on high-altitude block fields with a remarkable discrepancy between the ages of glacially transported erratic boulders versus parent bedrock or in situ weathered boulders (Fabel et al. 2002; Goehring et al. 2008; Marr et al. 2018), and the results of sophisticated high-resolution ice-sheet models for the Late Weichselian Scandinavian Ice Sheet (Patton et al. 2016) all indicate that Jotunheimen was likely covered by cold-based, low-velocity ice with very limited erosional potential. Only in topographically confined valleys draining towards the West and East ice-streams with higher erosional potential and different basal conditions existed that ultimately resulted in different geomorphological units. Unfortunately, there is very sparse information about the exact timing of Deglaciation in central Jotunheimen and, in particular, about the mechanisms involved. Whereas early work predicted middle- to late-Preboreal Deglaciation and thick ice during the Younger Dryas (Holmsen 1982; Sollid and Reite 1983; Sollid and Trollvik 1991), subsequent work opens for a more limited ice thickness during the Younger Dryas (Dahl et al. 1997; Lie et al. 2004). Currently there seems agreement about a final regional Deglaciation around 10 ka ago (Hughes et al. 2016; Stroeven et al. 2016; Patton et al. 2017), likely after the regional 'Erdalen event' at c. 9700 cal. a BP (Dahl et al. 2002; Matthews and Dresser 2008; Matthews et al. 2018). Matthews and Winkler (2011) date early Holocene moraines at a cirque glacier in western Jotunheimen (Austanbotnbreen) to the early (10.2 ka) or late Erdalen event (9.7 ka). This timing is in agreement with the proposed regional Deglaciation, but whether the widespread high-altitude plateaus outside these individual early Holocene glaciers possibly became ice-free already considerably (several centuries) earlier remains unknown. The lack of detailed age constraints for deglaciation of these widespread plateaus covered by periglacial landforms and inferred frozen-bed conditions during (at least) the Last Glacial Maximum significantly complicate any investigation on the development of permafrost and the onset of formation of the dominating patterned ground (Winkler et al. 2016).

The most relevant palaeoclimatic proxy-archive for an independent control on the Holocene climatic variability potentially underlying the development and distribution of periglacial landforms in Jotunheimen is the regional glacier chronology, mainly due to its natural proximity in actual location and altitudinal distribution. A few early Holocene glacier advances and periods of increased glacial activity rather connected to independent glaciers than remnants of the receding Scandinavian Ice Sheet have been identified in South Norway (Bakke et al. 2005; Nesje et al. 2008a; Nesje 2009). At Styggedalsbreen in the Hurrungane Massif of western Jotunheimen, early Holocene moraines outside the LIA-glacier foreland (Erikstad and Sollid 1986; Winkler 2001) likely date from the two-phased Erdalen event (Dahl et al. 2002; Matthews et al. 2008) and correspond to the accordingly dated moraines at Austanbotnbreen (Matthews and Winkler 2011). A few other potentially old pre-LIA moraines have already been identified by Matthews and Shakesby (1984) but lack numerical age constraints. Early Holocene ages have initially been suggested for high-altitude ice-cored moraine systems in parts of Breheimen and Jotunheimen (see review in Winkler 1996). Recent work has neither been able to confirm this theory nor to ultimately reject the possibility that some of the moraine complexes may originate from early Holocene glacier activity but have subsequently been substantially modified during the late Holocene (Shakesby et al. 2004; Matthews et al. 2014; see Sect. 8.4.3). The hypothesis that the younger 'Finse event' (the regional '8.2 ka-event'; see Nesje et al. 2008a; Nesje 2009 for details) was a period of substantial glacier activity in Jotunheimen has, however, been rejected by Matthews and Winkler (2011). This raises doubts about earlier assumptions that due to their altitudinal position early Holocene glacier activity in Jotunheimen continued considerably into the 'Holocene Thermal Maximum' (HTM; see Matthews and Karlén 1992; Karlén and Matthews 1992).

The HTM in South Norway is characterised by a lack of evidence for glacial activity and a climate apparently unfavourable for glaciers (Matthews and Dresser 2008; Nesje et al. 2008a; Nesje 2009). Jotunheimen does not constitute an exception from this general pattern. The HTM seems to have commenced immediately after the 8.2 ka BP-event, but detailed numerical control is scarce and it may well have started earlier (Velle et al. 2010). Even if some climatic variability has to be expected during the HTM, there are to date no confirmed cases of glaciers surviving during the entire Holocene including the HTM, not even high-altitude cirque glaciers in eastern Jotunheimen as the most likely suspects. Their ice-cored moraine complexes potentially initiated by early Holocene glacier activity could have survived the HTM even with the risen lower limit of permafrost responding to higher air temperatures (see Sect. 8.3.2) because permafrost conditions continued at the highest altitudes including these sites (Lilleøren et al. 2012). But any preserved moraine complex would have inevitably undergone substantial modification during subsequent late Holocene glacier advances (Matthews et al. 2014; see Sect. 8.4.3), meaning that these ice-cored moraines cannot be utilised as potential evidence for continuous glacier presence.

Stream-bank mires episodically flooded by glacial meltwater and lake sediment cores both contain deposits of glaciofluvial sediment that are accurately radiocarbon-datable by bracketing organic material. On that basis, Matthews and Dresser (2008) reconstruct a detailed record of glacial activity for the Smørstabbtindan Massif in western-central

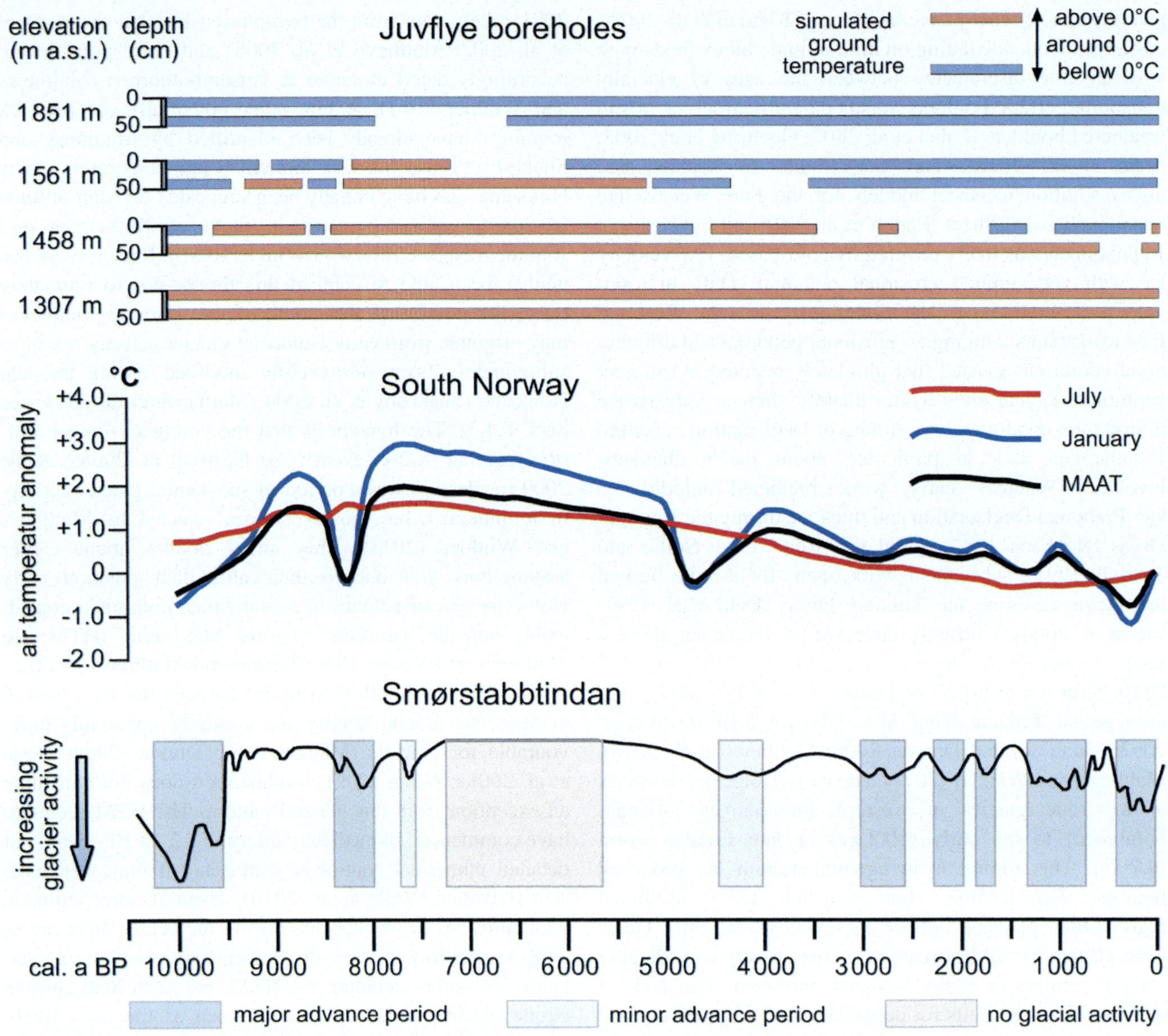

Fig. 8.4 Holocene palaeoclimatic data records from central Jotunheimen. Upper panel: Simulated ground temperatures for boreholes (at surface and 50 cm depth) at different elevations on Juvflye (generalised and modified from Lilleøren et al. 2012). Middle panel: Anomalies of January, July, and mean annual air temperatures in South Norway utilised for the simulation of ground temperatures as shown on upper panel (modified from Lilleøren et al. 2012). Lower panel: Glacier chronology for Smørstabbtindan in Jotunheimen (modified from Matthews and Dresser 2008)

Jotunheimen (Fig. 8.4), the best available regional glacier chronology for the Holocene that includes earlier work by Matthews et al. (2000, 2005). They outline seven millennial-scale neoglacial events called 'Smørstabbtindan events' with Smørstabbtindan I corresponding to the Erdalen event and Smørstabbtindan II to the Finse event, respectively. Their general subdivision of the Holocene displays three phases with the first (11,500–9,500 cal. a BP) characterised by regional/local re-advances of remnants of the Pleistocene ice-sheet and/or individual glaciers. The second phase (mid-Holocene; 9,500–5,000 cal. a BP) saw small or absent glaciers, whereas the last phase with enlarged and significantly fluctuating glaciers started around 5,000 cal. a BP. Apart from the abovementioned early Holocene Smørstabbtindan I and II events, all remaining millennial-scale glacial events occurred during that last phase: 4,800–3,900, 3,200–2,550, 2,350–1,700, 1,400–750, and finally 500–0 cal. a BP. Matthews and Dresser (2008) note that two or even more shorter, centennial-scale glacial events were nested within these millennial-scale events. The regional record of glacial activity for Jotunheimen corresponds well to other records published for South Norway (Nesje 2009), indicating that any occurring differences in glacier response are likely caused by different glaciological regimes (maritime versus continental) and the natural framework (altitude, glacier size/morphology, etc.).

As mentioned above, the ice-cored complex moraine systems at a number of mostly small, high-altitude cirque glaciers in Jotunheimen above both the current and the reconstructed limit of permafrost for the HTM only reveal inconclusive age constraints for their initial formation, thus cannot be utilised to date the regional onset of neoglaciation (Shakesby et al. 2004; Matthews et al. 2014). At a small polythermal cirque glacier in eastern Jotunheimen, Lie et al. (2004) present evidence for glacier reformation after 6,000 cal. a BP and a slow increase to approximate modern expansion until 3,800 cal. a BP (with reduced glacier activity around 5,200 cal. a BP). This earliest evidence for cessation of the regional HTM around 6,000 cal. a BP could be a model for similar glaciers in Jotunheimen, especially in higher eastern parts where these ice-cored moraines are concentrated.

The most severe climatic deterioration in Jotunheimen during the late Holocene occurred during the LIA when glaciers experienced their maximum extension. There is no historic evidence about the timing of the LIA-maximum in Jotunheimen except an eyewitness account noted by Øyen (1893) stating that Storbrean in Leirdalen had its largest expansion '100 years ago', i.e. at the end of the eighteenth century CE. This quote is, however, insufficient to act as reliable dating or fixed point for lichenometric dating curves. Whereas Erikstad and Sollid (1986) lichenometrically date the outermost terminal moraine of Storbrean to 1787 CE, Matthews (1974, 1975, 1977) demonstrates that detailed application of lichenometry favours a LIA-maximum ~1750, an age also suggested by Liestøl (1967) and Bogen et al. (1989). In a comprehensive study on the timing of the LIA-maximum in Jotunheimen, Matthews (2005) recognises a cluster of moraine formation between 1743 and 1750 as culmination of the major advance followed by three subsequent clusters during the next five decades (1762–1771, 1782–1790, 1796–1802). An initial slow and frequently interrupted retreat from the LIA-maximum position would easily explain the abovementioned sketchy eyewitness account as glaciers remained very close to their outermost position at least until around 1800. Matthews (2005) identified sub-regional differences in the timing of the LIA-maximum with glaciers in eastern Jotunheimen having their outermost LIA-moraines dated to the late eighteenth century. Winkler (1996, 2001) points out that outermost terminal moraines at some glaciers (e.g. Storbrean and Visbrean) are partly double-crested (see also Hoel and Werenskiold 1962). These double-crested moraines are exclusively concentrated in the western and western-central part of Jotunheimen, glaciers in eastern Jotunheimen show single ridges instead (Winkler 2002). Based on evidence about cold summer air temperatures during the second half of the eighteenth century (Nordli et al. 2003, 2005) this morphological difference can be interpreted in a way that while several western/central glaciers almost reached their previous maximum extent during one of the late eighteenth century re-advances and formed the inner crest of the outermost moraines (Winkler 2002), glaciers in eastern Jotunheimen concurrently responded stronger to low summer temperatures and overrode/destroyed any moraines formed around 1750. This difference in timing of the LIA-maximum reflects the climatological character of Jotunheimen as transitional zone and leaves glaciers in its western parts during the LIA responding mostly synchronous to the maritime glaciers of western South Norway (like Jostedalsbreen; see Chap. 3). Modelling the LIA-climate for Jotunheimen, Rasmussen et al. (2010) conclude that similar to maritime glacier regions in western South Norway (Nesje and Dahl 2003; Nesje et al. 2008b) winter precipitation was a major factor for the timing of the LIA-maximum in Jotunheimen. The relevance of these findings for permafrost and periglacial processes is that during the local culmination of the LIA not only lower air temperatures but also a potentially thicker and longer snow cover needs to be assumed.

Since culmination of the LIA, glaciers in Jotunheimen experienced an overall retreat barely interrupted by short re-advances, e.g. around c. 1920 (Winkler 1996, 2002; Andreassen et al. 2005). Between LIA-maximum and 2003 the glacier area in Jotunheimen shrank by 35%, their length by 34% (Baumann et al. 2009; Baumann and Winkler 2010; see Andreassen and Winsvold 2012 for the latest glacier inventory). The currently ongoing retreat from the LIA-maximum positions leaves considerable proportions of glacier forelands deglaciated and potentially affected by periglacial processes within the paraglacial process-system (see below).

8.3.2 Holocene Development of Permafrost

Whereas the current altitudinal and spatial distribution of permafrost in Jotunheimen is well studied (see Sect. 8.2.2), its development and variability during the Holocene remains mostly unknown. This constitutes a disadvantage as permafrost is decisive for periglacial processes and landforms, the latter owing potential as palaeoclimatic archives but experiencing general problems with the application of most numerical dating techniques. An attempt to reconstruct the altitudinal variability of permafrost in Jotunheimen and entire Norway has recently been undertaken by Lilleøren et al. (2012) utilising annual air temperature series for the past 10,000 years to drive a one-dimensional heat-flow model and a new equilibrium permafrost model (Gisnås et al. 2013) calibrated with existing borehole-temperature data (for example, Farbrot et al. 2011; Fig. 8.4). The palaeoclimatic temperature series are compiled from available proxy-data records like macrofossils and pollen from lake sediment cores, chironomid studies, or tree-line fluctuation data. The results have, therefore, an inferred character and cannot independently be directly re-assessed, just indirectly using periglacial landforms potentially affected by the temporal variability of permafrost, e.g. its altitudinal distribution and depths of active layer.

Lilleøren et al. (2012) conclude that at the highest altitudes in Jotunheimen permafrost existed since Deglaciation and though the entire Holocene. Large areas of permafrost subsequently degraded during the HTM but re-developed since with its largest mid-/late Holocene spatial expansion and lowest temperatures during the LIA, i.e. a development comparable to the regional Holocene glacier chronology (see above). At Juvflye, only sites above 1850 m a.s.l. should have experienced permafrost during the HTM, whereas in a lower zone stretching down to 1650–1700 m a.s.l. permafrost may have survived the HTM below superficial talks. Based on the morphology of its ice-cored moraine, Lie et al. (2004) conclude that the local lower limit of permafrost in the vicinity of a small polythermal cirque glacier in eastern Jotunheimen did not rise above 1,800 m a.s.l. even during warmer periods of the Holocene, a conclusion supported by the newer work of Lilleøren et al. (2012). By contrast, the altitudinal zone between 1600 and 1400 m a.s.l. at Juvflye in central Jotunheimen is expected to have developed permafrost post-HTM subsequently expanding in depth and towards lower altitudes until its peak during the LIA, but facing degradation since. Examining detailed results of the reconstructions for two individual boreholes at Juvflye used for calibration of the models utilised, Lilleøren et al. (2012) expect that at an altitude of c. 1560 m a.s.l. permafrost should have disappeared post-Deglaciation at 8.0–7.8 ka ago. This age corresponds remarkably well to the ages provided by Winkler et al. (2016) for the stabilisation of stone rings at the same location (8,240 ± 495 and 8,050 ± 560 ka ago for sites at 1500 and 1550 m a.s.l., respectively) obtained by SHD-measurements. It is, however, an interesting fact that the reconstructed re-development of permafrost and its subsequent expansion during the late Holocene does not show considerable impact on the patterned ground features investigated, thus opening for alternative interpretations of their age and formation processes (see Sect. 8.4.1).

8.4 Periglacial Landforms in Jotunheimen

8.4.1 Patterned Ground

Patterned ground with its different individual subtypes of landforms is indisputable the most prominent element of the periglacial process-system in Jotunheimen. It covers a considerable area of its surface in what Jotunheimen seems more comparable to Arctic uplands than to temperate mountain regions. Periglacial patterned ground is generally not considered a dominating landform in mountain environments and usually marginalised by more typical periglacial features like rock glaciers or different types of solifluction landforms. Jotunheimen constitutes an exception in this context and the dominance of patterned ground can easily be explained by its gross morphology (see Sect. 8.2.1). Its widespread high-altitude plateaus and low to moderately inclined slopes at summit level supporting rather rounded mountain tops facilitate the formation of patterned ground, which is virtually absent in the steep and rugged gross morphology of geologically young high mountain ranges. Furthermore, the terrain topographically suitable for the development of patterned ground is often covered by diamicton (in most cases local Weichselian till) promoting its development with heterogeneous grain size distributions required for landforms like sorted nets or stone rings. Most of the abovementioned plateaus are additionally located above the current lower limit of permafrost (Gisnås et al. 2013; see Sect. 8.2.2) generally seen as prerequisite for the formation of larger and well-established patterned ground features. As a consequence, various types of sorted patterned ground (circles, polygons, nets, steps, and stripes) cover a high percentage of the area above the lower limit of permafrost in Jotunheimen. Limitations to its distribution are mainly set by topography (slope angle), substrate (bedrock, homogeneous sediment/regolith), and snow/ice cover (distribution of modern glaciers and perennial snow beds). A number of studies have been conducted on patterned ground in Jotunheimen including detailed geomorphological mapping (Ødegård et al. 1987, 1988), micromorphological and –sedimentological analysis (Harris and Cook 1988), and attempts to obtain some age constraints and investigate their palaeoclimatic significance (Cook-Talbot 1991; Winkler et al. 2016). These studies focus on comparatively large patterned ground features at Juvflye and in eastern Jotunheimen (Figs. 8.5 and 8.6), whereas small-scale and obvious young patterned ground features reported from recently deglaciated glacier forelands (Fig. 8.9) are subject to different environmental conditions and probably affected by different driving factors.

Based on their detailed geomorphological mapping at Juvflye, Ødegård et al. (1987, 1988) report that 20–50% of the area with continuous till above 1750 m a.s.l. on slopes with less than 10° angles is covered by patterned ground. This percentage drops to less than 10% above 2000 m a.s.l. as a result of the lack of suitable surface material. In this context, Ødegård et al. mention a boundary between frost-sorted till and in situ weathered regolith at about 1950 m a.s.l. (lower at slopes steeper than 8–10°) that seems responsible for the described altitudinal distribution. The percentage of patterned

Fig. 8.5 Patterned ground on Juvflye: **a** Patterned ground, c. 1740 m a.s.l. **b** Stone ring with established vegetation in its fine centre, c. 1740 m a.s.l. **c** Patterned ground, c. 1830 m a.s.l. d Patterned ground, c. 1870 m a.s.l., Glittertinden in the far distance. **e** Patterned ground, c. 1850 m a.s.l. **f** Patterned ground, c. 1870 m a.s.l., unlike the other sites developed on in situ weathered regolith (Photos: Stefan Winkler, a–d 26.07.2018, e 19.08.2007, f 25.07.2018)

ground cover on the total area peaks between 1900 and 1950 m a.s.l. with up to 50% whereas it almost completely remains above 30% in the entire zone between 1800 and 2000 m a.s.l. (Figs. 8.3b, 8.5d and 8.6a). Identification and classification of patterned ground on the geomorphological map of Juvflye (Ødegård et al. 1987) follow Washburn (1956) with sorted polygons, nets, and circles typical for flat terrain (~0–6°) but sorted steps (2–11°), sorted stripes (3–17°), and boulder tongues (5–21°; average 13°) occurring on shallow and moderate slopes gradients (Ødegård et al. 1988; Fig. 8.7). Apart from the main processes involved in patterned ground formation on flat terrain such as frost heaving, cryoturbation and frost sorting, solifluction and creep are those additionally considered with the latter forms mapped on the slopes at Juvflye (see Washburn 1956, 1979; Goldthwait 1976; Harris and Cook 1988; Hallet 1990, 2015; Kessler et al. 2001;

Fig. 8.6 **a** Orthorectified aerial photo (2017) of patterned ground on Juvflye, c. 1840 m a.s.l. (adapted from www.norgeibilder.no, © Kartverket/www.kartverket.no). **b** View from the outer ridge of Gråsubrean's ice-cored moraine system (cf. Figure 8.11) onto distal flat terrain (Bukkeflye) with well-developed patterned ground. The perennial distal snow patch was very small during that summer and the patterned ground closest to the moraine shows brighter colours due to mostly lichen-free boulders previously covered by snow preventing lichen colonisation. **c** Close-up of patterned ground near moraine of Gråsubrean. Because there is no obvious morphological difference between (dark) heavily lichen-covered and (bright) lichen-free patterned ground the latter may have formed prior to it becoming covered by semi-perennial snow and indicates limited recent activity (Photos: Stefan Winkler, b, c 24.07.2013)

Matsuoka et al. 2003; French 2007; Ballantyne 2013a, 2018; Warburton 2013 for details). Ødegård et al. (1988) see a positive correlation between the mapped periglacial features and the occurrence of permafrost with water content, amount of fines, and slope gradient as decisive factors for the development of different surface features. According to their mapping active forms (with diameters of 0.3 to 1.5 m of their fine centres) are found above 1900 m a.s.l. only (Fig. 8.5f). Between 1700 and 1900 m a.s.l. fine centres of sorted circles and polygons typically exhibit diameters between 3 and 5 metres and are mostly free of vegetation indicating active frost processes. However, they point out that there is no evidence for activity of the borders/gutters or block-covered areas heavily covered by lichens. This 'fossil' appearance

Fig. 8.7 **a** Striped patterned ground on the northern slope of Juvflye, c. 1620 m a.s.l. **b** Striped patterned ground at Juvasshøe, c. 1870 m a.s.l. **c**, **d** Striped patterned ground on the southern slope of Juvflye, c. 1850/1800 m a.s.l. **e** Orthorectified aerial photo (2017) of striped patterned ground at approximate the site depicted on Fig. 8.7c (adapted from www.norgeibilder.no, © Kartverket/www.kartverket.no) (Photos: Stefan Winkler, a 18.08.2008, b 19.08.2007, c, d 08.08.1992)

(Ødegård et al. 1988) may suggest that the amount of fines is not sufficient to support continued active development of the features, a conclusion in its temporal context recently confirmed by Winkler et al. (2016). The morphology of the sorted circles at Juvflye is similar to those reported from comparable locations in eastern-central and eastern Jotunheimen, for example, from the Glittertinden Massif (circle diameters between 3.1 and 5.0 m, average 3.8 m) or the terrain distal to Gråsubrean's outermost (ice-cored) moraine (diameters 2.7–7.8 m with an average of 4.7 m—Harris and Cook 1988; Fig. 8.6b, c).

To relate the temporal development of patterned ground with the altitudinal fluctuations of the lower limits of permafrost during the Holocene and regional palaeoclimatic records some form of age constraint needs to be achieved notwithstanding the well-known issue to accurately date debris-rich landforms that develop over long periods of time. Cook-Talbot (1991) used lichenometry and Schmidt-hammer relative-age dating to obtain relative-age estimates for patterned ground in the Glittertinden Massif and its surroundings in eastern Jotunheimen. The application of established lichenometric growing curves revealed minimum ages of the sorted circles prior to the eighteenth century CE (i.e. prior to the regional LIA-maximum) with parts of their borders becoming stabilised since the twelfth century and smaller surface clasts within the fine centres since the LIA. It has, however, to be noted that for such apparent ages lichenometry is beyond its recommended time frame for reliable application (Matthews 1994, 2005). In theory, the exposure ages of clasts based on weathering criteria within patterned ground surfaces may provide insights into processes and dynamics of sorted patterned ground formation as well as chronological information about how long these clasts have been exposed at the surface and when the clast-rich margins became stabilised. Concurrently performed Schmidt-hammer relative-age dating revealed, therefore, not surprisingly that clasts in the borders probably had been exposed much earlier, at least 1500 years at some sites. Based on a comparison of these results to Schmidt-hammer data from terminal moraines in Jotunheimen reported by Matthews and Shakesby (1984), Cook-Talbot (1991) finally concludes that her evidence points towards an early Holocene origin of the sorted circles (excluding minor subsequent modification during the Late Holocene). This view is in agreement with the statement of Ødegård et al. (1988) about a 'fossil' appearance of sorted circles at Juvflye and supported by recent work of Winkler et al. (2016).

Winkler et al. (2016) apply SHD on sorted circles and polygons along an altitudinal profile at Juvflye within the area mapped by Ødegård et al. (1987). The sites located between 1500 and 1925 m a.s.l. indicate a distinct altitudinal gradient of increasing mean R-values with altitude (i.e. less weathered surfaces and decreasing surface-exposure ages) and exhibit R-value distributions with platycurtic modes, broad plateaus, narrow tails, and a negative skew. Using an established regional SHD-calibration curve for Jotunheimen (Matthews and Owen 2010) mean boulder exposure ages in the range 8,240 ± 495 to 6,910 ± 510 years ago are obtained. Despite some methodological uncertainty these SHD ages are assumed as indicative of the timing of patterned ground formation, representing minimum ages for active boulder upfreezing and maximum ages for the stabilisation of boulders in the encircling gutters.

The early Holocene ages for the patterned ground surface boulders yielded by SHD point towards an apparent cessation of major morpho-dynamic action during the HTM and lack of any significant increase of related activity during the late Holocene climate deterioration and permafrost re-development. Although the interpretation of these SHD ages from large-scale patterned ground remains challenging owing to their diachronous nature, a potentially complex history of formation, and influence of local non-climatic factors, they generally confirm the 'fossil' appearance of these features already noted by Ødegård et al. (1987). This can, furthermore, be supported by recent field observations that only minor secondary features like surface cracks and networks occur within the fine centres of large sorted circles and polygons that otherwise display no recent disturbance of clasts within the borders/gutters (Fig. 8.8a). The lack of evidence that could support reactivation of these 'relict' landforms despite lowering of the altitudinal limit of permafrost after the HTM or during the LIA allows more than one interpretation. Because the SHD ages for patterned ground are largely consistent with the establishment of regional permafrost shortly after Deglaciation in the early Holocene (~9,700 years ago) and its gradual degradation after the onset of the HTM would have affected the sites at lowest altitudes first (see Lilleøren et al. 2012), one could assume an initial period of efficient frost sorting during the Younger Dryas-Holocene transition with completion of sorted circle formation possibly ended even earlier than indicated by the SHD ages and the conventional timing of Deglaciation. The lack of detailed local ages for Deglaciation of Juvflye and comparable locations in eastern Jotunheimen (see Sect. 8.2.1) does, however, complicate any full assessment of such palaeoclimatic implications. If a comparison to micro-scale patterned ground developed on recently deglaciated ground (see below) is undertaken, moisture as crucial factor for patterned ground formation would be supported by slowly retreating glaciers (Matthews et al. 1998; Haugland 2006) as well as impermeable permafrost (Ødegård et al. 1988), the latter recently modelled to form very fast during Deglaciation on rock slopes in the nearby fjord region of western Norway (Steiger et al. 2016; Myhra et al. 2017). Water-soaked till is, therefore, likely promoting the development of large-scale sorted rings and

Fig. 8.8 **a** Initial secondary patterned ground formation within a larger, established stone ring at Juvflye, c. 1870 m a.s.l. **b** Small, initial patterned ground features on one of the outer ridges of the ice-cored moraine system at Gråsubrean (cf. Figure 8.11; see Matthews et al. 2014 for details) (Photos: Stefan Winkler, a 25.07.2018, b 24.07.2013)

polygons. However, as an alternative hypothesis patterned ground could also have formed prior to the Last Glacial Maximum and preserved beneath cold-based ice with subsequent emergence and completion of its current shape during Deglaciation. Although somehow less convincing, the latter hypothesis cannot fully be rejected based on currently available evidence.

Patterned ground features of smaller scales have been reported from recently deglaciated areas within LIA-glacier forelands at several glaciers in Jotunheimen (Fig. 8.9). Detailed studies on these features that include sorted nets, circles, and stripes which occur collectively with other periglacial forms like solifluction lobes, boulder-cored frost boils, ploughing boulders, and surface cracks have been conducted at Slettmarkbrean (Ballantyne and Matthews 1982; Haugland 2004, 2006), Storbrean (Ballantyne and Matthews 1983), Styggedalsbreen (Winkler 1996; Matthews et al. 1998; Haugland 2004, 2006), and Vestre Memurubrean (Haugland 2006). They are also reported from a number of other glaciers and don't seem to be uncommon in the region. It needs to be highlighted that apart from one site in the glacier foreland of Vestre Memurubrean ($\sim$1625 m a.s.l.) where permafrost according to the most recent models is possible (Lilleøren et al. 2012; Gisnås et al. 2013), all other sites studied in detail are located at altitudes at or below the current lower limit of permafrost (Styggedalsbreen as lowest site at 1270 m a.s.l.). Consequently, permafrost can be ruled out as contributing factor with patterned ground formation. All glaciers involved are generally warm-based and, therefore, exhibit different thermal regimes in their marginal glacier zones compared to high-altitude polythermal glaciers with ice-cored moraines concentrated in eastern Jotunheimen (see Sect. 8.4.3). Observations on recent moraine formation have, furthermore, confirmed that albeit subglacial freezing-on of sediment slabs by deep penetrating seasonal frost is a contributing process, there is no indication of any permafrost present or involved at these comparatively low-altitude glaciers (Matthews et al. 1995; Winkler 1996; Winkler and Matthews 2010; Hiemstra et al. 2015). As a consequence, factors other than permafrost need to be considered decisive for formation of the reported patterned ground features.

The reported sorted nets and circles occur on flat till surfaces such as fluted moraine surfaces or moraine crest, the latter mostly degraded minor (annual?) moraines that rise barely a few metres above the surrounding terrain like convex-shaped, flat mounds (Fig. 8.9a). Although Ballantyne and Matthews (1982) and Haugland (2006) report patterned ground from older terrain on glacier forelands within the LIA-maximum glacier extent, all studies highlight that patterned ground features are most active near the contemporaneous ice margin or even restricted to an area deglaciated since c. 1920/30 CE (Haugland 2004). Matthews et al. (1998) report that sorted nets at Styggedalsbreen stabilise on 60-year old terrain but show little difference to the same features on terrain deglaciated only 25 years ago. This observation could be confirmed by a field visit in 2018 that revealed no visible difference in the development of patterned ground during the past 20 years. By contrast, the development of vegetation has progressed since and older features are now less obvious than during previous field studies (Winkler 1996). Ballantyne and Matthews (1982) report a similar stabilisation of sorted circles achieved within a comparable time frame of 50 years after deglaciation at Slettmarkbrean. After first surface cracks of <20 mm depth in crack networks of 0.2–1.5 m length showed as early as 2–6 years after deglaciation at Storbrean their transition into mature sorted polygons in an exponential fashion was likely completed within 35 years after which they approached their end of development as indicated by the establishment of

Fig. 8.9 Patterned ground on recently deglaciated terrain in the glacier foreland of Styggedalsbreen: **a** Patterned ground on a degraded minor moraine dating from the 1920/30s CE. **b** Detailed view on patterned ground features, area deglaciated during the 1920/30s. **c** Initial patterned ground, area deglaciated during the 1960s. **d** Initial signs of patterned ground formation on the proximal slope of the recent terminal moraine system actively built up until c. 2000 (Photos: Stefan Winkler, a 04.09.1997, b 23.08.1992, c, d 24.07.2018)

vegetation (Ballantyne and Matthews 1983). Ballantyne and Matthews (1983) identify the initial surface cracks as desiccation cracks rather than seasonal frost or dilatation cracks. The subsequent formation of sorted circles is likely the result of frost sorting processes (Ballantyne and Matthews 1982). Water-soaked till and slow glacier retreat with its comparably colder and wetter conditions within the marginal proglacial zone are seen as major factors promoting rapid frost sorting and the related development of patterned ground according to Ballantyne and Matthews (1982). The abovementioned restriction of patterned ground to moraine mounds rising a few metres above the surrounding foreland is explained by their exposition prone to the effect of wind-blown snow in winter allowing deeper ground freezing. Ballantyne and Matthews (1982) see the decline of active processes and stabilisation of developed patterned ground after a few decades as the result of ground temperature amelioration (possibly also the absence of katabatic wind which only affects the proglacial area close to the glacier margin, Haugland 2006) and less moisture availability hampering processes like ice segregation, frost heave, or cryoturbation. Subsequent colonisation by plants or lichens may either be a cause or a consequence of this stabilisation.

The morphological parameters of patterned ground on such recently deglaciated ground are well documented (Ballantyne and Matthews 1982, 1983; Matthews et al. 1998). Individual sorted rings/polygons typically reach diameters between 30 and 75 cm (sometimes up to 1 m) in their mature stage and initial surface cracks usually do not exceed 50 mm depth (Winkler 1996). Their relationship to other periglacial forms and their place within the local micro- and meso-scale ecological system have been studied as well (Matthews et al. 1998; Haugland 2004, 2006). Summarising, the conditioning and controlling factors for patterned ground development as described from low- to mid-altitude glacier forelands in Jotunheimen are causally connected to the conditions within a rather narrow marginal zone (widths of up to 500 m) from of the retreating glaciers (especially moisture availability and specific micro-climatic conditions). Therefore, Matthews et al. (1998) rate the studied patterned ground formation as partly exogenous and paraglacial rather than endogenous and developmental like it is characteristic for most conventional models of patterned ground formation related to permafrost. This view is consistent to processes declining in their effectiveness with distance from the marginal zones and forms becoming stabilised after a short period of comparatively fast development already after several decades. Within this development the activity decreases first at the border/gutters, subsequently followed by the fine centres (Haugland 2004). Like a special case of patterned ground formation reported from an active supraglacial moraine at Storbrean (Ballantyne 1979) these forms are apart from their size most likely different in their mechanism(s) of formation to patterned ground on flat terrain at high altitude underlain by permafrost. Whether some factors responsible for their initial development in the marginal zones of retreating glaciers (e.g. moisture availability) also had an impact during the first stages of the development of the large-scale old patterned ground remains a hypothesis that definitely needs to be explored further.

8.4.2 Blockfields

Blockfields are common landforms at summit level in Jotunheimen and cover parts of the upper slopes and summit areas of such prominent peaks as Fannaråki (Fig. 8.2f), Galdhøpiggen (Fig. 8.10a), and Glittertinden. The lower boundary of blockfields reaches a maximum altitude of roughly 2000 m a.s.l. in central Jotunheimen (Nesje et al. 1988) but descends to altitudes several hundred metres lower towards both the West and East. This corresponds well to the local lower limit of in situ weathered regolith on Juvflye mapped at 1950 m a.s.l. by Ødegård et al. (1987, 1988). Given the extensive distribution of blockfields in Jotunheimen and surrounding mountain regions (Nesje et al. 1988) and the fact that they even appear on small-scale Quaternary geology maps (Klakegg et al. 1989; Sollid and Trollvik 1991) it seems surprising that no detailed local studies or systematic large-scale mapping attempts of these autochthonous blockfield have been conducted in Jotunheimen to date. By contrast, blockfields close to northeastern Jotunheimen (Tollan 1963; Nesje et al. 1994; Goehring et al. 2008; Marr et al. 2018) as well as in the fjord regions to the West and Northwest (Nesje et al. 1988, 1994; McCarroll and Nesje 1993) have been considerably investigated, primarily in the context of reconstructing the thickness and geometry of the Late Weichselian Scandinavian Ice Sheet (e.g. Nesje and Dahl 1990).

There seems consensus to use the term blockfield in a rather unspecific way to address the complexity and variability of these bouldery regolith covers related to past and/or present cold climate conditions (Rea 2013; Ballantyne 2018). Blockfields in Jotunheimen constitute typically coarse/bouldery regolith covers of mostly angular clasts free of fine sediment. Ballantyne (2018) uses the term openwork blockfields for such covers and connects them generally to resistant but well-jointed metamorphic (or igneous) rocks. This statement would fit the situation in Jotunheimen where the thickness of the coarse regolith cover overlaying a matrix of fine-grained weathering detritus can fall in the range between several decimetres and a few metres. Individual surface boulders on autochthonous blockfields have sizes of some tens of centimetres to one metre at average and are the result of in situ weathering without any major subsequent displacement. Despite the homogeneous bedrock in Jotunheimen (see Sect. 8.2.1) occasionally outcropping quartz veins still mirrored by the alignment of quartz blocks and boulders (see Fig. 8.10b) demonstrate that many blockfields in Jotunheimen belong to this type. Has the regolith considerably moved downslope and formed blockslopes or blockstreams, these deposits are defined as allochthonous (Ballantyne 2018). Almost all blockfields in Jotunheimen occur at altitudes where at least discontinuous permafrost has to be expected (see Sect. 8.2.2). For recent activity this may be of importance as frost wedging of jointed bedrock at the base of the active layer with dry permafrost is assumed to

Fig. 8.10 **a** Blockfield on the upper slope of Galdhøpiggen, view towards East. **b** Detail of in situ weathered quartz vein at the same site indicating no subsequent dislocation (Both photos: Stefan Winkler 29.08.1996)

be an efficient process of ongoing weathering beneath the regolith cover as the weathering front penetrates (Ballantyne 2018). Such a process would result in continuous or intermitted renewal as the regolith surface lowers by any form of denudation over time. It also opens for blockfields constituting quite old landforms with inherited characteristics on which certain hypotheses on their age and origin are founded (see below). On the other hand, the connection between blockfields and permafrost may coincidentally be caused by the location of suitable high-altitude topography in form of plateaus and shallow/moderate slopes and/or the local glaciation history with all terrain overridden by warm-based ice during the Last Glacial Maximum either covered by till or representing glacially eroded bedrock.

Two different models for autochthonous blockfield development have become established since the 1980s, the pre-Quaternary (preglacial) model focusing on intense chemical weathering on flat, preglacial land surfaces and the periglacial model emphasising the impact of mechanical/frost weathering under cold climate/periglacial conditions during the Quaternary cold periods (Ballantyne 2010; Rea 2013; Rixhon and Demoulin 2013). Although a number of arguments have been brought forward in support of the pre-Quaternary model including the location on flat or gently sloping land surfaces assumed to be of preglacial origin (see Sect. 8.2.1) or fine grain sizes and clay minerals supposedly related to effective chemical weathering (for details see Nesje et al. 1988), more recent research in Scandinavia and Scotland considered relevant for Jotunheimen because of comparable conditions strongly supports the periglacial model (Goodfellow et al. 2009, 2014; Ballantyne 2018). The pre-Quaternary model is additionally weakened by confirmed evidence that cold-based glaciers are generally able to protect relict landscapes and the fact that such cold-based conditions have been reconstructed for the former Scandinavian Ice Sheet over Jotunheimen (see Sect. 8.3.1). Morphological evidence supported by recent numerical dating has, furthermore, questioned the traditional interpretation of the lower boundaries of blockfields and glacial trimlines as actual upper limit of glacial erosion (e.g. Ahlmann 1922) in favour of more complex approaches considering these boundaries as potential boundaries between warm- and cold-based glacier conditions or the result of interaction between various processes (Ballantyne 2013b; McCarroll 2016).

Summarising, despite a lack of detailed local studies the most realistic scenario sees blockfields in Jotunheimen developing under cold but ice-free (i.e. periglacial) conditions during the Pleistocene (i.e. interglacials or interstadials). They most likely have been covered multiple times by the Scandinavian Ice Sheet, latest during the LGM, but cold-based glacier conditions protected them from destruction. One factor for their extensive distribution in Jotunheimen is related to the regional gross morphology in form of widespread high-altitude plateaus and shallow/moderate slopes at summit level. This fact seems coincidentally linking blockfields to the preglacial surface rather than any form of inheritance in the coarse-grained autochthonous regolith cover itself. Results from blockfields adjacent to Jotunheimen (e.g. Blåhø—see Marr et al. 2018 for details) indicate that they have been exposed earlier during Deglaciation than currently anticipated. Blockfields more distant to central Jotunheimen and its supposed remaining major ice culmination during Late Glacial may have experienced a similar development. Jotunheimen's blockfields definitely offer much to explore and future detailed research seems desirable to answer numerous remaining questions.

8.4.3 Ice-Cored Moraines

Although strictly speaking no periglacial landforms, the mostly multi-ridged and complex ice-cored moraine systems at several high-altitude glaciers (Figs. 8.11 and 8.12) are often included in studies focusing on the distribution of permafrost in Jotunheimen. Lilleøren and Etzelmüller (2011) define them as 'intact permafrost landforms' connected to glacial activity highlighting the abovementioned co-existence of glacial and periglacial process-systems in the region. They have first been investigated by Østrem (1961, 1963, 1964, 1965) who demonstrated the presence of ice cores beneath relatively thin debris covers. Furthermore, he emphasised major differences in morphology and scale to common moraine ridges formed by glaciers at lower altitude and specifically noted their spatial concentration in the northeastern part of Jotunheimen with its relatively high snow lines and continental climate (see also Baumann et al. 2009). He already anticipated their causal link to alpine permafrost without questioning their glacial origin. Østrem (1965) obtained radiocarbon dates for dust retrieved from the ice core at Gråsubrean that reveal decreasing ages from the inner towards the outer ridges, the latter apparently the youngest. His hypothesis for the origin of ice-cored moraines heavily relying on these radiocarbon dates is that the proximal ridges were deposited first and survived subsequent overriding by subsequent larger glacier advances. This explanation was challenged by Barsch (1971) who considered these features as (initial) rock glaciers. Unlike Østrem, Barsch interpreted the multi-ridged moraine morphology as rock glacier flow structures and saw these rock glaciers sourced by terminal moraines previously deposited by the glaciers. He accepted, however, that the outer ridges are the youngest and explained them by material being 'pressed-out' from the base under the rock glacier's own weight (Barsch 1977). In reply, Østrem (1971) argues that apart from detailed aspects of their morphology, the apparent lack of any movement of the ice-cored moraines and their location on level or near-level terrain do

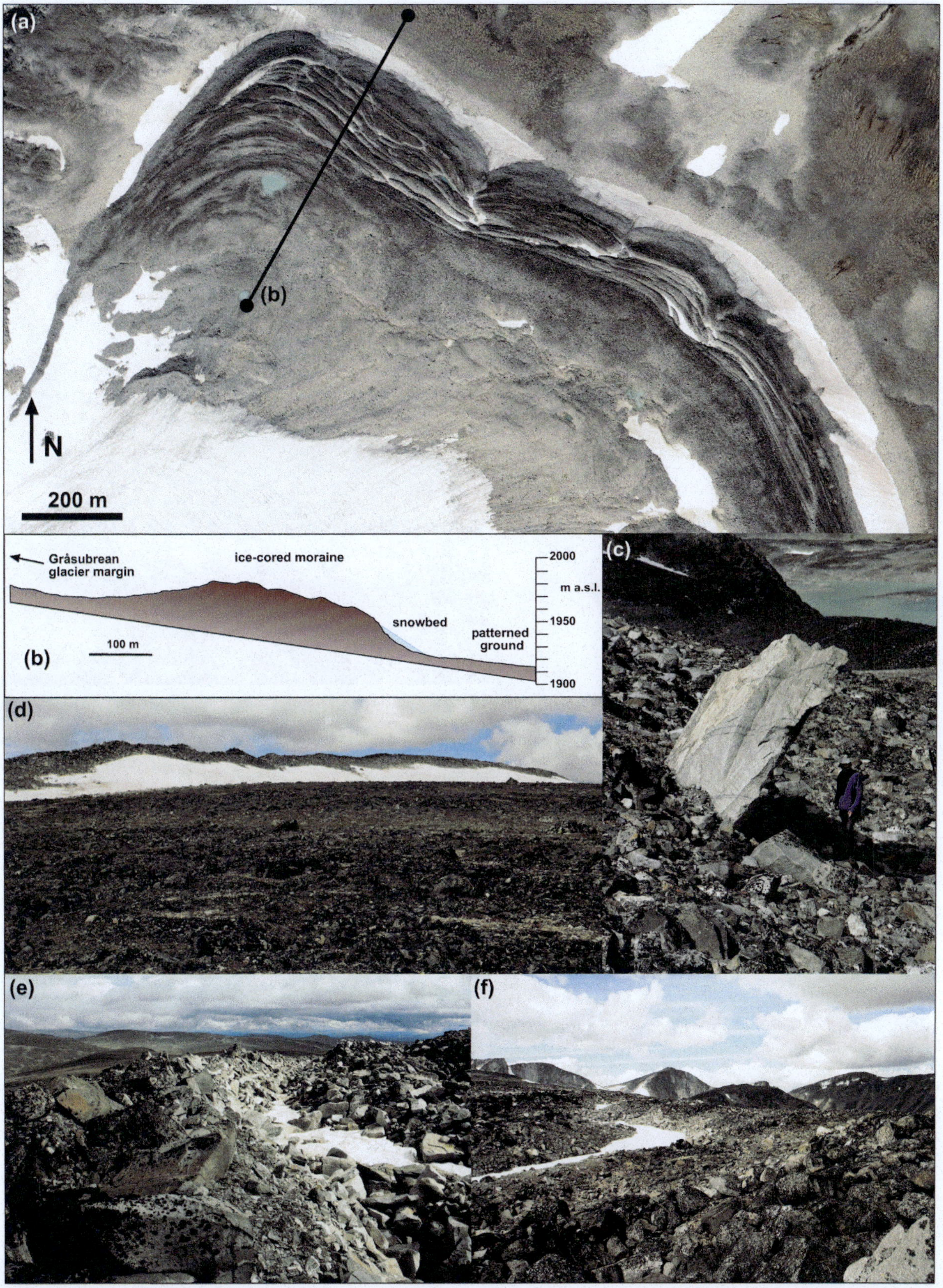

Fig. 8.11 Ice-cored moraine system at Gråsubrean: **a** Orthorectified aerial photo (2017) of its northeastern section, location of cross section indicated (adapted from www.norgeibilder.no, © Kartverket/www.kartverket.no). **b** Cross section of the multi-ridged ice-cored moraine system (modified from Matthews et al. 2014). **c** Up-thrust boulder on the proximal slope of the bench-like outermost moraine ridge. **d** Distal slope with typical semi-permanent distal snowbed; note patterned ground in foreground (cf. Figure 8.6b). **e**, **f** Multi-ridged surface of the ice-cored moraine (outer section), note furrows still partly covered by late-lying snow (Photos: Stefan Winkler, c, d, f 24.07.2013, e 01.08.2014)

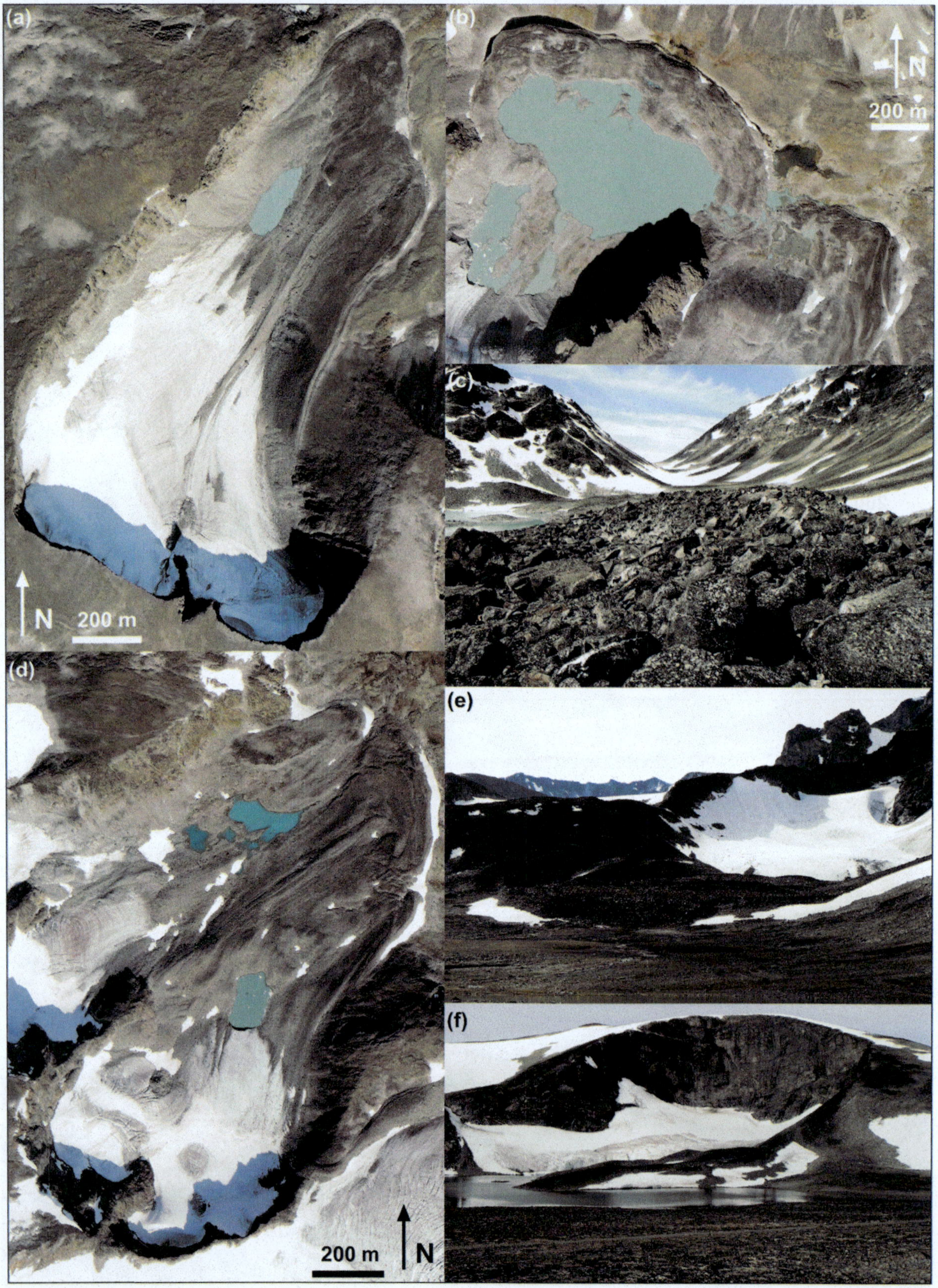

Fig. 8.12 Orthorectified aerial photos (2017) of ice-cored moraines: **a** Vestre Nautgardsbrean, **b** Østre Grotbrean, **d** Skautbreane (all adapted from www.norgeibilder.no, © Kartverket/www.kartverket.no). **c** Crest of ice-cored moraine at Østre Grotbrean. **e** Skautbreane with ice-cored moraine system. **f** Kjelen with ice-cored moraine (Photos: Stefan Winkler, c 15.07.2017, e 17.08.2008, f 18.08.2008)

not support any interpretation as rock glaciers. His view that these features are no rock glaciers has been accepted and retained since.

Matthews and Shakesby (1984) and Shakesby et al. (1987, 2004) designate the term 'push-deformation moraine' to these features and conclude that they likely are combinations of conventional push moraines and deformation structures of glacial origin, the latter being produced by the stresses associated with an advancing glacier in a permafrost environment. Etzelmüller et al. (2003) and Etzelmüller and Hagen (2005) reached similar conclusions in their comparison of ice-cored moraines in South Norway with landforms produced in the frontal ice-compression zone of polythermal glaciers in the Arctic. The most recent work of Matthews et al. (2014) interprets the ice-cored moraines as essentially glaciotectonic structures formed by the interaction of polythermal glaciers and alpine permafrost. Despite uncertainty about the precise mechanism of ridge formation, they conclude on basis of analogues from the Arctic and the Swiss Alps that a combination of subglacial, englacial, and proglacial creep with thrusting and folding in the zone of flow compression during relatively small glacial advances is involved. An obvious spatial concentration at high-altitude glaciers in eastern Jotunheimen seems, therefore, primary the result of the occurrence of small, polythermal glaciers in a more continental alpine permafrost environment. Glaciers in western and central Jotunheimen are, by contrast, generally warm-based. Their termini are located at lower altitude and their length changes in response to climate/mass balance variability are of higher magnitude, all factors restraining the formation of such ice-cored moraines. Matthews et al. (2014) reject the hypotheses of both Østrem and Barsch that the inner ridges are the oldest by application of SHD which reveals a reverse chronological order with the outer ridges constituting the youngest.

It seems very unlikely that proximal ridges would have survived multiple overriding by the glaciers given the delicate piles of boulders comprising most of the ridges and the lack of smooth profiles one would expect in such cases (Figs. 8.11e, f and 8.12c). Reasons for rejecting the original radiocarbon dates based on methodological uncertainties around the dated organic material of wind-blown origin have already been elaborated by Griffey and Worsley (1978). Both, ice and its organic content may have a considerable 'apparent age' when buried (see Matthews and Caseldine 1987). Østrem's (1961, 1963) initial crystallographic interpretation of the ice as originating from snowbeds rather than being glacier ice or ground ice does not necessarily support his theory. According to Matthews et al. (2014) the ice-cored moraine systems were essentially formed during the LIA from material deposited during multiple previous neoglacial events explaining the negatively skewed R-values of their SHD data and the considerable large 95% statistical confidence intervals, a pattern untypical for moraine ridges that were formed during one individual advance only but indicating considerable reworking. The small size of the glaciers involved supported their formation because once initially formed, any moraine ridge acts as obstacle and prevents complete overriding/destruction. Although moraine system formation surely started pre-LIA during older neoglacial events, they neither provide detailed evidence for their exact age nor help to decide whether some of those glaciers existed during the entire Holocene (including the HTM). Summarising, the ice-cored moraines in Jotunheimen should be classified as complex glaciotectonic terminal-moraine systems with their causal relationship to polythermal glaciers and permafrost constituting the obvious link to the alpine periglacial zone.

8.4.4 Rock Glaciers, Solifluction, and Other Periglacial Landforms

Rock glaciers are beyond controversy characteristic periglacial landforms of high mountain regions and widely utilised as indicators of former or current continuous permafrost (Barsch 1996; Janke et al. 2013). In Jotunheimen they are, however, rare features despite extensive occurrence of permafrost and only few rock glaciers have been mapped (Lilleøren and Etzelmüller 2011; Gisnås et al. 2013; Fig. 8.13). Within the entire mountain regions of South Norway, Lilleøren and Etzelmüller (2011) count only 6 talus-derived and 7 moraine-derived intact rock glaciers (but 40 ice-cored moraines) highlighting the general sparsity of these permafrost landforms. About half of these active forms are located in Jotunheimen, whereas some others have been reported and studied in Rondane to the East (cf. Barsch and Treter 1976; King 1984; Shakesby et al. 1987). Main reason for their scattered occurrence is the specific topography of Jotunheimen with steep slopes and rugged 'Alpine-type' landscapes largely absent at elevations where permafrost can be expected. Other potential locations for typical rock glacier formation like cirques or valley heads with steep headwalls are often still occupied by modern glaciers. Lilleøren and Etzelmüller (2011) also mention bedrock competence and debris availability as limiting factors, but the existence of ice-cored moraines clearly indicates that the low number of rock glaciers in Jotunheimen is predominately topographically controlled. The sole example of a rock glacier in Jotunheimen studied in greater detail is a moraine- or glacier-derived rock glacier (a debris-rock glacier following the classification of Barsch 1996) at Bukkeholsbrean that developed from a lateral moraine formed during the LIA-maximum (Vere and Matthews 1985; Fig. 8.11f). It is not clear if the small rock glacier of 100 m length and 130 m width is still active today, but ample debris supply at an

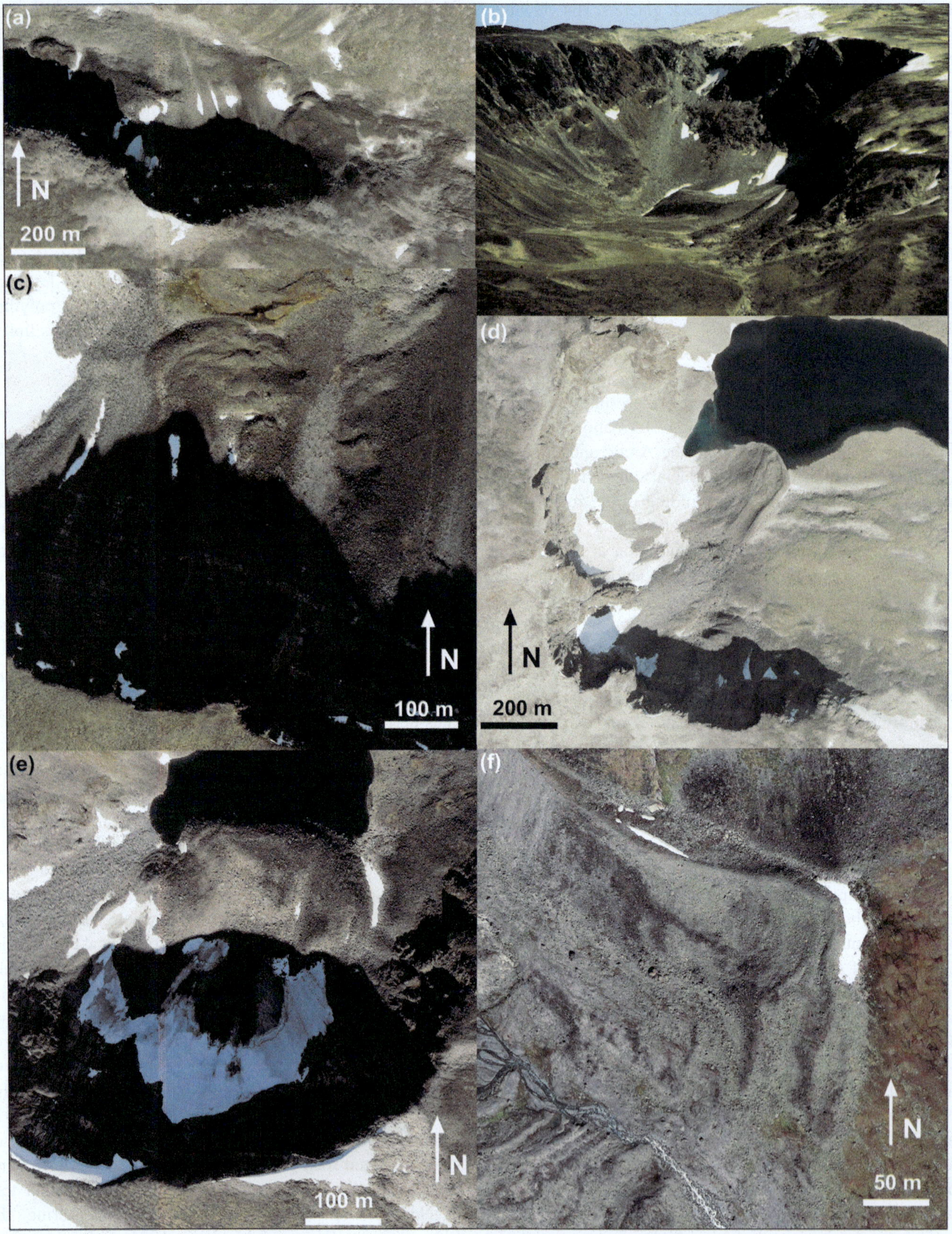

Fig. 8.13 Orthorectified aerial photos (2017) of rock glaciers: **a** (Initial) talus-derived rock glacier (Steindalen), **c** talus-derived rock glacier (Storglupen), **d** possible moraine-derived rock glacier (Hindholet/Nautgardstinden), **e** moraine-derived rock glacier east of Skauthøe on Skautflye, **f** rock glacier developed at the eastern lateral moraine system of Bukkeholsbrean (cf. Vere and Matthews 1985; all adapted from www.norgeibilder.no, © Kartverket/www.kartverket.no). **b** Talus-derived rock glacier at Glitterholet, west of Glittertinden; photo: Stefan Winkler, 26.07.2018)

altitude around the lower limit of permafrost may have been a decisive factor for its formation. A similar moraine-derived rock glacier developed from a lateral moraine is described by Matthews and Petch (1982) from Nordre Illåbrean as part of a broader study on the morphology and sedimentology of lateral moraines. For none of these two moraine-derived rock glaciers information about current internal dynamics, possible rates of movement, or the detailed relation to permafrost is available.

Various landforms of different morphology and scale related to solifluction have been reported from Jotunheimen, although no systematic review or comprehensive study on their spatial and altitudinal distribution has been undertaken yet. In general, solifluction is nowadays considered as an inclusive term describing mass movement of saturated material resulting from cyclic freezing and thawing of surface material by four different processes involved (Matsuoka 2001; Harris 2013; Millar 2013; Ballantyne 2018). The term gelifluction sometimes still used as a synonymous term needs to be identified as one of the four abovementioned different components of solifluction and defined as slow gravity-induced shear deformation associated with high pore-water pressure related to either seasonal freezing or the presence of permafrost. Detailed studies involving long-term monitoring (e.g. Harris et al. 2008) have revealed that the different processes involved with solifluction like frost creep, thaw consolidation, and the abovementioned gelifluction can best be described as form of viscous flow. Overall rates of downslope movements by solifluction are determined by a combination of factors like sediment properties, moisture content/pore-water pressure, snow distribution, or surface cover (vegetation cover). They are temporally and spatially highly variable. Measured rates of solifluction mostly fall in the range of some tens of millimeters per year, but can be significantly higher (Ballantyne 2018). By means of its relation to permafrost it must be noted that solifluction in periglacial environments does not require permafrost. Typical solifluction features may occur below the lower limit of permafrost and can, consequently, not be used for mapping regional or local permafrost distribution (Harris 1988; Lewkowicz 1988). The probably most common landforms produced by solifluction are either turf-banked or stone-banked (boulder-banked) lobes depending on whether the surface is covered by vegetation or not (Fig. 8.14). Recent research has shown that specific interactions between the vegetation cover or individual plant species and the morpho-dynamic processes at turf-bank solifluction lobes do exist (Draebing and Eichel 2017; Eichel et al. 2017). The initiation of solifluction is, furthermore, seen as palaeoclimatic indicator and has been studied in a number of mountain regions in Norway (Ellis 1979; Matthews et al. 1993; Nesje 1993; Worsley 1993; Elliott and Worsley 1999) and also in Jotunheimen (Matthews et al. 1986).

Matthews et al. (1986) investigate turf-banked solifluction lobes at an altitude of c. 1200 m a.s.l. in a northeasterly exposition in Leirdalen, central Jotunheimen. At this site below the lower limit of permafrost they examine both sedimentology and stratigraphy of a selected lobe and obtain several radiocarbon dates from organic content within. Based on this data the development started with an initial colluvial phase reflecting colder and wetter palaeoenvironmental conditions after about 3,500 to 3,000 cal. a BP with the gelifluction process itself commencing later after 1,000–800 cal. a BP, i.e. during the strongest climatic deterioration of the Late Holocene leading towards the LIA. By contrast to these solifluction lobes the sorted steps, sorted stripes, and boulder tongues mapped and described by Ødegård et al. (1987, 1988) at Juvflye constitute a combination of frost sorting processes and solifluction on shallow to moderate inclined slopes (Fig. 8.7). They seem, at least if such larger scale features are considered, bound to the

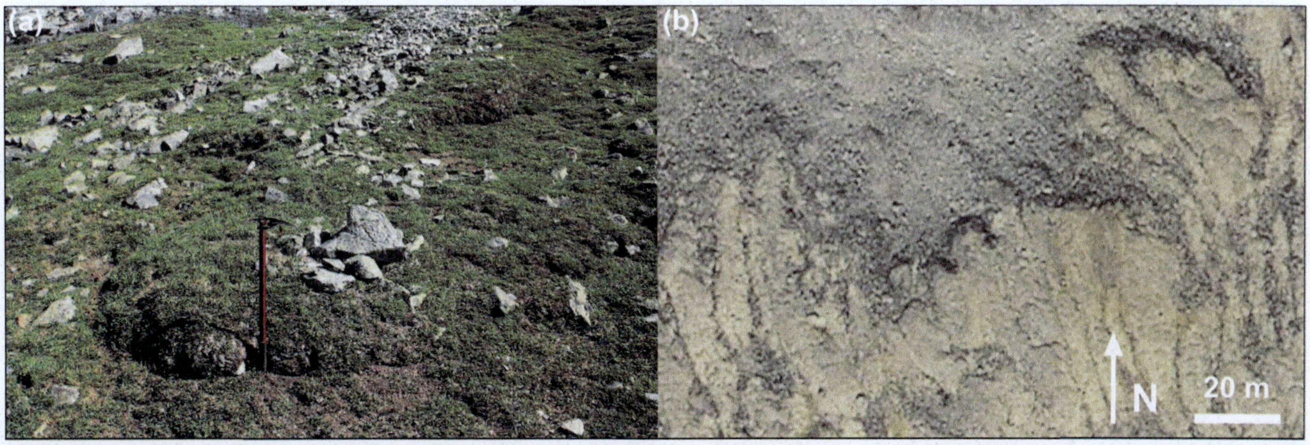

Fig. 8.14 a Turf-banked solifluction lobe in Leirdalen (Photo: Stefan Winkler 26.07.2008). **b** Orthorectified aerial photo (2017) of boulder tongues on the northern slope of Juvflye (adapted from www.norgeibilder.no, © Kartverket/www.kartverket.no)

occurrence of permafrost, but have not been studied in a chronologically context yet. Small-scale stripes are reported from recently deglaciated ground and do like adjacent patterned ground not require permafrost for their formation (Matthews et al. 1998). Within the small-scale geo-ecosystems on these recently deglaciated glacier forelands the family of small-scale periglacial landforms comprises some forms solely or partly influenced by solifluction, for example, solifluction areas encompassing numerous individual lobes with risers of up to 30 cm and ploughing blocks. The solifluction lobes are interpreted as having successively formed representing several generations and occur at average slope angles between 10° and 20°, but sometimes also steeper sections. At micro-scale, wrinkling of bryophite crust indicates the presence of solifluction that likely enhances the organic crust retaining moisture and increasing surface cohesion (Matthews et al. 1998).

Boulder-cored frost boils in different stages of their development have been reported from the glacier forelands of Bøverbrean, Storbrean and Styggedalsbreen (Harris and Matthews 1984; Winkler 1996; Matthews et al. 1998), all sites below the lower limit of permafrost (Fig. 8.15). The average diameters of these features are typically between 50 and 100 cm, but some reach diameters up to 150 cm or even 250 cm (Winkler 1996; Matthews et al. 1998). The height above ground does usually not exceed 50 cm and they are located in flat, poorly drained tills that have been deposited during the retreat from the LIA-maximum position (Fig. 8.15a). Some occur in areas that have been deglaciated after c. 1920/30 CE and are in close proximity to recently formed patterned ground, the latter, however, on moraine mounds rising a few metres above the wetter ground the former can be found on. During their development, boulder-cored frost boils are initially vegetation-capped but as they grow in height, this cap starts to tear up and fine sediment covers their surface (Harris and Matthews 1984; Fig. 8.15c). The development of these features is unanimously related to the vertical frost heaving of large, individual boulders within the initial till deposit (Fig. 8.15b) and well-developed boils show a shallow surrounding trough or moat, some exhibiting a concentration of gravel and stones. The sediment of the frost boils itself is not different to the till it has developed in.

Fig. 8.15 **a**, **b** Boulder-cored frost boils on the glacier foreland of Styggedalsbreen, area deglaciated during the early twentieth century CE (see Harris and Matthews 1984 and text for details). **c** Different stages of boulder-cored frost boils at Styggedalsbreen, the left one shows the final stage with the vegetation cover already torn up. **d** Earth hummock in upper Breiseterdalen at the site investigated by Ellis (1983). These features are located outside the glacier foreland of nearby Bøverbrean (Photos: Stefan Winkler, a, b 22.08.1996, c 05.08.1993, d 25.08.1994)

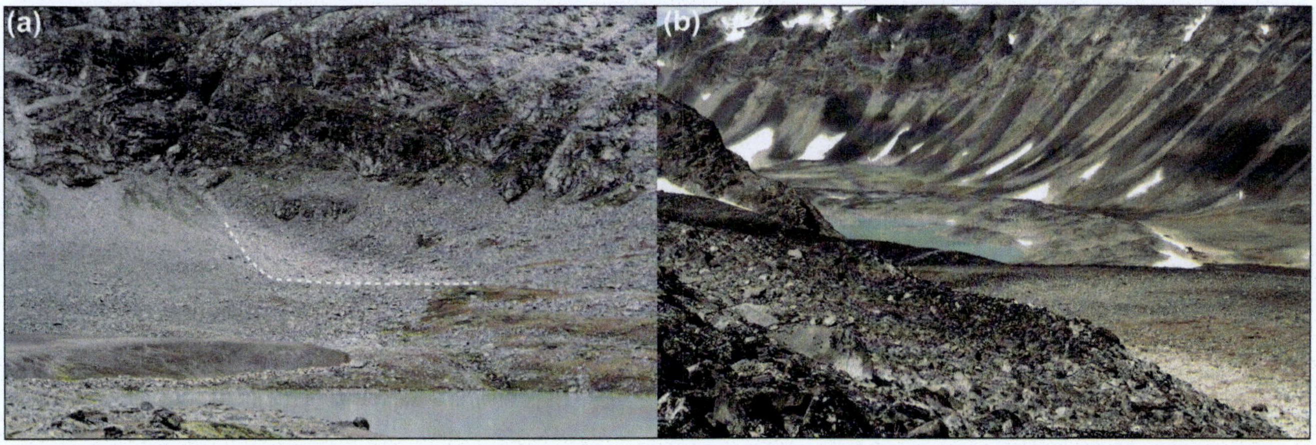

Fig. 8.16 **a** Pronival rampart (stippled line) near Styggedalsbreen in Hurrungane. **b** Avalanche boulder tongues at Trollsteinhøe as seen from Gråsubrean (Photos: Stefan Winkler, a 24.08.2006, b 24.07.2013)

By contrast to boulder-cored frost boils the earth hummocks described and radiocarbon dated by Ellis (1983) are neither located in Holocene glacier forelands nor have they developed in young till with large boulders. They are, however, also located below the lower limit of permafrost in Leirdalen and Breiseterdalen where cryoturbation is assumed to be the responsible process for their formation on imperfectly drained terrain at slope angles of 5° and 10°, respectively (Fig. 8.15d). They show a complex inner stratigraphy with organic-rich material, the latter despite all uncertainties involved with the dating of such material indicating their formation subsequent to the HTM as a consequence of climate change during the late Holocene.

Geomorphological processes and resulting landforms connected to snow avalanches are commonly considered as elements of slope processes within the periglacial process-system. Likewise nivation, a comprehensive term denoting various processes of mass wasting and weathering related to late-lying and perennial snow patches, is commonly dealt with in the context of periglacial geomorphology (e.g. Washburn 1979; Ballantyne 2018). Such landforms do also occur in Jotunheimen, but are generally infrequent with a scattered distribution. In this context, it seems interesting that observations by Liestøl (1974) in Leirdalen in western-central Jotunheimen established the 'avalanche plunge-pool effect' as one model for the formation of snow-avalanche impact ramparts and craters but subsequent recent research on snow-avalanche impact landforms have been conducted elsewhere in South Norway, in particular in Jostedalen and inner Sunnmøre (e.g. Matthews and McCarroll 1994; Owen et al. 2006; Matthews et al. 2017a). This concentration of research may reflect higher winter snow accumulation in (north)western South Norway, but a number of snow-avalanche impact forms can also be expected in western and western-central Jotunheimen in areas with a rugged 'alpine-type' morphology and steep valley slopes. An identical statement can be made concerning pronival ramparts, often developed as ridge-shaped debris accumulations at the foot of perennial snow patches that may form by an individual complex interaction of various processes (see Shakesby 1997; Matthews et al. 2011, 2017b; Hedding 2016). Small pronival ramparts in Jotunheimen can be found in the Hurrungane Massif (Fig. 8.16a) and do possibly also occur in other parts as these forms are not in any case easy to identify and can be confused with morphologically similar boulder accumulations of glacial, periglacial, or mass movement origin. Snow-avalanche boulder tongues sensu Rapp (1959) also occur in Jotunheimen, for example, north of Glittertinden in eastern Jotunheimen (Fig. 8.16b). Just as cryoplanation terraces these landforms related to snow avalanches and perennial or late-lying snow fields are still awaiting further exploration and have to date not received the attention they deserve, possibly because of the dominance of permafrost-related features like patterned ground or ice-cored moraines.

8.5 Altitudinal Distribution and Vertical Zonation of Periglacial Landforms

Investigating the altitudinal distribution of periglacial landforms allows identifying controlling factors for landform development within the periglacial process-system (Harris 1988). Outlining a vertical zonation for the distribution of individual periglacial landforms reveals if they are influenced by the occurrence of permafrost as well as which local environmental conditions like (micro-)topography, surface material/bedrock, moisture availability, or vegetation ground cover may additionally exert control. Whereas the altitudinal distribution of permafrost in Jotunheimen and its fluctuations

Table 8.3 Altitudinal distribution of selected periglacial landforms in Jotunheimen taken from existing literature (see text)

Landform	Altitude (m a.s.l.)	Location	Source
Blockfields (autochthonous)	1600–2100	Western Jotunhemen	Klakegg et al. (1989)
	1200–2300	Central/Eastern Jotunheimen	Sollid and Trollvik (1991)
	c. 2000 (cluster)	Jotunheimen	Nesje et al. (1988)
Patterned ground (large features)	1650–2000	Juvflye	Ødegård et al. (1987)
	1500–1925	Juvflye	Winkler et al. (2016)
	1085–2200	Northeastern Jotunheimen	Cook-Talbot (1991)
Patterned ground (small features)	1900–2250	Juvflye	Ødegård et al. (1987)
	1270–1625	Various glacier forelands	Haugland (2006)
Ice-cored moraines (intact)	1400–2000	Jotunheimen	Østrem et al. (1988)
	c. 1620 (mean)	South Norway[a]	Lilleøren and Etzelmüller (2011)
Rock glaciers (intact)	c. 1640/1685 (mean)[b]	South Norway	Lilleøren and Etzelmüller (2011)
Boulder-cored frost boils	1155–1375	Various glacier forelands	Harris and Matthews (1984)
Earth hummocks	1260–1350	Breiseter-/Leirdalen	Ellis (1983)

[a]Mean value for all features detected in South Norway
[b]First value for talus-derived, second for moraine-derived rock glaciers

during the Holocene is well documented (Lilleøren et al. 2012; Gisnås et al. 2013, 2017; see Sects. 8.2.2 and 8.3.2), the attempt to compile an altitudinal distribution of periglacial landforms on a regional scale for Jotunheimen based on existing literature proves to be challenging (Table 8.3). Lilleøren and Etzelmüller (2011) provide mean elevations for individual permafrost landforms (different rock glacier subtypes and ice-cored moraines, intact and relict features) summarised in their regional inventory for South Norway. In their glacier inventory for South Norway mainly based on 1970s aerial photography Østrem et al. (1988) provide additionally information on moraines at these glaciers including possible ice-cored moraines in Jotunheimen. The small-scale Quaternary geology maps of both Sollid and Trollvik (1991) and Klakegg et al. (1989) display blockfields in those parts of Jotunheimen their respective maps cover. But as with the blockfield studies of Nesje et al. (1988) and Nesje and Dahl (1990), the altitudinal and spatial resolution achieved is limited. Summarising, there seems to be a significant gap in scale considering the studies conducted so far. While some aim for a South Norwegian or Norwegian/Scandinavian overview, many other studies are restricted to small areas or even individual sites within Jotunheimen only (e.g. Juvflye or selected glacier forelands). They yield either more general statements regarding the altitudinal and spatial distribution of periglacial features or essentially provide insights into the occurrence of features at a few exemplary locations only. This makes every attempt to present a regional but nonetheless sufficiently detailed overview based on existing literature difficult if not impossible.

Because most available data on permafrost and periglacial landforms including detailed geomorphological mapping are centred on Juvflye and Galdhøpiggen/Visdalen, the attempt to graphically summarise the altitudinal distribution of individual landforms on Fig. 8.17 refers to this particular location in central Jotunheimen. The vertical zonation displayed may, therefore, deviate from conditions in western or eastern parts of Jotunheimen in their dependency on controlling factors like permafrost, climatic conditions, and topography in their particular local altitudinal variability. Additionally, the lower limit of permafrost as well as the lowermost glacier termini and steady-state glacier equilibrium lines schematically displayed on Fig. 8.17 apply for central Jotunheimen only. They do not take the West-East gradient of rising elevations into account. Autochthonous blockfields form a distinct cluster at around 2000 m a.s.l. at summit level throughout Jotunheimen and mainly cover an altitudinal range of roughly 1000 m between 1200 and 2300 m a.s.l. (Nesje et al. 1988; Klakegg et al. 1989; Nesje and Dahl 1990; Sollid and Trollvik 1991). They occur above the lower limit of permafrost and their upper limit constitutes an artefact of the local topography (summit height, slope angle) or perennial snow/glacier cover. Ice-cored moraines share an occurrence above the lower limit of permafrost and intact features are predominately reported from an altitudinal range between 1600 and 2000 m a.s.l. (Østrem 1964; Østrem et al. 1988; Lilleøren and Etzelmüller 2011; Matthews et al. 2014). Similar data are reported for intact rock glaciers with features assumed relict occasionally located slightly below the current lower limit of permafrost (Matthews and Petch 1982; Vere and Matthews 1985; Lilleøren and Etzelmüller 2011). The lowermost occurrence of large features of patterned ground usually coincide with the current lower limit of permafrost, e.g. at Juvflye at c. 1500 m a.s.l. (Ødegård et al. 1987; Winkler et al. 2016). It needs,

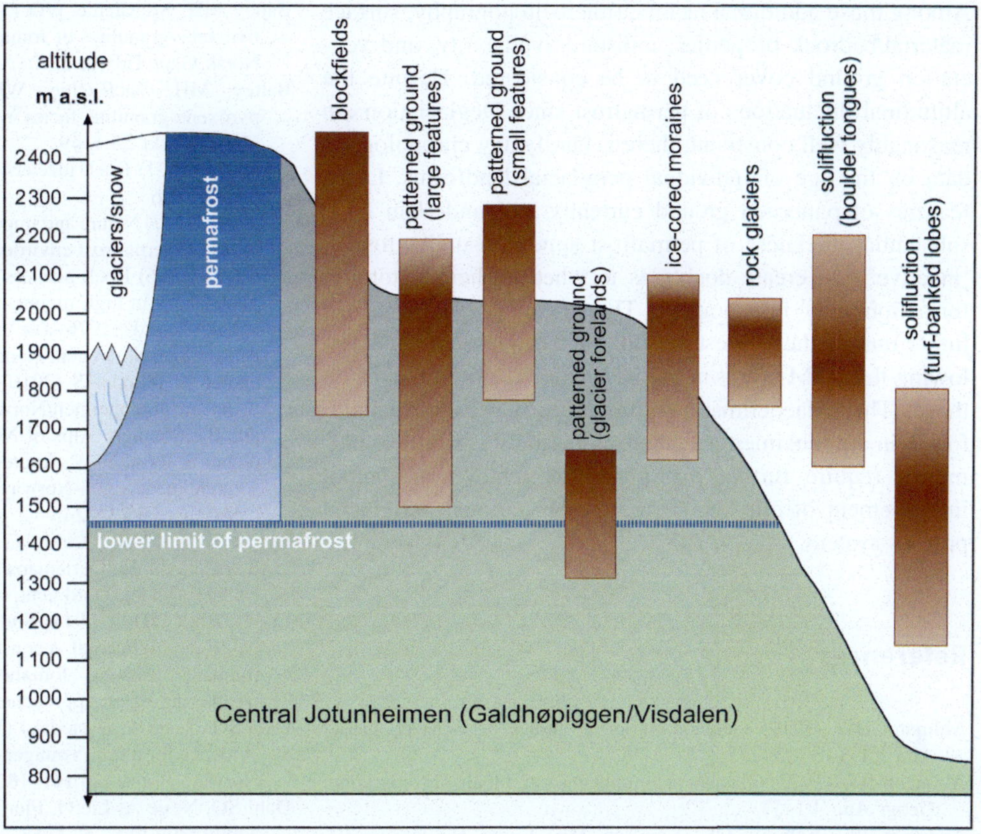

Fig. 8.17 Schematic graphic summary of the altitudinal distribution of periglacial landforms in central Jotunheimen based on existing literature and own observations (see text for details)

however, to be mentioned that (mostly small or secondary) features clearly showing signs of active processes can merely be found at around 1800 m a.s.l. and above (Cook-Talbot 1991; Winkler et al. 2016). The possible complex history of patterned ground formation (see Sect. 8.4.1) and local control by factors other than permafrost (surface material, moisture availability, deglaciation patterns, etc.) are responsible for this particular vertical zonation. Patterned ground as part of assemblages including various other small-scale periglacial landforms appears on recently deglaciated terrain on LIA-glacier forelands below or around the local lower limit of permafrost. Similar to the development of turf-banked solifluction lobes and few additional exceptions among intact/active periglacial landforms in Jotunheimen (Table 8.3), permafrost is not required for these small-scale patterned ground features.

8.6 Summary

Jotunheimen in central South Norway constitutes a major Scandinavian permafrost region and is characterised by a specific assemblage of periglacial landforms. The altitudinal distribution of individual periglacial features largely reflects the occurrence of permafrost as controlling factor for landform development within the periglacial process-system. Jotunheimen's periglacial landform assemblage needs, however, to be characterised as unique if compared on a global scale to other mid-latitudinal mountain ranges. Despite home of the highest peaks in Norway, Jotunheimen's gross morphology is distinguished by omnipresent high-altitude plateaus and shallow to moderate slopes at summit level. In interplay with permafrost this specific gross morphology provides a natural environment promoting extensive coverage of surfaces and slopes at higher elevations by different types of patterned ground. This is uncommon for the majority of alpine periglacial landsystems worldwide. By contrast, those periglacial features generally widespread at elevations above the lower limit of permafrost like rock glaciers are rare in Jotunheimen as a result of the combined effect of the lack of steep slopes, many potential locations (still) covered by modern glaciers, and bedrock properties limiting necessary debris supply. Blockfields and ice-cored moraines, the latter glaciotectonic terminal-moraine systems causally related to polythermal glaciers and permafrost, are other periglacial landforms that own their existence to some extent Jotunheimen's gross morphology. Similar to small-scale periglacial features developed on recently deglaciated glacier forelands below the local lower limit of permafrost, they provide evidence that factors other than permafrost may unfold influence on the altitudinal and spatial distribution of periglacial landforms.

Among those additional factors, (micro-)topography, surface material/bedrock properties, moisture availability, and vegetation ground cover need to be considered. Despite the altitudinal fluctuations of permafrost since Deglaciation are reasonably well constraint, there is hardly any chronological data on the age of individual periglacial landforms. Large features of patterned ground currently still underlain by a substantial thickness of permafrost appear visually 'fossil'/'inactive' and create doubts as to whether their history of formation may be complex. The lowering of permafrost limits during the coldest periods of the late Holocene following the HTM seems not to have exerted any influence on those. The palaeoclimatic significance of periglacial landforms in Jotunheimen and their altitudinal distribution ultimately require further exploration to achieve substantial improvement of our knowledge on the regional periglacial process-system.

References

Ahlmann HW (1919) Geomorphological studies in Norway. Geogr Ann 1:1–146/193–255

Ahlmann HW (1922) Glaciers in Jotunheimen and their physiography. Geogr Ann 4:1–57

Andersen JL, Egholm DL, Knudsen MF, Linge H, Jansen JD, Pedersen VK, Nielsen SB, Tikhomirov D, Olsen J, Fabel D, Xu S (2018) Widespread erosion on high plateaus during recent glaciations in Scandinavia. Nat Commun 9: article 830

Andreassen LM, Winsvold SH (eds) (2012) Inventory of Norwegian glaciers. NVE, Oslo, Rapport 38-2012

Andreassen LM, Elvehøy H, Kjøllmoen B, Engeset RV, Haakensen N (2005) Glacier mass balance and length variations in Norway. Ann Glaciol 42:317–325

Bakke J, Dahl SO, Nesje A (2005) Lateglacial and early Holocene palaeoclimatic reconstruction based on glacier fluctuations and equilibrium-line altitudes at northern Folgefonna, Hardanger, western Norway. J Quaternary Sci 20:179–198

Ballantyne CK (1979) Patterned ground on an active medial moraine, Jotunheimen, Norway. J Glaciol 22:396–401

Ballantyne CK (2010) A general model of autochthonous blockfield evolution. Permafrost Periglac 21:289–300

Ballantyne CK (2013a) Patterned ground. In: Elias SA, Mock CJ (eds) Encyclopedia of Quaternary science, 2nd edn, vol 3. Elsevier, Amsterdam, pp 452–463

Ballantyne CK (2013b) Trimlines and paleonunataks. In: Elias SA, Mock CJ (eds) Encyclopedia of Quaternary science, vol 1, 2nd edn. Elsevier, Amsterdam, pp 918–928

Ballantyne CK (2018) Periglacial geomorphology. Wiley Blackwell, Chichester

Ballantyne CK, Matthews JA (1982) The development of sorted circles on recently deglaciated terrain, Jotunheimen, Norway. Arctic Alpine Res 14:341–354

Ballantyne CK, Matthews JA (1983) Desiccation cracking and sorted polygon development, Jotunheimen, Norway. Arctic Alpine Res 15:339–349

Battey MH, Bryhni I (1981) Berggrunnen og landskapet. In: Garmo TT, Marker E (eds) Norges nasjonalparker 10 – Jotunheimen. Luther, Oslo, pp 21–33

Battey MH, McRitchie WD (1973) A geological traverse across the pyroxene-granulites of Jotunheimen in the Norwegian Caledonides. Norsk Geol Tidsskr 53:237–265

Battey MH, McRitchie WD (1975) The petrology of the pyroxene-granulite facies rocks of Jotunheimen, Norway. Norsk Geol Tidsskr 55:1–49

Barsch D (1971) Rock glaciers and ice-cored moraines. Geogr Ann A 53:203–206

Barsch D (1977) Nature and importance of mass wasting by rock glaciers in alpine permafrost environments. Earth Surf Proc 2:231–245

Barsch D (1996) Rock glaciers—indicators for the present and former geoecology in high mountain environments. Springer, Berlin

Barsch D, Treter U (1976) Zur Verbreitung von Periglazial-Phänomenen in Rondane/Norwegen. Geogr Ann A 58:83–93

Baumann S, Winkler S (2010) Parameterization of glacier inventory data from Jotunheimen/Norway in comparison to the European Alps and the Southern Alps of New Zealand. Erdkunde 64:155–177

Baumann S, Winkler S, Andreassen LM (2009) Mapping glaciers in Jotunheimen, South-Norway, during the 'Little Ice Age' maximum. Cryosphere 3:231–243

Bogen J, Wold B, Østrem G (1989) Historic glacier variations in Scandinavia. In: Oerlemans J (ed) Glacier fluctuations and climatic change. Reidel, Dordrecht, pp 109–128

Cook-Talbot JD (1991) Sorted circles, relative-age dating and palaeoenvironmental reconstruction in an alpine periglacial environment, eastern Jotunheimen, Norway: lichenometric and weathering-based approaches. Holocene 1:128–141

Dahl SO, Nesje A, Øvstedal J (1997) Cirque glaciers as morphological evidence for a thin Younger Dyras ice sheet in east-central southern Norway. Boreas 26:161–180

Dahl SO, Nesje A, Lie Ø, Fjordheim K, Matthews JA (2002) Timing, equilibrium-line altitudes and climatic implications of two early-Holocene glacier glacier readvances during the Erdalen event at Jostedalsbreen, western Norway. Holocene 12:17–25

Draebing D, Eichel J (2017) Spatial controls of Turf-banked solifluction lobes and their role for paraglacial adjustment in glacier forelands. Permafrost Periglac 28:446–459

Ebert K (2009) Terminology of long-term geomorphology: a Scandinavian perspective. Prog Phys Geogr 33:163–182

Egholm DL, Jansen JD, Brædstrup CF, Pedersen VK, Andersen JL, Ugelvig SV, Larsen NK, Knudsen MF (2017) Formation of plateau landscapes on glaciated continental margins. Nat Geosci 10:592–599

Eichel J, Draebing D, Klingbeil L, Wieland M, Eling C, Schmidtlein S, Kuhlmann H, Dikau R (2017) Solifluction meets vegetation: the role of biogeomorphic feedbacks for turf-banked solifluction lobe development. Earth Surf Proc Land 42:1623–1635

Elliott G, Worsley P (1999) The sedimentology, stratigraphy and 14C dating of a turf-banked solifluction lobe: evidence for Holocene slope instability at Okstindan, northern Norway. J Quaternary Sci 14:175–188

Ellis S (1979) Radiocarbon dating evidence for the initiation of solifluction ca. 5500 years B.P. at Okstindan, North Norway. Geogr Ann A 61:29–33

Ellis S (1983) Stratigraphy and 14C dating of two earth hummocks, Jotunheimen, south central Norway. Geogr Ann A 65:279–287

Erikstad L, Sollid JL (1986) Neoglaciation in South Norway using lichenometric methods. Norsk Geogr Tidsskr 40:85–105

Etzelmüller B, Hagen JO (2005) Glacier-permafrost interaction in Arctic and alpine mountain environments with examples from southern Norway and Svalbard. In: Harris C, Murton JB (eds) Cryospheric systems: glaciers and permafrost. Geological Society of London Special Publication 242, Bath, pp 11–27

Etzelmüller B, Berthing I, Sollid JL (2003) Aspects and concepts on the geomorphological significance of Holocene permafrost in southern Norway. Geomorphology 52:87–104

Etzelmüller B, Romstad B, Fjellanger J (2007) Automatic regional classification of topography in Norway. Norw J Geol 87:167–180

Fabel D, Stroeven A, Harbor J, Kleman J, Elmore D, Fink D (2002) Landscape preservation under Fennoscandian ice sheets determined from in situ produced ^{10}Be and ^{26}Al. Earth Planet Sc Lett 201:397–406

Farbrot H, Hipp TF, Etzelmüller B, Isaksen K, Ødegård RS, Schuler TV, Humlum O (2011) Air and ground temperature variations observed along elevation and continentality gradients in southern Norway. Permafrost Periglac 22:343–360

Fossen H, Pedersen R-B, Bergh S, Andresen A (2008) Creation of a mountain chain. The building up of the Caledonides, about 500-405 Ma. In: Ramberg I, Bryhni I, Nøttvedt A, Rangnes K (eds) The making of a land—geology of Norway. NGF, Trondheim, pp 178–231

French HM (2007) The periglacial environment, 3rd edn. Wiley, Chichester

Gisnås K, Etzelmüller B, Farbrot H, Schuler TV, Westermann S (2013) CryoGRID 1.0: permafrost distribution in Norway estimated by a spatial numerical model. Permafrost Periglac 24:2–19

Gisnås K, Etzelmüller B, Lussana C, Hjort J, Sannel BK, Isaksen K, Westermann S, Kuhry P, Christiansen HH, Frampton A, Åkerman J (2017) Permafrost map for Norway, Sweden and Finland. Permafrost Periglac 28:359–378

Gjessing J (1967) Norway's paleic surface. Norsk Geogr Tidsskr 21:69–132

Gjessing J (1978) Norges landformer. Universitetsforlaget, Oslo/Bergen/Tromsø

Goehring BM, Brook EJ, Linge H, Raisbeck GM, Yiou F (2008) Beryllium-10 exposure ages of erratic boulders in southern Norway and implications for the history of the Fennoscandian ice sheet. Quaternary Sci Rev 27:320–336

Goldthwait RP (1976) Frost-sorted patterned ground: a review. Quaternary Res 6:27–35

Goodfellow BW, Stroeven AP, Hättestrand C, Kleman J, Jansson KN (2008) Deciphering a non-glacial/glacial landscape mosaic in the northern Swedish mountains. Geomorphology 93:213–232

Goodfellow BW, Fredin IO, Derron M-H, Stroeven AP (2009) Weathering processes and Quaternary origin of an alpine blockfield in Arctic Sweden. Boreas 38:379–398

Goodfellow BW, Stroeven AP, Fabel D, Fredin O, Derron M-H, Bintanja R, Caffee MW (2014) Arctic–alpine blockfields in the northern Swedish Scandes: late Quaternary—not Neogene. Earth Surf Dynam 2:383–401

Griffey NJ, Worsley P (1978) The pattern of Neoglacial glacier variations in the Okstindan region of northern Norway during the last three millennia. Boreas 7:1–17

Hall AM, Ebert K, Kleman J, Nesje A, Ottesen D (2013) Selective glacial erosion on the Norwegian passive margin. Geology 41:1203–1206

Hallet B (1990) Self-organization in freezing soils—from microscopic ice lenses to patterned ground. Can J Phys 68:842–852

Hallet B (2015) Stone circles: form and soil kinematics. Philos T R Soc A 371:20120357

Harris C (2013) Slope deposits and forms. In: Elias SA (ed) Encyclopedia of Quaternary science, vol 3, 2nd edn. Elsevier, Amsterdam, pp 481–489

Harris C, Cook JD (1986) The detection of high altitude permafrost in Jotunheimen, Norway using seismic refraction techniques: an assessment. Arctic Alpine Res 18:19–26

Harris C, Cook JD (1988) Micromorphology and microfabrics of sorted circles, Jotunheimen, southern Norway. In: Senneset K (ed) 5th international conference on permafrost proceedings, vol 1. Tapir, Trondheim, pp 776–783

Harris C, Matthews JA (1984) Some observations on boulder-cored frost boils. Geogr J 150:63–73

Harris C, Kern-Luetschg M, Smith M, Isaksen K (2008) Solifluction processes in an area of seasonal ground freezing, Dovrefjell, Norway. Permafrost Periglac 19:31–47

Harris SA (1988) The alpine periglacial zone. In: Clark MJ (ed) Advances in periglacial geomorphology. Wiley, Chichester, pp 369–413

Hauck C, Isaksen K, Vonder Mühll D, Sollid JL (2004) Geophysical surveys designed to delineate the altitudinal limit of mountain permafrost: an example from Jotunheimen, Norway. Permafrost Periglac 15:191–205

Haugland JE (2004) Formation of patterned ground and fine-scale soil development within two late Holocene glacial chronosequences: Jotunheimen, Norway. Geomorphology 61:287–301

Haugland JE (2006) Short-term periglacial processes, vegetation succession, and soil development within sorted patterned ground: Jotunheimen, Norway. Arctic Antarc Alp Res 38:82–89

Hedding DW (2016) Pronival ramparts: a review. Prog Phys Geog 40:835–855

Hiemstra JF, Matthews JW, Evans DJA, Owen G (2015) Sediment fingerprinting and the mode of singular and composite annual moraine ridges at two glacier margins, Jotunheimen, southern Norway. Holocene 25:1772–1785

Hipp T, Etzelmüller B, Farbrot H, Schuler TV, Westermann S (2012) Modelling borehole temperatures in Southern Norway—insights into permafrost dynamics during the 20th and 21st century. Cryosphere 6:553–571

Hoel A, Werenskiold W (1962) Glaciers and snowfields in Norway. Norsk Polarinstitut Skrifter, Oslo, p 114

Holmsen P (1982) Jotunheimen - beskrivelse til kvartærgeologisk kart M 1:250.000. Norges Geologiske Undersøkelse 374 Skrifter 37, Trondheim

Hughes ALC, Gyllencreutz R, Lohne ØS, Mangerud J, Svendsen JI (2016) The last Eurasian ice sheets—a chronological database and time-slice reconstruction, DATED-1. Boreas 45:1–45

Isaksen K, Holmlund P, Sollid JL, Harris C (2001) Three deep alpine-permafrost boreholes in Svalbard and Scandinavia. Permafrost Periglac 12:13–25

Isaksen K, Hauck C, Gudevang E, Ødegård RS, Sollid JL (2002) Mountain permafrost distribution in Dovrefjell and Jotunheimen, southern Norway, based on BTS and DC resistivity tomography data. Norsk Geogr Tidsskr 56:122–136

Isaksen K, Heggem ESF, Bakkehoi S, Ødegård RS, Eiken T, Etzelmüller B, Sollid JL (2003) Mountain permafrost and energy balance on Juvvasshoe, southern Norway. In: Phillips M, Springman SM, Arenson LU (eds) Permafrost, 2 vols. Swets and Zeitlinger, Lisse, pp 467–472

Isaksen K, Sollid JL, Holmlund P, Harris C (2007) Recent warming of mountain permafrost in Svalbard and Scandinavia. J Geophys Res 112:F92S04

Isaksen K, Ødegård RS, Etzelmüller B, Hilbich C, Hauck C, Farbrot H, Eiken T, Hagen JO, Hipp TF (2011) Degrading mountain permafrost in Southern Norway: Spatial and temporal variability of ground temperatures, 1999–2009. Permafrost Periglac 22:361–377

Janke JR, Regmi NR, Giardino JR, Vitek JD (2013) Rock glaciers. In: Shroder J, Giardino R, Harbor J (eds) Treatise on geomorphology, vol 8. Glacial and periglacial geomorphology. San Diego Academic Press, San Diego, pp 238–273

Karlén W, Matthews JA (1992) Reconstructing Holocene glacier variations from glacial lake sediments: studies from Nordvestlandet and Jostedalsbreen-Jotunheimen, southern Norway. Geogr Ann A 74:327–348

Kessler MA, Murray AB, Werner TB, Hallet B (2001) A model for sorted circles as self-organized patterns. J Geophys Res Sol-EA 106:13287–13306

King L (1984) Permafrost in Skandinavien: Untersuchungsergebnisse aus Lappland, Jotunheimen und Dovre/Rondane. Universität Heidelberg, Heidelberger Geographische Arbeiten 76, Heidelberg

Klakegg O, Norddahl-Olsen T, Sønstegaard E, Aa AR (1989) Sogn og Fjordane fylke - kvartærgeologisk kart M 1:250.000. Norges Geologiske Undersøkelse, Trondheim

Kleman J, Stroeven AP, Lundqvist J (2008) Patterns of Quaternary ice sheet erosion and deposition in Fennoscandia and a theoretical framework for explanation. Geomorphology 97:73–90

Kleman J, Fastook J, Ebert K, Nilsson J, Caballer R (2013) Pre-LGM northern hemisphere ice sheet topography. Clim Past 9:2365–2378

Klemsdal T, Sjulsen E (1988) The Norwegian macro-landforms: definition, distribution and system of evolution. Norsk Geogr Tidsskr 42:133–147

Lewkowicz AG (1988) Slope processes. In: Clark MJ (ed) Advances in periglacial geomorphology. Wiley, Chichester, pp 325–368

Lidmar-Bergström K (1988) Preglacial weathering and landform evolution in Fennoscandia. Geogr Ann A 70:273–276

Lidmar-Bergström K, Ollier CD, Sulebak JR (2000) Landforms and uplift history of southern Norway. Global Planet Change 24:211–231

Lie Ø, Dahl SO, Nesje A, Matthews JA, Sandvold S (2004) Holocene fluctuations of a polythermal glacier in high-alpine eastern Jotunheimen, central-southern Norway. Quaternary Sci Rev 23:1925–1945

Liestøl O (1967) Storbreen glacier in Jotunheimen. Norsk Polarinstitutt, Oslo, Skrifter, p 141

Liestøl O (1974) Avalanche plunge-pool effect. Norsk Polarinstitutt Årbok 1972, Oslo, pp 179–181

Lilleøren KS, Etzelmüller B (2011) A regional inventory of rock glaciers and ice-cored moraines in Norway. Geogr Ann A 93:175–191

Lilleøren K, Etzelmüller B, Schuler TV, Gisnås K, Humlum O (2012) The relative age of mountain permafrost—estimation of Holocene permafrost limits in Norway. Global Planet Change 92–93:209–223

Lutro O, Tveten E (1996) Geologisk kart over Norge, berggrunnskart Årdal M 1:250.000 Årdal. Trondheim, Norges Geologiske Undersøkelse

Mangerud J, Gyllencreutz R, Lohne Ø, Svendsen JI (2011) Glacial history of Norway. In: Ehlers J, Gibbard PL, Hughes PD (eds) Quaternary glaciations—extent and chronology. Elsevier, Amsterdam, pp 279–298

Marr P, Winkler S, Löffler J (2018) Investigations on blockfields and related landforms at Blåhø (Southern Norway) using Schmidt-hammer exposure-age dating (SHD): palaeoclimatic and morphodynamic implications. Geogr Ann A 100:285–306

Matsuoka N (2001) Solifluction rates, processes and landforms: a global review. Earth Sci Rev 55:107–134

Matsuoka N, Abe M, Ijiri M (2003) Differential frost heave and sorted patterned ground: field measurements and a laboratory experiment. Geomorphology 52:73–85

Matthews JA (1974) Families of lichenometric dating curves from the Storbreen gletschervorfeld, Jotunheimen, Norway. Norsk Geogr Tidsskr 28:215–235

Matthews JA (1975) Experiments on the reproductibility and reliability of lichenometric dates, Storbreen gletschervorfeld, Jotunheimen, Norway. Norsk Geogr Tidsskr 29:97–109

Matthews JA (1977) A lichenometric test of the 1750 end-moraine hypothesis: Storbreen gletschervorfeld, southern Norway. Norsk Geogr Tidsskr 31:129–136

Matthews JA (1994) Lichenometric dating: a review with particular reference to 'Little Ice Age' moraines in southern Norway. In: Beck C (ed) Dating in exposed and surface contexts. University of New Mexiko Press, Albuquerque, pp 185–212

Matthews JA (2005) 'Little Ice Age' glacier variations in Jotunheimen, southern Norway: a study in regionally controlled lichenometric dating of recessional moraines with implications for climate and lichen growth rates. Holocene 15:1–19

Matthews JA, Caseldine CJ (1987) Arctic-alpine brown soils as a source of palaeoenvironmental information: further ^{14}C dating and palynological evidence from Vestre Memurubreen, Jotunheimen, Norway. J Quaternary Sci 2:59–71

Matthews JA, Dresser PQ (2008) Holocene glacier variation chronology of the Smørstabbtindan massif, Jotunheimen, southern Norway, and the recognition of century- to millennial-scale European Neoglacial events. Holocene 18:181–201

Matthews JA, Karlén W (1992) Asynchronous Neoglaciation and Holocene climatic change recon-structed from Norwegian glaciolacustrine sedimentary sequences. Geology 20:991–994

Matthews JA, McCarroll D (1994) Snow-avalanche impact landforms in Breheimen, southern Norway: origin, age, and paleoclimatic implications. Arctic Alp Res 26:103–115

Matthews JA, Owen G (2010) Schmidt hammer exposure-age dating: development linear age-calibration curves using Holocene bedrock surfaces from the Jotunheimen-Jostedalsbreen regions of southern Norway. Boreas 39:105–115

Matthews JA, Petch JR (1982) Within-valley asymmetry and related problems of Neoglacial lateral moraine development at certain Jotunheimen glaciers, southern Norway. Boreas 11:225–245

Matthews JA, Shakesby RA (1984) The status of the 'LIttle Ice Age' in southern Norway: a relative-age dating of Neoglacial moraines with Schmidt hammer and lichenometry. Boreas 13:333–346

Matthews JA, Winkler S (2011) Schmidt-hammer exposure-age dating (SHD): application to early-Holocene moraines and a reappraisal of the reliability of terrestrial cosmogenic-nuclide dating (TCND) at Austanbotnbreen, Jotunheimen, Norway. Boreas 40:256–270

Matthews JA, Harris C, Ballantyne CK (1986) Studies on a gelifluction lobe, Jotunheimen, Norway: 14C chronology, stratigraphy, sedimentology and palæeoenvironment. Geogr Ann A 68:345–360

Matthews JA, Ballantyne C, Harris C, McCarroll D (1993) Solifluction and climatic variation in the Holocene: discussion and synthesis. In: Frenzel B (ed) Solifluction and climatic variations in the Holocene. Fischer, Paläoklimaforschung 11, Suttgart, pp 339–361

Matthews JA, McCarroll D, Shakesby RA (1995) Contemporary terminal-moraine ridge formation at a temperate glacier: Styggedalsbreen, Jotunheimen, southern Norway. Boreas 24:129–139

Matthews JA, Shakesby RA, Berrisford MS, McEwen LJ (1998) Periglacial patterned ground on the Styggedalsbreen glacier foreland, Jotunheimen, southern Norway: micro-topographic, paraglacial and geoecological controls. Permafrost Periglac 9:147–166

Matthews JA, Dahl SO, Nesje A, Berrisford MS, Andersson C (2000) Holocene glacier variations in central Jotunheimen, southern Norway based on distal glaciolacustrine sediment cores. Quaternary Sci Rev 19:1625–1647

Matthews JA, Berrisford MS, Dresser PQ, Nesje A, Dahl SO, Bjune AE, Bakke J, Birks HJB, Lie Ø, Dumayne-Peaty L, Barnett C (2005) Holocene glacier history of Bjørnbreen and climatic reconstruction in central Jotunheimen, Norway, based on proximal glaciofluvial stream-bank mires. Quaternary Sci Rev 24:67–90

Matthews JA, Shakesby RA, Schnabel C, Freeman S (2008) Cosmogenic ^{10}Be and ^{26}Al ages of Holocene moraines in southern Norway I: testing the method and confirmation of the date of the Erdalen event (c. 10 ka) at its type-site. Holocene 18:1155–1164

Matthews JA, Shakesby RA, Owen G, Vater AE (2011) Pronival rampart formation in relation to snow-avalanche activity and Schmidt-hammer exposure-age dating (SHD): three case studies from southern Norway. Geomorphology 130:280–288

Matthews JA, Winkler S, Wilson P (2014) Age and origin of ice-cored moraines in Jotunheimen and Breheimen, Southern Norway: insights from Schmidt-hammer exposure-age dating. Geogr Ann A 96:531–548

Matthews JA, Owen G, Winkler S, Vater AE, Wilson P, Mourne RW, Hill JL (2016) A rock-surface microweathering index from Schmidt hammer R-values and its preliminary application to some common rock types in southern Norway. CATENA 143:35–44

Matthews JA, Owen G, McEwen LJ, Shakesby RA, Hill JL, Vater AE, Ratcliffe AC (2017a) Snow-avalanche impact craters in southern Norway: their morphology and dynamics compared with small terrestrial meteorite craters. Geomorphology 296:11–30

Matthews JA, Wilson P, Mourne RW (2017b) Landform transitions from pronival ramparts to moraines and rock glaciers: a case study from the Smørbotn cirque, Romsdalsalpane, southern Norway. Geogr Ann A 99:15–37

Matthews JA, Winkler S, Wilson P, Tomkins M, Dortch J, Mourne R, Hill J, Owen G, Tomkins J, Vater A (2018) Small rock-slope failures conditions by Holocene permafrost degradation: a new approach and conceptual model based on Schmidt-hammer exposure-age dating in Jotunheimen, southern Norway. Boreas 47:1144–1169

McCarroll D (2016) Trimline Trauma: the wider implications of a paradigm shift in recognising and interpreting glacial limits. Scot Geogr J 132:130–139

McCarroll D, Nesje A (1993) The vertical extent of ice sheets in Nordfjord, western Norway: measuring degree of rock surface weathering. Boreas 22:255–265

Millar S (2013) Mass movement processes in the periglacial environment. In: Shroder J, Giardino R, Harbor J (eds) Treatise on geomorphology, vol 8. Glacial and periglacial geomorphology. San Diego Academic Press, San Dieogo, pp 374–391

Moen A (1998) Nasjonalatlas for Norge: Vegetasjon, 2nd edn. Statens Kartverk, Hønefoss

Myhra KS, Westermann S, Etzelmüller B (2017) Modelled distribution and temporal evolution of permafrost in steep rock walls along a latitudinal transect in Norway by CryoGrid 2D. Permafrost Periglac 28:172–182

Nesje A (1993) Neoglacial gelifluction in the Jostedalsbreen region, western Norway: evidence from dated buried soils. In: Frenzel B (ed) Solifluction and climatic variations in the Holocene. Fischer, Paläoklimaforschung 11, Stuttgart, pp 37–47

Nesje A (2009) Late Pleistocene and Holocene alpine glacier fluctuation in Scandinavia. Quaternary Sci Rev 28:2119–2136

Nesje A, Dahl SO (1990) Autochthonous block fields in southern Norway: implications for the geometry, thickness, and isostatic loading of the Late Weichselian Scandinavian ice sheet. J Quaternary Sci 5:225–234

Nesje A, Dahl SO (2003) The 'Little Ice Age'—only temperature? Holocene 13:139–145

Nesje A, Whillans IM (1994) Erosion of Sognefjord, Norway. Geomorphology 9:33–45

Nesje A, Dahl SO. Anda E, Rye N (1988) Block fields in southern Norway: significance for the Late Weichselian ice sheet. Norsk Geol Tidsskri 68:149–169

Nesje A, Dahl SO, Valen V, Ovstedal J (1992) Quaternary erosion in the Sognefjord drainage basin, western Norway. Geomorphology 5:511–520

Nesje A, McCarroll D, Dahl SO (1994) Degree of rock surface weathering as an indicator of ice-sheet thickness along an east-west transect across southern Norway. J Quaternary Sci 9:337–347

Nesje A, Bakke J, Dahl SO, Lie Ø, Matthews JA (2008a) Norwegian mountain glaciers in the past, present and future. Global Planet Change 60:10–27

Nesje A, Dahl SO, Thun T, Nordli Ø (2008b) The "Little Ice Age" glacial expansion in western Scandinavia: summer temperature or winter precipitation? Clim Dynam 30:789–801

Nordli PØ, Lie Ø, Nesje A, Dahl SO (2003) Spring-summer temperature in western Norway 1734–2003: a data-synthesis approach. Int J Climatol 23:182–1841

Nordli PØ, Lie Ø, Nesje A, Benestad R (2005) Glacier mass balance in southern Norway modelled by circulation indices and spring-summer temperature AD 1781–2000. Geogr Ann A 87:431–445

Ødegård RS, Sollid JL, Liestøl O (1987) Juvflya – Kvartærgeologi og geomorfologi M 1:10.000. Geografisk Institutt, Universitetet I Oslo, Oslo

Ødegård RS, Sollid JL, Liestøl O (1988) Periglacial forms related to terrain parameters in Jotunheimen, southern Norway. In: Senneset K (ed) 5th international conference on permafrost proceedings, vol 3. Tapir, Trondheim, pp 59–61

Ødegård RS, Sollid JL, Liestøl O (1992) Ground temperature measurements in mountain permafrost, Jotunheimen, southern Norway. Permafrost Periglac 3:231–234

Ødegård RS, Hoelzle M, Johansen KV, Sollid JL (1996) Permafrost mapping and prospecting in southern Norway. Norsk Geogr Tidsskr 50:41–53

Ødegård RS, Isaksen K, Mastervik M, Billdal L, Engler M, Sollid JL (1999) Comparison of BTS and Landsat TM data from Jotunheimen, southern Norway. Norsk Geogr Tidsskr 53:226–233

Østrem G (1961) Breer og morener i Jotunheimen. Norsk Geogr Tidsskr 17:210–243

Østrem G (1963) Comparative crystallographic studies on ice from ice-cored moraines, snow-banks and glaciers. Geogr Ann 45:210–240

Østrem G (1964) Ice-cored moraines in Scandinavia. Geogr Ann 46:282–337

Østrem G (1965) Problems of dating ice-cored moraines. Geogr Ann 47:1–38

Østrem G (1971) Rock glaciers and ice cored moraines, a reply to D. Barsch. Geogr Ann 53 A:207–213

Østrem G, Dale Selvig K, Tandberg K (1988) Atlas over breer i Sør-Norge. NVE, Oslo, Meddelelser fra Hydrologisk Avdeling 61

Owen G, Matthews JA, Shakesby RA, He X (2006) Snow-avalanche impact landforms, deposits and effects at Urdvatnat, southern Norway: implications for avalanche style and process. Geogr Ann A 88:295–307

Øyen PA (1893) Isbræstudier i Jotunheimen. Nytt Magasin for Naturvidenskabere 31:26–27

Patton H, Hubbard A, Andreassen K, Winsborrow M, Stroeven AP (2016) The build-up, configuration, and dynamical sensitivity of the Eurasian ice-sheet complex to Late Weichselian climatic and oceanic forcing. Quaternary Sci Rev 153:97–121

Patton H, Hubbard A, Andreassen K, Auriac A, Whitehouse PL, Stroeven AP, Shackleton C, Winsborrow M, Heyman J, Hall AM (2017) Deglaciation of the Eurasian ice sheet complex. Quaternary Sci Rev 169:148–172

Rapp A (1959) Avalanche boulder tongues in Lappland. Geogr Ann 41:34–48

Rasmussen LA, Andreassen LM, Baumann S, Conway H (2010) 'Little Ice Age' precipitation in Jotunheimen, southern Norway. Holocene 20:1039–1045

Rea BR (2013) Blockfields (Felsenmeer). In: Elias SA, Mock CJ (eds) Encyclopedia of Quaternary science, vol 3, 2nd edn. Elsevier, Amsterdam, pp 523–534

Reusch H (1901) Nogle bidrag till forstaaelsen af hvorledes Norges dale og fjelde er blevne til. Norg Geol Unders 32 (Aarbog 1900):124–263

Rixhon G, Demoulin A (2013) Evolution of slopes in a cold climate. In: Shroder J, Giardino R, Harbor J (eds) Treatise on geomorphology, vol 8. Glacial and periglacial geomorphology. San Diego Academic Press, San Diego, pp 392–415

Roberts D, Gee DG (1985) An introduction to the structure of the Scandinavian Caledonides. In: Gee DG, Sturt BA (eds) The Caledonide Orogen—Scandinavia and related areas, vol 1. Wiley, Chichester, pp 55–68

Rudberg S (1988) Gross morphology of Scandinavia—six complementary ways of explanation. Geogr Ann A 70:135–167

Shakesby RA (1997) Pronival (protalus) ramparts: a review of forms, processes, diagnostic criteria and palaeoenvironment implications. Prog Phys Geog 21:394–418

Shakesby RA, Dawson AG, Matthews JA (1987) Rock glaciers, protalus ramparts and related phenomena, Rondane, Norway: a continuum of large-scale talus derived landforms. Boreas 16:305–317

Shakesby RA, Matthews JA, Winkler S (2004) Glacier variations in Breheimen, southern Norway: relative-age dating of Holocene moraine complexes at six high-altitude glaciers. Holocene 14:899–910

Sigmond EMO, Gustavson M, Roberts D (1984) Berggrunnskart over Norge 1:1 milion. Norges Geologiske Undersøkelse, Trondheim

Sollid JL, Reite JA (1983) The last glaciation and deglaciation of Central Norway. In: Ehlers J (ed) Glacial deposits of North-West Europe. Balkema, Rotterdam, pp 41–59

Sollid JL, Trollvik JA (1991) Oppland fylke, kvartærgeologi og geomorfologi 1:250.000. Universitetet I Oslo, Institutt for naturgeografi, Oslo

Steiger C, Etzelmüller B, Westermann S, Myhra KS (2016) Modelling the permafrost distribution in steep rock walls in Norway. Norw J Geol 96:1–13

Stroeven AP, Hättestrand C, Kleman J, Heyman J, Fabel D, Fredin O, Goodfellow BW, Harbor JM, Jansen JD, Olsen L, Caffee MW, Fink D, Lundqvist J, Rosqvist GC, Strömberg B, Jansson KN (2016) Deglaciation of Fennoscandia. Quaternary Sci Rev 147:91–121

Tollan A (1963) Trekk av isbevegelsen og isavsmeltningen i Nordre Gudbrandsdalens fjelltrakter. Nor Geol Unders 223:328–345

Velle G, Bjune AE, Larsen J, Birks HJB (2010) Holocene climate and environmental history of Brurskardstjørni, a lake in the catchment of Øvre Heimdalsvatnet, south-central Norway. Hydrobiologia 642:13–34

Vere DM, Matthews JA (1985) Rock glacier formation from a lateral moraine at Bukkeholsbreen, Jotunheimen, Norway: a sedimentological approach. Z Geomorphol 29:397–415

Washburn AL (1956) Classification of patterned ground and review of suggested origins. Geol Soc Am Bull 67:823–865

Washburn AL (1979) Geocryology, 2nd edn. Arnold, London

Warburton J (2013) Patterned ground and polygons. In: Giordano JR, Harbour JM (eds) Treatise on geomorphology, vol 8. Glacial and periglacial geomorphology. Elsevier, Amsterdam, pp 298–312

Winkler S (1996) Frührezente und rezente Gletscherstandsschwankungen in Ostalpen und West-/Zentralnorwegen. University of Trier, Trierer Geographische Studien 15, Trier

Winkler S (2001) Untersuchungen zur Klima- und Morphodynamik in skandinavischen Gebirgsregionen während des Holozän - ein Vergleich ihrer Wechselwirkungen und Prozeßsysteme im überregionalen Kontext kaltgemäßigter maritimer Gebirgsregionen. Habilitation thesis (unpublished), University of Trier, Trier

Winkler S (2002) Von der "Kleinen Eiszeit" zum "globalen Gletscherrückzug" - eignen sich Gletscher als Klimazeugen? Akademie der Wissenschaften und der Literatur, Mainz, Abhandlungen der Mathematisch-naturwissenschaftlichen Klasse 3

Winkler S, Matthews JA (2010) Observations on terminal moraine-ridge formation during recent advances of southern Norwegian glaciers. Geomorphology 116:87–106

Winkler S, Matthews JA (2014) Comparison of electronic and mechanical Schmidt hammers in the context of exposure-age dating: are Q- and R-values interconvertible? Earth Surf Proc Land 39:1128–1136

Winkler S, Matthews JA, Mourne RW, Wilson P (2016) Schmidt-hammer exposure ages from periglacial patterned ground (sorted circles) in Jotunheimen, Norway, and their interpretative problems. Geogr Ann A 98:265–285

Worsley P (1993) Holocene solifluction at Okstindan, northern Norway: a reassessment. In: Frenzel B (ed) Solifluction and climatic variations in the Holocene. Fischer, Paläoklimaforschung 11, Stuttgart, pp 49–57

www.høydedata.no—open portal of Kartverket/www.kartverket.no. Last accessed 23 Dec 2018

www.met.no—open portal of Meteorologisk insitutt/www.met.no. Last accessed 19 Dec 2018

www.norgeibilder.no—open portal of Kartverket/www.kartverket.no. Last accessed 23 Dec 2018

www.senorge.no—open portal in collaboration between Norges Vassdrags- og Energidirektoratet/www.nve.no, Meteorologisk institutt/www.met.no, and Kartverket/www.kartverket.no. Last accessed 11 Dec 2018

Characterization of Scree Slopes in the Rondane Mountains (South-Central Norway)

9

Dominique Sellier and Riwan Kerguillec

Abstract

The Rondane massif, lying in South-central Norway, near the 62nd North parallel and the 9th meridian, is one of the highest mountains in Scandinavia, with Jotunheimen, Dovrefjell and Kebnekaise, rising to 2178 m at Rondslottet. The massif is entirely composed of quartzite rocks, known for their high frost susceptibility, particularly to macro frost shattering. Then, quartzite tends to produce coarse angular fragments in cold regions, and is especially suitable for block fields and scree. The current morphogenetic conditions in the Rondane are those of a relatively dry periglacial-type climate. The slopes, rising more than 1000 m in height, but devoid of glaciers, are entirely within the active periglacial zone (about 1200 m, thus one of the most extended in height in Europe). The Rondane scree slopes present the most typical periglacial forms of quartzite mountains in north-western Europe and the massif offers exemplary terrain for studying scree slopes in the Nordic environment. This chapter has three main purposes. The first one is to characterize the properties of scree slopes in quartzite, using the Rondane massif as an example. The second is to present a morphological and dynamic typology of scree slopes. The last one is to relate the scree process to the kinds of slope and the local morphodynamic and morphostructural parameters.

Keywords

Central norway • Rondane • High atlantic mountain • Quartzitic relief • Periglacial processes • Scree slopes

D. Sellier (✉) · R. Kerguillec
University of Nantes, Nantes, France
e-mail: dominique.sellier@univ-nantes.fr

A. A. Beylich (ed.), *Landscapes and Landforms of Norway*, World Geomorphological Landscapes,
https://doi.org/10.1007/978-3-030-52563-7_9

9.1 Introduction

The Rondane mountains, rising to 2178 m (Rondslottet) and lying near the 62nd North parallel and the 9th meridian, form part of the Scandes chain. They are among the highest mountains in Scandinavia after Jotunheimen (Galdhøpiggen, 2469 m) and Dovrefjell (Snøhetta, 2286 m).

The Rondane massif has three fundamental properties (Kerguillec and Sellier 2015). The first relates to their overall relief, which is similar to that of high alpine-type mountains, with peaks above 2000 m in altitude and mighty glacial relicts (cirques and U-shaped valleys). The second pertains to the particular conditions determining the morphogenetic processes at work, with active periglacial processes above 1000 m in altitude causing minor sorted features (earth patches and small sorted circles) and the first sporadic permafrost lenses (King 1983, 1984, 1986). The low precipitation levels explain why there is no genuine glacio-nival zone, with two consequences. First, the slopes, rising more than 1000 m in height, are entirely within the active periglacial zone. Second, the current morphogenetic conditions are those of a relatively dry periglacial-type climate. The third fundamental property of the Rondane massif is structural. The chain is composed entirely of quartzite rocks, known for their frost susceptibility. These rocks are thus especially prone to producing coarse homometric angular frost-shattered debris.

The Rondane scree slopes present the most typical periglacial forms of quartzite mountains in north-western Europe, with the chain offering exemplary terrain for studying scree slopes in the Nordic environment (Sellier 2002). The purpose of this chapter is thus to help characterize the properties of scree slopes in quartzites, using the Rondane massif as an example. A second purpose is to relate the forms of scree process to the types of slope and the local morphodynamic and morphostructural parameters. It discusses the general nature of the relief, the morphological properties of the slopes, and the conditions for scree processes in the periglacial environment specific to the Rondane area, before defining the morphometric and sedimentological properties of the scree slopes on the basis of a series of transects surveyed in the range.

9.2 A Dry Periglacial High Quartzite Mountain

9.2.1 A High Mountain

The Rondane mountains (*62°03′N–61°46′N*) are to the north-east of the Jotunheimen range (Galdhøpiggen, 2469 m) and south-east of the Dovrefjell range (Snøhetta, 2286 m). Lying about 200 km from the Norwegian coast, they are the most continental of the Norwegian mountains. The range runs for about 30 km north-south and east-west. It culminates at 2178 m on the peak of Rondslottet (*61°54′58″ N/9°51′15″E*). It dominates the landscape, rising above the plateaus and low mountains of Hövringen fjell to the north-west and the Ringebu fjell to the south-east. It is thus an example of a "mountain on a high surface" (Sellier 2002). It is flanked by the large U-shaped valleys of Atnadalen to the east and Gudbrandsdalen to the south-west (Fig. 9.1).

The relief illustrates a characteristic feature of southern Scandinavian landscapes, associating high mountain ranges (of over 2000 m), vast sub-horizontal plateaus (the fjell, at around 1000 m), and glacial U-shaped valleys, the bottoms of which lie at about 300 m in altitude (Peulvast 1985).

The fjell (1000–1200 m) was ice scoured by Pleistocene glaciers, and in places resembles the relief of Scottish knobs and lochans.

Large U-shaped valleys have been carved out in its surface: Atna valley (300–400 m deep), and especially Gudbrandsdalen valley (700–800 m deep).

The Rondane massif stands above this surface (Fig. 9.2). It includes a ring of peripheral medium mountains with rounded summits rising to between 1600 and 1850 m, whose flanks are indented by widening cirques. It is comprised especially of central massifs of analogous volume and altitude, separated by the U-shaped valleys of Døralen to the north, Illmandalen to the south, and Rondvassdalen in the center. These three central massifs have summits of almost equal altitude, all rising to over 2000 m (2015–2018 m for Smiubelgen to the south-east, 2015–2114 m for Høgronden to the north, and 2044–2178 m for Rondslottet to the south-east). They are primarily distinguished by their high mountain landforms, characterized by the many broads, deep glacial cirques, divided by narrow ridges. The relief also relates to the morphology of the slopes, characterized by heights of over 600–800 m, as well as by the large size of the rocky free faces and scree slopes. The content of this chapter thus focuses mainly on these "central massifs".

9.2.2 A Quartzite Mountain

The Rondane mountains present homogenous structural properties, explaining the simplicity and unity of their relief. The massif is composed entirely of quartzite rocks enriched with feldspars and micas, belonging to the *sparagmite* group (Oftedahl 1950), but retaining the fundamental characteristics of pure quartzites (siliceous recrystallized stratified and intensely jointed rocks of detrital origin). The sparagmites are macro frost-susceptible and tend to produce coarse angular fragments of the sort comprising block fields and especially scree.

9.2.3 A Dry Periglacial Mountain

Data about the climate properties of the massif is provided by the Venabu meteorological station on Ringebu fjell 30 km south of the center of the Rondane massif, at an altitude of 930 m, hence similar to the base of the mountains (Fig. 9.1).

9.2.3.1 Thermal Data at the Base of the Massif

At Venabu, the warmest monthly average occurs in July (+10.4 °C) and the coldest in January (−9.7 °C) (Norsk Meteorologisk Institutt 2018). The average annual thermal range is thus 20.1 °C, a characteristic attributable to the continentality in comparison to the massifs further west (Jotunheimen and Dovrefjell). There are six months with an average above 0 °C at Venabu, from May to October. Average monthly temperatures are below 0 °C the rest of the year, but only exceed +10 °C in July, a typical feature of sub-polar climates (Péguy 1970; Kerguillec 2013). The thermal properties of the Rondane area are thus harsh, despite marked interannual variations. At the Venabu station at around 1000 m in altitude, there are on average at least 84 frost/thaw cycles per year (Kerguillec 2015a). Freeze frequency is influenced by the North Atlantic Oscillation (NAO), as elsewhere in Norway. At Venabu, years with fewer frost/thaw cycles correspond to negative NAO phases, and those with more cycles occur during positive phases (Kerguillec 2015a).

9.2.3.2 Precipitation

The amount, nature, and regime of precipitation are indicative of the continental tendency of the Rondane climate. The massif is among the driest in Scandinavia. Venabu receives on average 660 mm per year, due both to its distance from the sea and its shelter position in relation to Dovrefjell and Jotunheimen, which stand, respectively, 100 and 300 m higher to the north-west and west. Additionally, annual total precipitation is 440 mm on average at Hövringen (935 m in altitude) immediately west of the Rondane massif, and only

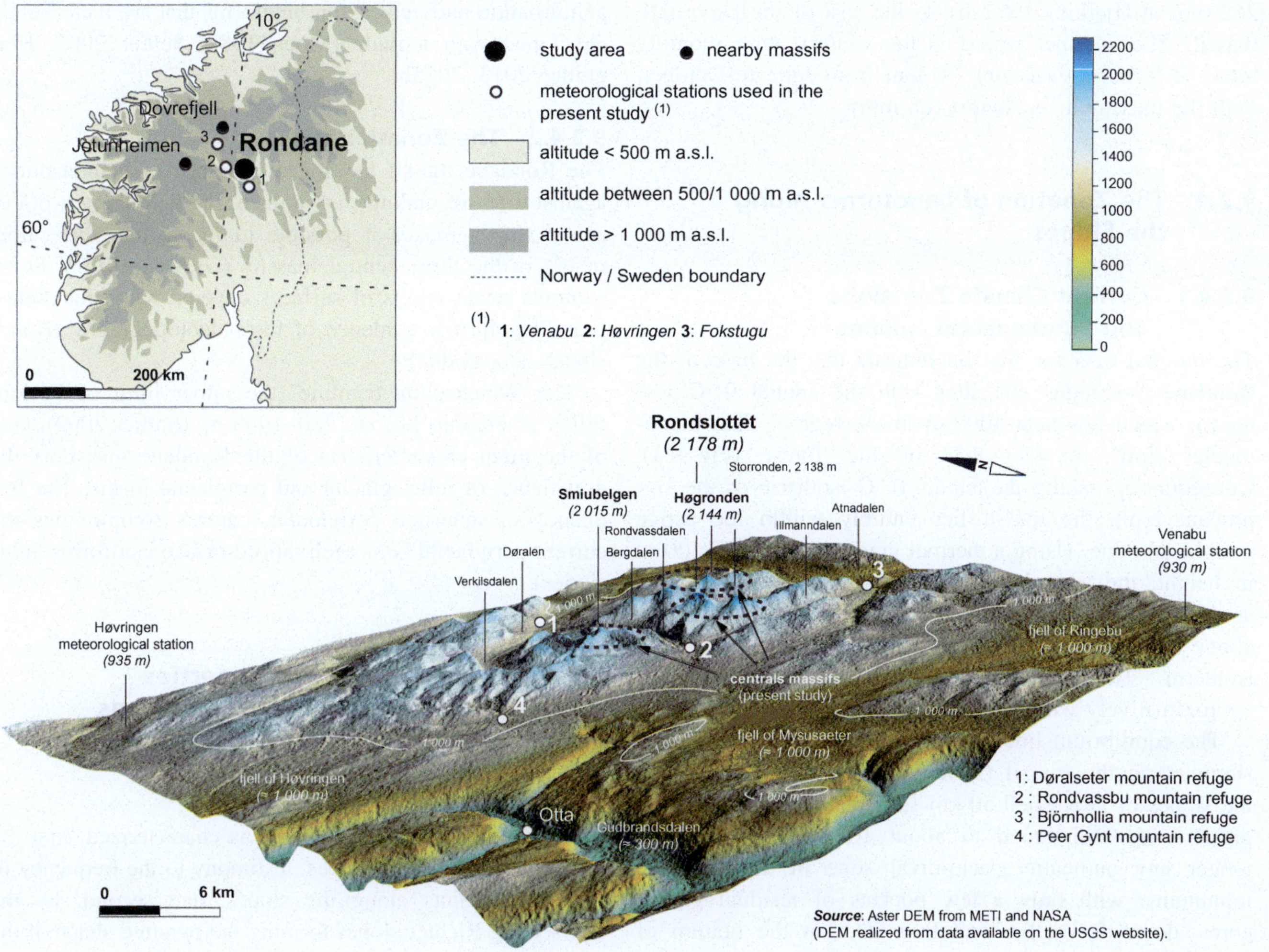

Fig. 9.1 Location map and digital elevation model of the study area

Fig. 9.2 Southern part of Rondane mountains, Rondslottet (2178 m) and Storronden (2138 m) from the fjell (1000 m)

252 mm at Hjerkinn (953 m), to the east of the Dovrefjell massif. The summer period is the wettest, with monthly totals at Venabu exceeding 70 mm from June to October, with the maximum in August (88 mm).

9.2.4 The Zonation of Landforms Along the Slopes

9.2.4.1 Current Climate Zonations and Fundamental Isolines

The thermal data for Venabu indicate that the base of the Rondane mountains coincides with the annual 0 °C isotherm, which lies near 1000 m in the region. This fundamental limit is also that of the forest (Fig. 9.1). Consequently, taking the annual 0 °C isotherm as the significant limit, the massif lies entirely within the active periglacial zone. Using a thermal gradient of 0.6 °C/100 m in height, the annual −2 °C and −5 °C isotherms lie at around 1300 and 1800 m, respectively, with the summits above 2000 m being immediately above the annual –6 °C isotherm. The summit of Rondslottet thus coincides approximatively with the annual −7.5 °C isotherm.

The equilibrium line altitude (ELA) is near the highest summits on the massif (2000–2200 m), whereas it is on the slopes of Dovrefjell 50 km further north, where cirque glaciers are still found to about 1650 m. There is no longer any authentic glacio-nival zone in the Rondane mountains, with only a few pockets of residual glacial cores along the gullies and free faces at the bottom of cirques, on the ubac at altitudes above 1500 m. This singularity in comparison to other massifs of above 2000 m (Jotunheimen, Snøhetta) is due to the relative dryness of the Rondane mountains, itself determined by their continentality and low level of snow accumulation. Under these circumstances, the difference in attitude between the annual 0 °C isotherm and the ELA is about 1000–1200 m, explaining why the Rondane massif lies entirely within the periglacial zone, which has the greatest altitude range in Scandinavia, and Europe even (Sellier 2002, 2006; Kerguillec 2013).

The Rondane are thus a typical example of a “high dry periglacial mountain”. All the massif lies within the mountain permafrost domain. According to the theoretical limits established by King for Scandinavia (1983, 1984, 1986), the sporadic permafrost zone (corresponding to the lower alpine zone of mountain tundra) lies between 1000 and 1450 m, sparse discontinuous permafrost (corresponding to the mid-alpine zone with more open vegetation) lies between 1500 and 1650 m, and extensive discontinuous permafrost (corresponding to the upper alpine zone) lies above 1650 m.

The Rondane mountains also contain several successive active periglacial sub-zones, distinguished by a series of geliturbation and gelifluxion landforms that are increasingly developed with attitude (Moen 1987; Sellier 2002; Kerguillec 2013, 2015b).

9.2.4.2 The Zonation of Relicts

The Rondane massif has the appearance of a monadnock against the surrounding slaty units of the fjell. Its network of paleic slopes makes it possible to reconstitute the initial reliefs in the three central massifs (Gjessing 1967). Some summits retain relicts of surfaces covered with blockfields, providing further evidence of the paleotopography (Rondslottet, Storronden).

The Weichselian trimline (the upper limit of Weichselian glaciation) lies at 1750–1900 m. (Sellier 2002). One of the main characteristics of the Rondane massif is the prevalence of relict glacial and periglacial forms. The frequency of inherited periglacial features complexifies the current periglacial belts (gelivation surface landforms, talus slopes).

9.3 The Morphological Properties of the Slopes and the Conditions Governing the Scree Processes in the Quartzites

The relief of the Rondane massif is characterized, first, by the frequency of scree slopes, testimony to the frequency of periglacial relicts along the slopes, and second, by the intersecting Richter slopes forming the pyramid shape of the mountains.

9.3.1 Outer Slopes and Inner Slopes

Presentation and classification of the scree slopes need to distinguish between paleoreliefs (old lands) and “alpine” type reliefs (glacial cirques and valleys) (Gjessing 1967; Rudberg 1965–1966). These two types of relief determine the two main categories of slopes: the outer paleic slopes, and the inner rocky slopes of glacial cirques.

The outer slopes lie around the periphery of the massif. They correspond to paleic slopes of insufficient incline to generate fall or scree processes. They are thus outside the scope of this study.

The inner slopes lie in the three central massifs of the Rondane mountains and are mainly cirque slopes. They contain relicts of glacial erosion (rocky free faces) and periglacial erosion effects (scree slopes). Their initially steep incline, which caused the subsequent scree processes, was produced by glacial ablation. The positioning of slopes of glacial origin thus determined where the scree slopes and Richter slopes now lie.

9.3.2 Inner Slope Properties: Free Faces and Scree Processes

The inner slopes have profiles or profile segments belonging to four categories:

- rocky summit slopes resulting from strips of paleic landforms, similar to the abovementioned paleic slopes,
- rocky free faces, which are specially developed in quartzites and characteristic of inner slopes,
- scree slopes (either continuous talus slopes or series of contiguous scree cones).
- uniform Richter slopes partially covered in debris.

9.3.2.1 The Morphology of the Free Faces

"Free faces" is here used to refer to large rocky escarpments making up over half a slope profile, in which they differ from "ledges" which take up less than half the profile and are interrupted by benches. The Rondane quartzite free faces are remarkable for their great vertical and lateral continuity. They sometimes form scarps rising without interruption over heights of several hundreds of meters (nearly 600 m on the north of Rondslottet), and tend to be north-facing. Their general incline is above 45°, and in all cases above 35°, but rarely 75°. Since they cut across beds, free faces running contrary to the dip need to be distinguished from those following the dip, for this influences the conditions and forms of scree process.

The free faces originate in the actions of cirque glaciers, and appeared once the latter melted. However, many retain the effects of postglacial frost shattering and have scree slopes at their base, which may be classified into several types depending on their degree of development.

9.3.2.2 The Frost Susceptibility of Quartzites and Frost Shattering of Free Faces

Quartzites such as those of the Rondane mountains are extremely susceptible to frost shattering, particularly macro frost shattering, because they are fine-grained homometric siliceous rocks, by definition wholly recrystallized, hence very hard and mechanically very fragile and very breakable. These rocks contain many discontinuity planes, especially as the dense, regular joint planes are accompanied by bedding planes. As outcrops, these rocks are very permeable and macro frost-susceptible, and thus conducive to the production of coarse, angular, homometric fragments. Quartzites are thus ideal for producing scree.

On the other hand, and for the same reasons, quartzites are very resistant to chemical weathering, and thus tend not to disintegrate. They have low porosity, all those sampled being impermeable, and are thus not generally prone to micro frost action or the production of fine material.

In conclusion, quartzites produce fragments of contrasting size (with large quantities of coarse fragments, and small quantities of sands and pelites). The regoliths are prone to sieve effects and openwork formations on their surface.

9.3.2.3 Scree Processes in Quartzites

Scree processes result from a mechanism of progressive, primarily individual fragmentation (primary rockfalls) (Rapp 1960; Francou 1988). These processes may produce additional fragments when impacting on benches, in gullies, or on the scree slopes beneath (secondary rockfalls). In cold environments, fragmentation results primarily from frost shattering (a periglacial process), in the wake of postglacial decompression effects occurring along the slopes after the ice retreated (a paraglacial process). Scree processes thus result primarily from fall processes (simple gravity scree). In fact, they result from complex phenomena combining the actions of avalanches, run-off, gelifluction, and the flowage of ice lenses contained between fragments. These processes, the effects of which characterize assisted gravity scree, trigger the production of fragments. Above all, they determine the transit of fragments and the reworking of scree. They thus modify the profile and even the overall form of scree slopes (Francou and Hetu 1989).

In addition to their exceptional extent, quartzite scree presents various properties that are immediately noticeable in the field. Quartzite scree is composed of larger fragments than scree in other rocks. The material size is bimodal, and may be classified into coarse fragments of the size of blocks, and a fine, mainly sandy fraction. They have an openwork texture, with an often unstable surface. Fragments tend to have sharp angles and flat faces.

9.3.3 The Inner Slopes: Type A Slopes and Type B Slopes

Nearly all the inner slopes of the Rondane mountains have scree slopes composed of debris resulting from the postglacial frost shattering of the free faces, and partly from postglacial decompression. Their typology depends on the relationship between the free faces and the scree slopes. The most widely used ratio to account for the proportions of each section of a slope with a rockface above and scree slope below is Ho/Hi, as defined by Statham (1976), where Ho represents the height of the scree slope and Hi that of the slope as a whole (Fig. 9.3). We shall differentiate between type A slopes, with a preponderant rockface and a Ho/Hi ratio below 0.50, and type B slopes with a preponderant oblique slope, similar to Richter slopes, and Ho/Hi ratio above 0.50.

On this basis, four categories of type A slopes and two categories of type B slopes may be differentiated (Fig. 9.4).

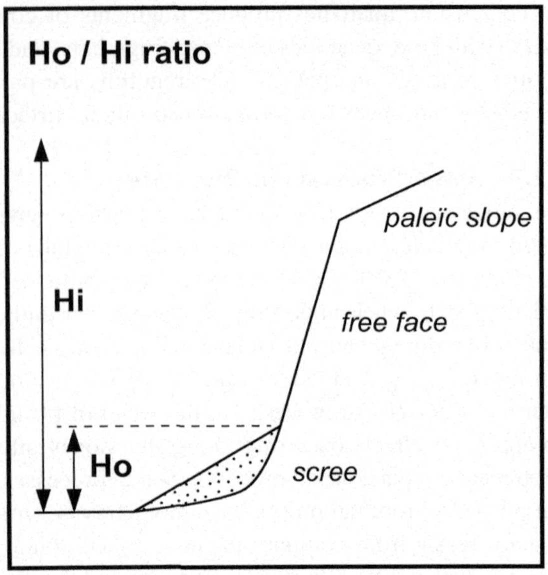

Fig. 9.3 The ratio H0/Hi (from Statham, 1976)

Among type A slopes with a preponderant free face (Ho/Hi < 0.50):

- P0 free face slopes, with a large rocky face and no scree processes, a theoretical type rarely encountered in the Rondane mountains
- P1 free face slopes, with a continuous free face above primitive talus slopes
- P2 free face slopes indented with gullies, and primitive scree cones beneath
- P2R ruined free face slopes indented with gullies, and avalanche cones and composite cones beneath (scree cones reworked by avalanches).

Among type B slopes with a Ho/Hi ratio between 0.50 and 1:

- VT slopes, on which only relict ledges remain, with Richter slope sections standing above large developed talus slopes
- VR slopes or Richter slopes, once again a theoretical type rarely encountered here.

9.3.4 Type A and Type B Slopes and Their Associated Scree Slopes

9.3.4.1 Type A Slopes and Their Types of Scree Slope

P0 free face slopes without scree slopes

Such slopes are rare in the Rondane mountains. They are characterized by their great height (500–600 m), sharp incline (60°–80°), and longitudinal and lateral continuity (with no gullies). They are north-facing and probably attributable to late ice retreat. In principle, the free faces occupy

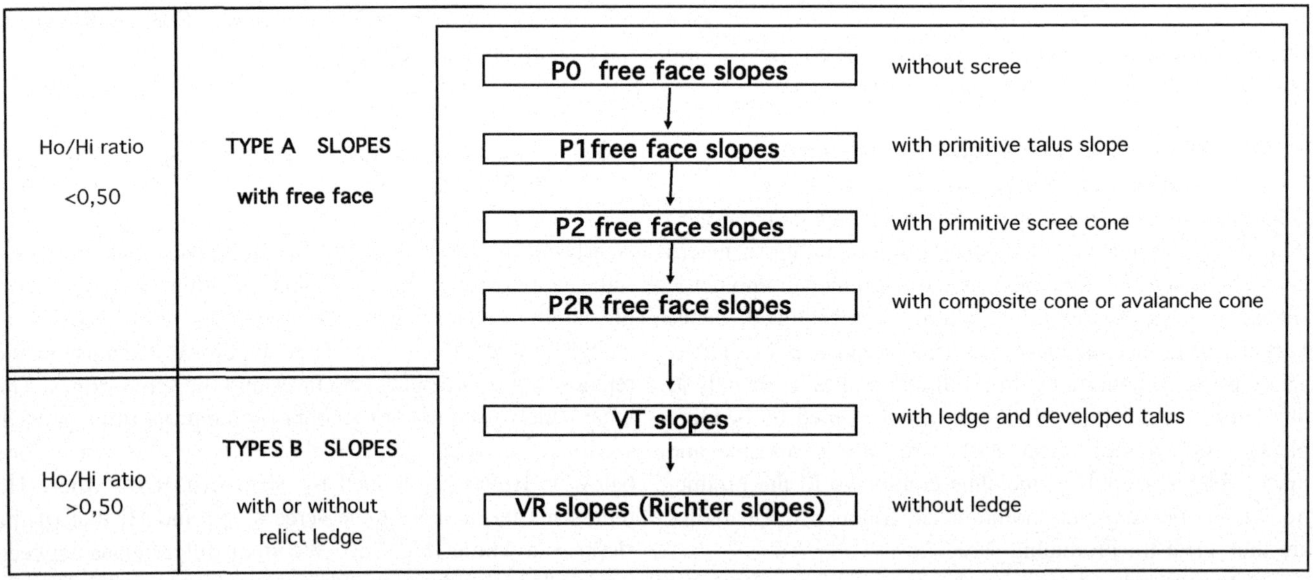

Fig. 9.4 Types of slopes in Rondane

the entire profile and have no scree (with a Ho/Hi ratio close to 0). P0 free faces are characteristic of the actions of glaciers. They have smooth surfaces and decompression slabs corresponding to the initial stages of their degradation. They sometimes have debris at their base issuing from reduced attacks on the rocky surfaces above.

P1 free face slopes with primitive talus slopes
These slopes are found beneath the highest summits, in ubac position or above cirques. They have large, relatively regular free faces and still rise several hundred meters upwards, but with a less sharp incline (50°–75°). Decompression marks are rare or have been effaced by frost shattering. Cutting is reduced and there are no gullies. P1 free faces stand above continuous short talus slopes of reduced height, referred to here as *primitive talus slopes*.

P2 free face slopes with primitive scree cones
There are many such slopes in the Rondane mountains. P2 free faces are not as high as the previous ones. They have a lesser incline (50°–70°) due to the more intense effects of frost shattering, and decompression slab surfaces now give way to rougher surfaces. The principal characteristic of P2 free faces is their lateral discontinuity. They are cut by gullies with scree cones beneath, referred to here as *primitive cones*. These gullies result from differential postglacial frost shattering attacking intensely fractured zones. They produce large quantities of fragments that are then channeled. Consequently, when a gully originates in a P1 type free face, primitive talus slopes are replaced by scree cones, and the Ho/Hi ratio increases, but without reaching 0.50. The remaining space between the gullies has the form of trapezoidal facets, sometimes reduced to triangular facets or pinnacles. Obviously, many possible intermediary stages exist between P1 free face slopes and P2 free face slopes (P1-2 free face slopes).

P2R free face slopes
P2R slopes are P2 free faces that have been ruined by frost shattering. The free face has been replaced by rocky sections which still take up over half of the slope, but reduced to a series of ledges and relict crags. Their incline varies between 50° and 35° and tends to be similar to that of Richter slopes. The relative indentation of gullies is progressively attenuated by the lowering of the remaining rocky facets between them. Consequently, scree cones take up more place, and the Ho/Hi ratio nears 0.50. Their profiles become concave under the additional action of avalanches. The space between cones is progressively filled by the product of scree processes on the abovementioned rocky facets, tending to result in talus slopes with irregular longitudinal and transversal forms.

9.3.4.2 Type B Slopes

Straight slopes with ledges and developed taluses (VT slopes)
Such slopes are also common in the Rondane mountains, especially on the adret. They have a Ho/Hi ratio of above 0.50. The free faces are replaced by relict ledges, isolated crags, or straight rocky segments comparable to *Richter segments* as defined by Francou and Hetu (1989). There are no longer any gullies, and cones are replaced by large talus slopes, referred to here as developed taluses. In general, the incline of such slopes, including their ledges, is around 35°. These straight slopes with ledges and taluses (VT slopes) mark an important stage in the development of inner slopes. They have a preponderance of periglacial relicts, giving rise to two hypotheses. The first consists in considering VT slopes as indicative of particularly rapid and effective postglacial frost shattering and scree processes. The second, on the contrary, consists in considering them as relatively preserved forms of glacial erosion during the latest glaciation or glaciations. This hypothesis raises the issue of the effectiveness of glacial erosion in certain cirques and the conditions in which they were glaciated.

Richter slopes (VR)
A different type of rectilinear slope differs from the previous ones in having no ledges or continuous taluses. Their incline ranges between 30° and 35° and their Ho/Hi ratio is close to 1. Certain have a rocky segment in their upper section and are thinly covered in angular debris. These slopes are found alongside and on the same sites as VT slopes, hence mainly on the adret of cirques. They may be assimilated to Richter slopes due to their profile and incline, although perfect Richter slopes with an entirely rocky surface and no cover are very rare in the Rondane mountains.

In conclusion, the typology put forward here is based on the profile of the slopes, and the proportion occupied by scree slopes (their Ho/Hi ratio) (Fig. 9.5). It is composed of five categories of fundamental forms. The first three, of type A (the P1, P2, and P2R free face slopes), are characterized by the preponderance of the free face in their profile, with a Ho/Hi ratio < 0.50. The other two, of type B (the VT and VR slopes), are characterized by their straight profiles, with or without ledges, and have an incline of less than 35° and Ho/Hi ratio > 0.50. This classification accounts for the glacial and periglacial relicts and the inversely proportional operative periglacial modeling. They give rise to a theoretical succession of landforms revealing the series of processes at work, to be specified by analyzing the scree slopes.

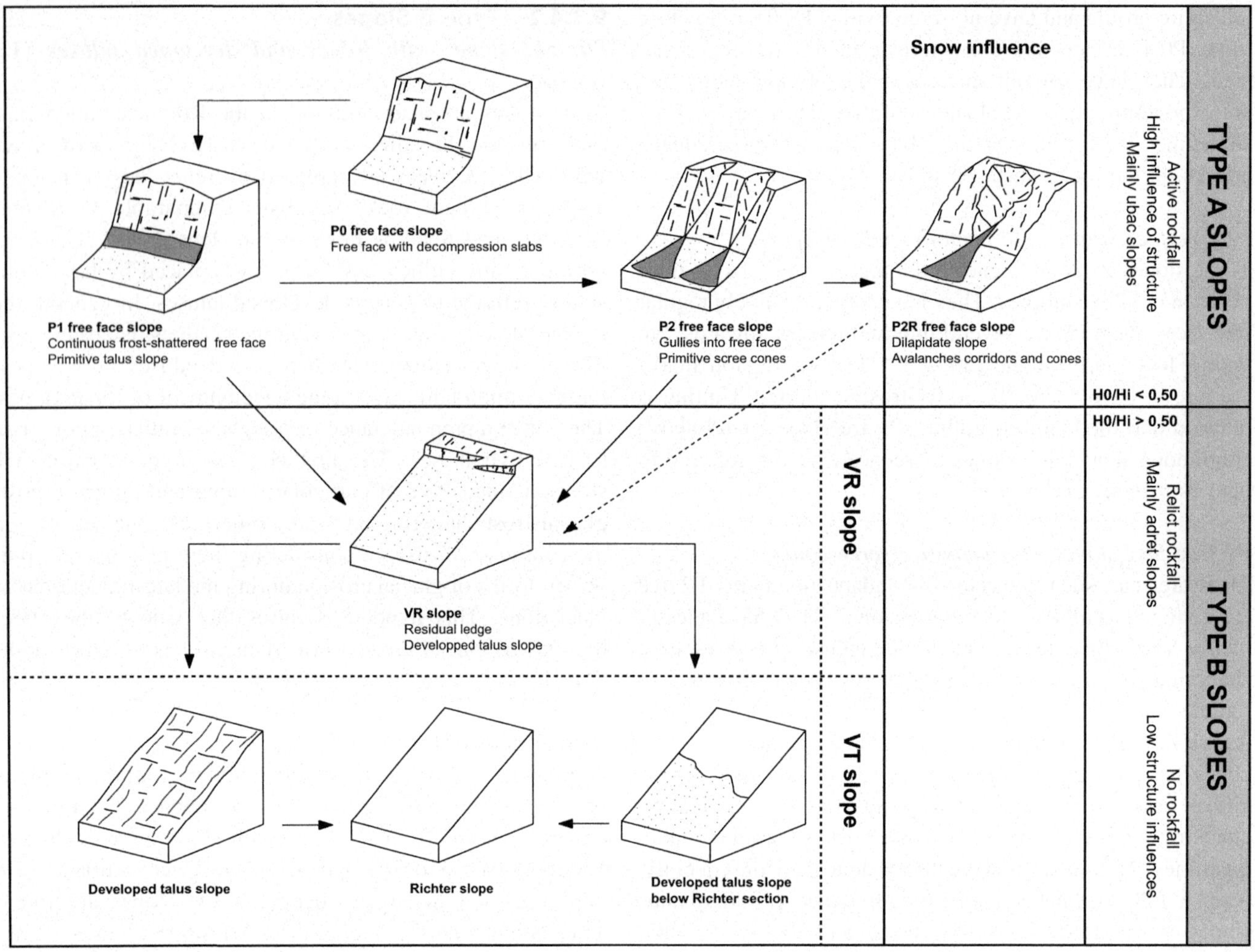

Fig. 9.5 Types of slopes (A and B) and scree slopes in Rondane

9.4 The General Properties of Scree Slopes in the Rondane Mountains

Scree is a deposit of rocky fragments progressively built up by *scree processes* on a slope (Francou and Hetu 1989). Scree processes result from the successive detachment of fragments from a free face or a rocky slope with an incline of 30°–35° by frost shattering or decompression, followed by the fall, transit, and accumulation of these fragments at the base of a slope. The *scree slope* designates the forms resulting from scree processes (s*cree cones* and *talus slopes*). A scree slope is characterized by its dimensions, its height (Ho), its incline, and its profile.

9.4.1 Study Sites and Choice of Transects

Thirty transects were surveyed in the Rondane massif, taking into account the environment of the slopes (the types of cirques), how they lay in relation to the dip (following or running contrary to it), the exposition (adret, ubac, or west- or east-facing, acting respectively as adret or ubac), the altitude of the apex of the scree slopes (1250–1660 m) and of the summit of the slopes (1400–2000 m), the types of free face (P1, P2, P2R), and the nature of the scree slopes (whether taluses or cones). The transects finally selected lie on ten slopes, representing the different types of scree slopes present in the Rondane massif, considering solely simple gravity slopes and composite cones (scree slopes assisted by run-off

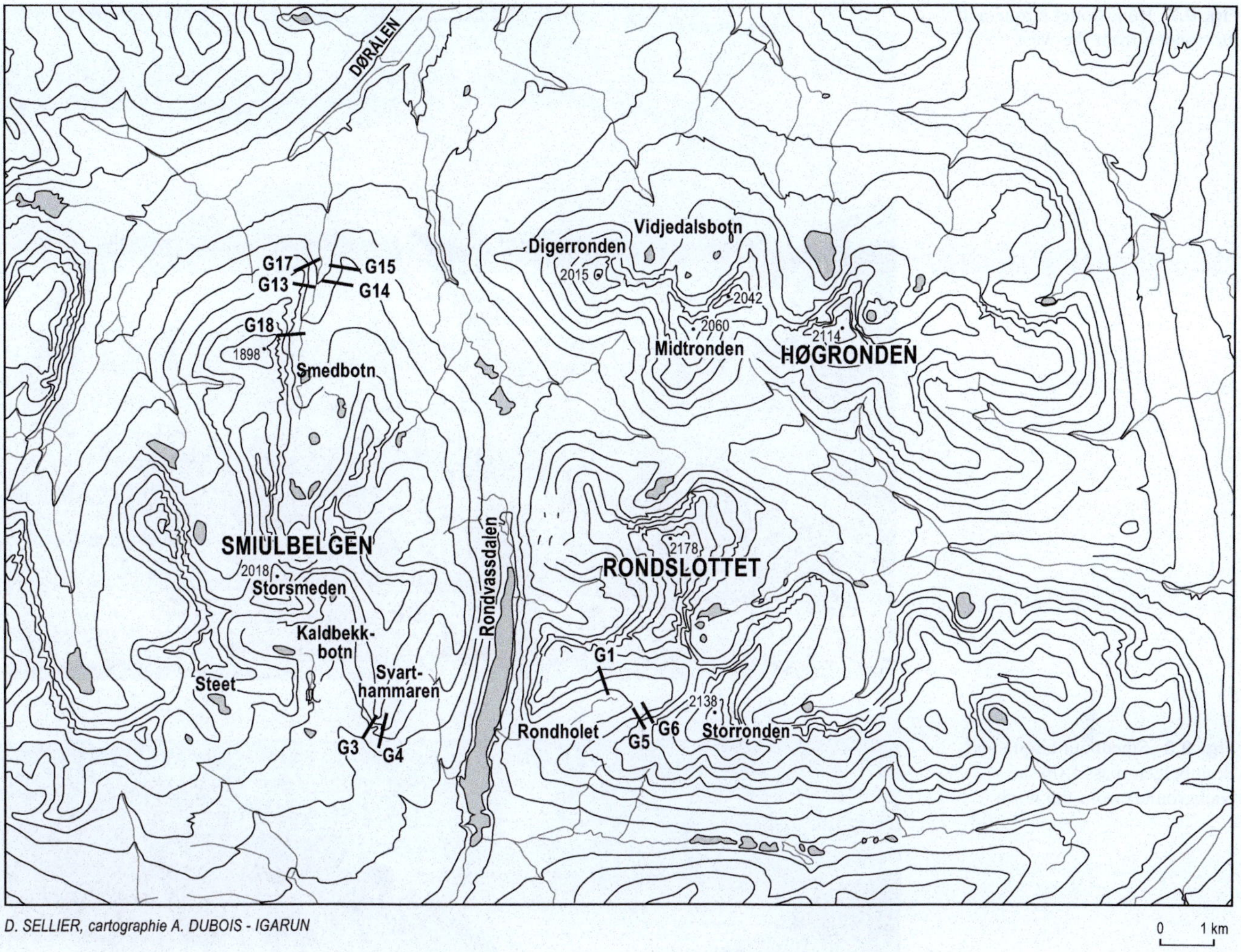

Fig. 9.6 Location map of surveyed transects in Rondane

and reworked by flowage were excluded from the survey). The surveys were conducted on three sites (Fig. 9.6).

To the south of Rondslottet, Rondholet (1500 m) is a cirque running east-west, whose base is at around 1500 m. The ubac slope follows the dip and rises to about 1800 m, with a P2 free face of 150–200 m in height and incline of around 75°. This free face is scored by gullies with a series of six primitive scree cones beneath (transect G5), separated by intermediary talus slopes (transect G6). The adret lies contrary to the dip, and corresponds to a straight slope with a P2R face or slope with ledges and developed taluses (VT slope) (transect G1) (Fig. 9.7).

South of Smiubelgen, Kaldbekkbotn (1400–1600 m) is an elongated cirque that opens to the south onto a large south-facing spur (Svart Hammaren). The tip of this spur, contrary to the dip and in adret position, has P1-2 free faces rising above talus slopes (transect G4). They are indented by a few fault gullies, the largest of which is above a composite cone (transect G3) (Fig. 9.8).

To the north of Smiubelgen, Smedbotn is a cirque running 5 km north-south. Five transects were surveyed in its northern part. The steep west slope, running contrary to the dip, rises to 1780 and has solely type A slopes. There are primitive talus slopes beneath P1 free faces in the higher sectors with fewer faults (transect G18), while series of primitive cones coalesce beneath P2 free faces with gullies (transect G13). There are also several composite cones, particularly at the entrance to the cirque (G17). The east slope, running contrary to the dip and with an attenuated profile, has solely type B slopes. These include straight slopes with ledges and developed taluses (VT) (transects G14 and G15) (Fig. 9.9).

This sample contains the main types of scree slope found in the Rondane mountains: three "primitive" talus slopes and two "primitive" cones, depending primarily on gravity mechanisms, and two "composite" cones and three "developed" taluses (Fig. 9.10).

Fig. 9.7 Rondslottet cirque and Storronden from the west

Fig. 9.8 Smedbotn cirque and Smiubelgen massif (in the background) from the north

9.4.2 The Morphometric Properties of the Scree Slopes

The morphometric characterization of the scree slopes is based on information from topographical transects surveyed at 10 m intervals, measuring the incline between each station. The length of the scree slope transects ranged from 90 to 210 m, and their height from 40 to 110 m. Their Ho/Hi ratio (ranging from 0.15 to 0.85) is an essential criterion for distinguishing between type A slopes, with minority scree slopes (Ho/Hi < 0.50), and type B slopes, with majority scree slopes (Ho/Hi > 0.50). The average incline of the transects ranged from 23.7° to 31.9°. The maximal incline of the transects (the steepest 10 m segment) ranged from 32.5° to 42°. Their longitudinal profiles were straight or concave. The straightest and steepest are characteristic of scree slopes

Fig. 9.9 Kaldbekkbotn cirque and Svart Hammaren spur (on the right side) from the south

Cross sections	Slope type / dip direction	exposure	altitude of the slope summit (m)	Hi (slope elevation) (m)	type of slope	type of talus slope	basal altitude (m)	summital altitude (m)	scree elevation (H0)	Hi / H0 ratio	length of the scree	average inclination	maximum inclination
Rondholet													
G1	opposite	adret	1,660	150	P2R-VT	evolved talus slope	1,510	1,580	70	0.46	150	24.3	35
G5	conform	ubac	1,760	230	P2	primitive cone	1,530	1,595	65	0.28	150	23.7	33
G6	conform	ubac	1,780	240	P1-2	talus slope	1,540	1,595	55	0.23	90	24.6	32.5
Kaldbekkbotn													
G3	opposite	adret	1,660	240	P1-2	composite cone	1,420	1,500	80	0.33	170	27.41	35
G4	opposite	adret	1,660	230	P1-2	primitive talus slope	1,430	1,470	40	0.17	90	27.55	34
Smedbotn													
G17	opposite	eastward	1,470	200	P1-2	composite cone	1,270	1,335	65	0.32	140	26	36
G13	opposite	eastward	1,520	220	P2	primitive cone	1,300	1,385	85	0.38	160	31.46	39
G18	opposite	eastward	1,780	360	P1	primitive talus slope	1,420	1,475	55	0.15	100	31.85	35
G15	opposite	westward	1,410	130	VT	evolved talus slope	1,280	1,390	110	0.85	210	30.83	42
G14	opposite	westward	1,405	95	VT	evolved talus slope	1,310	1,375	65	0.68	135	29.25	36

Fig. 9.10 Morphometric characteristics of selected slopes

depending more directly on simple gravity mechanisms (taluses and primitive cones).

Various scholars have devised graphics to present minor variations in incline along transects. Basing his work on Rapp (1960), Francou (1988) devised a "graph of the fractional distances to the apex" model, placing the incline of each segment on the x-axis, and the fractional distance of each station from the apex on the y-axis. The method consists in reducing the length of each segment to its dimension in comparison to the transect as a whole, taking the apex of the scree slope as point of reference. It thus reduces all the graphics to the same scale. "Morphometric graphs", inspired by Francou's model, were compiled for each transect surveyed in the Rondane mountains. Here, however, the inclines relative to each segment are placed on the y-axis and the proportional distance between each station on the x-axis, making it possible to read the graph in the same direction as the transect profile. Thus a straight section on the scree slope appears as a set of horizontal segments on the graph, a concave section as a series of ascending segments, a convex section as a series of descending segments, and a drop as a peak. A reference line corresponding to 30° incline facilitates comparison between the graphs.

Correlations between the Ho/Hi ratio and the length of scree slopes (Fig. 9.11), together with correlations between this ratio and the average incline of slopes, justify the distinction into four types of scree slopes differentiated in the field: primitive taluses (G18, G4, G6), primitive cones (G13, G5), "composite" cones (intermediary between primitive cones and avalanche cones, G3, G17) and "developed" taluses (G1, G14, G15). Primitive taluses (G18, G4, G6) are remarkably homogenous in character. Cones are found in more scattered positions due to the respective influences of gravity (primitive cones) and snow (composite cones). Developed taluses have the highest Ho/Hi ratios.

9.4.2.1 Primitive Talus Slopes (G18, G4, G6)

These scree slopes built up beneath P1 rock faces are the shortest (90–100 m), straightest, and among the steepest. They have a low Ho/Hi ratio (0.15–0.23). They correspond to the most elementary forms of scree process and result from simple gravity. The G18 transect is the most representative of this category (with a length of 100 m, Ho/Hi ratio of 0.15, average incline of 31.85°, and a very regular profile) (Fig. 9.12). Transect G4 already presents signs of flowage (length 100 m, Ho/Hi ratio 0.17, and a slightly convex proximal section with distal concavity).

Many of the scree slopes in the Rondane mountains present marks of flowage in the form of minor modifications in incline, particularly toward their base. Many taluses also have ridges in their distal sections, accompanied by a convexo-concave profile on the lower third or quarter of the transect. Transect G6, surveyed between cones on the Rondholet ubac, is of similar dimensions to the previous ones (90 m in length, 55 m in height, and Ho/Hi ratio of 0.23), but with a slightly lesser incline (25°–32° on average, 32°–34° at most), and a composite profile characterized by a protalus ridge.

9.4.2.2 Primitive Scree Cones (G13 and G5)

These cones develop beneath P2 free faces. Their morphometric properties are similar to those of primitive talus slopes because they too result from the preponderant actions of simple gravity. They form, however, a distinct dynamic and morphological entity, despite being the next stage in slope development after primitive taluses. They are longer than primitive talus slopes since they extend down the gullies indenting the free faces, which also supply them with debris. Their Ho/Hi ratio is consequently higher. Their maximum incline, always at the apex, is steeper. They have more varied longitudinal and transversal profiles, depending upon the stage of evolution of the free faces. Scree cones are more voluminous than taluses, because they have more extensive contributing areas, are found on denser fragmentation sites, and result from more sizeable accumulations. Transect G13 is an example of this type of scree slope (Fig. 9.13). It is long (160 m) and high (85 m), with a Ho/Hi ratio of 0.38. Its average incline is 31.5°, reaching 39° at its steepest.

Some of the scree cones have protalus ridges which are suggestive of flowage phenomena, especially as they extend to the base of adjacent intercone taluses. This is the case of cone G5, surveyed on the ubac of Rondholet cirque, which is comparable to a flowed primitive talus such as talus G6. Cone G5 is semicircular, and despite having only a slightly larger Ho/Hi ratio of 0.28 (as against 0.23),

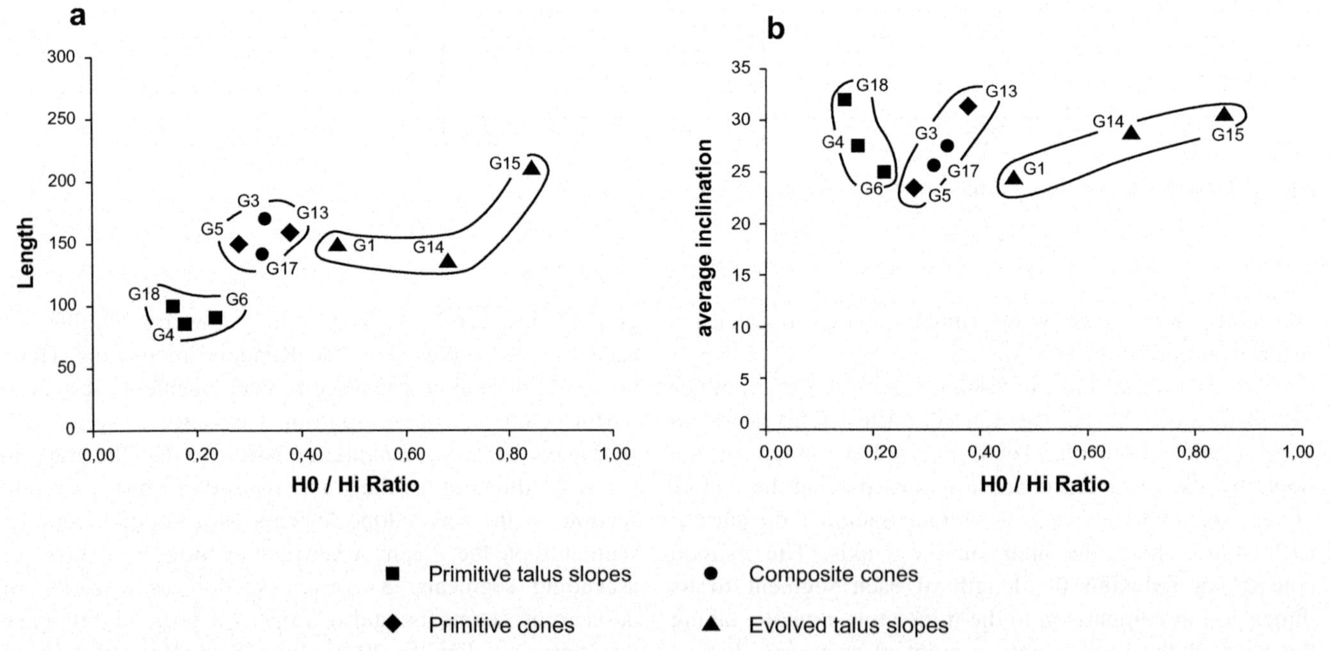

Fig. 9.11 Correlations between scree slope length and H0 /Hi ratio of the ten transects surveyed in Rondane (A) and between mean inclination and H0 /Hi ratio of the same transects (B)

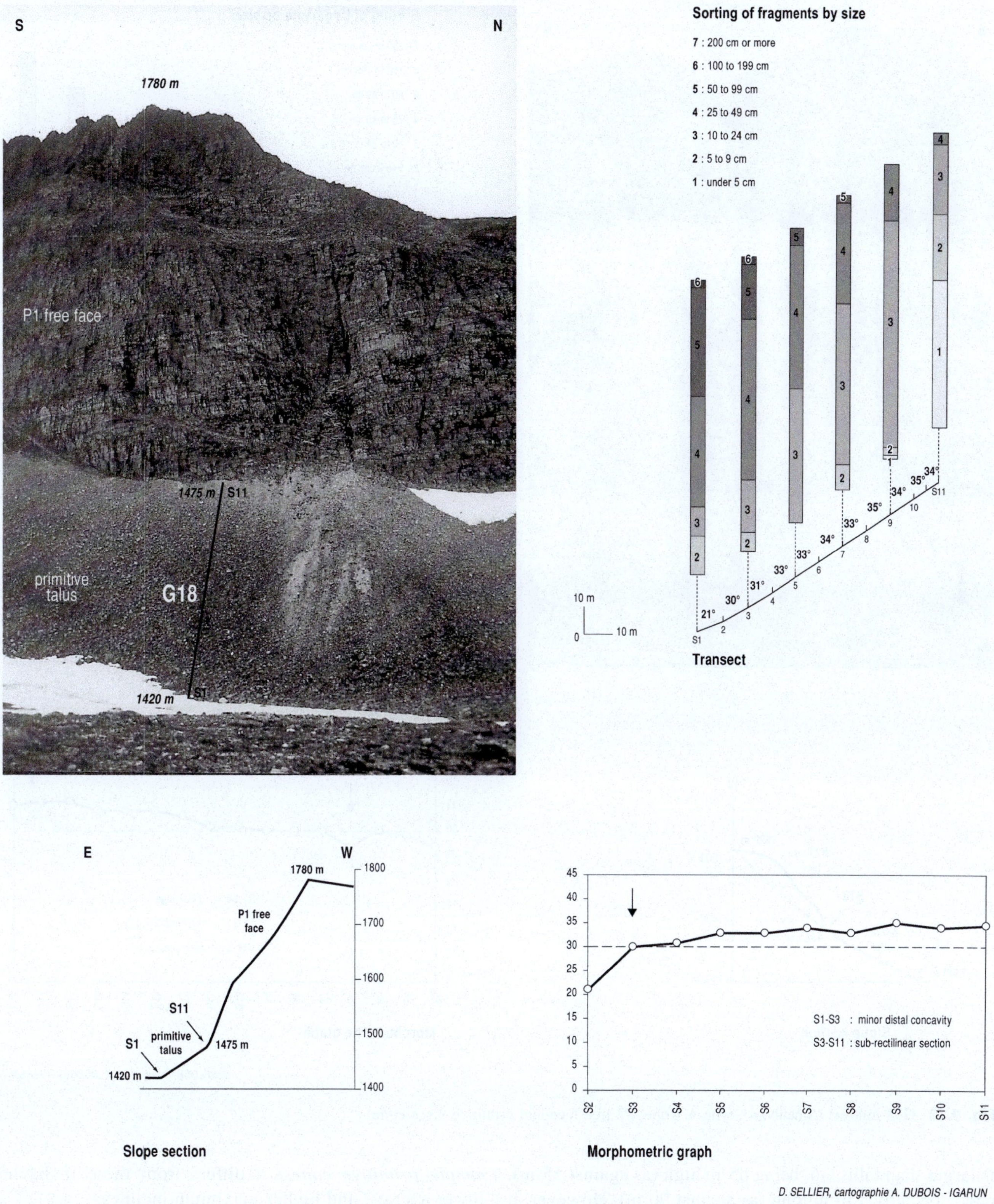

Fig. 9.12 G18 transect (Smedbotn): type A slope, P1 free face and primitive talus slope

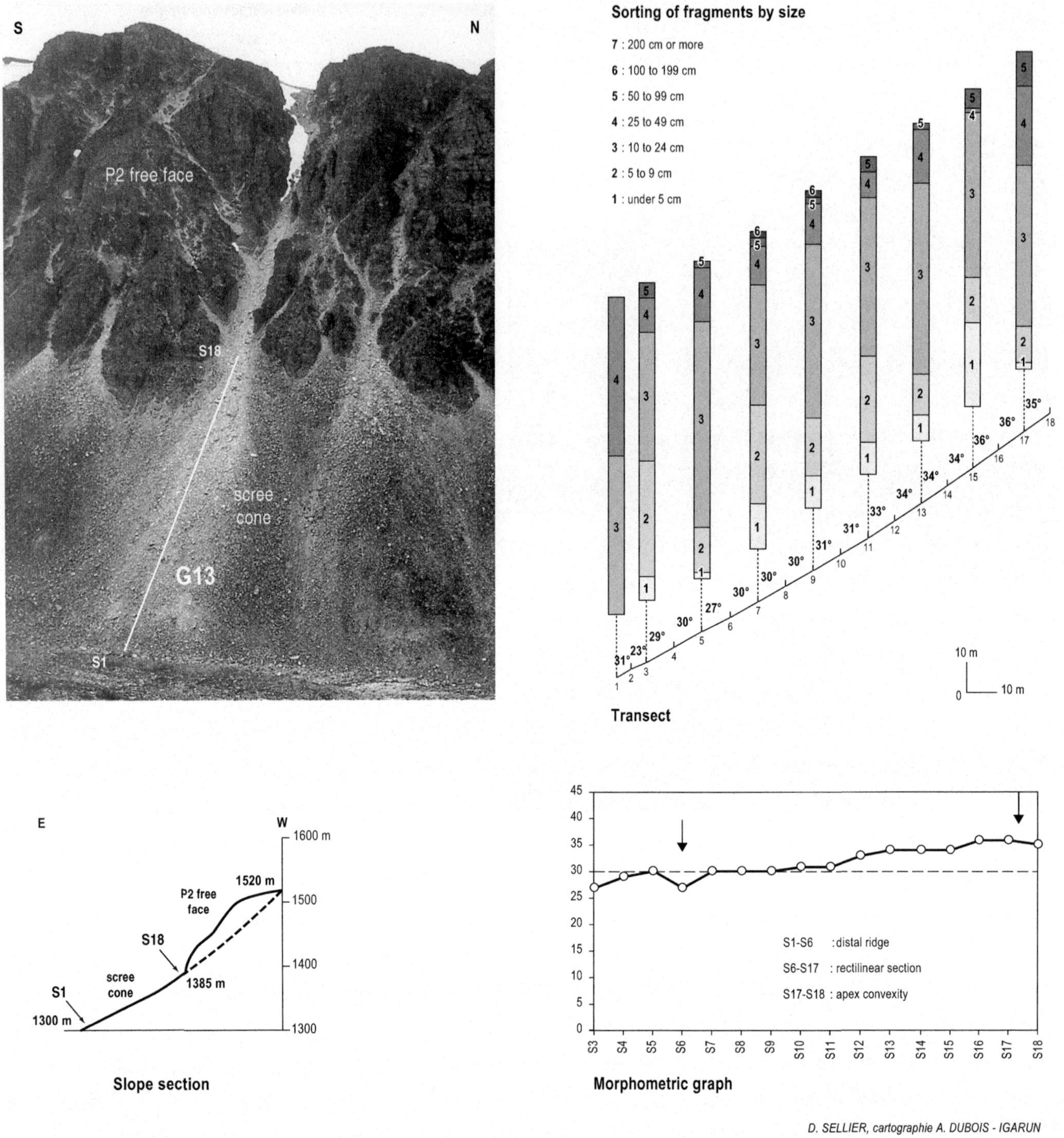

Fig. 9.13 G13 transect (Smedbotn): type A slope, P2 free face and primitive scree cone

is larger than talus G6, being 65 m high (as against 55 m), and especially 150 m long (as against 90 m). However, it has an analogous profile and incline, with an average slope of 23.7° (as against 24.6°), and maximum incline of 33° (as against 32.5°). Its profile is similar to that of talus G6, despite having a less marked distal ridge. While the dimensions of cone G5 are compatible with those of *simple primitive cones*, it differs from these in having lower average and higher maximum inclines.

9.4.2.3 Composite Scree Cones (G3 and G17)

These scree cones reveal reworking by avalanches. They are found beneath various types of free face (P1-2, P2, and P2R). The size of their contributing areas is between that of scree cones and that of avalanche cones, supplying them with frost-shattered debris from the surrounding free faces and material swept into gullies by avalanches. Two similar examples bring out the main characteristics, namely G3 in adret position to the south of Svart Hammaren spur (Fig. 9.14), and cone G17 facing north-east at the entrance to Smedbotn. These cones are similar in terms of their length (170 and 140 m), average incline (27.4° and 26°), and

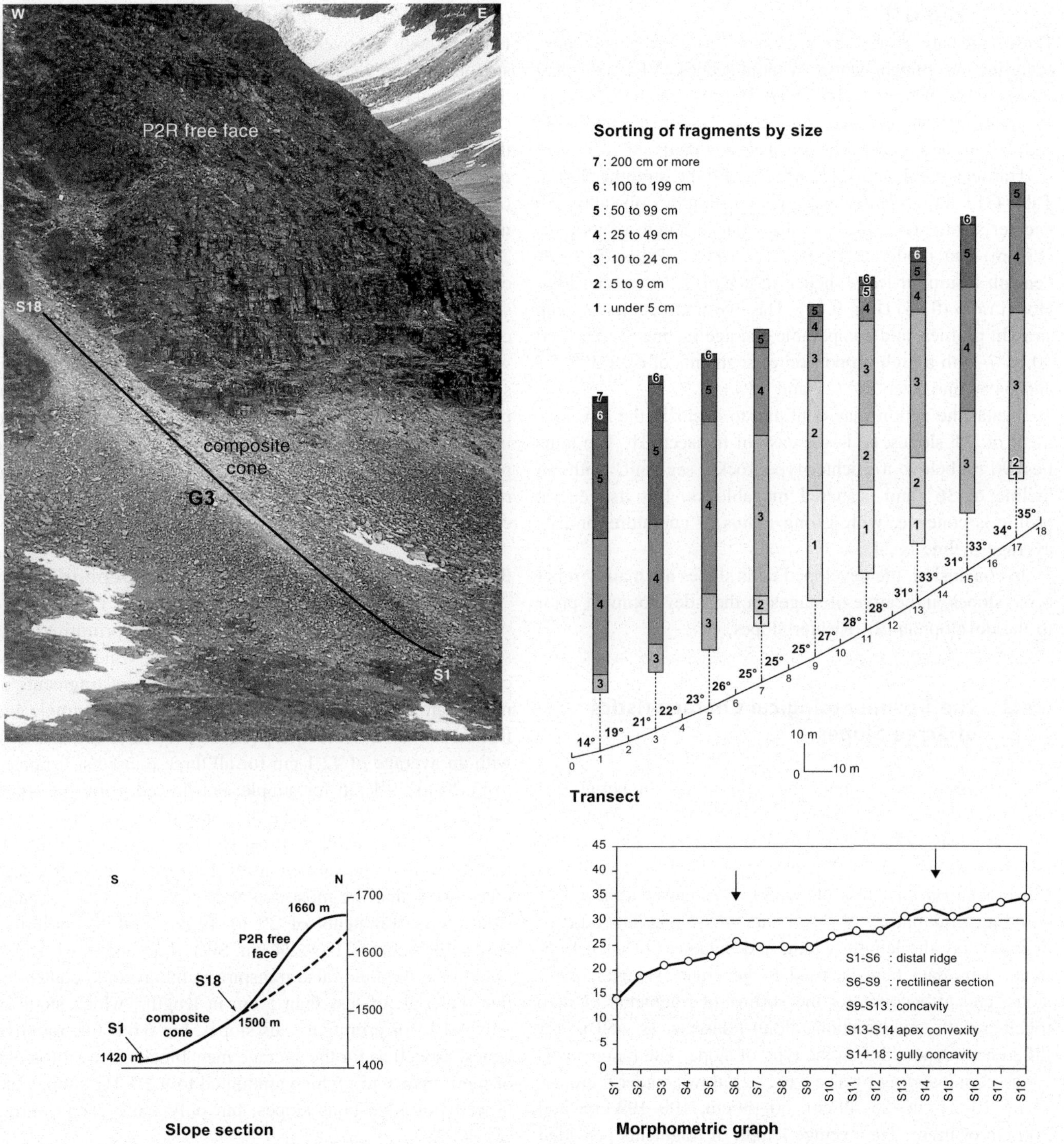

D. SELLIER, cartographie A. DUBOIS - IGARUN

Fig. 9.14 G3 transect (Kaldbekkbotn-Svart Hammaren): type A slope, P2R free face and composite scree cone

maximum incline (35° and 36°). They are concave but characterized by having a relatively high proportion of segments of over 30°, and their morphometric graphs are convex in shape.

9.4.2.4 Developed Talus Slopes (G14, G15, and G1)

These large talus slopes are common in the Rondane mountains, occurring on straight slopes without ledges (VR), or, more frequently, with relict ledges (VT). These slopes have inclines of around 35° and are characterized by their great length (135 and 210 m) and a Ho/Hi ratio of between 0.50 and 1.

Transects G14 and G15 are of VT type straight slopes. Talus G14 sits beneath a ledge with a straight slope above (a Richter segment), explaining the reduced length of its talus (135 m) and moderate Ho/Hi ratio (0.68). Talus G15 lies beneath a summit ledge, hence its length (210 m) and high Ho/Hi ratio (0.85) (Fig. 9.15). The two transects have steep straight profiles, and comparable average incline (29.25° and 30.83°), with a high proportion of segments of over 30° (57 and 64%) and over 35° (21 and 38%).

Avalanche action may continue to occur in the development of VT slopes, as in the case of transect G1. The talus lies 90 m below a Richter-type rocky segment, with an incline of 36° and scattered megablocks. But its general profile is concave, with a long radius of curvature, and its average incline is 24.3°.

In conclusion, the developed talus slopes are mainly relict scree slopes, indicative of stages in their development prior to the development of Richter slopes.

9.4.3 The Sedimentological Characteristics of Scree Slopes

The recording, processing, and definition of material size characteristics for the ten scree slopes examined in the Rondane massif (G1, 3, 4, 5, 6, 13, 14, 15, 17, and 18) pertained to surface sedimentology. Three measurements were taken on the basis of 30–50 fragments within a 1 m counting grid placed at 20 m intervals along the transect axis, namely the length, width, and thickness of each fragment. This data may be used to generate average dimensions, and to account for the sorting of fragments by size together with any variations along transects, as well as for differences depending on the type of slope. The ranges used for assessing sorting by size were as follows: under 5 cm, 5–9 cm, 10–24 cm, 25–49 cm, 50–99 cm, 100–199 cm, and 200 cm or more. The average length of fragments provided below (recorded as Lm on the diagrams) only relate to fragments of at least 5 cm in size. Nevertheless, nearly 2/3 of the counting stations included no fragment of less than 5 cm, and only 17% of these stations had more than 10% of "fine fragments", already a good indication of the overall size properties of local scree material (Fig. 9.16).

9.4.3.1 Coarse-Grained Scree

The coarseness of fragments is a specificity of quartzite scree in general. The average length (by station) of those measured on the surface of the Rondane transects ranged from 18.2 cm to 85.9, with an overall average of 41.5 cm. For 77% of stations, material size classes in the 10–50 cm range were in the majority. Fine material was very much in the minority (5.5% on average of the examined scree slope surfaces). On the other hand, scree containing blocks with dimensions of more than 1 m or more occurred in varying proportions (ranging from 0.5 to 30% of the surface fragments). In the majority of cases, these megablocks were between 1.50 and 2.50 m in length, but could measure 3.50 m in places. These characteristics explain the openwork texture of the surface scree, except at the apex, where the texture was semi-open or closed. The majority of fragments were thus contiguous but without any matrix. They were often unstable due to their size and shape. These properties reflect the regular joint planes found in sparagmites (thin beds that are easily divided into slabs). The scree material size reveals the effects of the processes by which they were formed (gravity processes) or reworked (flowage, avalanches, run-off, slope wash, or sieve effects), and hence their degree of development.

9.4.3.2 Differences in Material Size Depending Upon Scree Slope Type (Fig. 9.17)

A first distinction may be made between primitive talus slopes and primitive or composite cones, which have less coarse material overall. The average length of fragments of more than 5 cm ranged from 27.6 to 74.5 cm for simple and flowed primitive taluses (respectively, G18 and G4, and G6), with an average of 42.1 cm for all three transects. It ranged from 20 to 29.1 cm for simple and flowed primitive cones (respectively, G13 and G5), dropping to an overall average of 25.5 cm, diminishing further still away from the bisecting axis, a common phenomenon for this type of form. It is also noteworthy that the most common class for primitive talus slopes was fragments of 25 to 49 cm, and for primitive cones fragments of 10 to 24 cm. Such differences, which are obvious in the field, also transpire in the average content of fine material (of less than 5 cm in length), which stood at only 1.1% for primitive talus slopes, but 6.6% for primitive cones, as well as for the average megablocks content (blocks of more than 1 m), which amounted to 9.2% for simple and flowed primitive talus slopes, but only 2.6% for primitive cones.

The average material size for composite cones (G3, G17) is influenced by avalanches, and while fairly similar to that for primitive scree cones is in fact even smaller. The average length of fragments of more than 5 cm ranged from 18.2 to

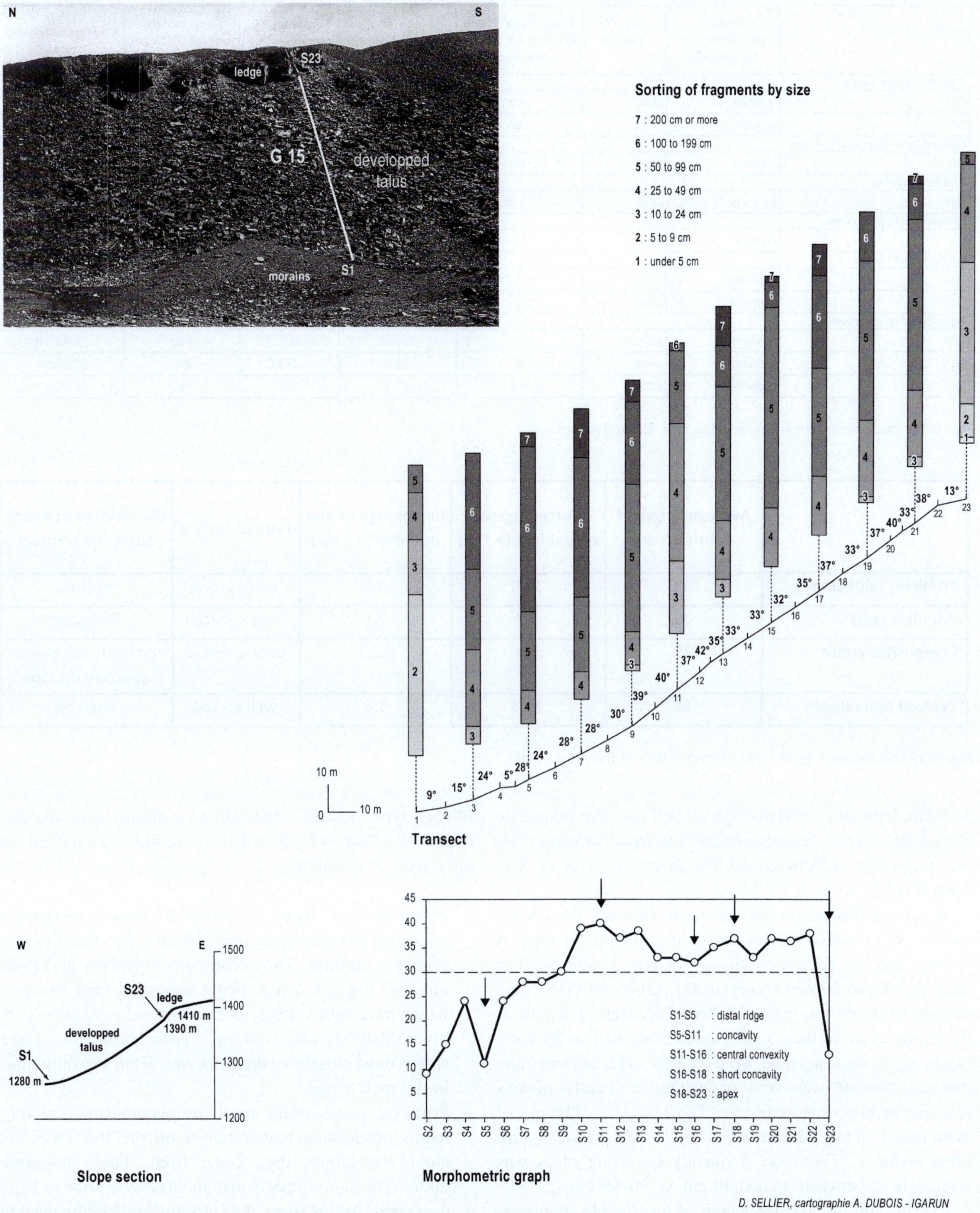

Fig. 9.15 G15 transect (Smedbotn): type B slope, VT slope, developed talus slope

	Slope : dip conform or dip opposite	Aspect	Percentage of fine material < 5 cm per transect	Average length of debris > 5 cm per transect	Percentage of megablocks > 1 m	Length of the longest block per transect	Main granulometric class
Primitive talus slopes							
G18	opposite	ubac	4.2	27.6	1	134	25-49 cm
G4	opposite	adret	0	74.5	30	275	50-99 cm
Flowed primitive talus slopes							
G6	conform	ubac	3.3	31.4	5.3	250	10-24 cm
Primitive cones							
G13	opposite	ubac	9.5	20	0.6	180	10-24 cm
Flowed primitive cones							
G5	conform	ubac	6.7	29.1	3.3	235	10-24 cm
Composites cones							
G3	opposite	adret	10.9	34.6	4	225	10-24 cm
G17	opposite	ubac	24.2	18.2	0.5	176	5-9 cm
Developed talus slopes							
G14	conform	adret	1.7	40.5	8.5	270	25-49 cm
G15	conform	adret	0	85.9	28.4	322	50-99 cm
G1	opposite	adret	2.8	57.4	9.2	350	25-49 cm

Fig. 9.16 Sedimentological characteristics of selected slopes

	Average length of debris > 5 cm	Percentage of megablocs (> 1 m)	Percentage of fine material (< 5 cm)	Type of sorting	Distribution of debris along the transect
Primitive talus slopes	45	11.5	1.5	well sorted	normal
Primitive cones	26	2.7	7.5	poorly sorted	irregular
Composites cones	25	2.4	12.5	poorly sorted	normal downslope opposite upslope
Evolved talus slopes	61	15.3	1.5	well sorted	anarchic

Fig. 9.17 Sedimentological mean characteristics of the scree slopes

34.6 cm, with an overall average of 24.9 cm. The frequency of megablocks was broadly similar. The most common class was fragments of between 10 and 24 cm (G3), or of 5 to 9 cm (G17).

A second distinction may be made between the above-mentioned primitive taluses and cones (found on type A slopes), and the developed taluses (of type B slopes). The three developed taluses surveyed (G1, G14, and G15) were notable for their very coarse debris. The average length of fragments of more than 5 cm ranged from 40.5 to 85.9 cm (with an overall average of 61.2 cm). The surface fine material content was zero or negligible, nearly always resulting in an openwork texture. The frequency of blocks of more than 1 m ranged from 8.5 to 28.4%. These blocks were often unstable. The most frequently occurring class was fragments of between 25 and 49 cm, or 50–99 cm.

To sum up, the average length of measurable fragments by transect was about 40 cm for primitive talus slopes, about 20–30 cm for primitive and composite cones, but about 60 cm for developed talus slopes. The specific material size of each type of scree slope follows naturally from the conditions in which rocky fragments were initially supplied and the effects of reworking.

- Primitive talus slopes are caused by gravity mechanism effects on frost-shattered debris from a free face that is in principle uniform. The constituent fragments underwent minimal impacts before being deposited, and are those which have subsequently been reworked most slowly (by dry avalanches and flowage). Their dimensions correspond most closely to their distance from discontinuities in the rock above.
- Primitive cones result from the accumulation of fragments undergoing longer transit on the free faces and along the gullies they come from. These fragments underwent more impacts and subdivisions prior to being deposited. At this stage, they are smaller than the material found on neighboring taluses. Nevertheless, the phenomenon is still partially subject to initial determining structural factors. The siting of gullies and hence cones is

often dictated by the greater density of joints or faults, which consequently produce fragments of smaller size.

- In the case of composite cones, avalanches provide additional material by sweeping along gullies and redistributing the material already deposited on the surface of cones. This results in the brutal displacement of material, with increased shocks and fragment comminution.
- The case of developed slopes differs from the previous ones due to their particularly voluminous debris. Their material size results firstly from the conditions in which fragments were supplied. At this stage in slope evolution, the remaining ledges reduce the height of fall, thus tending to produce large blocks. In any event, they reduce rockfall dynamic, ordinarily the major process along a talus, and thereby promote sieve effects, which are especially prevalent given the already coarse size of the fragments.

9.4.3.3 Differences in Sorting

Longitudinal sorting may be analyzed in two different ways: the distribution of average debris lengths of 5 cm or more per station along the transects, and the distribution of fragments by class of material size for each station.

Most of the scree slopes displayed decreasing material size toward the apex with the average size of fragments diminishing as one moves upwards, a phenomenon known as "normal sorting", and relating to the kinetic energy of fragments as a function of their mass (Rapp 1960; Francou and Hetu 1989). Certain scree slopes display "inverse sorting", characterized by larger sized fragments being found as one moves upwards. Others displayed no sorting, suggesting the influence of assisted scree processes (Fig. 9.18).

Normal sorting was only found on primitive talus slopes (G18, G4). Talus G18, in particular, displayed a typical reduction in material size following a linear-type relationship between size of fragment and distance from apex. This type of organization confirms the direct relationship between reduced material size toward the apex and simple gravity scree processes.

The distribution of fragments along primitive cones was more complex. Thus cone G13 did not present any obvious sorting, though the largest fragments were found at its top and bottom. The reduction in material size moving upwards was masked by the accumulation at the apex of coarse fragments from gullies. This tendency (decreasing material size at the base, increasing material size toward the top) was more obvious along composite slopes (G3, G17), due to the greater action of assisted scree processes by avalanches from gullies.

The distribution of fragments by size along developed talus slopes (G1, G14, and G15) differed from these previous models. Each case presented a different type. Any initial sorting determined by rockfall dynamic has since been reworked and mixed by flowage, snow actions, and sieve effects, reorganizing fragments along the slopes which, ultimately, are more like debris rectilinear slopes than scree slopes.

The distribution of fragments by class of material size confirms and supplements the previous distinctions (Fig. 9.19).

9.5 Summary and Conclusion. The Genetic Relationship Between Types of Scree Slope

As the Rondane slopes show one of the most typical periglacial forms of quartzite mountains in north-western Europe, the massif offers exemplary terrain for studying scree slopes in the Nordic environment. This chapter leads to characterize the properties of scree slopes in quartzite, using the Rondane massif as an example, to present a morphological and dynamic typology of scree slopes, and to relate the scree process to the kinds of slope and the local morphodynamic and morphostructural parameters.

The Rondane mountains are a massif where scree is particularly extensive due to structural reasons (the thickness and homogeneity of the quartzite substratum) and to the climate conditions (a periglacial zone with an extremely large altitude range and the presence of permafrost on all the slopes). On the basis of the morphometric and sedimentological analysis presented in this chapter, it is possible to set out the defining properties of the scree slopes in the Rondane massif, thereby helping to specify the properties of quartzite scree slopes in general. Such scree slopes are characterized primarily by their coarse homometric material and openwork texture.

The scree slopes under consideration derive mainly from postglacial (and in places supraglacial) relicts, and hence have been progressively built up since the Weichselian ice melt. It is probable that, here as elsewhere, the main stages date from periods immediately after deglaciation, despite a recrudescence of episodic periglacial activity due to recent climate fluctuations. Most of the scree are fossils, as indicated by their proportion of lichen coverage. The scree is currently being reworked by dry avalanches, flowage, avalanches, and run-off, even though it is directly observable that scree processes remain active.

Irrespective of this, the cirques in the Rondane mountains associate various types of slopes. P0 free faces, virtually intact glacial relicts, are found in those cirques where the ice cover disappeared most recently. P1 free face slopes correspond to an initial stage in effective periglacial erosion, and remain abrupt and uniform. P2 free face slopes result from selective periglacial erosion of the free faces from gullies, determined by the fracturing of the rock. VT type slopes bear testimony to generalized periglacial erosion. VR type slopes result from completed periglacial actions operative on cirque slopes.

Glacial relicts thus predominate along the profiles of type A slopes, and periglacial relicts along those of type B slopes.

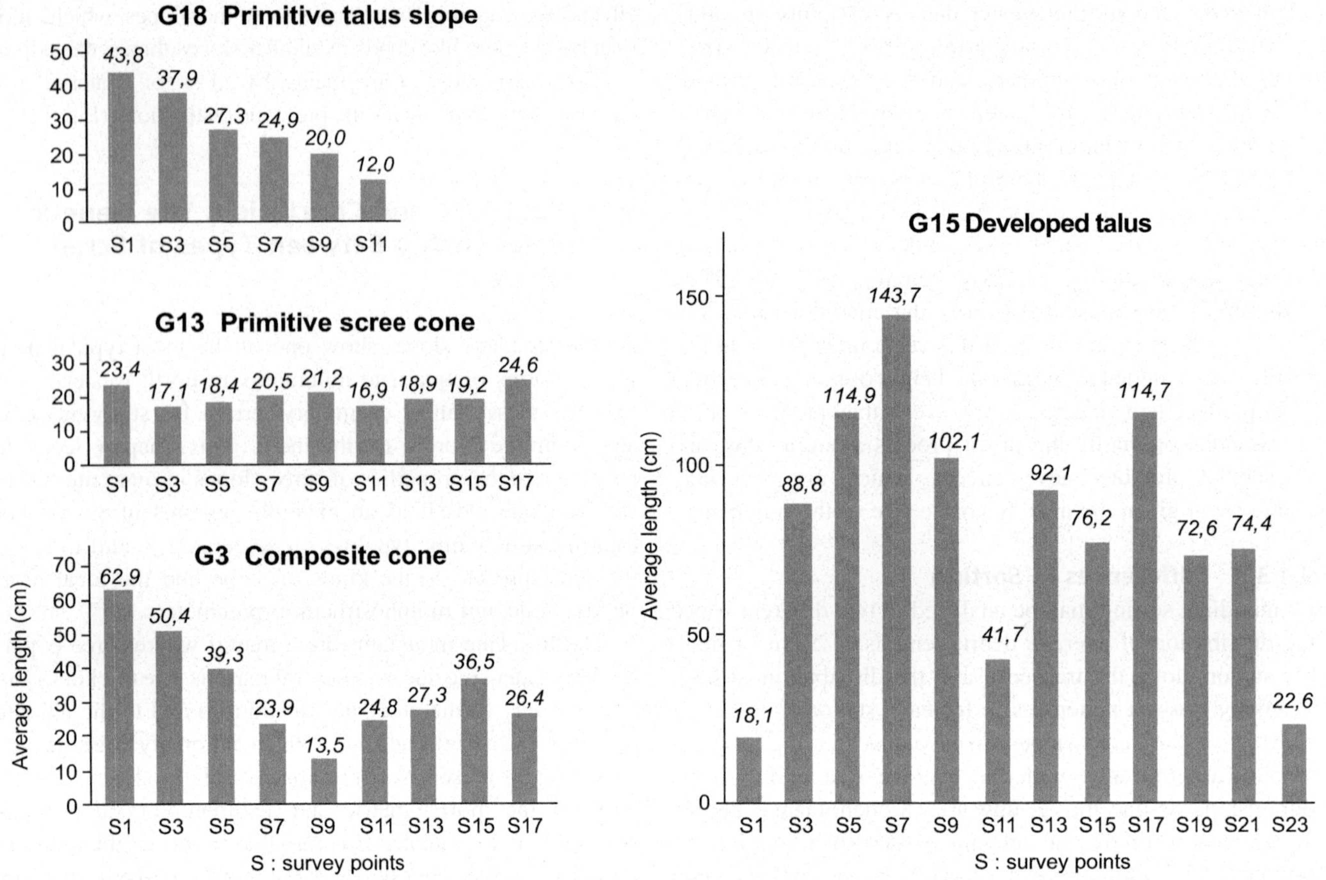

Fig. 9.18 Fragment average length at each survey point along scree slopes

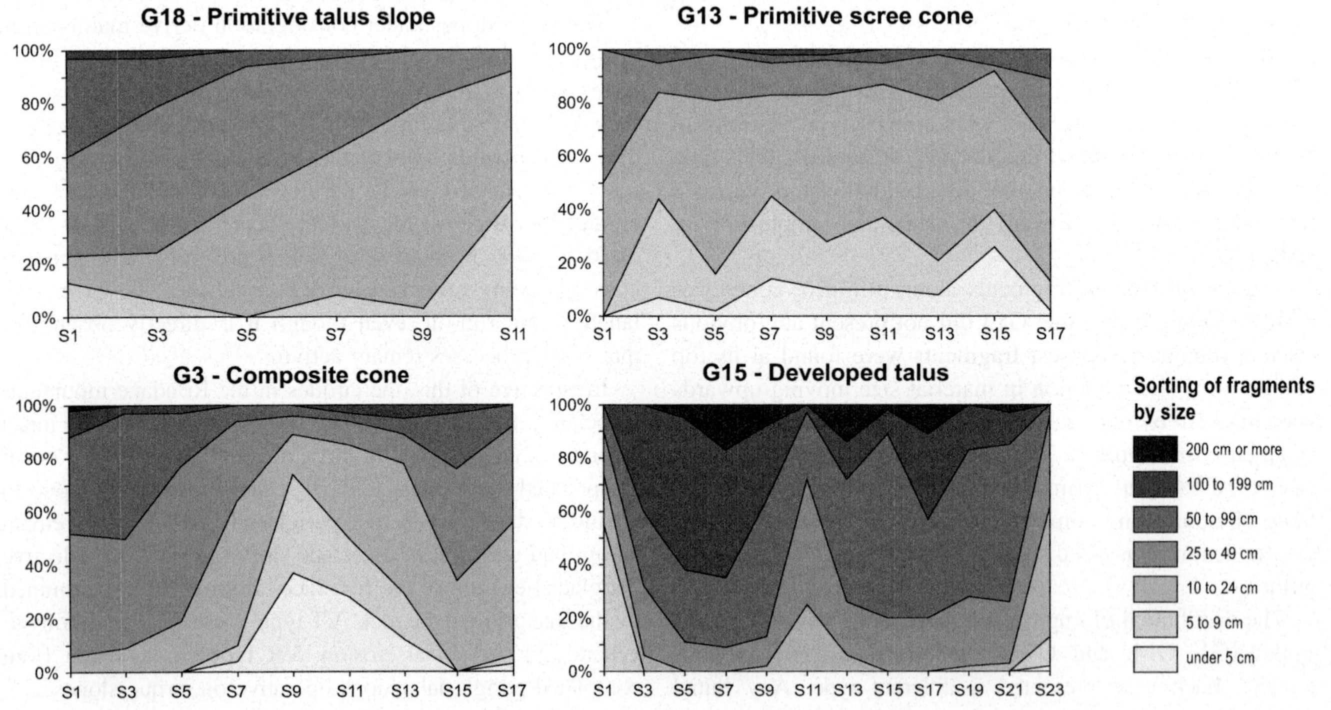

Fig. 9.19 Scree size sorting at each survey point

This classification is based on the inverse proportionality between glacial and periglacial relicts. It suggests a tendency and morphogenetic continuum even between each type of slope and its corresponding scree slope defined by its morphometric and sedimentological characteristics. P2 free face slopes and primitive cones necessarily follow on from P1 free face slopes and primitive taluses. Richter slopes present in each case the most fully developed forms of scree processes.

Nevertheless, this classification also indicates by its very principle that the effect of interglacial stages in shaping the inner slopes has been at least equivalent to the effect of glacial stages in certain places. It thus does not imply any systematic chronology within a given cirque. It does not mean that the straight slopes within cirques are more recent than the P2R, P2, or P1 slopes, nor a fortiori that they result exclusively from postglacial morphogenesis. The distribution of slope type by region and by type of cirque, together with the coexistence of various types of slopes inside these same cirques, does suggest on the other hand that the development of these slopes may have proceeded from initially different forms, occurring at different rates, developing to different stages of evolution, and perhaps resulting in different states today.

Additionally, the shift from primitive talus slopes beneath P1 free faces to developed talus slopes on VT and VR slopes, hence from type A slopes to type B slopes, may follow different models of development, by the retreat and concomitant lowering of free faces and the transformation of these free faces into relict ledges and/or Richter segments—without necessarily passing via the stage of primitive cones beneath P2 free faces when the structural properties do not favor the digging out of deep individual gullies. Under these circumstances, the talus slopes are neither "primitive" nor "developed", but are continually growing in length and surface area, corresponding to all the degrees of development, ranging from primitive talus slopes beneath P1 free faces, to developed talus slopes beneath ledges (VT slopes), and large developed talus whose ledges have disappeared (VR slopes). At the same time, the free faces, originally type P1, are progressively reduced to discontinuous ledges, and tend toward straight segments.

Lastly, the distribution of the landforms most clearly revealing glacial action (P0 and P1 free faces) helps position the Weichselian trimline at between 1750 and 1900 m in the Rondane mountains (Sellier 2002). It also indicates that the glaciation of cirques was highly variable from place to place, depending on the altitude of their base, on that of the surrounding summits, and especially on their aspect.

Acknowledgements This chapter is dedicated to Andrée Dubois, geographer, cartographer at the Institut de Géographie de l'Université de Nantes, passed away in June 2019, who drew the figures and innumerable others. The authors would like to thank the Institut de Géographie de l'Université de Nantes and the Laboratoire LETG Nantes (CNRS) for founding the translation of the manuscript.

References

Francou B (1988) L'éboulisation en Haute montagne. Thèse d'Etat, Centre de Géomorphologie du CNRS, Caen

Francou B, Hetu B (1989) Eboulis et autres formations de pente hétérométriques. Contribution à une terminologie géomorphologique, Comité national français de Géographie, Commission pour l'étude des phénomènes périglaciaires, Notes et Comptes Rendus du Groupe de Travail « Régionalisation du périglaciaire » XIV:11–69

Gjessing J (1967) Norway's paleïc surface. Nor Geogr Tidsskr 21 (2):69–132

Kerguillec R (2013) Les dynamiques périglaciaires actuelles dans un milieu de haute montagne atlantique: parcs nationaux du Oppland et du Sør-Trøndelag, Norvège centrale. PhD thesis, Université de Nantes

Kerguillec R (2015a) Seasonal distribution and variability of atmospheric freeze/thaw cycles in Norway over the last six decades (1950–2013). Boreas 44(3):526–542

Kerguillec R (2015b) Characteristic and altitudinal distribution of periglacial decay phenomena in the massif of Rondane, central Norway. Geogr Ann 97(2):299–315

Kerguillec R, Sellier D (2015) Selection and promotion of geomorphosites in the Rondane National park (Central Norway): landforms, heritages and geodynamics. Géomorphologie: relief, processus, environnement 21(2): 131–144

King L (1983) High mountain permafrost in Scandinavia. In: Proceedings of the 4th international conference permafrost. National Academy Press, Washington, pp 612–617

King L (1984) Permafrost in Skandinavien. Heidelberger Geographische Arbeiten 76:174

King L (1986) Zonation and ecology of high mountain permafrost in Scandinavia, Geografiska Annaler A 68(3):131–139

Moen A (1987) The regional vegetation of Norway: that of central Norway in particular. Nor Geogr Tidsskr 41(4):179–226

Norsk Meteorologisk Institutt (2018). http://www.eklima.no. Accessed Jan 2019

Oftedahl C (1950) Petrology and geology of the Rondane area. Nor Geol Tidsskr 28:199–225

Péguy CHP (1970) Précis de Climatologie. Masson, Paris

Peulvast JP (1985) Relief, érosion différentielle et morphogenèse dans un bourrelet montagneux de Haute latitude: Lofoten-Vesterålen et Sogn-Jotun (Norvège). Thèse d'Etat, Université de Paris I

Rapp A (1960) Recent development of mountain slopes in Kärkevagge and surroundings, Northern Scandinavia. Geografiska Annaler XLII (2–3):71–200

Rudberg S (1965–1966) Reconstruction of polycyclical relief in Scandinavia. Norsk Geografisk Tidsskrift XX(3–4):65–73

Sellier D (2002) Géomorphologie des versants quartzitiques en milieux froids: l'exemple de montagnes d'Europe du nord-ouest. Thèse d'Etat, Université de Paris I, p 1888

Sellier D (2006) Les limites de l'étage périglaciaire fonctionnel dans les montagnes atlantiques de l'Europe: éléments d'identification à partir de marqueurs morphologiques. Environ Périglaciaires 13:41–59

Statham I (1976) A scree slope rockfall model. Earth Surf Proc Land 1:43–62

10 Morphological Description of Erosional and Depositional Landforms Formed by Debris Flow Processes in Mainland Norway

Lena Rubensdotter, Kari Sletten, and Gro Sandøy

Abstract

The predominantly alpine landscape of Norway is situated along the Atlantic Ocean and most of Norway has today a relatively maritime climate with high both summer and winter precipitation. This is a generally favourable setting for precipitation induced debris flows given that there are sediments available for mobilization. This condition is also fulfilled since the last ice age left a landscape with valley sides covered with extensive deposits of glacial till. In addition to the till, extensive glaciofluvial deposits are found in the lower parts of the valleys and bedrock in high elevation areas are often draped with in situ weathering material. Weathering, rock falls and snow avalanches are active processes which continue to contribute more sediment to the slopes systems, available for remobilization by different types of debris flows. The varying morphological imprint of debris flow processes are readily identifiable in the landscape and have influenced the general development of valley sides and bottoms since the last ice age. Debris flows initiate and develop differently depending on the large-scale morphology of the slopes, the properties of the available sediments and the hydrological condition at initiation. Debris flow terminology is complex and may be based on several factors relating to release mechanism and type of movement within the flow. Here we use primarily morphological characteristics and thus describe and divide the Norwegian debris flows into two major types: A. *Open-slope debris flows* and B. *Fluvial channel-dependent debris flows*. The first type is further subdivided into: *A1, channel forming debris flows*; *A2, widening debris flows* and *A3, high-viscosity debris flows*. In addition, we mention and discuss larger scale landforms where debris flow processes play a significant role: *C. Debris flow fan systems* and *D. Multi-process debris flow fans*. We describe the general setting, release and movement mechanisms for the different types and the resulting morphological characteristics, using illustrated examples from around Norway.

Keywords

Debris flow • Slope process • Landforms • Geomorphology • Mass movement

10.1 Introduction

Most of the Norwegian landscape is alpine and has an Atlantic maritime climate setting (Seppälä 2005, Fig. 10.1, left panel). This setting, together with the general Quaternary sediment cover of thick and thin till and other sediment types on the slopes (Fig. 10.1, right panel), form favourable conditions for several types of mass movements. Debris flow processes are readily identifiable in the landscape due to their characteristic morphology with vertically extensive erosion tracks, ridges, lobes and large fan-shaped deposits and have significantly influenced the morphology of most valley sides and bottoms since the last ice age. Differences in local climate, topography, hydrology and sedimentology influences which sub-types of debris flow processes are active, and these variations are detectable in the resulting morphological forms. Here we introduce the main types of debris flow deposits in Norway and sub-divide them based on their morphological characteristics.

10.2 Terminology and Preconditions for Debris Flows

The terminology of mass movement processes and deposits is complex (Hungr et al. 2001) and may vary depending if focus is on release processes, mechanical mass movements

L. Rubensdotter (✉) · K. Sletten · G. Sandøy
Geological Survey of Norway (NGU), Postal Box 6315 Torgarden
NO-7491 Trondheim, Norway
e-mail: lena.rubensdotter@ngu.no

A. A. Beylich (ed.), *Landscapes and Landforms of Norway*, World Geomorphological Landscapes,
https://doi.org/10.1007/978-3-030-52563-7_10

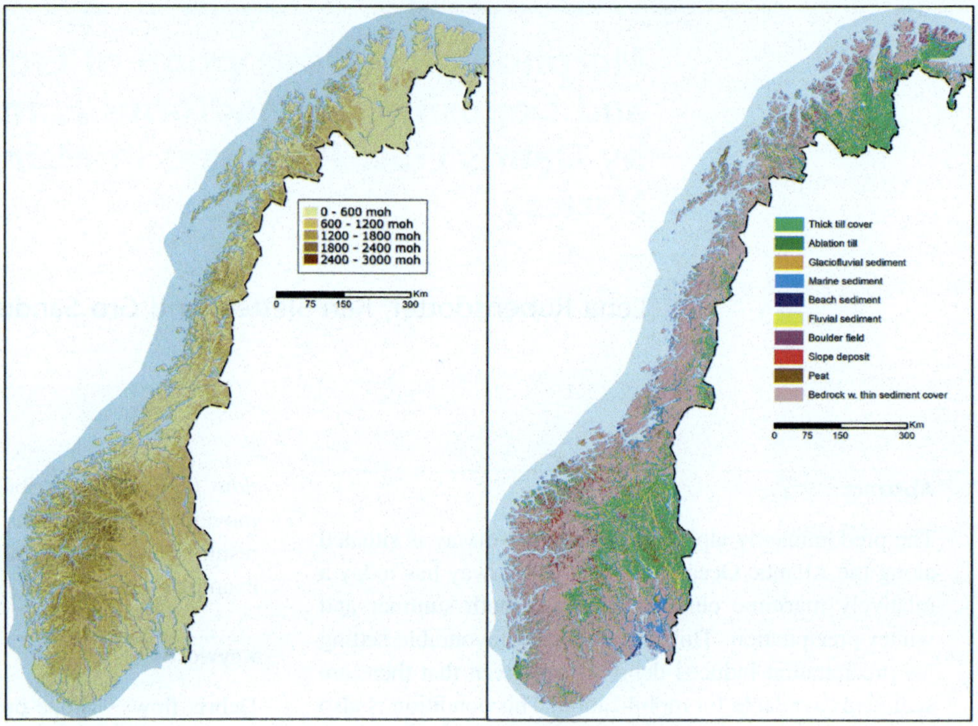

Fig. 10.1 Maps over mainland Norway, showing overall topography as elevation segments (left) and quaternary sediment cover (right, very generalized). Note that the pink unit, e.g. Bedrock w. thin sediment cover, encompasses both bare bedrock and areas where a thin cover of different types of sediments, e.g. till, peat or weathering material, cover the bedrock. Data: the Norwegian mapping authority (left map) and The Geological Survey of Norway (right map)

types or on erosional and depositional forms (morphology). Here we use an approach focussing on the resulting landforms, thus not subdividing types depending on minor variations in the movement or water content. Some of the debris flows described here could therefore also be called debris avalanches and to some degree debris flood/debris torrent and earth flow/mud flow.

The pre-requisite of debris flows is sediments that may be eroded, and therefore the geological history of Norway with repeated glaciations is crucial for the processes today. The Quaternary ice age period in Scandinavia is often imagined as long time periods with a continuous km-thick ice cover over the whole land-mass. This is however not the whole story, and it was only during the maximum glaciations that all land was covered by glaciers. This means that most of the time the ice sheet was existing only as separate ice sheets with the areas between the domes characterized by valley-glaciers networks, like the situation around the Jostedalsbreen ice-field today. Research have shown that the thermal regime of the glaciers also varied in time and space, with cold-based ice at higher elevation not eroding but rather preserving older landscapes and sediments (Kleman 1994; Davis et al. 2006). This means that the sediment cover of Norway has a long and complex history and, in more detail, encompasses large areas with in situ weathering material of varying thickness over bedrock, as well as large areas of thin or discontinuous till cover and smaller areas with glaciofluvial deposits. Since the earth's crust was pressed down under the last big ice sheet, the landscape has risen many meters (varying between some tenths to above 200 m) in relation to the sea since the deglaciation. This in turn means that marine deposits have been raised over the sea level.

In addition to the large lateral variations in the topographically dissected landscape of Norway, the sediment covers also often vary with depth. It is not uncommon with, for example, a stratigraphy with peat on top of a marine deposit, which in turn overlays a till unit. Another common situation is that a loosely consolidated, sandy till unit is draped on top of a compact, silty-clayey, sub-glacial till unit. The regional variation of surficial sediment cover and composition is one explanation for the varying debris flow activity and varying morphological imprint we see in the landscape today.

10.3 Debris Flows in Norway Since the Last Deglaciation

Over the post-glacial time scale (the Holocene) there has been both a paraglacial landscape- and sediment adjustment period (Ballantyne 2002; Ballantyne and Benn 2004) and a subsequently continued landscape development through long-term weathering, mass wasting and fluvial processes (Blikra and Nemec 1998). Since debris flows to a very high degree are a precipitation induced processes, climate variations over the Holocene has also been of importance for this development, both directly and indirectly. The climatic conditions in Norway today are very varied depending on both latitude and relation to the sea in the west. Most of the country has, however, a maritime climate with relatively high precipitation around the year. The exceptions are the eastern parts of south Norway and the inner Finnmark area in the far north, towards the Finnish border, which have a more continental climate. Large parts of these continental areas also have a lower relative relief than much of the rest of the country (see Fig. 10.1, left panel), which combined with relatively dry climate results in a lower general debris flow activity here. However, since debris flows are dependent on both sediment and precipitation, some maritime areas that have already experienced many mass movement events may have fewer debris flows today due to a lack of sediment to erode. Such debris flow systems could be classified as sediment-restricted, as opposed to the more common precipitation restricted systems.

Indications are that there was a high debris flow activity directly after deglaciation, when vegetation and soil-forming processes had not yet stabilized newly deposited glacial sediments on slopes (the paraglacial period, Ballantyne 2002; Ballantyne and Benn 2004). This is not reliably dated for many Holocene slope systems but is a logical assumption and is supported by studies of recently deglaciated valleys (Fig. 10.2 and Laute and Beylich 2012). During the Holocene climate optimum (Jansen et al. 2009) temperatures were generally higher and vegetation cover was established over most areas. Although the reginal climate variations are not well known for the Holocene in Norway it can be assumed that debris flow activity was generally lower and more sporadic than directly after the deglaciation, due to the combined effect of more stable soils and a warmer climate with less precipitation. This may, however, have varied significantly over the landscape and the general slope geomorphology in some coastal and fjord areas with large debris flow fans could be interpreted as a sign of regular debris flow activity throughout the post-glacial period (Fig. 10.3).

The climate deterioration after the Holocene climate optimum has not been defined clearly on a regional scale, but most archives indicate onset of colder and wetter climate sometime 5000–3000 BP and up until today (Schweinsberg et al. 2017 and references therein). The most marked period during the late Holocene is the so-called Little Ice Age (LIA) during medieval times when historical records tells about much colder climate and events of catastrophic episodic debris flow events over large areas in Norway. The best documented of these events is the Storofsen event, when large parts of the Gudbrandsdalen valley was so heavily affected by debris flows and floods that large areas were declared uninhabitable and unfit for farming and whole communities were abandoned (Sletten and Blikra 2007). It is said that almost all arable soil was eroded or covered with course-grained debris. The morphological traces of this and other similar large events are visible today as highly eroded hillslopes with numerous debris flow tracks, but most of the deposits are today not visible due to surface planation by farmers and society (Fig. 10.4). Large parts of the valley-floor deposits were also low-viscosity flows and travelled into the river channels where it was subsequently eroded and transported away, leaving no traces today (Sletten and Blikra 2007).

10.4 Schematic Morphological Description of Debris Flows

The Norwegian debris flows can be described as having three main morphological parts, linked to the processes from (i) starting/release, (ii) transport/erosion and (iii) deposition (Fig. 10.5). The two latter can occur together as deposition of ridges along a track with erosion in the central part of the track. At the terminus of the track we find the main depositional area.

(i) The starting zone is often, but not always, linked to a back-scarp of some form, either on an open slope or as a scoured eroded area inside a stream channel.
(ii) The transport/ erosion zone is mostly elongated and situated along the steepest part of the flow pathway. It is often possible to distinguish morphologically an upper part with only traces of erosion from a lower part where simultaneous deposition occurs along the sides of the flow path. The relative expansions of these parts cannot be defined, and it is not uncommon to see traces of some deposition almost at the release area, depending on the variability in slope angle and morphology, over which the debris flow travels.
(iii) The depositional zones are found both as elongated ridges (levées) along the flow path and as larger deposits at the lower end of the flow. Levées are formed by debris in the lateral parts of the flow that have stopped to move due to friction against material and vegetation outside the track. Levées can have very varied compositions, but when they are not complicated by large debris such as tree trunks, they are often described as having an imbrication of larger particles directed away from the main direction for flow. The main deposits are found at the lower end of the track where the energy of the flow decrease, which is usually coinciding with a marked reduction in the angle of slope. The morphology of this part is much dependent on the viscosity of the flowing mass (e.g. water content) and may form convex lobes or tongues, or more fan-shaped forms with a dispersed form and fluvial-type patterns on the surface (Fig. 10.6).

Fig. 10.2 Examples of paraglacial debris flows associated with recently deglaciated valley sides in Bødalen, Stryn, western Norway. Note the sparse vegetation cover and large amounts of material eroded, many places down to the underlying bedrock. Encircled person for scale. Photograph Lena Rubensdotter, NGU

10.5 General Types of Debris Flows in Norway and Their Geomorphological Traits

To describe the geomorphological types of debris flow tracks and deposits in Norway, it is natural to include some parts of the release and transport mechanisms in the description because this often influence the resulting landforms.

Since debris flows consists of a mix of debris and water moving through turbulent flow down a slope, it is logical that the proportions of debris and water, as well as the type of incorporated debris, will affect how the landscape is formed by the process. Higher water content in the moving mass reduces internal friction and keeps more energy in the flow. This in turn causes a longer run-out distance. The water content is also related to the more precise internal movement of particles in the moving mass, which in turn influences the resulting morphology. Generally, larger debris particles are pushed laterally in the flow and stopped and deposited along the path due to the rapid increase in friction to surrounding masses when moving towards the sides/edges of the flow. If the water content is high enough, this pattern may be almost fluvial, with only the larger particles being stacked against each other and fines sediments being removed with the water. When the water content is lower it may be observed larger rocks and boulders that seems to be floating on top of an otherwise finer grained flow.

Debris flow landforms in Norway can be roughly divided into two primary types, A and B, with three sub-types of A, based on a combination of process and releasing mechanism

Fig. 10.3 Illustration of the morphology and position of a debris flow fan partly incised into a rock fall dominated talus slope on Flakstadøya, Lofoten, north-western Norway. Note that the toe of the fan is deposited down to the sea-shore and has been eroded by wave action (seen as a straight shelf, or scarp, in the lowermost part of the fan, white arrow). The inset image shows more details of the debris flow erosion track and the levées and deposits forming the fan below the track. The 3D landscape depiction was created by a draping a hillshade image over a digital elevation model (DEM) in a GIS environment. Data from the Norwegian mapping authority

and resulting morphology. In addition to this, combinations of the different types of debris flows alone or together with other slope processes forms larger debris flow fans, which are also discussed as type C and D.

A. *Open-slope or hillslope debris flows, subdivided into*
   - *A1, Channel forming debris flows,*
   - *A2, Widening debris flows*
   - *A3, High viscosity debris flows*

B. *Fluvial channel-dependent debris flows.*

C. *Debris flow fan systems*

D. *Multi-process debris flow fans*

10.6 Open-Slope or Hillslope Debris Flows (A)

The term open-slope debris flow is here used for debris flows occurring outside pre-existing fluvial channels or tracks. This type is found in many different types of settings and may be developed in different sediment types and can also occur where there is only a thin vegetation cover on bare bedrock. The morphological features characterizing this type of debris flows are a relatively distinct starting point or zone, an erosion and transportation zone, which is quite often

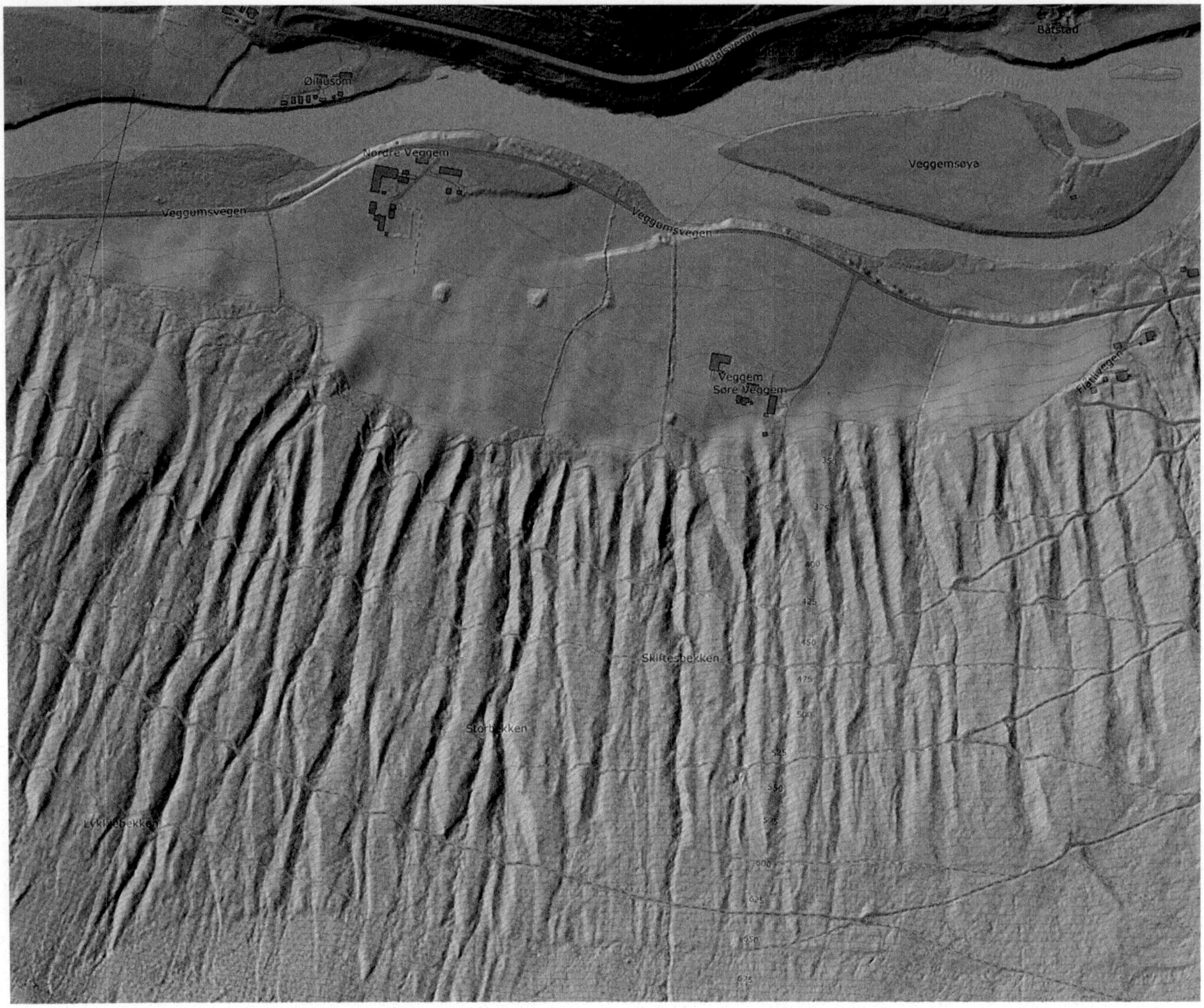

Fig. 10.4 Illustration of a north-facing hillslope with numerous debris flow channels at Veggem, Otta, south-central Norway. The foot of the slope is farmed and the deposits from the debris flows are thereby obscured in the morphology. Some of the debris from the events travelled all the way to the river in the upper edge of the image. The image is made through draping a hillshade image over a map in a GIS environment. Data from the Norwegian mapping authority

slightly winding in form, and a deposition zone. The starting zone or release area is often preserved as a depression in the sediment cover with a distinct upper edge (Fig. 10.7). The depth of the depression might vary but is often somewhere 0,5 to 1,5 m deep and the width is usually only some meters or tens of meters wide. The continued debris flow track can then either be channelized or of a widening type.

10.6.1 Channel-Forming Debris Flows (A1)

The channel forming type of open-slope debris flows erodes and creates its own channel (Fig. 10.8). This type may occur as a singular form but is most often found as sets of channels along a slope, where geological and meteorological conditions are similar. The starting zones are often clustered at similar elevations on the slope, indicating that precise angle of slope and/or small variations in sediment and ground properties are important for the release of events (Figs. 10.4 and 10.6).

The erosion and transport zone of the channel forming debris flows are incised into the substrate and forms a long narrow channel. The channels, or tracks, are often slightly winding, resembling fluvial channels on lower angle slopes (Figs. 10.8, 10.9 and 10.10). The winding may be a result of adapting to the pre-existing topography but is also induced by the turbulent flowing movement within the debris flow itself. Channel width is usually very constant along the track and the depth varies from some decimetres to several meters,

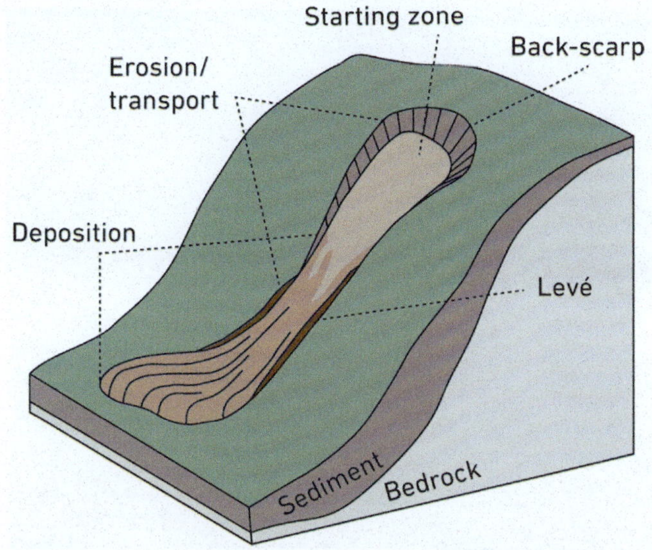

Fig. 10.5 Schematic illustration of a single debris flow with the morphological features and zones named (Modified after Nettleton et al. 2005). This schematic debris flow is shortened and disproportioned in comparison with most of the Norwegian type debris flows, which typically occur on long slopes with a very high length-to-width ratio of the path

but is usually relatively similar along a track, unless it passes over a protruding bedrock unit. Channel-forming debris flows often have levées developed along much of the track (Figs. 10.8, 10.9 and 10.10), and the winding morphology is sometimes accentuated by irregularities in width height and thickness of the levées. The winding may also be further accentuated if separate pulses or events re-use the same track since levées are often more pronounced in the outer swings of the channel where a secondary flow pulse is pushed over the channel boundaries, depositing sediments. The depth of a channel may also seem deeper than it is since the levées are built up on the sides along the track.

The deposition areas may have quite variable morphology depending on if it is one or repeated debris flows reusing the same track. A single event usually results in one tongue-shaped deposit with a convex surface in the outermost part and sometimes a more or less well-developed depression in the middle of the lobe. It is also common that the original lobe-shape is broken through at one point by a small fluvial erosion track. A secondary, finer grained, fluvial deposit is often found connected to this break in the lobe. This is the result of secondary fluvial activity released water from the eroded track or originating from small streams that was cut off by the debris flow. Several debris flows coming down the same channel often form separate lobes and a slightly dendritic pattern in the deposition area. Both thickness and form of the levées and lobes may vary between flows, sometimes taping out to very thin narrow tracks with hardly any erosion and just small levées draped over the pre-existing surface (see the small deposit by the arrow in Fig. 10.10).

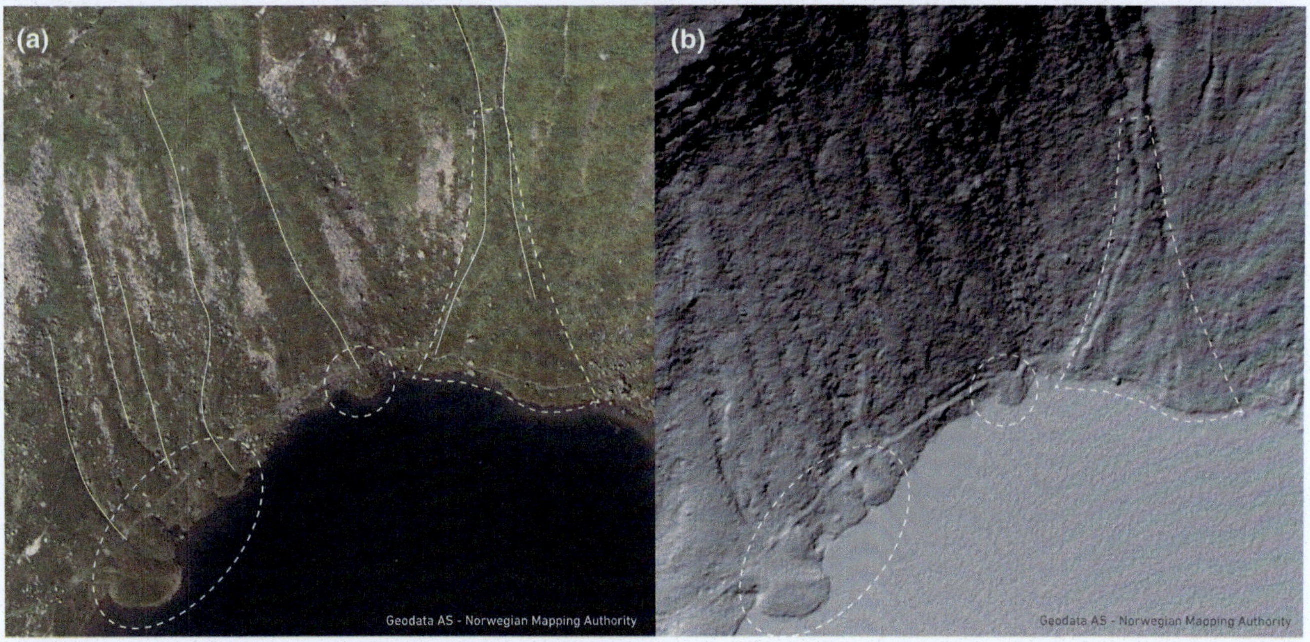

Fig. 10.6 Two images showing the same landforms on Flakstadøya, Lofoten, north-western Norway. In the left part of the images four small debris flow lobes are seen developed out into a small lake (encircled with dashed lines). They are formed by open-slope debris flows (white lines) with relatively short run-out and convex lobes as deposits. To the right in the images is marked the outline of a small debris flow fan (dashes line) with one clear track on the left side of the fan. Data from Geodata AS, Norwegian mapping authority

Fig. 10.7 Combined photos showing the starting point of an open-slope debris flow of the widening type in Vatne, Ørsta, western Norway. Note the fact that the starting zone is not eroded down to bedrock, but that a sliding surface has been developed ca 0,5 m down in a relatively unsorted sediment (till). In the upper part of the image is seen the lowermost part of the debris flow track, which continued down to a lake. Photograph Gro Sandøy, NGU

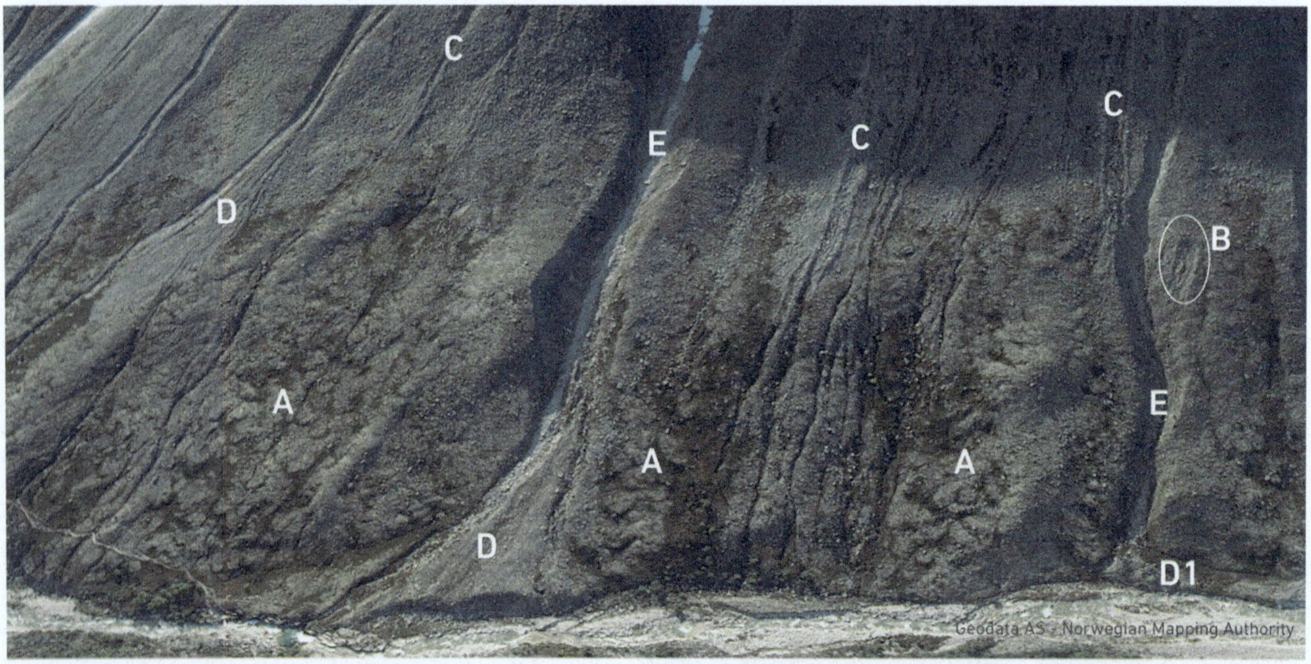

Fig. 10.8 Slanted image of a slope with multiple different debris flows in the continentally set in the Rondane mountain massif, central south Norway. A, regular slope surface with numerous active solifluction lobes. B, Starting point of open-slope debris flow that continues into a larger pre-existing debris flow track (E). C, Channel forming single debris flows with clearly visible levées. D, Debris flow fans. D1 is un-proportionally small and most of the deposits have been transported away by the river in the lower edge of the image. E, Over-deepened channels formed by repeated debris flows. Data from Geodata AS, Norwegian mapping authority

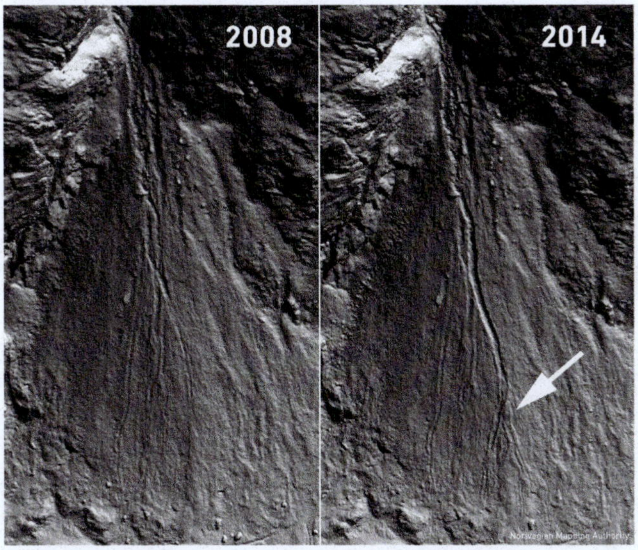

Fig. 10.9 Hillshade illustrations of a complex debris flow fan in two separate years (2008 and 2014), Drivdalen, central south Norway. This fan is also affected by both by snow avalanches and rock falls. Winding debris flow tracks with channels, levées and lobes can be seen in both images and the new events in the 2014 images show a pattern with several consecutive flows. Looking at the new debris flow tracks occurring only in the 2014 image, the second last flow or pulse has formed a plug in the pre-existing channel (white arrow). This plug has then deflected the last flow to make a left turn, into a new part of the lower fan. The same fan is also shown in aerial photographs in Fig. 10.10. Data from the Norwegian mapping authority

Fig. 10.10 Aerial photographs of the same complex debris flow fan as in Fig. 10.9. Small inset in the middle is an enlarged portion of the 2014 image with the most recent channel forming debris flow events clearly visible. Several debris flow paths are diverting from the active central channel and can be seen tapering out in different directions. This fan is also affected by both snow avalanches and rock falls

10.6.2 Widening Debris Flows (A2)

The widening type of open-slope debris flows may be called Debris avalanches and is in Norway often termed Triangular debris flows (Fig. 10.11). They initiate similarly to many channel forming debris flows, but instead of incising a channel they widen along the flow path, often reaching more than a hundred meters width at the bottom. It goes without saying that this type debrisof flows is one of the most destructive for society and infrastructure.

The starting zone is usually rather small, some 5–10 meters wide and some 0,5 meters deep (Fig. 10.7) and initiates like a slide with a distinct lower sliding surface. This surface is often not an underlying bedrock surface, but more typically a harder or denser layer in a sediment stratigraphy (Sandøy et al. 2017). The starting zones are thus not dissimilar to the channel forming open-slope debris flows. Once the slide is in motion it transitions into a turbulent flow and erodes and incorporates both vegetation and sediment along the path.

The widening of the flow path is the result of a shallow cover of erodible material over a harder surface (which can be bedrock, but also consolidated till) that is has a very low permeability and thus do not drain water out of the debris flow mass. These factors combine to contribute in two ways to the widening of the track; Firstly, the underlying hard surface means no deeper channel incision can be mechanically made. The incorporation of surface sediment and vegetation thus only increases the mass and potential energy of the debris flow, without losing much energy through erosion of underlying sediment. This means that the energy of the flow is upheld and increasing along the path, which widening in size as material is incorporated and therefore eroding out to the sides even more effectively. The widening becomes a self-feeding process as the material eroded also incorporates whole trees and larger vegetation, since they also are more easily eroded if standing on a compact bedrock surface shallowly covered by sediment. The non-permeable subsurface secondly reduces the loss of water from the flow mass through drainage into underlying sediment. This too contributes to maintaining and increasing the potential energy of the flow mass at the same time as keeping the water content reduces internal friction and induces more turbulent flow, resulting in even more effective lateral erosion. All factors together lead to continued increasing width of the debris flow along its path, as long as the flow is unable to erode into the subsurface material. The underlying hard surface in the track is mostly consisting of glacially polished compact bedrock (e.g. gneiss or granitic composition), but sometimes it seems that compact clay and silty compacted till can act in a similar way, if covered with a thinner cover of more loosely consolidated sediment.

Fig. 10.11 Photo looking up a large widening debris flow track in Årsetdalen, Ørsta, Western Norway. Both bare bedrock and a residual sediment cover is seen inside the path, while the visible levées (to the right in the photo) are mostly composed of large tree trunks and organic debris. The deposition zone is not shown on the photo but is situated behind the photographer. Photograph Lena Rubensdotter, NGU

The morphological imprint of a widening debris flow is a triangular scar in the sediment starting from a point and widening downslope and often exposing bedrock in the erosion part of the path (Fig. 10.11). It is not uncommon with levées along the lower part of the flow path, but they are morphologically insignificant in relation to the total width of the path. Since widening debris flows usually erode large areas of vegetation, the deposits often has a high content of tree trunks and other organic content (see lower right side of the debris flow in Fig. 10.11). This obscures the lasting morphology of the events as the organic material to a large extent is decomposed and removed over time, reducing the relative size of both levées and valley-bottom deposit. The major deposit may form a rather undefined lobate accumulation of debris on the valley bottom (Fig. 10.7). However, the high water content and thus energy often drives the more fine-grained part of the mass flow all the way down to a river or water body lowermost in the valley bottom (see upper part of Fig. 10.7), where much of the material is dispersed and removed over time. The long-term morphological imprint is, therefore, surprisingly insignificant given the destructive nature of these types of debris flows when they occur.

10.6.3 High-Viscosity Debris Flows (A3)

High-viscosity debris flows have a relatively high debris-to-water ratio, which influences both the type of movement and the total outrun length of the flows. This type of debris flow may occur both as open-slope or fluvial-track flow, however, the open-slope type is not so common in Norway. The high-viscosity flows mostly start as a slump on relatively low gradient slopes. High-viscosity debris flows are affected by high friction both internally, within the moving mass, and externally, between the moving mass and the underlying material on the slope. The open-slope types usually leave a distinct back-scarp, even when the subsequent erosion/transport zone is relatively short. The low water content and high friction results in a relatively short run-out and conditions the morphology of the deposit, resulting in a usually evenly thick, or convex, long lobe (Fig. 10.12). This type of debris flow thus results in a well contained, tongue-formed deposit with often the same width as the track and no lateral widening. Another morphological distinction in the open-slope type is that they often do not develop clear levées. From a sedimentological point of view

Fig. 10.12 Photos of an open-slope high-viscosity debris flow track and deposit, near Harstad, Troms, North-western Norway. The deposit is old and vegetated, but else composed of a previous till cover. The inset photo shows the front of the debris flow tongue, where is seen a unsorted distribution of grain sizes (white sheep for scale). Note the lack of levées and the long extent and even width of the deposit. Photograph Kari Sletten, NGU

particles in the deposit may be slightly imbricated, but more common is an unsorted matrix with no clear orientation or forting of particles. Single open-slope high-viscosity debris flows could from a morphological view be confused with solifluction lobes, which are generally thinner and shorter without a starting back-scarp. They might also resemble inactive rock glaciers, which however are generally of larger size and originate in a rock-scarp or near vertical rock slope.

10.7 Fluvial Channel-Dependent Debris Flows (B)

Debris flows initiated in a pre-existing fluvial channel are very common in the alpine Norwegian landscapes. The size of the channel may vary significantly from a small seasonally active stream to large v-shaped channels with significant year-round discharge. This type of fluvially dependent debris flows will have some similar traits as the channel forming open-slope debris flows but are usually strongly affected by the higher amount of available water for incorporation in the mass flow (Fig. 10.13).

Since debris flows inside fluvial channels are directly associated with increased water discharge during snow melt or precipitation events, direct fluvial processes will also affect the landforms before, during and after the debris flow event. This results in a residual morphology in which the deposition zone is more fluvial in pattern and extension (Fig. 10.13). Most fluvial channel-dependent debris flows are initiated when anomalously high discharge mobilize and erode sediment inside the stream channel itself, often starting as a flash flood type of event that transforms into a debris torrent/debris flood, and later debris flow, as more solid debris is incorporated into the flow. The precise starting point might, therefore, be difficult to pin point, as it often does not leave any lasting morphological traces in the channel. The increased mass of the flow makes it continuously more erosive downstream in the channel, at the same time as it incorporates water from the fluvial flow. Fluvial channel dependent debris flows may also initiate or be

Fig. 10.13 Photo of a fluvial channel-dependent debris flow with very low viscosity in Kirkeselva, Troms, Northern Norway. The high water content and pulsating nature of the process has resulted in multiple flow paths spread out on the flat valley bottom. Note the fine-grained sediments in the outermost parts of the flow paths and the fluvial winding tracks on the surface resulting in very relief deposits. Closer to the spreading point at the trees, further up the slope, larger boulders and stones are deposited, although not in any clear pattern. Photograph Odd-Arne Mikkelsen, NVE

increased in size through lateral slumping from the channel sides, in which case the resulting scarps of these slumps will usually also leave a morphological imprint (Fig. 10.14). Lateral slumps may also occur because of stream bank erosion caused by the debris flow itself.

Levées are often formed also along fluvial channel-dependent debris flows. Depending on the topography and depth of the pre-existing channel they may not be deposited as clear ridges outside the actual channel but may in larger channels often seen as relatively short-lived "trim-lines" of debris on the channel sides. When the debris flow reaches lower angle slopes along the channel it is more common to see more morphologically significant levées (Figs. 10.3, 10.9 and 10.10).

Fluvial channel-dependent debris flows generally have a long run-out due to the high water content and the depositional zone may continue out on flat or almost flat valley bottoms (Figs. 10.15 and 10.16). The location within an active fluvial channel increases the tendency for the debris flow to come in "pulses" during an event, which may stretch over hours or even days. The high water content affects the morphology and sorting of the deposits, which often includes imbrication to the larger particles and a tendency for fluvial-type lateral grain-size sorting in the depositional area. It is not uncommon that different flows have a distinct grain-size sorting in the deposits (Fig. 10.15). Large particles like boulders and stones are typically deposited first, while the finer grain sizes are washed further and deposited in a more low-angle fan-shape further out on the valley bottom (Fig. 10.13). Depending on the precise water-to-sediment ratio the deposits might form sheets, thin lobes or lobes with a depression in the middle. The occurrence of separate debris pulses during the same event results in multiple lobes and deposits, often differing slightly in grain size composition between the lobes and stretching out in several directions. It is also common with a very gradual transition from actual debris flow deposition over to a more common fluvial flood deposit over the valley-floor. The common fine-grained nature of the outermost deposit (Fig. 10.13) also contributes to the disappearance of the morphological traces over time, since it often is relatively rapidly re-moulded by farming activities into an even valley bottom. As with the channelized debris flows, this may result in the only morphological traces being overly incised channels on the slopes (see Fig. 10.4).

10.8 Debris Flow Fans–Composite Landforms of Different Viscosity Flows (C)

Most debris flows in Norway are of the fluvial channel-dependent type (B), but the channels have often originally formed as an open-slope debris flow of type A1. Since channels thus are contributing factors to debris flow

Fig. 10.14 Composite of two photos showing different parts of the same channel dependent low-viscosity debris flow in Liadalen, Ørsta, western Norway. This event might have been more of a debris flood/debris torrent that a classical debris flow, since the water content was very high. The two images show the higher reaches of the debris flow, near the starting point in the right image and a kilometre downflow in the left image, but still several km from the final exit from the channel into a fjord. The deposits in the channel is relatively well sorted with only the larger particles remaining (stones and boulders) indicating a very high degree of fluvial sorting in the end of the event. The actual process. Note the secondary slumps from the channel brinks, leaving concave scars in the channel sides. Photographs Lena Rubensdotter, NGU

initiation it is unsurprising that fans resulting from repeated debris flows are common forms in the landscape (e.g. Fig. 10.3 and 10.8).

A large debris flow system with repeated activity from the same track over many years may release flows of different viscosity over time, depending on fluvial activity and variations in debris availability. It is, therefore, not uncommon to find depositional lobes from high-viscosity flows relatively high up on a debris flow fan, inside older tracks.

This nature of repeated debris flows from the same channel is the key to the development of larger fan-shaped debris flow deposits. The higher viscosity flows stop higher up the slope than the low viscosity ones and the plugs formed will block the channel. The next debris flow, hours, days or years after, might not have enough energy to pass the plug, but will rather spill over the channel wall and form a completely new track down a different part of the slope (see Fig. 10.9, at the arrow). Repetition of this sequence leads to the build-up over time of the triangular shaped fans. Debris flow fans may vary greatly in size, depending on the type and number of events building the deposit, but are usually from some tens of meters up to several hundred meters wide.

Many of the debris flow fans are today still connected to active starting zones. In areas with an originally thin sediment cover the available debris might be totally depleted over time and the fans thus becomes inactive with continuous vegetation cover, etc. (e.g. Figure 10.6). Sediment replenishment usually occurs though weathering of bedrock and

Fig. 10.15 Photo showing part of a multi-process fan in Erdalen, Stryn, western Norway. A low-viscosity debris flow leading to the left of the fan forms a deposit towards the camera position. The debris flow track is winding, and the deposit has a seemingly sorted and relatively fine-grained grain size. Around the debris flow deposit are large boulders and stones resulting from rock falls. The profile of the fan in the background is shaped by multiple events of snow avalanches, of which some snow from the last event is still sitting in the track on the fan. Photograph Lena Rubensdotter, NGU

subsequent rock fall, or snow avalanche erosion. The material ends up in depressions in the landscape—often channels, where it is available for debris flow erosion. The speed of this sediment replenishment varies a lot due to primarily varying weathering resistance of different types of bedrock. Hard bedrocks, such as granites and some types of gneisses, are most resistant, and therefore we find more inactive debris flows fans in granitic areas. If a debris flow fan becomes inactive, other processes such as solifluction (the slow deformation and downslope movement of soils on a slope) may over time "whisk out" the debris flow morphology (levées, lobes and plugs) and form a smoother surface (Figs. 10.6 and 10.8). If the fan itself is in the depositional zone for rock falls and snow avalanches, this too might contribute to reducing the debris flow morphology (Figs. 10.11, 10.15 and 10.16).

10.9 Multi-process Debris Flow Fans (D)

Since debris flows are occurring on slopes, it comes naturally that they will occur in combination with other slope processes. The most important ones in the Norwegian landscapes are rock falls and different types of snow avalanches (Fig. 10.15). Larger rock avalanches, although relatively common in the landscapes, usually greatest very coarse-grained deposits which are not so consistent with debris flows.

Rock falls interact with debris flows in several ways, of which the most important is through the deposition of "new" material down the slopes, where it is available for subsequent debris flows. The rock fall taluses vary in grain size composition depending on both bedrock type, distance from the source and subsequent weathering conditions of the talus itself. Sediment grain size is generally smaller higher up on a talus, and it is therefore often here the debris flows initiate because of the water-retaining ability of finer sediments. Smaller grain sizes can also be a secondary development due to in situ weathering of the talus material, to the degree where it retains enough water to release open-slope debris flows (type A1 or A2). In Fig. 10.9, the process interaction is demonstrated as fine-grained talus is reused by repeated channel forming debris flows. In Fig. 10.15, large rock-fall boulders are dispersed around and below a low-viscosity debris flow deposit.

Snow avalanches are also closely linked to debris flow activity and morphology. Dry snow avalanches contribute to

Fig. 10.16 Photo showing a multi-process fan in Erdalen, Stryn, western Norway. This fan is smoothed by snow avalanche deposits and erosion. Some relatively recent debris flow events have deposited thin convex channels and lobes towards the camera position, which can be observed as pebbly unvegetated patches of slightly convex ground. The long run-out, irregular morphology without clear levées and grain-size sorting hints as slush avalanches as initiating process for the low-viscosity flows. Photograph Lena Rubensdotter, NGU

bringing debris onto the slope, similarly to rock fall (Figs. 10.15 and 10.16). Repeated avalanching also serves to remobilize and deform debris flow deposits and morphology and bring material further out from the slope. If dry snow avalanches are active on a fan this often results in just the most recent debris flows being distinguishable in the fan surface morphology.

Wet snow avalanches and slush avalanches (oversaturated snow that collapses and goes over into turbulent flow) often transforms into debris flows if they are erosive enough to start to incorporate solid debris such as rocks, sediments and vegetation, into the flow. Slush avalanches almost always undergo this transition into debris flows, and it is difficult to distinguish any specific traits in the debris flows starting that way. From a morphological point of view the high content of slush and water in the flow will increase the tendency of fluvial-type grain-size sorting, and the resulting debris flow deposits might sometimes seem almost depleted of finer grain sizes (silt and clay, see Figs. 10.14, 10.15 and 10.16). Slush-avalanche initiated debris flows may, if not erosive in the lower parts, also deposit the flow mass on top of a snow cover in the valley bottom. When the snow under and inside the primary deposit melts, the resulting debris flow tracks and lobes may transition into a weakly convex morphology of very low ridges "overprinted" on the underlying landforms (Fig. 10.16).

10.10 Summary

Debris flows are important geomorphological agents in most parts of the alpine landscapes of Norway. They are primarily precipitation driven gravitational slope processes and the resulting morphology can be found in both small scale, formed by single events, and as more complex landform systems covering much of the lower slopes in some valleys. The alpine topography and mostly maritime climate in Norway, together with extensive but varying sediment cover of glacial till, in situ weathering material, rock fall talus and more, sets the conditions for a range of different debris flow process types and the resulting erosional and depositional landforms.

Using morphology as a diagnostic one may subdivide and describe the characteristics of two main types of debris flows in Norway: *A. Open-slope debris flows* and *B. Fluvial channel-dependent debris flows*. Type A can be further subdivided into: *A1, channel forming debris flows*; *A2, widening debris flows* and *A3, high-viscosity debris flows*, and the major morphological characteristics of these types has been described and illustrated. In addition, two larger landform types dependent on debris flow activity and common in many parts of the Norwegian alpine landscape have been described: *C. Debris flow fan systems* and *D. Multi-process debris flow fans*.

References

Ballantyne CK (2002) A general model of paraglacial landscape response. Holocene 12(3):371–376

Ballantyne CK, Benn DI (2004) Paraglacial slope adjustment and resedimentation following recent glacier retreat, Fåbergstølsdalen, Norway. Arct Alp Res 26(3):255–269

Blikra LH, Nemec W (1998) Post-glacial colluvium in western Norway depositional processes, facies and palaeoclimatic record. Sedimentology 45:909–959

Davis PT, Briner JP, Coulthard RD, Finkel RC, Miller GH (2006) Preservation of Arctic landscapes overridden by cold-based ice sheets. Quatern Res 65:156–163

Hungr O, Evans SG, Bovis MJ, Hutchingson JN (2001) A review of the classification of landslides of the flow type. Environ Eng Geosci 7 (3):221–238

Jansen E, Andersson-Dahl C, Moros M, Nisancioglu K, Nyland B, Telford R (2009) The early to mid-Holocene thermal optimum in the North Atlantic. In: Natural climate variability and global warming: a holocene perspective, pp 123–137

Kleman J (1994) Preservation of landforms under ice sheets and ice caps. Geomorphology 9(1):19–32

Laute K, Beylich AA (2012) Influences of the little ice age glacier advance on hillslope morphometry and development in paraglacial valley systems around the Jostedalsbreen ice cap in Western Norway. Geomorphology 167–168:51–69

Nettleton IM, Martin S, Hencher S, Moore R (2005) Debris flow types as mechanisms. In: Winter MG, Mcgregor F, Shackman L (eds) Scottish road network landslide study. The Scottish Executive, Edinburgh, pp 45–67

Sandøy G, Rubensdotter L, Devoli G (2017) Trekantformede jordskred —studie av fem skredhendelser i Norge. Geological Survey of Norway (NGU) Report no 2017.017

Seppälä M (ed) (2005) The physical geography of fennoscandia. Oxford University Press, Oxford

Sletten K, Blikra LH (2007) Holocene colluvial (debris-flow and water-flow) processes in eastern Norway: stratigraphy, chronology and palaeoenvironmental implications. J Quat Sci 22:619–635

Schweinsberg AD, Briner JP, Miller GH, Bennike O, Thomas EK (2017) Local glaciation in West Greenland linked to North Atlantic Ocean circulation during the Holocene. Geology 45(3):195–198

Part III

The Status and Value of Geomorphological Heritage in Norway

Landforms and Geomorphosite Designation on Mount Gausta (Telemark)

11

Dominique Sellier and Riwan Kerguillec

Abstract

Mount Gausta is the highest point in Telemark. It stands above a vast plateau lying at 1000 m or so in altitude (a fjell), where the large Vestfjorddalen valley is entrenched. There are two parts to it, with a pyramid peaking at 1881 m to the north (Gaustatoppen), and a long plateau at between 1500 and 1600 m (Gaustaråen). The relief is characteristic of north European quartzite mountains because of the large Richter and scree slopes around Gaustatoppen, together with a large blockfield (or felsenmeer) on the Gaustaråen plateau. Thanks to the altitudes and absence of glaciers, it is possible to observe the vertical zonation of periglacial features and glacial relict landforms over nearly 900 m in elevation. The properties of the felsenmeer help determine the altitude of the Weichselian trimline and confirm there was a paleonunatak where Gaustatoppen stands. In addition to presenting the main geomorphological properties of Mount Gausta, the purpose of this chapter is to select geomorphosites, that is say sites of specific interest for visitors, using a deductive method based on defining the major geomorphological components (fjell, Gaustaråen, Gaustatoppen, Vestfjorddalen), and then the types of relief (geomorphotypes) and most significant sites for these reliefs (geomorphosites).

Keywords

Telemark • Mount Gausta • Quartzite • Monadnock • Blockfield • Geomorphosites

11.1 Introduction

Telemark lies mainly south of the 60th parallel. Its relief, typical of southern Norway, is defined by three essential characteristics: the surface of the fjell, with mountain ranges above, and large valleys below (Peulvast 1985, 1987). The term fjell designates large plateaus at an altitude of around 1000 m (in the strict meaning of the term), though it is also sometimes used to refer to uplands composed of these plateaus together with the mountainous relief above (fjell in the broad meaning of the term). The mountainous relief in the form of freestanding chains such as Gausta in Telemark, or large chains elsewhere in Norway (such as Snöhetta, Rondane, and Jotunheimen), rises to between 1,800 and 2,400 m in altitude. Large valleys of glacial origin have been carved out several hundred meters below the fjell (Vestfjorddalen in Telemark, and Gudbrandsdalen in Oppland).

High plateaus are thus a fundamental feature of the Telemark landscape (fjells in the strict meaning of the term). These plateaus, at an altitude of between 1,000 m and 1,200 m, are incised by deep glacial U-shaped valleys, such as Heddal valley to the north of Lifjell, Lake Tinsjö valley to the west of Blefjell, and especially Vestfjorddalen valley to the north of the Gausta range. Freestanding mountains rise above them. Mount Gausta is the highest of these mountains. Its peak, Gaustatoppen (59°51′20″N, 8°39′00″E), stands at 1,881 m, 120 km away from the sea. It is thus typical of quartzite relief in Scandinavia.

This chapter takes a heritage approach. Its purpose is to present the geomorphological properties of Mount Gausta and its surrounding area in terms accessible to a visiting public, using a popularization method based on analyzing the relief features on several different scales. Its plan ties in with this method.

Popularizing geomorphology entails presenting noteworthy relief features, termed geomorphosites. *Geomorphosites*, initially defined by Panizza (2001) and Reynard (2009a), are relief features which, due to their scientific and

D. Sellier (✉) · R. Kerguillec
University of Nantes, Campus Tertre, IGARUN,
chemin de la censive du tertre, BP 81 227,
44 312 Nantes Cedex 3, France
e-mail: dominique.sellier@univ-nantes.fr

A. A. Beylich (ed.), *Landscapes and Landforms of Norway*, World Geomorphological Landscapes,
https://doi.org/10.1007/978-3-030-52563-7_11

educational interest, merit commentary addressed to a general public and to specialists. *Geomorphosites* are thus objects for observation that may be seen from appropriate viewing points, here referred to as *observation stations* (viewing points and belvederes lying at varying distances from the *geomorphosites* under observation).

Popularization thus entails choosing the sites as a first step. Several non-exclusive methods may be used, particularly a *selective method* in which sites of geomorphological interest are inventoried and assessed using pre-established scientific and educational criteria, and a *deductive method* leading to the selection of *geomorphosites* based on analysis of the relief at various scales, the method adopted here. It comprises several stages.

- The first is to recognize the main geomorphological properties of the space being popularized (the study area), in this instance Mount Gausta and its environs.
- Then identifying the major geomorphological components, which entail subdividing the study area into several subsets of comparable dimension and complementary properties following standard geomorphological criteria (topographic, structural, hydrographic, and paleogeographic factors).
- Defining the elementary relief features, referred to here as *geomorphotypes*, in terms of their fundamental components, based on their morphostructural properties and their morphological and dynamic relations.
- Selecting geomorphosites to represent each *geomorphotype*, in accordance with scientific criteria (representativity, educational interest, and so on). G*eomorphosites* are objects of observation (places observed) whose scale is such that the sites may be photographed.
- Determining the *observation stations* providing a view of each *geomorphosite*, taking into account tourism criteria (accessibility and readability). In addition to individual observation points, the stations may take the form of itineraries along a slope, crest line, or valley. They act as privileged places for imparting knowledge about the geomorphosite, and on occasions about where the station lies via subsidiary observation of minor relief features (such as the top of a slope, outcrop, section, and the vertical zonation of landforms and relict landforms).

Each *geomorphosite* thus belongs to a series of taxons of comparable scientific and educational interest. The series of *geomorphosites* issue from a joint process of multiscalar analysis of the relief and of deductive selection, hence from integrated analysis. This method has already been applied to Norway (Kerguillec and Sellier 2015) after being tested several times in France (Sellier 2010, 2013). It may be applied to Mount Gausta to present its relief features and select a series of significant *geomorphosites*.

11.2 The General Geomorphological Properties of the Study Area

The study area lies in the north of Telemark. It covers Mount Gausta and its environs extending to Blefjell to the east and Lifjell to the south (Fig. 11.1). This area presents topographical contrasts not found anywhere else in central or southern Norway.

11.2.1 A Freestanding High Mountain in Southern Norway

11.2.1.1 Relief Culminating in a Pyramid

Mount Gausta stands out in the landscape due to its size and shape. It is used as a landmark throughout southern Norway. Lying 100 km south of Hallingskarvet, the nearest landform of similar altitude, Mount Gausta is the highest point in the territory between the 60°30′N parallel and the southern coastline of Norway (58°N), where the highest peaks do not exceed 1400 m to 1700 m. It is made up of a chain with parallel flanks running north–south-east, 10 km long by 3 km wide. Its pyramidal landform, in its northerly part (Gaustatoppen) and surrounded by large rectilinear slopes, rises to nearly 1900 m above the Vestfjorddalen valley to the north over 1600 m below (Fig. 11.2).

11.2.1.2 An *Open Mountain*

Mount Gausta is what is called an *open mountain*, a term corresponding to large freestanding chains dominating the surrounding surfaces lying all around them, as opposed to *closed mountains* which are ranges dug out by cirques and vallies (such as Jotunheimen, Dovre, and the Rondane mountains,) (Sellier 2002). The distinction between open and closed mountains is essential. It helps determine the presence and form of the glacial relief features (particularly cirques), and consequently the type of slopes. The slopes of open mountains such as Gausta have only undergone the effects of ice-sheet glaciation, and thus differ from the slopes of closed mountains which additionally have cirque slopes.

11.2.1.3 Vertically Zoned Relief

The relief of Mount Gausta and its region is typical of the tripartite topographical configuration set out in the introduction (being made up of mountains, plateau, and valleys) (Fig. 11.3).

Mount Gausta includes the Gaustatoppen pyramid on its northern edge (1881 m) together with an elongated plateau, Gaustaråen (1621 m on its north-west tip, 1506 m on its south-east tip) (Fig. 11.4). Gaustaråen plateau may be the vestige of a planation level that also formed some of the adjacent relief and, further afield, the table summits of

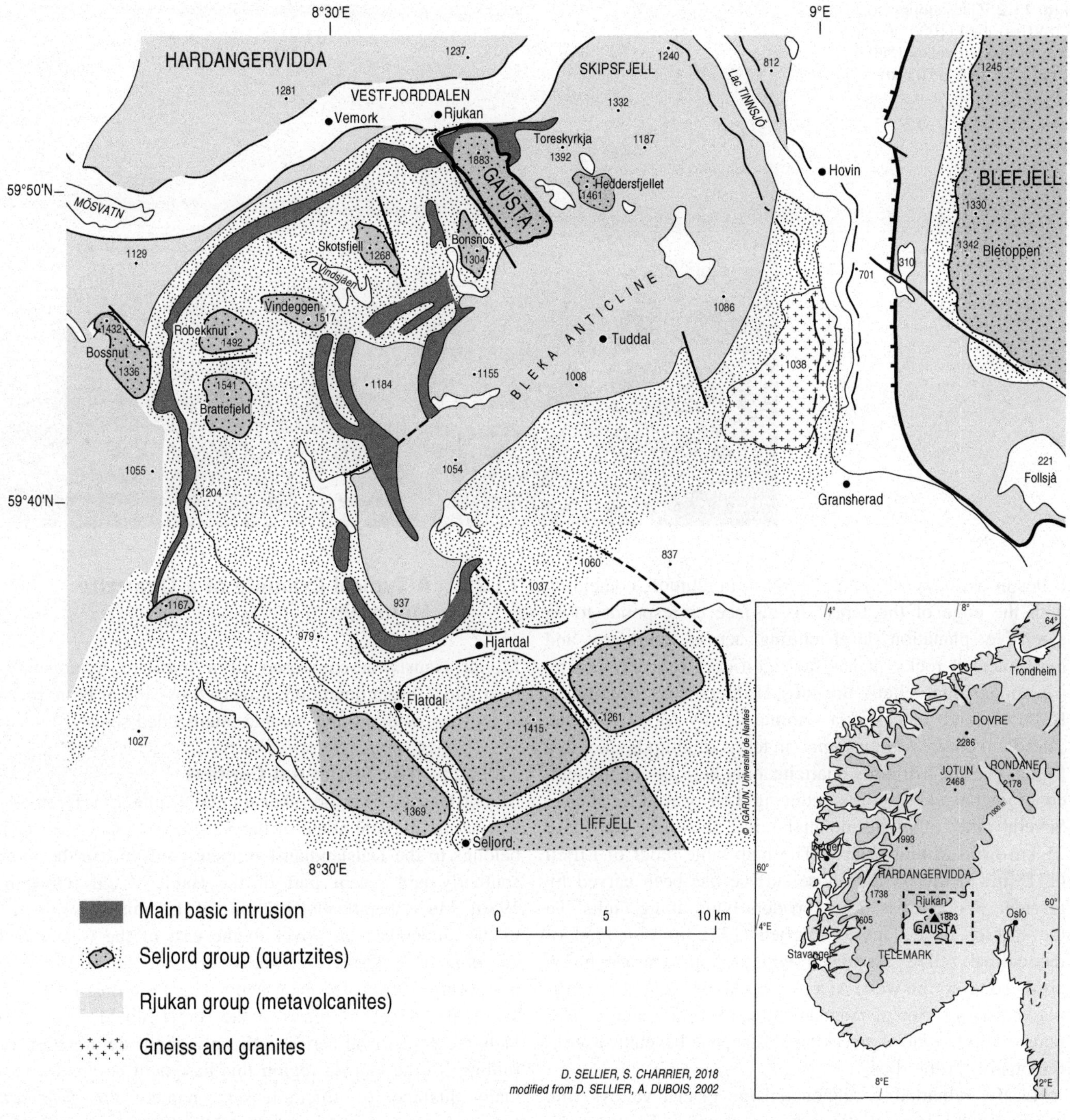

Fig. 11.1 Location of Mount Gausta (Telemark)

Blefjell (1342 m) and Lifjell (1415 m). It is transversely and longitudinally incurved. At its southerly end are two mounds, Lille Gaustakne (1443 m) and Store Gaustakne (1523 m), so named because of their knee shape. The plateau contains one of the largest blockfields in Scandinavia. Gaustatoppen and Gaustaråen are separated by a col (1580 m). Around lie large rectilinear slopes of periglacial origin such as those surrounding Gaustatoppen (Richter slopes), or concave slopes of primarily glacial origin such as those around Gaustaråen. There are two large glacial transfluence ways on either side of the Gausta chain—Gausdalen to the west and Heddersvatn to the east—lying along parallel fault lines with strings of lakes.

The fjell beneath is the extension of the vast Hardangervidda platform, dipping toward southern Norway (Gjessing 1967; Peulvast 1978; Klemsdal and Sjulsen 1988).

Fig. 11.2 Gaustatopppen (1881 m), fjell (1000 m) and Vestfjorddalen valley (300 m) from Skipsfjell (1103 m)

It lies at between 950 m and 1200 m in altitude, coinciding with the edge of the forest. Its surface corresponds to an extensive planation level cutting across quartzites and metavolcanic rocks. It has fault corridors re-excavated by ice, and residual relief—Bonsnos, Heddersfjellet, Bossnuten, Brattfjell, and Toreskyrkja—dominated by Mount Gausta. These *intermediary mountains* (in terms of altitude) are, like Mount Gausta, primarily quartzite, only here associated with conglomerates and schists, reducing their overall resistance. Several other quartzite mountains rise above the fjell over 25 km from Mount Gausta. The high table relief of Lifjell (1413 m) to the south, whose surface has been carved by glaciers, is crisscrossed by corridors lying along faults. To the east of Lake Tinnsjö, Blefjell (1342 m) is a kind of monoclinal relief, associated with a fault scarp running north–south to the west. At a more detailed scale, the fjell is marked by a series of mounds and lakes. The outcrops display multiple glacial traces and disappear beneath discontinuous till in the dips.

Lastly, remarkably calibrated large glacial valleys sink nearly 1000 m beneath the surface of the fjell. To the north, Gaustatoppen stands above Vestfjorddalen, the northern limit of the Hardangervidda plateau and one of the most emblematic examples of a large glacial valley in Scandinavia due to its size and steep slopes (Dons 1960a, 1961, 1972; Gjessing 1966, 1987). This impressive place is home to the town of Rjukan, and was the site of the "heavy water battle" about the Vemork hydro-electric power plant.

11.2.2 A Typical Example of a Quartzite Mountain

Mount Gausta is composed of a mass of homogenous quartzite (Gausta quartzite), traversed by sills of basic rocks at its base, without any notable interbedded schists.

11.2.2.1 The Structural Context

Mount Gausta, like the other quartzite reliefs in Telemark, lies on the internal slope of the Scandes Mountains. It also belongs to the Dalslandian basement, and thus to the most southerly and recent part of the Baltic shield (Oftedahl 1980). More specifically, it is one of the units incorporated in the autochthonous cover to the east of the Caledonian orogenic belt. These units thus form the foreland to the periphery of the Caledonian nappes, and are thus related to the "Caledonianized" parts of the Baltic Shield. They also include supracrustal series with only slight deformation or folding. In the Gausta region this basement comprises two major juxtaposed structural sets, namely the *Telemark Supracrustal Suite* to the north and the *Southwestern Gneiss Region* to the south (Verschure et al. 1990).

The plateaus and mountains in north Telemark—Gausta, Blefjell, Lifjell, and their neighbors—are composed of thick formations of Precambrian quartzites lying over the Baltic Shield, which have been related to Jotnian sediments. These supracrustal formations are among the oldest in Fennoscandia (between 1500 and 1100 million years old).

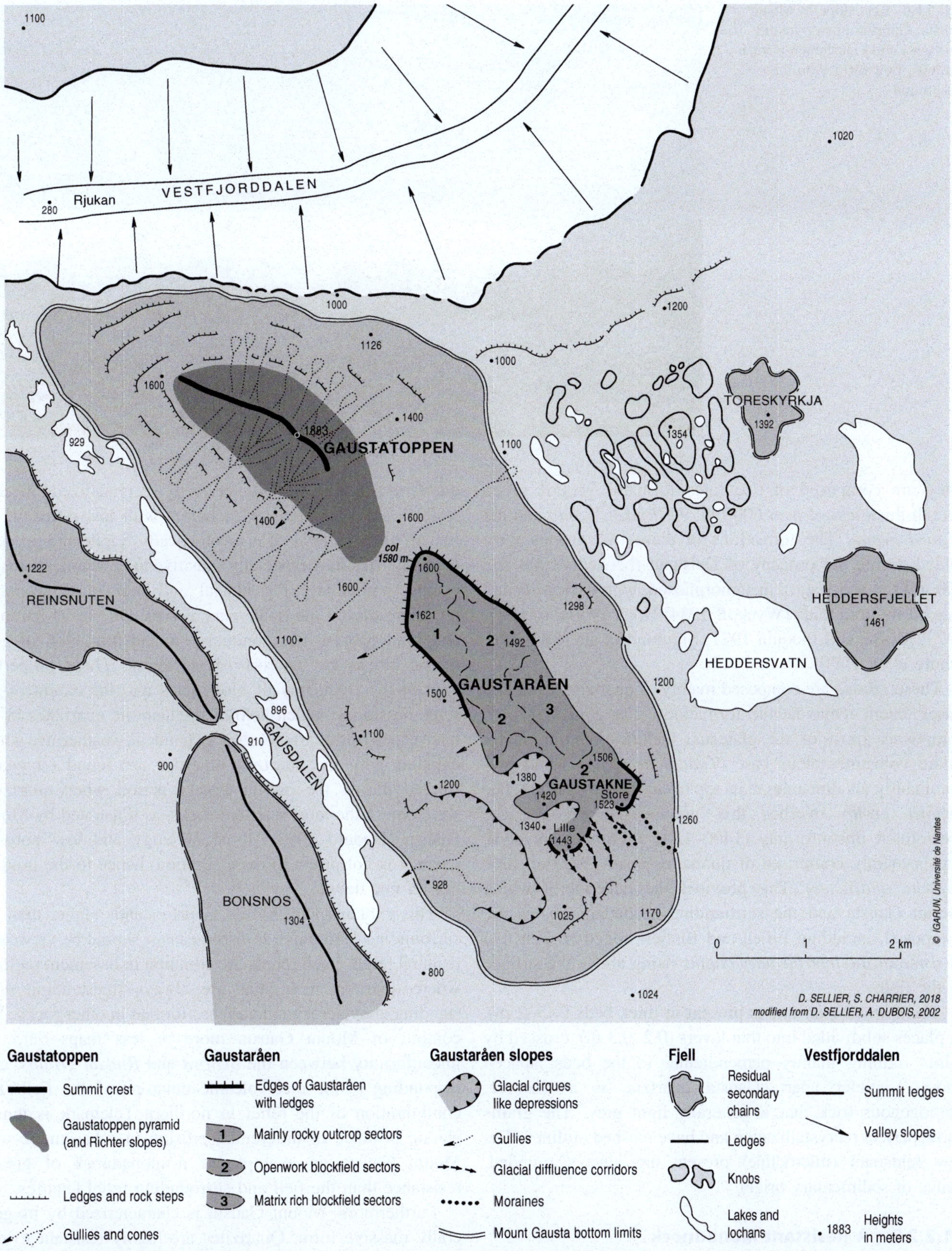

Fig. 11.3 Geomorphic map of Mount Gausta

Fig. 11.4 East slope of Mount Gausta: Gaustatoppen pyramid (on right) and Gaustaråen plateau (on left), fjell surface on the foreground

They are composed of two unconformable groups of an overall thickness of over 6000 m: the *Rjukan group* and the *Seljord group*. The formations of these two groups were folded during the orogeny of Dalsland (between 1000 and 900 million years ago), metamorphosed, with basic sills and pierced by granitoids (Wyckoff 1934; Dons 1960a, b, 1961, 1972; Moine and Ploquin 1971; Sigmond et al. 1984; Verschure et al. 1990).

These groups are composed mainly of quartzites, and to a lesser extent metavolcanic formations. The *Rjukan group,* composing most of the plateaus, includes predominantly acidic volcanites at its base (*Tuddal Formation*) and predominantly alkaline ones at its top (*Vemork Formation*). The *Seljord group* overlies this previous group via the pre-Jotnian unconformity (1400–1300 million years ago), and is mainly comprised of quartzites about 2000 m thick (*Gausta quartzites*). This provides the material of which Mount Gausta and the surrounding secondary chains are composed, including Lifjell and Blefjell, together with that of some of the *intermediary chains* rising above the surface of the fjell.

The *Gausta quartzite* is present in thick beds (0.5–2 m), in places subdivided into thin layers (0.2–0.3 m), crossed by joints running mainly perpendicular to the beds, spaced about a meter apart. Gausta quartzite is a massive, homogenous rock that is generally light grey. The grains show perfect recrystallization and have meshed outlines. The few feldspars (microcline) present are always rounded, hence of sedimentary origin.

11.2.2.2 A Resistant Monadnock

The properties of quartzite relief are in all cases wholly conditioned by the relationship between two factors: quartzite is a homogenous, perfectly recrystallized siliceous rock, hence very hard and coherent with low porosity, yet like all hard materials it is mechanically fragile and prone to fissuring. It is thus especially sensitive to mechanical erosion agents, particularly the glacial and periglacial processes prevalent during the quaternary period. Due to its hardness and joint pattern it is conducive to the formation of geometric blocks by macro-frost shattering. These properties explain the frequency of blockfields and the extent of the scree on slopes, which help distinguish the quartzites in the landscape. But quartzite resists chemical weathering, which explains why the quartzite outcrops are found on peaks, formed during the pre-quaternary period when quartzites were one of the most resistant rocks, as illustrated by Mount Gausta (Godard 1993). Its mineralogy and low porosity mean it is not prone to disintegration, hence to the production of fine debris.

This explains why Mount Gausta stands higher than the surrounding landscape. It also means it should be viewed as residual relief. Such reliefs are common in basement regions, where quartzite mountains are always freestanding units standing above erosional surfaces formed in other rocks. The contour of Mount Gausta more or less maps onto the unconformity between the *Seljord* and *Rjukan groups*, corresponding to the metavolcanic outcrops on the fjell. The configuration of the relief in northern Telemark is thus a classic instance of differential erosion in a basement, with Mount Gausta an example of a monadnock of greater resistance than the fjell and surrounding relief features.

Furthermore, Mount Gausta is characterized by its generally massive form. Quartzites are highly resistant to dissection (Godard 1965). The densely present joints facilitate surface water infiltration and impede the concentration of water on the surface. The gullies are shallow and spaced out with little branching, except around the peaks, such as Gaustatoppen where they help form an alpine landscape. The quartzite slopes, however, present remarkable lateral continuity. Furthermore, there are few or no cirques along *open mountains* as their altitude and configuration prevent the formation of the appropriate types of glaciers.

11.2.2.3 The Vestiges of Appalachian-Type Relief

Mount Gausta and the neighboring chains present syncline flexures together with tilted fault blocks. These mountains, which are currently in a high position, are thus inverse relief, comparable to Appalachian-type reliefs. There are two categories of deformation. The first concerns major folds running north-east–south-west, centered on the Bleka anticline, which is flanked by two syncline axes containing *Seljord quartzites*. Mount Gausta belongs to the more northerly of the two. The second category of subaltern deformations runs north-west–south-east, taking the form of transversal faults and flexures.

Mount Gausta displays this double series of deformations. The quartzite strata are folded longitudinally (north-east–south-west). They also reveal a synclinal trough along the Gaustatoppen–Gaustaråen axis (north-west–south-east). As a result, the Gaustaråen plateau takes the form of an upstanding syncline. This overall framework conditions the forms of its bordering slopes, which all run counter to the dip. The other reliefs associated with the Heddersfjellet–Brattfjell axis correspond to monoclinal blocks bordered by transversal faults. This results in dissymmetrical chains and pyramids, such as Bonsnos and Bossnuten.

11.2.3 A Periglacial Mountain

11.2.3.1 Vegetation Tiering

One of the most obvious facts is the contrast between the forest covering the slopes of Vestfjorddalen and the zones above covering the fjell and slopes of Mount Gausta (Fig. 11.5). To the north of Mount Gausta, the edge of the

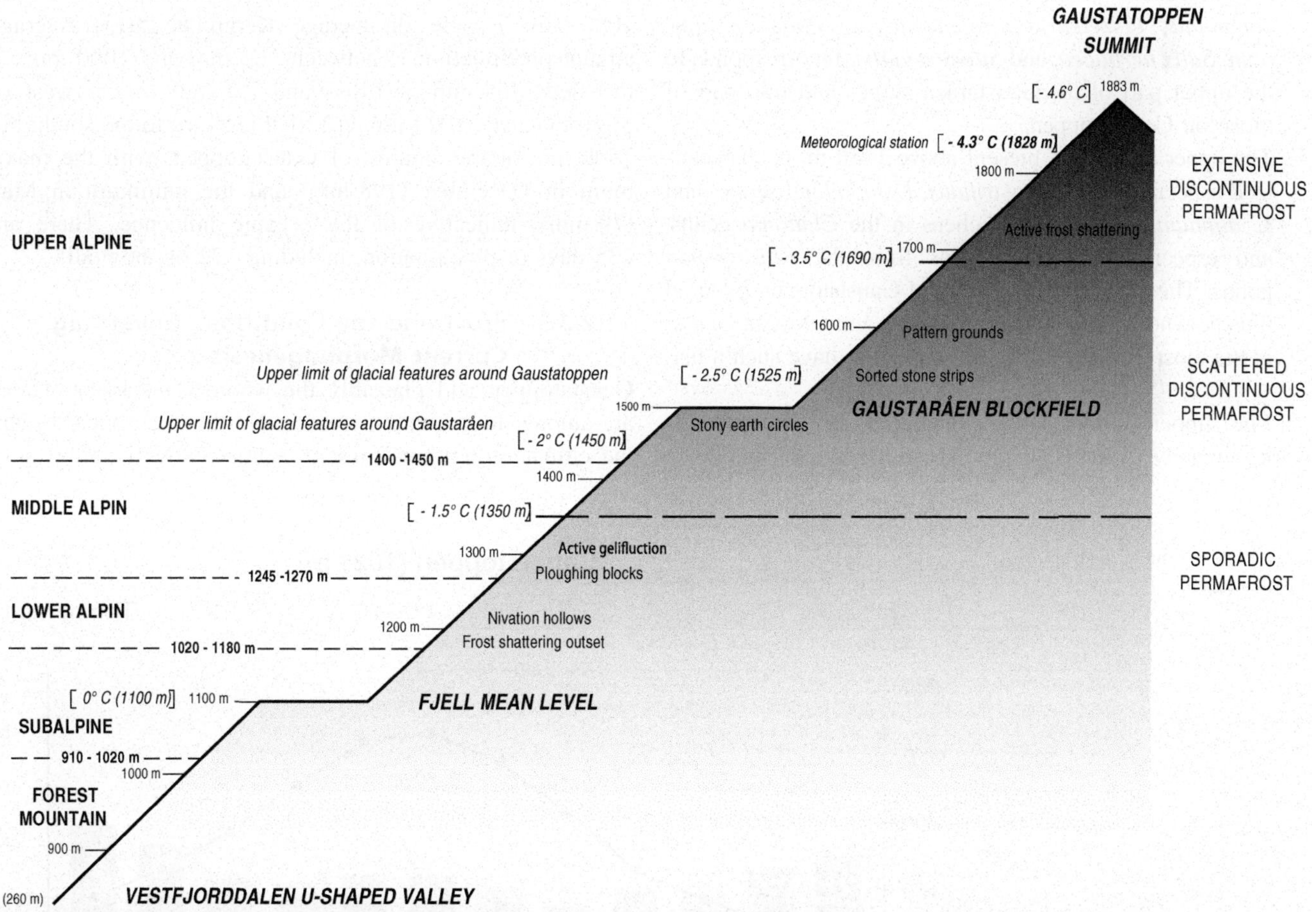

Fig. 11.5 Biogeographical and morphoclimatological tiers on the slopes of Mount Gausta. –Altitudes of Mean Annual Temperatures in °C, extrapolated from records of Gaustatoppen meteorological station, 1828 m (Norsk Meteorologisk Institutt). –Permafrost distribution, from L. King zonation (1984). –Vegetation tiers and lower limits of active periglacial features, from field surveys

forest (of Scotch pine, spruce, birch, and rowan) coincides nearly everywhere with the edge of the fjell (910–1020 m). Above this, and up to 1180 m, thus across much of the fjell, is a subalpine zone as defined for Scandinavia (Moen 1987), corresponding to a mixed forest—high tundra ecotone dotted with atrophied trees and bushes (willow, twisted birch, juniper, and Scotch pine).

Moving up the slopes on Mount Gausta we may detect three tiers.

- The lower alpine tier (1020–1270 m) of high altitude tundra, characterized primarily by *Alchemilla alpina, Betula nana, Carex* sp., *Empetrum hermaphroditum, Eriophorum scheuchzeri, Juncus trifidus, Loiseleuria procumbens, Lycopodium selago, Phyllodoce caerulea, Rubus chamaemorus, Salix* sp., *Vaccinium myrtillus, V. uliginosum,* and *V. vitis-idaea.*
- The mid-alpine tier (1270–1450 m) with the disappearance of *Betula nana* and fewer *Empetrum* and *Vaccinium.* It is characterized primarily by *Arctostaphylus alpinus, Carex bigelowii, Cassiope hypnoides, Cryptogramma crispa, Deschampsia flexuosa, Juncus trifidus, Juniperus communis, Loiseleuria procumbens, Lycopodium alpinum, Salix herbacea,* and *Silene acaulis.* It corresponds to the upper part of the Gaustaråen slopes and mid part of those on Gaustatoppen.
- The upper alpine tier, present above 1450 m, is characterized especially by *Juncus trifidus, Lycopodium selago* and *L. alpinum,* with mosses, lichens in the *Cladonia* genus, and especially crustose lichens from the *Rhizocarpon* genus. The Gaustaråen plateau and Gaustatoppen pyramid thus lie wholly within the upper alpine zone. Mount Gausta is the most southerly relief in Norway to have such a tier.

The upper limit of the subalpine tier corresponds approximately to the 0 °C annual isotherm. This implies a short growing season, probably less than 140 days per year if defined as the number of days with an average temperature of 6 °C or more (Nordseth 1987). It also coincides with the appearance of sporadic permafrost (King 1984, 1986), thus the bottom of the operative periglacial zone. Both here and in most of southern Norway this limit corresponds to the edge of the fjell. All the mountains rising above the fjell are within the alpine tiers, with high altitude tundra giving way to rocky outcrops toward the peaks.

11.2.3.2 Climate Data

There is a meteorological station at the summit of Gaustatoppen (1828 m), where the average annual temperature is −4.3 °C, the warmest monthly average is +4.9 °C, and the coldest monthly average −11.7 °C (Fig. 11.6). The average annual range is 16.6 °C. The average is above 0 °C for just four months of the year. There are 286 days per year when the average is below 0 °C, with 109 when the average is below −10 °C (*Norsk Meteorologisk Institutt*). The 0 °C average annual isotherm has been calculated to be around 1100 m and the +10 °C summer isotherm near 1050 m, where the number of frost-thaw cycles is between 48 and 105, with 77 cycles on average (Kerguillec 2015). Average annual precipitation is between 700 mm and 1000 mm on the fjell: 734 mm at Mösstrand (948 m) to the west of Mount Gausta, 1024 mm at Lifjell (354 m) to the south, but 1509 mm at the summit of Gaustatoppen, with the maximum in December (198 mm) and the minimum in May (79 mm), indicative of the oceanic influence. There are 235 days of precipitation, including 172 of snowfall.

11.2.3.3 Frost and the Conditions Governing Current Morphogenesis

Gaustatoppen and plausibly the whole Gaustaråen plateau are among the mountain areas in Northern European where the climate conditions present characteristics nearing those

Fig. 11.6 Climatic diagram of Gaustatoppen

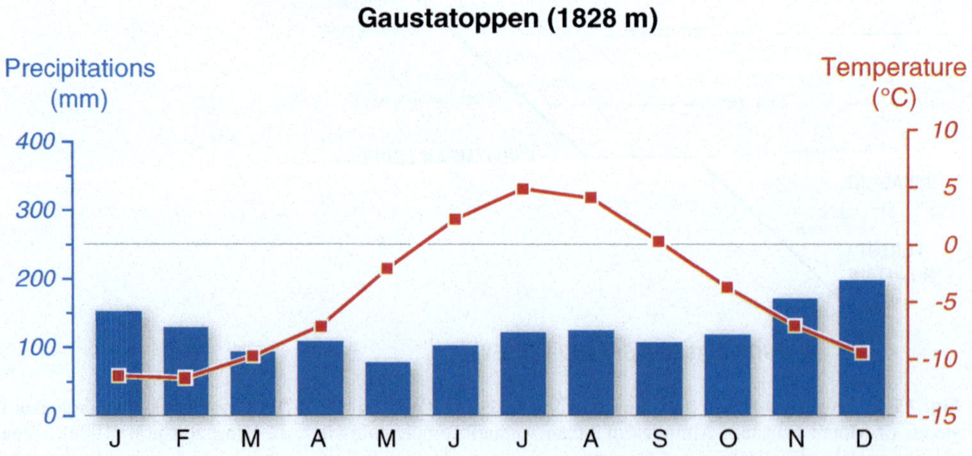

found in oceanic polar regions due to the influence of the latitude, altitude, and comparative proximity to the coast (negative average annual temperatures, four months with an average above 0 °C, summer averages below +10 °C, annual temperature range of 16 °C, and total annual precipitation of around 1500 mm).

There is no permanent snowline cutting across the topography. In principle, there is no snow on the Gaustatoppen peak in summer. Current morphogenesis is thus largely determined by the actions of freezing and meltwater.

Taking a gradient of 0.6 °C per 100 m, and working on the basis of the 0 °C average annual isotherm being near 1100 m, the annual average at the level of Gaustaråen plateau would be −2.5 °C, and −4.6 °C on the Gaustatoppen summit. All of Mount Gausta would thus lie within the active periglacial zone. Taking the thermal thresholds used to establish the limits of the various types of permafrost in Scandinavia (King 1984, 1986), the line of contact between sporadic and sparse discontinuous permafrost (MAT −1.5 °C) would be around 1350 m on Mount Gausta, the lower limit of extensive discontinuous permafrost (MAT −3.5 °C) somewhere around 1690 m, but the continuous permafrost (MAT −6 °C) altitude would be higher than the summit of Gaustatoppen. Most of Mount Gausta would thus be within the discontinuous permafrost zone, with it being sporadic on Gaustaråen plateau and extensive on the Gaustatoppen pyramid.

The effects of processes due to the cold are, however, reduced, and this for three reasons. Mount Gausta is characterized by a comparatively damp periglacial system without continuous permafrost, lying below 1900 m and in a latitude below the 60 °N parallel. The properties of the local quartzites, characteristically broken into large fragments with little fine debris, hinder geliturbation and gelifluction. The types of slopes (glacially formed and Richter slopes) limit geliturbation and rockfall, and hence scree processes.

The active periglacial phenomena are of three elementary types. The phenomena of geliturbation and gelifluction are distributed as follows. The first ploughing blocks are encountered above 1280 m, hence within the mid-Alpine zone, followed by terracettes and gelifluction lobes between 1320 and 1390 m. At 1500 m, hence within the upper alpine zone, stony earth circles and small sorted circles start to make their appearance, and the marks left by gelifraction increase in scale. Above 1520 m, sorted nets with multiple stony earth circles appear, as do sorted stone strips. Some active gelifluction lobes may be observed on the slopes of Gaustaråen above 1400 m. However, the slopes do not display any generalized gelifuction. Frost shattering is operative from the base of the slopes, hence the lower alpine zone, around particular sites such as nivation hollows, and becomes the dominant process operating above 1650 m.

11.2.3.4 The Problem of the Upper Limits of Weichselian Ice Cover

Due to its location and morphology, Mount Gausta is of particular interest for the ice cover of southern Norway. During the Weichselian, the ice divide lay where Hardangervidda plateau is, and the ice in Telemark was generally moving south-east (Vorren 1977). In the late Preboreal, the edges of the ice sheet were still near Mount Gausta (Sollid and Torp 1984; Andersen and Karlsen 1986; Lundqvist 1987). Wille (1905), Ahlmann (1919), followed by Wyckoff (1934) and Linton (1949), realized early on that Gaustatoppen had been a nunatak during the latest glaciation, due to its overall form, the profile of its slopes, and the absence of any glacial traces at altitude.

The zonation of the glacial and periglacial relict landforms along the slopes around Mount provides a first set of indications about the upper limits to ice coverage in the region. These slopes still have a lower section with glacial relief features and an upper section with periglacial ones, as often in places once covered in ice.

The slopes rising about 500 m around Mount Gaustaråen are characterized by having been fashioned primarily by glacial action, transpiring in the concave profile slopes, ice-molded rocks, crescentic gouges or striations, and sporadic morainic deposits up to 1400 m in altitude. The periglacial relief of the upper sections has been caused by the frost shattering of free faces, resulting in talus slopes and scree cones spilling down as far as the lower sections. The slopes of Mount Gaustatoppen are over 800 m in height, and include a lower section whose glacial forms are broadly comparable to the previous ones, and an upper section with alpine-type periglacial relief at altitudes above 1500 m on the north-east and above 1350 m on the south-east.

Gaustaråen plateau is 3.5 km long and 2 km wide, rising along its edges (to 1621 m to the north and 1523 m to the south). Its expense is mainly flat, and provides a second series of indications about the extent of ice cover and the existence of a nunatak where Mount Gausta stands given that its altitudes are close to the abovementioned limits. The plateau is covered by a block formation, which in terms of its extent, location, dimensions, angularity, and interlocking of its material meets the definition of a blockfield (Rudberg 1977; Washburn 1979).

11.3 The Major Geomorphological Components and Geomorphotypes

Mount Gausta and its environs are in fact made up of four major components: the plateau (or fjell), the Gaustaråen plateau, the Gaustoppen pyramid, and Vestfjorddalen valley. Each major component includes two geomorphotypes that

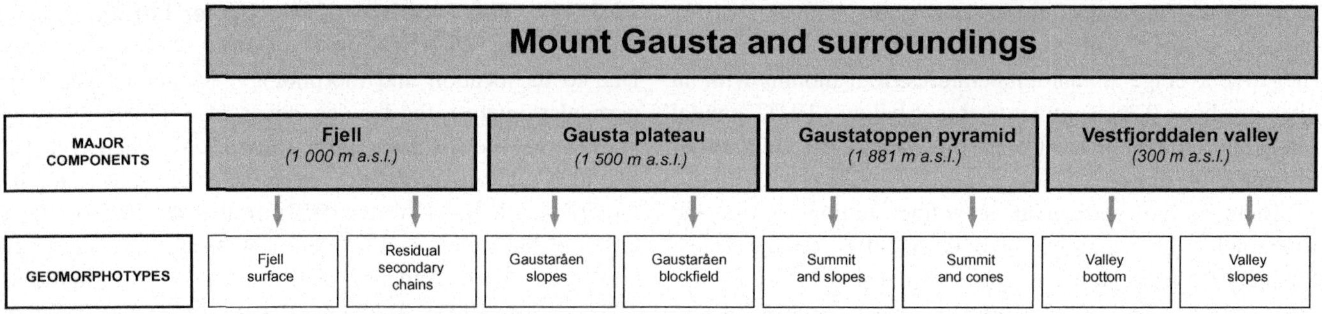

Fig. 11.7 Geomorphotypes of mount Gausta and surroundings

may be directly discerned by people visiting the landscape (Fig. 11.7).

11.3.1 The Fjell

The fjell *stricto* sensu, which is very extensive in Telemark, is in places topped by mountain chains such as Gausta. It lies at between 1050 m alongside Vestfjorddalen and 950 m to the south, where it does not rise above the tree line. Its overall topography results from pre-quaternary planation cutting across the quartzites and metavolcanites. Two geomorphotypes may be noted here.

11.3.1.1 The Surface of the Fjell

Its monotony is characteristic of basement areas subjected to the actions of ice sheets (Rudberg 1962; Peulvast 1985; Benn and Evans 2002). One of the first notable characteristics is that it is punctuated here and there by freestanding rocky mounds worn by glaciers, less than 100 or so meters in height, whose summits are at approximately the same altitude, issuing from planation levels. They are aligned along the most resistant outcrops. They reveal the repeated actions of differential glacial erosion and bear the marks of ice channeling. In places they disappear beneath morainic deposits. These alternate with lakes, some of which are interconnected, often following faults or points of lithologic contact. These lakes sometimes give way to marshes and cottongrass bogs. The landforms overall resemble the *knobs and lochans* of Scottish and Finnish platforms. The relief is in fact conditioned by a ribboned structure determined by the roots of the Dalslandian folds in the *Seljord group* quartzites to the west, and by orthogonal faults in the *Rjukan* group metavolcanics to the east.

11.3.1.2 Intermediary Residual Chains

In addition to this, the fjell is punctuated by distinct chains: Brattefjellet (1541 m), Vindegen (1517 m), Heddersfjellet (1461 m), Bossnuten (1432 m), Toreskyrkja (1392 m), Nutan (1313 m), Bonsnos (1304 m), Store Björndalsnuten (1293 m), and Diplanuten (1277 m).

The highest of these, to the west of Mount Gausta, are composed of *Seljord group* quartzite. Their shape is characterized by ridges and elongated pyramids. They rise above corridors dug into the schists or basic intrusions. They are residual relief derived from folded structures. Those to the east and south of Mount Gausta are primarily composed of metavolcanics from the *Rjukan group*. They have flat summits and massive forms with geometrical contours determined by intersecting fault lines.

These chains were entirely covered by ice sheets during the most recent glaciations, as indicated by the erratic blocks still found on the taller ones and the roches moutonnées on their slopes. There are no cirques however. Such reliefs are typical of the scouring of periglacial landforms in the basement, and are common in Scandinavia, where Rudberg (1973, 1988) refers to them as *flyggbergs* (Fig. 11.8).

11.3.2 Gaustaråen Plateau

Mount Gausta is the major landform in Telemark in terms of its size, altitude, and the expanse of its surrounding slopes. Gaustaråen plateau, corresponding to just two-thirds of the chain, and Gaustatoppen, dominating its north-east, are nevertheless two distinct entities due to their respective altitudes (1500–1600 m for the former, 1881 m for the latter), forms (the first being a plateau, the second a pyramid), and their slope systems (Fig. 11.9).

11.3.2.1 The Mainly Glacial Forms of the Gaustaråen Slopes

These slopes correspond to the edges of an upstanding syncline, hence inclined contrary to the dip. They enable analysis of the glacial landforms along the comparatively high slopes (900–1200 m at their base, and 1500–1600 m at their summit) in an area subject exclusively to ice-sheet glaciation. The slopes are characterized by the length of their

Fig. 11.8 West slope of Gaustatoppen (ravines and cones) in the foreground, Bonsnos (intermediary mountain) and fjell surface (knobs and lochans lanscape) in the background

Fig. 11.9 Gaustaråen plateau and Gaustatoppen pyramid to the north

bare rocky outcrops, the lack of incisions, and the thin and sporadic nature of deposits, within which there are morainic elements than scree. They are comprised of a long concave incline of glacial origin with a short ledge at the top of periglacial origin.

The concave incline (30° at the top, and 10–20° on the lower section) accounts for most of the overall slopes. It displays deep scaring of the quartzite beds and many signs of glacial ablation. The heterogeneous and sporadic deposits have been rearranged by avalanches and meltwater, and occupy slopes up to 1400 m in altitude. They are comprised of large blocks of sub-angular quartzite, together with erratic blocks of granite and basic rocks, and angular fragments caused by the frost shattering of the summit ledge. The latter is discontinuous and low, only a few dozen meters at most. It bears the signs of frost shattering, the effects of which amplify only above 1400 m. It is uneven, with funnels produced by frost action, and scree cones and slopes at some places beneath it. Run-off has caused debris flows. It is currently the most active process on the slopes.

The minor marks left by the ice may be divided into two categories. The first is ice-molded rocks, characterized by multiple convexities and minor cavities on the quartzite beds. The second are polished surfaces, striations, crescentic gouges, and hyperbolic crack strings, showing the direction the ice moved in, in this instance south-east. All these landforms progressively disappear with altitude, while the effects of frost shattering become increasingly present. The highest marks of glacial ablation are to be found on the edge of the plateau at around 1480 m. The upper parts of the concave inclines are thus covered in places by coarse frost-shattered blocks (between 0.30 and 2 m long), with sharp and often unstable angles, combined with what may be the result of postglacial decompression, as suggested by the fissures present on the ledge running above.

11.3.2.2 The Gaustaråen Blockfield

The Gaustaråen plateau has one of the most remarkable blockfields in northern Europe (Dahl 1955; Rudberg 1984; Jansen 1986; Nesje 1989; Nesje et al. 1990; Sellier 1995). Landforms known as *felsenmeers*, *blockfields*, or *mountain top detritus,* are characteristic features of the most tabular-shaped British and Scandinavian mountains, where their presence was noted as of the early twentieth century and attributed to the action of frost processes (Wille 1905; Högbom 1914; Ahlmann 1919). They are generally composed of coarse angular blocks, which may be openwork or else wedged in an interstitial matrix of fine elements (matrix-rich) (Ballantyne 1998). The many questions they raise relate to the respective role played by glacial and periglacial processes in their elaboration, together with how they relate to the upper levels of quaternary glaciations, and hence their interest for identifying paleonunataks. These issues pertain to their type, evolution, and age, all of which clearly vary (Fig. 11.10).

The Gaustaråen felsenmeer is made up of three features, each corresponding to a geomorphosite: sporadic block rocky outcrops, openwork blockfields, and matrix-rich blockfields.

The *rocky outcrops* are found on the highest points, lying north and south of Gaustaråen, on the edge of the blockfields. They are comprised of slabs rising at angles of 1–5° near the edges of the plateau: *attenuated slabs* at a lesser angle than the dip, and *exaggerated slabs* at a greater angle than the dip. They form steps, the backslopes of which favor the breaking off and slippage of blocks. Some bear relict or operative frost splitting marks, such as fissures several centimeters wide with angular edges (Fig. 11.11).

Disintegration has blunted the edges of most fissured slabs and resulted in the presence of *pseudo-pebbles*, rocky fragments measurable in tens of centimeters, whose exposed faces have become rounded and rough, while the faces in contact with the rocky outcrop have remained smooth and flat. The depressions formed between the attenuated slabs are covered by pebble areas of several square meters, composed of rocky fragments, pseudo-pebbles, and a sand-pebble matrix. Some slabs have wide, shallow weathering basins which are clearly operative, as shown by the absence of lichen and concentration of sand at the bottom.

In addition to this, the rocky outcrops have glacial relict forms which may also be found on the neighboring blockfields. These relict forms are three types. Crescentic gouges and hyperbolic crack strings are common. The crescent-shaped gouges may occur in isolation or lines, with their

Fig. 11.10 General view of Gaustaråen plateau from Gaustatoppen pyramid, to the south

Fig. 11.11 Quartzite rocky outcrops (attenuated slabs), north-west of Gaustaråen plateau

convexity pointing against the direction of flow, with the upstream edge being steepest, and the horns here measuring between 3 and 60 cm. They result from ice-plucking of rocky fragments due to the intermittent pressure of indenter elements (Laverdière et al. 1968). The *hyperbolic crack strings* are curved fissures, whose concavity this time points in the direction of flow, with horns of between 3 and 8 cm. They result from the intermittent pressure of an indenter without rock plucking. They occur up to the top of Gaustaråen accompanied by striations. The glacially truncated sides produced on the edge of the slabs, on the upstream side, often form a series of coalescing arcs, whose horns measure between 10 and 70 cm. Traces of polishing are more frequent, and may be observed especially near Gaustakne, where they indicate the particularly intense action of the ice.

In addition to this, blocks of varying size, sometimes 2–3 m long, are littered over the slabs. The density naturally increases in zones of passage with blockfields. Most of these blocks are autochthonous and were moved by slippage on the outcrops. Some are erratics, here scarce dispersed granite blocks smaller in size than the surrounding quartzite elements (30–80 cm), and fragments of basic rocks of even smaller size (20–30 cm). These blocks are always very rounded. There are also allochthonous quartzite elements. Some stand out for their petrography, such as the purple quartzite elements found here and there on the autochthonous grey quartzites. Others may be recognized due to the way they are lying, such as the superposed blocks and *tripod perched blocks*. The former are small blocks lying on top of larger blocks, which in turn lie on rocky slabs. The latter are on the contrary large blocks (up to 2 m long) resting on at least three blocks (of 10–20 cm), which rest in turn on slabs. All these blocks are similar to forms known elsewhere under the generic term of *perched blocks* and considered as erratics (Whittow 2000), though they are in fact a key feature for interpreting the genesis of the Gaustaråen felsenmeer. Some have traces of glacial erosion on their sides or lower face, with rounded edges, truncated sides, and multidirectional crescentic gouges and hyperbolic crack strings. The presence of such blocks (of which there are a remarkably large number to the north of the plateau and around Gaustakne) is indicative of the sub-glacial or post-glacial outwash leading to the tables being wedged where they current lie.

The openwork blockfields are the major feature on the felsenmeer. They occur on horizontal or slightly sloping areas where they entirely mask the rocky outcrops. They are composed of large interlocking blocks without any interstitial matrix on the surface, and never form sorted nets. Blocks at any given site are homometric (0.5–2 m). The largest is over 4 m long and 2.5 m wide. They are colonized by lichens, mainly in the *Rhizocarpon* genus. Many are unstable, even the largest of them (Fig. 11.12). The effects of frost have resulted in clean-edged fissures which may run through blocks several tens of centimeters thick. Disintegration has not affected the large blocks, which are still very angular, but transpires in the presence of a few weathering basins. There are common glacial marks: truncation chippings on the edges of certain blocks, hyperbolic crack strings and crescentic gouges on the surface, polishing marks, and relicts of abrasion antecedent to the dislocation of the substratum. The openwork blockfields include erratics of granite and basic rocks, and especially of dark quartzites mixed with clear

Fig. 11.12 Openwork blockfield, north-east of Gaustaråen plateau

Fig. 11.13 Tripod perched block (2 m high) on Gaustaråen plateau in an openwork blockfield sector

quartzites. They include *superposed blocks*, including *tripod perched blocks* 1–3 m long, resting on small elements lying in turn on large blocks (Fig. 11.13).

The matrix-rich blockfields are characterized by the presence of blocks of varying size (0.2–3 m) wedged against each other, with their bases surrounded by a mix of sand and small fragments. There are fewer giant blocks. Disintegrated forms are more developed. There are few notable marks of frost shattering. The glacially truncated sides, crescentic gouges, and hyperbolic crack strings present the same properties as in the openwork blockfields. Erratics of granite, basic rocks, and quartzite are mixed with autochthonous blocks of the same proportions (Fig. 11.14).

The interstitial matrix is thus the distinctive characteristic of this type of blockfield. The matrix is composed of sands and fragments of between a few centimeters and several decimeters in size. It contains few fines, due to the rarity of feldspars in the local quartzite. It is composed of periglacial and glacial fine debris, the results of micro-frost shattering, disintegration, and postglacial illuviation. All this material was buried by run-off, particularly snowmelt. This bimodal granulometric classification, due to the low proportion of

Fig. 11.14 Matrix-rich blockfield in the central part of Gaustaråen plateau, pebble area in the foreground

intermediary elements between the large blocks and their matrix, may be attributed to the dual action of macro-ice shattering and disintegration. The matrix-rich blockfields further stand out for the small number of superposed blocks and absence of tripod perched blocks, which are incompatible with this type of formation. The pebble and pseudo-pebble areas are numerous, and accompanied by patches of fruticose lichen, clumps of *Juncus trifidus*, and cushions of *Loiseleuria procumbens*.

The texture of the matrix-rich blockfields has enabled sorted circles to develop, with cells measuring between 0.5 and 2 m in diameter. These are relicts. There are, however, active stony earth circles, together with active *sinking* and *flowing blocks* of between 0.5 and 1.8 m long. The stony earth circles are between 10 and 30 cm in diameter, and formed of fines and mobile material, without any form of plant colonization. The *sinking blocks* result from sinking in the active layer. They are surrounded by rims of fines composed of sand and pebbles colonized by *Juncus trifidus*, *Lycopodium selago*, and *L. alpinum*. The *flowing blocks* are associated with pebble areas at an angle of 2–3°. Their upper ends have sunk under the effect of minor movements, while there are rims around their lower ends.

In conclusion, the rocky outcrops are found at the highest and most sloping places, the openwork blockfields tend to occur in sectors with little incline, while the matrix-rich blockfields occupy flatter and lower places in the center of the plateau. The complementary distribution of these three types of landscape is due to the topography and site conditions. It is also informative about their origins.

The three things we can learn from the Gaustaråen felsenmeer

First it tells us about the glaciation of Gaustaråen during the quaternary due to the presence of several relict landforms, such as the glacially truncated sides, crescentic gouges, hyperbolic crack strings, and erratics, together with the diffluence corridors and morainic ridges present to the south of the plateau beneath Gaustakne (Fig. 11.15). The distribution of these relict landforms shows that the whole plateau was covered in ice up to an altitude of at least 1620 m.

Some of these relict landforms do not have any precise chronological signification and may be the result of glaciations prior to the Weichselian, as are the granite and basic rocks. The dispersal across the plateau and on its slopes does however indicate that Gaustaråen as a whole once lay under ice, and the glacial dynamic was capable of lifting granite fragments over an elevation of nearly 300 m in the event of there being no granite outcrops further to the known sites beneath the plateau.

Yet certain phenomena suggest the Gaustaråen plateau was completely ice covered during the latest glaciation. This is suggested by the fresh glacial marks on the blocks and rocky outcrops, such as the crescentic gouges and hyperbolic crack strings, even though quartzites retain this type of mark better than other rocks do at a given altitude or site. In conjunction with this, there are very few forms of weathering here, often considered as indicative of the age of a felsenmeer (Dahl 1966a, b; Ives 1958, 1966; Nesje 1989; Nesje et al. 1988). The effects of apparently recent disintegration are limited to pseudo-pebbles and round-edged slabs.

Fig. 11.15 Glacially truncated side produced on the edge of a quartzite block in the central part of Gaustaråen plateau

The large blocks still have angular shapes, smooth surfaces, and sharp edges. The presence of superposed blocks and especially of tripod blocks still on their perches despite the many risks of destabilization is especially revelatory of the recent glaciation of the plateau. The marks left by glaciation on Gaustaråen result from the advance of ice toward the south-south-east, which fits in with the general movement of ice in Telemark during the latest glaciation (Vorren 1977; Sollid and Torp 1984). *Hence the preceding observations indicate that Gaustaråen was entirely covered by ice during the Weichselian, but that it may retain traces of earlier glaciations, to which may be attributed some of the erratics and hollow terrain (vestiges of border cirques and diffluence corridors).*

The properties of the felsenmeer also indicate that glacial erosion played a direct part in its genesis, at different places and different times. The structural slabs are dislocated by large fissures without any apparent vertical displacement. Some straight or broken slabs between 3 and 15 m long, with openings between 20 and 60 cm wide, often perpendicular to the direction of the ice, are particularly indicative of plucking. The largest of them, partially encumbered with angular debris, are over 3 m deep and several beds thick. In places their sides have been glacially truncated and have hyperbolic crack strings, suggesting that sub-glacial plucking was responsible for the intermittent dislocation of the rocky substratum and for carrying blocks to the felsenmeer.

The quartzite erratics show that the action of the ice was accompanied by the transport of material. The superposed blocks and tripod blocks indicate displacement and that there was sub-glacial or postglacial outwash of a sand and pebble matrix still present at their base in places. Nevertheless, there are no real moraines on the surface of the plateau, and *it would be mistaken to view the Gaustaråen felsenmeer as just an outwashed morainic mantle rearranged by periglacial processes.*

Many of the marks associated with the blocks are indicative of pressure and crushing. Certain striations and series of hyperbolic crack strings run in multiple directions on a given block, or else are curved, correlating to the rotation of the block in question. The preponderance of non-continuous marks (crescentic gouges and hyperbolic crack strings), the scarcity of polished surfaces, and the mainly angular contours of the blocks similarly suggest the influence of transport over very short distances. Some of the forms observed, such as the erratics and tripod perched blocks, further prove that the *current configuration of the felsenmeer does not result simply from the destruction by postglacial periglacial processes of a rocky surface previously marked by glacial imprints.*

Some of the phenomena observed on the surface of the Gaustaråen plateau thus indicate deep digging of the substratum under the effect of the differential localized movement of a glacier that was momentarily temperate, though largely immobile overall. Limited movements at places where there are rocky outcrops and open blockfields could have sufficed to cause blocks to be dislodged and move them short distances over great expenses by breaking up the substratum. These processes may have been aided by the specific fissuring of the quartzites, and by the substratum having been prepared by pre-glacial periglacial processes. They may have been assisted by the depressed form of the plateau, favoring the concentration, slippage, and overflow of ice. Admitting that the plateau was occupied by a cold-based glacier for most of the last glaciation, such processes are conceivable during a cataglacial phase, when the ice coverage is thinner and acquiring the properties of a warm-based glacier (Sellier 1995, 2002, 2007).

The properties of the felsenmeer are nevertheless indicative of postglacial periglacial rearrangement. Frost shattering, which is still active, led to the comminution of blocks during the Holocene. The fractures of the larger blocks reveal macro-frost shattering. Gelifluction has occurred around the plateau, and geliturbation within the matrix-rich blockfields. Postglacial periglacial processes are clearly partly responsible for the configuration of the felsenmeer. This has been accompanied by snowmelt run-off in a site prone to immobile snow slabs. This process has caused the least coarse material to be drawn away, thereby leading to certain blocks being destabilized and re-mobilized. This has contributed to material sorting, as indicated by the presence of tripod perched blocks. The relationship between the rocky outcrops, openwork blockfields, and matrix-rich blockfields is suggestive of analogous sorting right across the plateau, with outcrops having been exhumed in places and open blockfields subjected to sieving. Nevertheless, *the action of Holocene periglacial processes alone does not suffice to explain the overall configuration of the felsenmeer, and nor does the action of any supraglacial periglacial processes that may have occurred.*

The rate of lichen cover and diameter of certain *Rhizocarpon* individuals, sometimes nearly 20 cm, suggest limited frost partitioning on the felsenmeer. Furthermore, the large blocks remain major features of the landscape. In any case, there is less frost shattering than on Gaustaråen plateau, lying in the zone of sporadic discontinuous permafrost, or on Gaustatoppen above the level of extensive discontinuous permafrost. This difference pertains both to the current period and for successive ones since deglaciation. The relative scarcity of postglacial forms of scree processes on the Gaustaråen slopes in comparison to those of Gaustatoppen also illustrates the reduced role of frost shattering, both past and present, at the altitudes of the felsenmeer. The same holds for the forms of geliturbation and gelifluction. There are several indications that the material has been stable for a long time. The pseudo-pebbles have disintegrated solely on the face exposed to the weather, indicating that they have not been turned for a considerable length of time. The fact that tripod perched blocks are still to be found suggest the sectors where they lie have been stable since deglaciation. While frost action has operated in certain places under favorable circumstances, as may have occurred immediately after the withdrawal of the ice sheets, frost action alone cannot explain the interlocking of blocks observable in numerous sectors. *Hence it would be a further mistake to view the Gaustaråen felsenmeer as just a postglacial periglacial mantle.*

The blockfields on the surface of Gaustaråen plateau retaining the properties of in situ formations seem to result from changes caused by multiple factors. They do not correspond to a morainic deposit that was subsequently outwashed, or to a field of Holocene periglacial debris, even less to an older periglacial mantle that may or may not have been covered by a cold-based glacier.

They result mainly from the products of localized glacial deep digging having been rearranged by frost action and run-off during the Holocene, without excluding the momentary influence of supraglacial periglacial phenomena, or the reworking of earlier materials. Gaustaråen felsenmeer would appear to have experienced varied morphogenic conditions, though long near to the *Weichselian trimline* or else a bit lower (thin glaciation due to the proximity of the upper margin of the ice sheet) or a bit higher (supraglacial periglacial morphogenesis). It thus displays the characteristics of a supraglacial margin *felsenmeer* or a "trimline" felsenmeer (Sellier 1995, 2007).

Yet it is difficult to determine the relative share of glacial and postglacial processes in the genesis of the felsenmeer. For instance, the most extensive rocky outcrops are to be found near the bordering edges and on the steepest inclines, where there is also the clearest evidence of plucking and quarrying, hence the most developed ablation marks. The openwork blockfields below have the coarsest material and the largest number of tripod perched blocks. The largest matrix-rich blockfields with the highest number of indications of periglacial rearrangement lie in the flattest areas on the center of the plateau. This configuration is testimony to how glacial and periglacial processes complement each other in two ways: first, predominant ablation to the north-west, with reduced transport and the accumulation of blocks toward the center and south, and second, instances of outwash in the highest and steepest sectors, with the concentration and infilling of fine material in the lowest and most tabular sectors. It is thus necessary to emphasize the determining influence that snow and meltwater have played in differentiating the textures and current appearance of the felsenmeer.

11.3.3 The Gaustatoppen Pyramid

The part of Mount Gausta corresponding to Gaustatoppen is typical of quartzite mountain pyramids. It is visible from everywhere in Telemark and surrounded by large rectilinear slopes whose profile resembles Richter slopes, rising 800–1000 m above the fjell. It stands 300 m above the surface of Gaustaråen but over 1500 m above the bottom of

Vestfjorddalen. The slopes need to be examined zone by zone, for this determines the proportion of glacial and periglacial relict landforms along their profile, raising the question of glaciation levels once again. They also need to be considered transversally, contrasting the large rectilinear inclines forming their armature (*divide slopes*) to the funnels and channels dividing the current relief into triangular facets. The morphology of these slopes thus records the action of glaciers (basal concavities), periglacial processes (Richter slopes and scree), and gullying (cones and ravines).

11.3.3.1 The Top and the Divide Slopes

Not all pyramid peaks and divide ridges are quartzites, but quartzite ones are characterized presence of quartzites at the top and the intersection of large rectilinear slopes resulting from the convergence of large Richter slopes. The top of Gaustatoppen is in fact a ragged divide ridge with tower-like rocks ruined by macro-frost shattering. This crest peaks at 1881 m and drops 1700 m in altitude over one kilometer. It is comparable to the landscapes of Scandinavian high mountains, though without the marks of such active periglacial morphogenesis as found on Snöhetta or in the Rondane mountains. There is no "periglacial summit" zone here since the altitude is insufficient, as is volume of the summit. The landscape displays characteristics of "summit" relief but in fact corresponds to a "periglacial sub-summit" zone (Fig. 11.16).

The profiles of the Gaustatoppen slopes are bipartite, with their type depending on their aspect and extent of dissection. These slopes, necessarily issuing from pre-quaternary landforms, have a concave lower section of glacial origin with few incisions, and an upper section of periglacial origin with a rectilinear profile derived from Richter slopes. These two sections intersect above 1450 m, toward the top of the Gaustaråen slopes.

The lower section is concave and at an angle of less than 30°, indicating even more powerful ice-sheet scouring than around Gaustaråen. Minor figures correlating to the recent passage of ice (polishing, striations, glacial truncated sides, and crescentic gouges) may be found on outcrops up to 1530 m, resulting from the ice advancing toward the south-east, conforming once again to the Weichselian ice movement. In places these slopes are covered by thin sporadic heterometric blocks mainly of morainic origin, sometimes up to an altitude of 1575 m.

The upper part intersects with quartzite beds partially covered by openwork coarse scree that is stabilized and contains solely periglacial relict landforms. To the north-east is a rectilinear rocky incline 350 m to 450 m high, topped by discontinuous jagged corniches. Its slope, mainly 30–35°, increasing to 36–37° at the top, indicates that it issues from old Richter slopes gashed by branching gullies. Dissection is such that it is reduced to large triangular facets, giving way to simple divide crests toward the top. To the south-west the slopes have a similar profile but face the sun (adret slopes), resulting in more intense scree process and avalanche activity.

Unlike other quartzite mountains, the slopes of Mount Gausta display few instances of pure gravity scree processes. Admittedly, conditions in Telemark are favorable to scree processes (homogenous quartzite slopes, a periglacial zone over an elevation of nearly 1000 m, and active frost shattering over the last 9000 years or more). Nevertheless, the profiles and angles of the slope are not compatible with ongoing scree processes. The upper section of the Gaustatoppen slopes lying

Fig. 11.16 Crest peak of Gaustatoppen (1881 m), ruined by active macro-frost shattering of quartzite

above 1350–1500 m derive from Richter slopes, the angle of which naturally limits scree processes dynamics even though frost shattering continues to very operative at the summit. Equidistant funnels to the north-east mean that these slopes have been reduced to triangular facets deeply gullied to the south-west by corridors running more or less parallel. The debris covering them has been colonized by lichen over 20–60% of its surface. Furthermore, these typical open mountain slopes—having no cirques—have solely undergone the action of ice sheets, resulting in insufficiently steep profiles for scree processes to have occurred during the Holocene. Lastly, the presence of a nunatak and the effects of supraglacial scree processes have helped to lastingly regularize the upper slopes, which were spared any Holocene scree processes other than in the corridors.

Mount Gausta is thus a quartzite massif incompatible with any sustained scree processes action other than on its summit crest and in its ravines, which have been periodically swept by avalanches and torrent processes. Its periglacial morphogenesis did not maintain any real scree slopes or cones around Gaustatoppen. It did however contribute to building up large debris cones deposited in funnels spreading out around the base of the slopes.

The slopes of Mount Gausta display contact at around 1440–1530 m on the north-east of Gaustatoppen, at about 1400 m on its south-west, and at 1350–1470 m around Gaustaråen. Beneath this, the glacial imprints have been retained, but above they have been replaced or else destroyed by marks resulting exclusively from frost shattering, associated with the alpine topography there. This contact suggests in turn a trimline marking the separation between a periodically glaciated zone and a supraglacial zone subjected over time to periglacial processes. This would confirm the existence of a nunatak where Gaustatoppen now stands during the Weichselian and previous glaciations.

However, observations of the glacial marks along the slopes of Gaustatoppen and Gaustaråen shows that the contact between the zone displaying glacial imprints and that with only paraglacial marks occurs at altitudes of around 1300 m up to altitudes of over 1500 m. It is thus plausible that there were glacial relict landforms above the contact under consideration, which were subsequently erased by frost shattering, and thus that the true upper limits of previous glaciations are in places concealed within the post-glacial periglacial zone.

11.3.3.2 Gullies and Their Related Landforms

Gaustatoppen provides remarkable examples of hierarchized gullies cut into the remains of Richter slopes. The alpine relief characterizing the upper part of the slopes largely results from the cutting of corridors and the isolation of intermediary rocky crests with jagged profiles. All the indications are that the dissection of the Richter slopes is a relatively old phenomenon. The gullies are deep (between about 10 and 30 m) and long (400–600 m) with little branching. They are similar to *slush channels*, even though frost has helped widen them and they have been cleaned by run-off as indicated by *debris flows*. Above, the cones start at around 1450 m, spread to 1300 m, before descending to 900 m. Some thus reach the subalpine or forest zone. The largest are still operative to 950 m. Yet their elongation, flat transversal profile, and concave longitudinal profile (18° at the base and 25° at the summit) are characteristics of composite cones, resulting from predominant scree processes and surface rearrangement by avalanches (Fig. 11.17).

Fig. 11.17 Ravines and scree cones on north-east slope of Gaustatoppen

The north-east slope, as seen from the point at 1126 m (*geomorphosite 12*), looks the least dissected. The narrow, shallow corridors gullying it from its summit lead to half a dozen cones starting toward 1450–1500 m and stretching down to 1325 m. The most representative of the best preserved of these cones is 260 m in length, running north-east–south-west, with an elevation of 90 m from 1395 to 1485 m in altitude. Its slope is 19.5° on average, with a maximum of 25°. Its longitudinal profile is concave, and transversal profile relatively flat. Fragments of over 5 cm observed at 14 stations have an average length of 17.7 cm. Across the whole cone the largest do not exceed 1.50 m. Fragments under 5 cm occur at a rate of just 5.7%. The grano-classification of the debris is normal in the distal section, then inverted in the proximal section. Everything thus suggests that this cone was caused by avalanche processes. Pellicular avalanches may be observed in its proximal part during the summer, and it is likely that this cone, and its neighbors, continue to operate under the combined effects of scree processes along the corridors and slush flows. Its current morphological properties are similar to those of composite cones (Sellier 2002).

Because of its aspect, the south-west slope has been subject to more active gullying and scree processes. The ravines are longer (1000–1500 m), deeper (10–50 m), and more hierarchized than on the north-east. They give way to funnels which have dismantled the rocky facets and replaced them by ruined spurs and pinnacle crests contributing to the alpine nature of the relief. Above, the cones—though still concave with flat transversal profiles—are wider and longer. They descend to the bottom of the Gausdalen corridor (905 m) and thus reach the forest zone. The largest of these cones, lying between 920 and 1275 m to the north of Aslaktaul (*geomorphosite 13*), at the bottom of a slope with an elevation of over 960 m, is 1700 m long and over 800 m wide. It follows on from a 1300 m-wide and 1500 m-long funnel running up over an elevation of more than 600 m to the summit of Gaustatoppen. The cone slopes at 12° at the bottom and 28° at the top. It is composed of angular, heterometric openwork blocks (between 50 and 150 cm long at the base, and 20–80 cm at the top). The volume of material suggests intense initial frost shattering operating from corridors continuing once the slope had been regularized. The blocks on the lower part of the cone are currently wholly colonized by crustose lichen. There is less colonization on the upper part, indicating avalanche deposits, in all likelihood with a lengthy recurrence interval. The corridors cutting into the upper part of the slope are the marks left by rainwater and snowmelt torrential erosion, resulting from the high volumes of local precipitation. This cone would thus appear to be a vast polygenic cone, combining products from frost shattering and from torrential run-off, as well as from heavily laden avalanches. This is further suggested by the regular longitudinal profiles, marked by wavy-edged transversal rims. There are fairly high levels of avalanche activity in Telemark due to the form of the slopes and the long periods of snow cover.

The volume of the cones on the south-west slope of Gaustatoppen and their position at altitude are thus indicative of greater postglacial activity than on its north-east slope. Nevertheless, as in the previous case, these forms result from processes driven by torrents and avalanches that have taken over from scree processes. The cones may be categorized as avalanche cones, though probably derive from old scree cones and are merely the most recent landform associated with the lengthy evolution of the slopes. They are fairly thin, as their profile indicates, and in their current state have lesser volume than those in the corridors above, suggesting that the slopes of Gaustatoppen underwent major supraglacial frost shattering during the Weichselian, but that some of the scree processes material was carried along by the glaciers below, in a process similar to the *transferred cones* at Spitsberg (as described in André 1991). The volumes of the scree cones accumulated at the foot of the Gaustatoppen pyramid seem undersized in comparison to those of the funnels extending above. Under such conditions, the cones would be composed solely of the products of postglacial scree processes and associated avalanche processes, whereas the corridors above bear the marks of older ablation, some during the latest glaciation. They thus provide indications about the position of the Weichselian *trimline* and the evolution of the Richter slopes above.

In conclusion, the entire form of Mount Gausta, rising nearly 900 m above the surface of the fjell like a gigantic profiled block, confirms the action of ice sheets moving in a south-easterly direction, even though most of its appearance results from its past as residual relief, and despite its slopes having been reworked by periglacial, supraglacial, and postglacial processes. The north-eastern tip of the mountain,

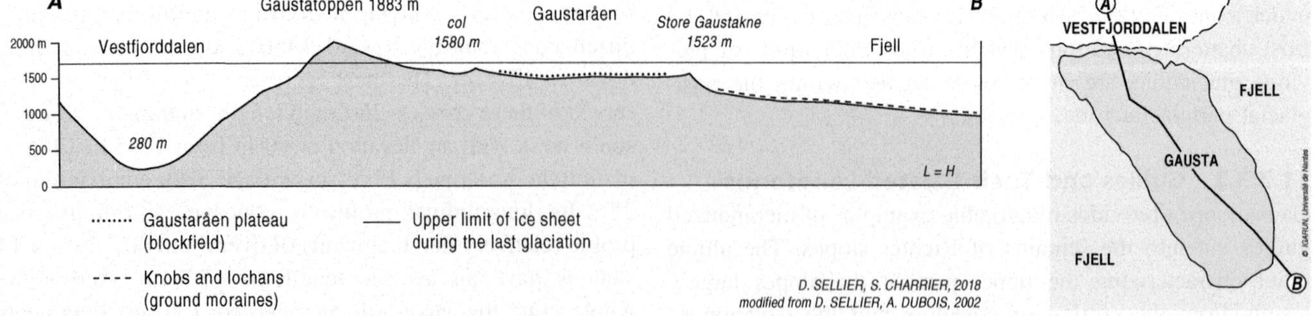

Fig. 11.18 Gausta topographic profile and assumed upper limit of Weischelian ice sheet

which divided and deviated the ice flows, has a characteristic prow shape. Its flanks, which channeled the ice, retain a conclave profile over some of their length. The appearance of its south-eastern tip (marked by the "knees" of Store Gaustakne and Lille Gaustakne, separated by diffluence corridors and basins resulting from former cirques) is still wholly fashioned by the ice. Several arguments (such as the large frost action funnels and transferred cones) confirmed that Gaustatoppen was a nunatak during the Weichselian, and that the rectilinear upper sections of its slopes evolved in a supraglacial environment. Their degradation may in itself be considered as indicative of supraglacial evolution. Ahlmann (1919), followed by Linton (1949), interpreted it long ago as indicating there had been a nunatak where Gaustatoppen lies (Fig. 11.18).

11.3.4 The Rjukan Valley (Vestfjorddalen)

Thanks to its dimensions and relief features, Vestfjorddalen is one of the great *glacial valleys*, like Drivdalen to the east of Dovrefjell, and especially Gudbrandsdalen to the west of the Rondane mountains. These are among the largest U-shaped valleys in basement areas in the world, due to the rise of the Scandes Mountains where Norway lies. They record major channeling of the ice, but the fact they are part of hierarchized systems and their relation to V-shaped valleys higher up suggest they result from periglacial and interglacial river systems (Gjessing 1966).

11.3.4.1 The Overall Shape and Landforms

Vestfjorddalen runs for 35 km across Telemark beneath Mount Gausta, from Lake Mösvatn to the west (Hardangervidda) to Lake Tinnsjö to the east (Blefjell), thus deviating around quartzites like all great glacial valleys. It is wide (3 km between its upper edges), drops 900 m beneath the surface of the fjell, and though regularly calibrated is sinuous in shape. It has a classic parabolic transversal profile. Its bottom is flat, wide (500 m at Rjukan), and covered by morainic and alluvial upbuild, where the River Måna traces free-swinging meanders before emptying into Lake Tinnsjö to the east.

11.3.4.2 The Slopes

The valley is flanked by large, uniform, continuous, symmetrical slopes which are among the highest in Norway. They have concave profiles with a long radius of curvature at the bottom, rectilinear sections in the middle, where the average slope can exceed 45° over an elevation of 500 m, and free faces and ledges at the top. The overall shape results from Pleistocene glacial processes. The surface is entirely rocky with the exception of the occasional scree and rockslide at the bottom, and wooded when the slope permits it. It is smooth and striated. The scarcity of postglacial scree processes may be explained by the properties of the substratum, formed mainly of metavolcanites which are less prone to frost action than quartzites, and by the altitudes, which are below those where periglacial processes operate. The conservation of many slabs, indicating postglacial decompression, is testimony to the accumulation and release of constraints along the slopes by glaciation then deglaciation, and to the degradation of the walls of a valley that channeled powerful ice flows.

11.4 The Geomorphosites and Observation Stations

11.4.1 Selection Methods

Each of the geomorphotypes is illustrated by an exemplary geomorphosite observable from a station, or by several complementary geomorphosites.

11.4.1.1 Geomorphosite Selection

The scientific criteria employed to choose geomorphosites differ depending on the author and subject. Several overviews have been published on this topic (Reynard 2009a, b; Reynard et al. 2009). The criteria selected here relate to

- the capacity of the selected geomorphosite to exemplify the geomorphotype as defined within the study area,
- the scientific and educational value of the geomorphosite above and beyond the study area,
- the state of conservation of the geomorphosite and any protection measures in place,
- the aesthetic qualities of the geomorphosite (particularly its photogenic properties) together with any cultural value (relating to history, literature, accounts, painting, and imagery).

11.4.1.2 Observation Station Selection

The criteria retained here for selecting observation stations pertain to

- the interest of the station as a viewing point for the preselected geomorphosite,
- the interest of the station site itself (the presence of any outcrops, sections, or any minor relief),
- accessibility and conditions for halting at the station,
- any risk of damage caused by excessive frequentation, considered as an eliminating factor.

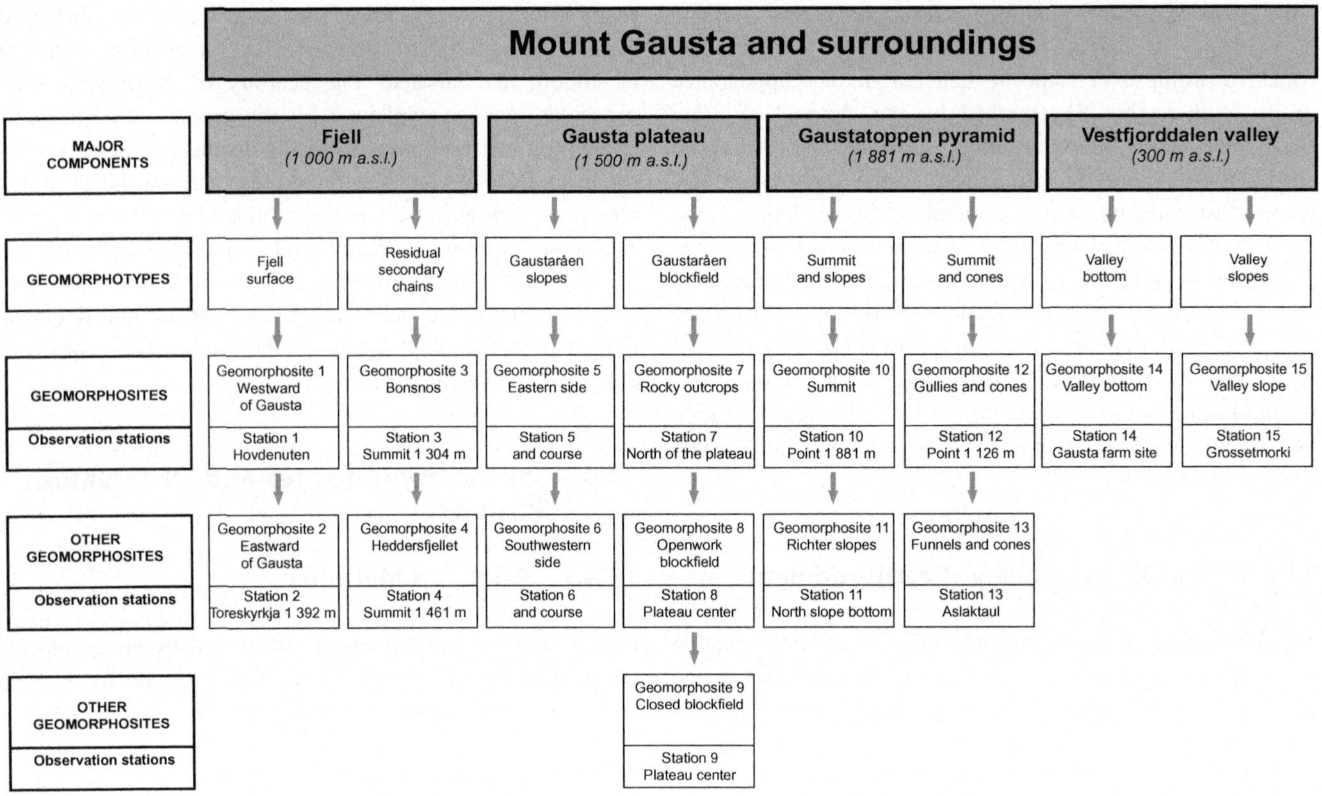

Fig. 11.19 Geomorphosites of mount Gausta and surroundings

11.4.2 Proposed Geomorphosites (Figs. 11.19 and 11.20)

11.4.2.1 On the Fjell

– **On the surface of the fjell (geomorphotype)**

Geomorphosite 1: to the west of Mount Gausta, panorama westwards from the bottom of the west slope of Gaustatoppen, at Hovdenuten (station 1). The surface of the fjell cuts across the *Seljord group* quartzites in the area around Bjorndalen-Gausdalen. The terrain is uneven with mounds and basins forming a string of lakes (*knob and lochan* type landscape). The relief is conditioned by a ribboned structure determined by the roots of faults.

Geomorphosite 2: to the east of Mount Gausta, 360° view from the 1392 m summit (station 2), near Toreskyrkja. The mounds here are less ordered but there are many lakes and corridors because the *Rjukan group* metavolcanites, through which the fjell surface cuts, are less fractured, and thus ultimately less permeable than the quartzites. The relief is conditioned by the system of orthogonal faults.

– **At the site of the intermediary residual chains (geomorphotype)**

Geomorphosite 3: to the west of Mount Gausta, and 360° view from the 1304 m summit of Bonsnos (station 3) over the monoclinal chain formed of quartzites, and panoramic view over the west flank of Mount Gausta.

Geomorphosite 4: to the east of Mount Gausta, 360° view from the 1461 m summit of Heddersfjellet (station 4) over the chain formed of metavolcanites, entirely shaped by ice, and panoramic view over the east flank of Mount Gausta.

11.4.2.2 Along the Gaustaråen Plateau

– **On the slopes bordering Gaustaråen (geomorphotype)**

Geomorphosite 5: on the east slope of Gaustaråen, itinerary from Stavronuten et via the 1492 m pointed toward Store Gaustakne (station 5), and panoramic view over Heddersfjellet. The slope is wholly marked by glacial relict landforms (in terms of its slope profile and minor forms).

Geomorphosite 6: on the west slope of Gaustaråen, itinerary from Myklestul via the summit (1500 m) toward Lille

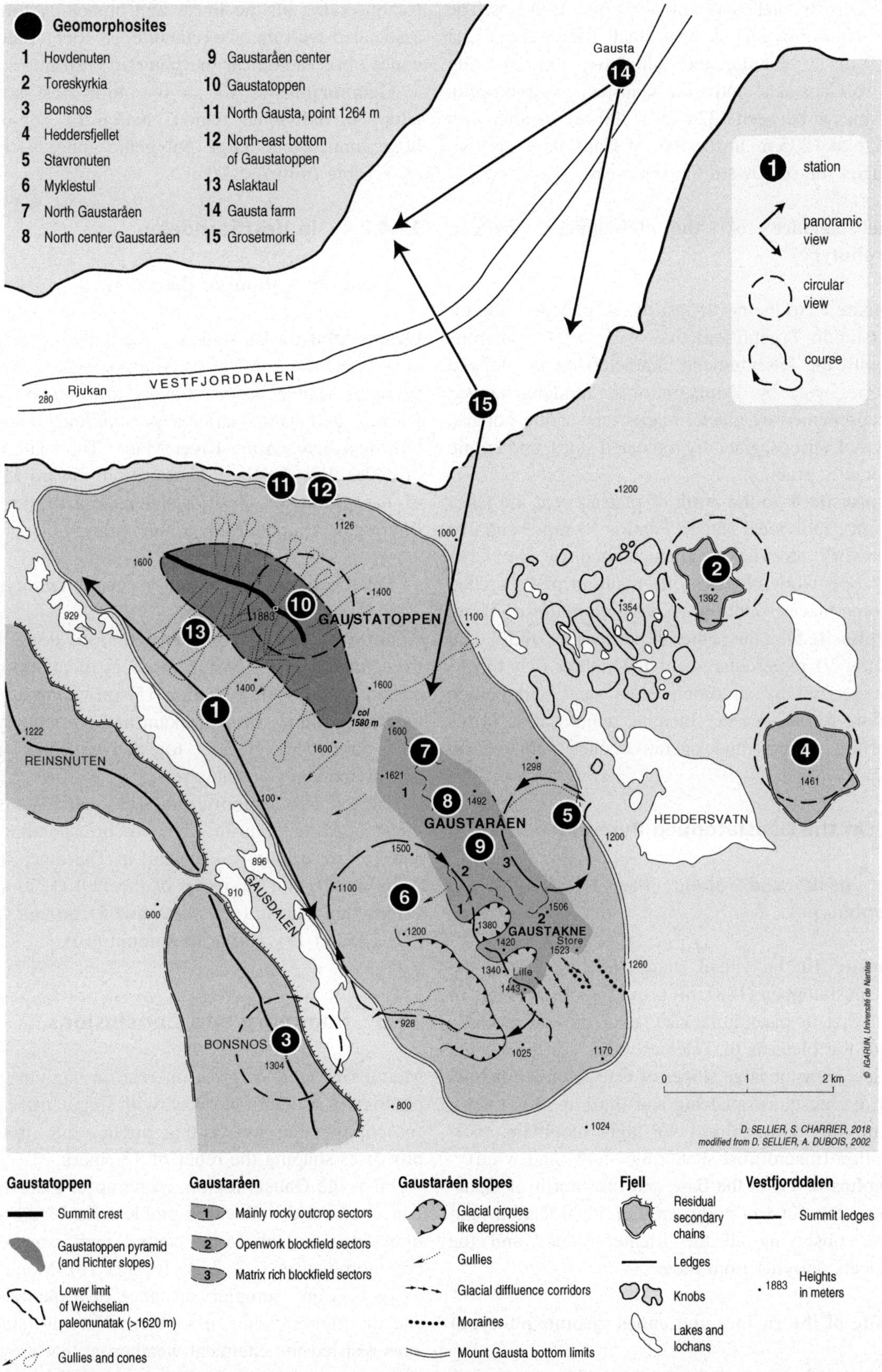

Fig. 11.20 Location map of geomorphosites on mount Gausta and surroundings

Gaustakne (station 6), and panoramic view over Bonsnos. The glacial relict landforms include two small flared cirques, at around 1350 m in altitude, and diffluence corridors dug between Store Gaustakne and Lille Gaustakne (Arvedalen). Lateral moraines at between 1320 and 1200 m, together with a frontal ridge at 1295 m, at the foot of Gaustakne, are evidence of glacial run-off toward the south-east.

– **On the surface of the felsenmeer plateau (geomorphotype)**

Geomorphosite 7: to the north-east of the plateau, individual station (station 7) representative of the rocky outcrops associated with the felsenmmeer, characterized by stepped quartzite slopes worn by disintegration or rendered jagged by frost, accompanied by glacial marks (crescentic gouges, hyperbolic crack strings, glacially truncated sides, and erratic blocks, especially granite).

Geomorphosite 8: to the north of plateau near the lakes (1492 m point), individual station (station 8) representative of the openwork blockfields, characterized by the large interlocking openwork blocks with many tripod perched blocks, in some places twenty or so within a radius of 50 m.

Geomorphosite 9: at the center of the plateau, individual station (station 9) representative of the matrix-rich blockfields, characterized by an omnipresent sand and pebble matrix, leading to the presence of relict geliturbation forms (fossil-patterned ground) and operative ones (stony earth circles).

11.4.2.3 On the Gaustatoppen Pyramid

– **At the peak and along the Richter slopes (geomorphotype)**

Geomorphosite 10: individual station (station 10) at the summit of Gaustatoppen (1881 m) providing three levels of observation: first, a panoramic view over the entire chain and, beyond, the plateaus of Telemark and Vestfjorddalen; second, a view over the large slopes of Gaustatoppen below, and the entire Gaustaråen plateau; and third, a view over a quartzite summit crest subject to high mountain morphogenic action (macro-frost shattering, snow, and wind).

Geomorphosite 11: at the base and turn north of Gaustatoppen, near the 1264 m point, an individual station (station 11) for observing all the Richter slopes and the triangular facets deriving from them.

– **At the site of the ravines and cones (geomorphotype)**

Geomorphosite 12: at the north-east base of Gaustatoppen, near the preceding station, an individual station (station 12) for observing all the linear and branching forms of ravine associated with the abovementioned Richter slopes, and the cones observable near the point at 1126 m.

Geomorphosite 13: at the north-west base of Gaustatoppen, individual station (station 13) for observing the large funnels and large polygenic cones such as the one observable from Aslaktaul.

11.4.2.4 In Vestfjorddalen

– **Along the bottom of the valley**

Geomorphosite 14: station at the bottom of Vestfjorddalen, at a dwelling known as Gausta, where Mount Gausta becomes visible as one travels from Lake Tinsjö toward Rjukan, and named after a person said to have been the ferryman here on the River Måna. The station (station 14) provides three levels of observation: toward Mount Gausta, of the profile of Vestfjorddalen, and of the bottom and fluvio-glacial upbuilding of the valley.

– **On the slopes of the valley (geomorphotype)**

Geomorphosite 15: from Grosetmorki, along the road over the south slope of Vestfjorddalen, at about 900 m in altitude, an individual station (station 15) providing a view over all the valley slopes from Rjukan, and particularly of the free faces formed by glaciers, together with the Richter slopes and ravines on Gaustatoppen.

The last two geomorphosites (14 and 15) provide general views of Mount Gausta. The two corresponding stations (14 and 15) are on the way toward it. The other stations could form circuits, for example, of the fjell (1, 3, 4, and 2), on Gaustaråen plateau (6, 7, 8, 9, and 5), around Gaustatoppen (11, 12, and 13), and at its summit (10).

11.5 Summary and Conclusions

Mount Gausta is of great interest for studying the geomorphology of southern Norway, with singularities that are ideal for popularizing for visiting publics the current and past processes shaping the relief of Telemark.

First, the Gausta region, made up of glacial fjell, mountain chain, and valleys, is emblematic of the general configuration of relief in central and southern Norway (geomorphosites 1, 2, 3, 4, 10, and 14). Mount Gausta, for its part, is an exemplary instance of a quartzite mountain, and the highest relief in southern Norway (since its quartzites resisted the chemical weathering that prevailed during pre-quaternary times). It also stands out for its abundance of rocky debris, and especially the presence of one of the most remarkable felsenmeers in Norway (this time due to the sensitivity of quartzites to the mechanical erosion that prevailed in quaternary times) (geomorphosites 7, 8, and 9). Mount Gausta is a simple chain comprised of large slopes intersecting at the summit of Gaustatoppen pyramid, and of lower slopes flanking the Gaustaråen plateau (geomorphosites 5, 6, 10, 11,12, and 13). Its simplicity derives from its being an "open mountain" rising above the fjell, and without any large glacial cirques.

In addition to these general characteristics, Mount Gausta has two major properties worthy of visitors' attention. The first pertains to the tiering of its vegetation and its current and relict landforms, displaying the influence of climate and erosion over an elevation of nearly 900 m. The second is the presence of felsenmeer, making it possible to define the characteristics of "trimline felsenmeers". These two properties help confirm that Gaustatoppen was the site of a paleonunatak during the Weichselian, at a time when Mount Gausta lay between the ice divide and the edge of the ice sheet.

The current zonation of the terrain is one of the most obvious features of the landscape. In the absence of any nivo-glacial zone, the slopes of Mount Gausta in their entirety lie within the operative periglacial zone that rises nearly 900 m above the level of fjell, passing through the 0 °C annual isotherm. It is thus one of the thickest operative periglacial zones in Scandinavia after the Rondane mountains. The major active periglacial phenomena is frost shattering, present in places on the surface of the fjell, and increasing above 1500 m (the top of the Gaustaråen slopes), and then again above 1700 m (the lower limit of extensive sporadic permafrost). At the summit, micro-frost shattering helps maintain the alpine landscape (with ruined tower-like crests, pinnacles, and corridors). Three sets of features need considering.

- On the lower sections of the slopes, current periglacial morphogenesis is determined by the climate characteristics specific to lower and mid-alpine zones. Apart from frost shattering, the main active periglacial phenomena are terracettes and small gelifluction lobes.
- On Gaustaråen plateau, between 1500 and 1650 m, thus at the base of the upper alpine zone, the only active periglacial phenomena are stony earth circles generally occurring in the center of fossil-patterned ground, where there are increased marks of frost shattering.
- On the inclines around Gaustatoppen pyramid, those deriving from Richter slopes are subject to more intense frost shattering above 1650 m, but periglacial activity is still selective (at the summit crest and in the bottom of corridors).

Concerning the zonation of relict landforms, the base slopes of Mount Gausta (around Gaustaråen and the lower section of Gaustatoppen) correspond to slopes fashioned by glacial ice, while the upper slopes (the upper part of Gaustatoppen), where ravining has taken over from scree processes, correspond to ones derived from old Richter slopes.

Several indications of the presence of a Weichselian trimline may furthermore be identified. The first is the erratic blocks present on the peaks of the surrounding chains (Bonsnos, Heddersfjellet), the slopes of Gaustaråen and Gaustatoppen, and the Gaustaråen *felsenmeer.* The fact that there are none on the Gaustatoppen pyramid supports the idea that the Weichselian *trimline* was somewhere around 1600 m. The second set of markers relates to the distribution of elementary glacial erosion marks on the slopes. These relict landforms become fewer and then disappear at between 1400 and 1530 m beneath Gaustatoppen.

Several other properties of the Gaustatoppen slopes confirm the presence of a Weichselian paleonunatak above this altitude. The main indication is the forms of ravining found on Gaustatoppen. These are present mainly above 1600 m and are disproportionate to the scale of the cones beneath, supporting the hypothesis of a partially supraglacial dissection.

Detailed examination of the Gaustaråen plateau shows the additional interest of analyzing slopes and felsenmeers. There are several features (marks left by deep glacial digging, tripod perched blocks, and so on) suggesting it be regarded as testimony to pellicular ice cover and an example of a *"trimline" felsenmeer*, whose prophecies suggest the Weichselian *trimline* was somewhere around 1620 m on Mount Gausta during the latest glaciation.

The block arrangements and marks they bear indicate that this felsenmeer is of composite type, resulting from the influences of initial deep digging by the differential, localized, restricted movements of a warm-based glacier, in all likelihood mostly during the cataglacial period, followed by rearrangement by postglacial frost action agents, together with sorting of the finest fragments by gravity and snowmelt outwash. These processes probably reworked an old stock of material prepared by earlier paleogeographic sequences. The case of Gaustaråen suggests that the evolution of a felsenmeer may, in certain circumstances, coincide with warm-based glaciation, drawing on preparation by deep glacial digging (Sellier 2007).

If the Gaustaråen plateau was indeed covered by ice during the Weichselian, the problem is to assess its extent together with the duration of glaciation.

The slopes of Gaustatoppen derive from old Richter slopes and were glaciated up to at least 1530 m, as indicated by the altitude of glacial imprints still present, and up to at least 1620 m if we look at Gaustaråen. These upper inclines have been degraded by frost shattering, gullying, and avalanches. This degradation may be considered as indicating supraglacial evolution, hence the presence of a paleonunatak where Gaustatoppen lies.

Furthermore, the scree cones present at the base of the Gaustatoppen pyramid look relatively small and ultimately undersized in comparison to the frost action funnels and corridors above, indicating that some of the scree material was carried away by glaciers beneath, in a process comparable to *cone transfer*. They would thus result solely from postglacial scree processes, while the corridors above record the partially earlier effects of ablation.

In addition to this, the Gaustaråen plateau displays many marks of glacial erosion extending up to its highest parts, suggesting the existence of sufficiently thick ice cover, mobile at least some of the time. Yet the Gaustatoppen pyramid peaks just 300 m above the highest edges of the Gaustaråen plateau and does not in its current state present any traces of glaciation.

The hypothesis that Gaustaråen was covered by a thin, momentarily warm-based icecap is therefore plausible. The Gaustatoppen pyramid would thus always have been near the upper limits of ice cover during the latest glaciation. In any case, Mount Gausta provides an opportunity to observe the geomorphological properties of a mountain that is the sole example of a paleonunatak in Telemark.

Acknowledgements The authors would like to thank Simon Charrier, cartographer, who redrew the maps, and the Institute of Geography of Nantes University and the Laboratory Géolittomer-Nantes (CNRS) for founding the translation of the manuscript.

References

Ahlmann HW (1919) Geomorphological studies in Norway. Geogr Ann 1:1–148, 193–252

Andersen BG Karlsen M (1986) Glasialkronologi-Isfrontens Tilbaketrekning, Nationalatlas for Norge, Staten Kartverk

André MF (1991) Dynamique actuelle et évolution holocène des versants du Spitsberg. Thèse d'Etat, University of Paris I

Ballantyne CK (1998) Age and significance of mountain top detritus. Permafrost Periglac Process 9:327–345

Benn DI, Evans D (2002) Glaciers and glaciations. Arnold, London

Dahl E (1955) Biogeographic and geologic indications of unglaciated areas in Scandinavia during the glacial ages. Bull Geol Soc Am 66:1499–1519

Dahl R (1966a) Blockfields, weathering pits and tor-like forms in the Narvik Mountains, Nordland, Norway. Geogr Ann 48A:55–85

Dahl R (1966b) Blockfields and other weathering forms in the Narvik Mountains. Geogr Ann 48(A):224–227

Dons J (1960a) Telemark supracrustal and associated rocks, in geology of Norway. Norsk Geologisk Undersökelse 208:49–58

Dons J (1960b) The stratigraphy of supracrustal rocks, granitization and tectonics in the Précambrian Telemark area. Norsk Geologisk Undersökelse 212:2–27

Dons J (1961) Rjukan, Geologisk Kart, Scale 1/100 000. Norges Geologisk Undersölkelse

Dons J (1972) The Telemark area, a brief presentation. Sciences de la Terre XVII 1–2:23–29

Gjessing J (1966) Some effects of ice erosion on the development of Norwegian valleys and fjords. Nor Geogr Tidsskr 20(8):273–299

Gjessing J (1967) Norway's paleic surface. Nor Geogr Tidsskr 21 (2):69–132

Gjessing J (1987) Geomorphology of Norway. In Varjo U, Tietze W, Norden (eds) Man and environment. Gebrüder Borntraeger, Berlin-Stuttgart, pp 90–103

Godard A (1965) Recherches géomorphologiques en Ecosse du Nord-Ouest. Thèse d'Etat, Lettres, Paris, Publications de la Faculté des Lettres de l'Université de Strasbourg

Godard A (1993) Rhythms in the geomorphological evolution and time-space related scale levels in the high latitudes. Zeitschrift für Geomorphologie, Suppl. 93:61–67

Högbom B (1914) Über die geologische Bedeutung des Frostes. Bull Geol Inst Univ Uppsala, XII, pp 257–389

Ives JD (1958) Mountain-top detritus and the extent of the last glaciation in Northeastern Labrador-Ungava. Can Geogr 12:25–31

Ives JD (1966) Blockfields, associated weathering forms on mountain tops and the Nunatak hypothesis. Geogr Ann 48A(4):220–223

Jansen IJ (1986) Kvartaergeologi, Jord og Landskap i Telemark gjennom 11000 ar. Institutt for Naturanalyse

Kerguillec R (2015) Seasonal distribution and variability of atmospheric freeze/thaw cycles in Norway over the last six decades (1950–2013). Boreas 44(3):526–542

Kerguillec R, Sellier D (2015) Selection of geomorphosites in the Rondane National Park (central Norway): landform and popularization. Géomorphol Relief Process Environment 21(2):131–144

King L (1984) Permafrost in Skandinavien. Heidelberger geographische Arbeiten 76

King L (1986) Zonation and ecology of high mountain permafrost in Scandinavia. Geogr Ann 68A(3):131–139

Klemsdal T, Sjulsen OE (1988) The Norwegian macro-landforms: definitions, distribution and system of evolution. Nor Geogr Tidsskr 42:133–147

Laverdière C, Bernard C, Dionne JC (1968) Les types de broutures glaciaires; 1, Classification et nomenclature franco-anglaise. Revue géographique de Montréal XXII 1:21–33

Linton D (1949) Unglaciated areas in Scandinavia and Great Britain. Irish Geogr 2:25–33

Lundqvist J (1987) Late weichselian glaciation and deglaciation in Scandinavia. In Sibrova V, Bowen DQ, Richmond GM (eds) Quaternary glaciation in Northern Hemisphere. Pergamon, Oxford, pp 269–292

Moen A (1987) The regional vegetation of Norway; that of central Norway in particular. Nor Geogr Tiddskiftt 4:179–226

Moine B, Ploquin A (1971) Caractères géochimiques des séries de roches supracrustales du Telemark, Précambrien épizonal de la Norvège du Sud. Comptes-rendus de la Société géologique de France 3:139–141

Nesje A (1989) The geographical and altitudinal distribution of blockfields in southern Norway and its significance to the Pleistocene Ice sheet. Zeitschrift für Geomorphologie, Suppl. 72:41–53

Nesje A et al (1990) Autochthonous blockfields in southern Norway: implications for the geometry, thickness, and isostatic loading of the Late Weichselian Scandinavian ice sheet. J Quat Sci 5(3):225–234

Nesje A, Dahl SO, Anda E, Rye N (1988) Block fields in southern Norway: significance for the Late Weichseilian ice sheet. Nor Geol Tidsskr 68:149–169

Nordseth K (1987) Climate, hydrology and biogeography of Norway. In: Varjo U, Tietze W, Norden (eds) Man and environment. Gebrüder Borntraeger, Berlin-Stuttgart, pp 159–169

Oftedahl C (1980) Geology of Norway. Norges Geologiske Undersökelse, Universitetsforlaget 54(356):3–115

Panizza M (2001) Geomorphosites: concepts, methods and examples of geomorphological survey. Chin Sci Bull 46(1):4–6

Peulvast JP (1978) Le bourrelet scandinave et les Calédonides: un essai de reconstitution des modelés de la morphogenèse en Norvège. Géographie physique et Quaternaire XXXII 4:295–320
Peulvast JP (1985) Relief, érosion différentielle et morphogenèse dans un bourrelet montagneux de haute latitude: Lofoten-Vesteralen et Sogn-Jotun (Norvège). Thèse d'Etat, Université de Paris I
Peulvast JP (1987) Surfaces d'aplanissement étagées dans les Scandes: principes et implications de l'étude des paléosurfaces dans une montagne de haute latitude. Revue de Géomorphologie Dynamique XXXVI 3:83
Reynard E (2009a) Geomorphosites: definition and characteristics. In: Reynard E, Coratza P, Regolini-Bissig G (eds). Friedrich Pfeil, Munich, pp 9–20
Reynard E (2009b) The assessement of geomorphosites, In: Reynard E, Coratza P, Regolini-Bissig G (eds) Friedrich Pfeil, Munich, pp 63–71
Reynard E, Coratza P, Regolini-Bissig G (2009) Geomorphosites. Friedrich Pfeil, Munich
Rudberg S (1962) Geology and morphology of the "fjells". Biul Peryglac 11:173–186
Rudberg S (1973) Glacial erosion forms of medium size—a discussion based on four Swedish case studies. Zeitschrift für Geomorphologie, Suppl. Band 17:33–48
Rudberg S (1977) Periglacial zonation in Scandinavia. In: Poser H (ed) Formen, Formengesellschaften und Untergrenzen in den heuteigen periglazialen Höhenstufen der Hochgebirge Europas und Afrikas zwischen Arktis und Äquator. Bericht über ein Symposium, Akad. Wiss. Göttingen Abh., Math-Phys., Kl. Folge E 31, 92–104
Rudberg S (1984) Fennoscandian shield: Finland, Sweden and Norway. In: Embleton C (ed) Geomorphology of Europe. MacMillan Publishers, Londres, pp 55–74
Rudberg S (1988) Gross morphology of Fennoskandia. Six complementary ways of explanation. Geogr Ann 70A(3):135–167
Sellier D (1995) Le felsenmeer du mont Gausta (Telemark, Norvège): Environnement, caractères morphologiques et significations paléogéographiques. Géog Phys Quatern 49(2):185–205
Sellier D (2002) Géomorphologie des versants quartzitiques en milieux froids : l'exemple des montagnes de l'Europe du Nord-Ouest. Thèse de doctorat d'Etat, Université de Paris I Panthéon-Sorbonne
Sellier D (2007) Le rôle des relais de processus glaciaires et périglaciaires dans la genèse des felsenmeers. Du continent au bassin versant: théories et pratiques en géographie physique (Hommage au Professeur Alain Godard). Presses Universitaires Blaise Pascal, Collection Nature et Société, Clermont-Ferrand, pp 355–374
Sellier D (2010) L'analyse intégrée du relief et la sélection déductive des géomorphosites: application au relief de la Charente-Maritime (France). Géomorphol Reliefs Process Environ 2:199–214
Sellier D (2013) Le relief de la Loire-Atlantique: patrimoine géomorphologique et géomorphosites. In: Morice JR, Saupin G, Vivier N (eds) Les nouveaux patrimoines des Pays de la Loire, Presses universitaires de Rennes, pp 236–255
Sigmond EMO, Gustavson M, Roberts D (1984) Berggrunnskart over Norge, Nasjonalatlas for Norge
Sollid JL, Torp B (1984) Glasialgeologisk Kart over Norge, 1:1 000 000, Nasjonalatlas for Norge, Geografisk Institutt, Universitet i Oslo
Verschure RH, Maijer C, Andriessen PAM (1990) Isotopic age determination in South Norway: II. The problem of errorchron ages from Telemark rhyolites. Nor Geologisk Undersökelse 418:47–60
Vorren TO (1977) Weichselian ice movements in South Norway and adjacent areas. Boreas 6:247–257
Washburn AL (1979) Geocryology, a survey of periglacial processes and environments. Arnold, Londres
Wille N (1905) Om innvandringen av det arktiske floraelement til Norge. Nyt Magazin for Naturvidenskap 43
Wyckoff D (1934) Geology of the Mt. Gausta region in Telemark, Norway. Nor Geologisk Tidsskrift 13:1–72
Whittow J (2000) Dictionary of physical geography. Penguin Books, 590 pp

Selection of Geomorphosites in the Dovrefjell National Park (Central Norway)

12

Riwan Kerguillec and Dominique Sellier

Abstract

Located in central Norway, the Dovrefjell is part of the Scandes Mountains and is one of the highest mountain of Scandinavia, rising to 2 286 m at the Snøhetta. The massif has been a National Park since 1974. Dovrefjell-Sunndalsfjella National Park covers more than 4 000 km^2, making this one of the largest protected areas in Norway. The massif presents a high mountain landscape characterized by sharp summits, extended slopes and large glacial cirques that conserve small glaciers. Dovrefjell has rich geomorphological potential dues to its lithology, it is clearly developed mountainous belts and the prevalence of active/inherited reliefs, and this potential could lead to a project of the promotion of relief/topography and landforms of scientific and educational interest. Such valuation mostly refers to the notion of geomorphosites and this concept requires an inventory of sites of geomorphological interest before the selection of priority sites. In this context, several selection methods are commonly used. The major aim of this case study is to increase awareness of the landforms of Dovrefjell in the perspective of their valuation in the National Park. A new way to select geomorphosites is also proposed in this chapter by applying the two main used methods jointly ("deductive method" and "selective method").

Keywords

Dovrefjell • Central norway • High atlantic mountain landscapes • Selection of geomorphosites • Reliefs and landforms valuation

12.1 Introduction

The Norwegian massif of Dovrefjell has been a National Park since 1974, a little later than the massif of Rondane (1962) but 6 years before the Jotunheimen (1980). The Dovrefjell-Sunndalsfjella National Park was established in 2002 to replace and enlarge it. The massif is part of the Scandes Mountains and is situated near the 62nd north parallel and the 9th east meridian. Its relief has three main properties. First, the Dovrefjell is a quartzitic mountain and this homogeneous lithology plays a major role regardless of the scale level. Secondly, the massif extends from the Norwegian fjell (900–1 000 m a.s.l.) to 2 286 m a.s.l at the Snøhetta and belongs to the higher massifs of Scandinavia as well as remaining among the most continental high mountains of Norway. This location and its corresponding climate involve clearly developed mountainous belts above the timberline, including one of the thickest periglacial belts in Europe and one of the most continental glacio-nival belts of central Norway. Finally, the quartzitic structure and the climate of the mountain area allow widespread active and inherited periglacial features, as well as numerous other types of landforms such as pre-Quaternary, glacial, and paraglacial landforms. For all these reasons, the massif of Dovrefjell has a high geomorphological value which could lead to a promotion of relief representative of a Nordic quartzitic high mountain. The first objective of such valuation would be to communicate scientific knowledge to a non-specialist public, the second to provide this public with the major principles to observe the reliefs by themselves, to interest them in relief evolution and to help them make sense of the spatial and temporal relationships between these reliefs.

The promotion of relief is commonly based on geomorphosites descriptions. Geomorphosites can be defined as landforms of different scales characterized by scientific, cultural and historical, aesthetic and social/economic values (Panizza 2001). Using this concept requires an inventory of sites of geomorphological interest, before targeting the

R. Kerguillec (✉) · D. Sellier
University of Nantes, Campus Tertre, IGARUN, Chemin de La Censive Du Tertre, BP 81 227, 44 312 Nantes cedex 3 Nantes, France
e-mail: riwan.kerguillec@univ-nantes.fr

A. A. Beylich (ed.), *Landscapes and Landforms of Norway*, World Geomorphological Landscapes,
https://doi.org/10.1007/978-3-030-52563-7_12

priority sites for the valuation of relief. Two main selection methods are available in this perspective. The "selective approach" involves elaborate assessment methods and is especially used when an area comprises numerous sites of geomorphological interest. This method has been previously implemented in Alpine and Mediterranean regions. One of the most prevailing issues in geomorphological heritage research is the development of methodologies that focus on this estimation of different types of value present in geomorphosites. Several assessment methods have been proposed (Grandgirard 1999; Reynard 2005; Reynard and Panizza 2005; Reynard et al. 2007; Pereira and Pereira 2010). These methods select priority geomorphosites according to scientific parameters (e.g., exemplarity, educational interest, representativeness) and several additional parameters (e.g., accessibility, legibility). Described as a "deductive method", the second method corresponds to an integrated approach guided by deductive reasoning and based on the prior multi-level analysis of the relief. It leads to a selection of geomorphosites representative of several geomorphotypes and eventually of their minor components. It focuses on the properties of the space concerned by the valuation and is suitable for various types of reliefs (e.g., mountains, plains, continental, coastal…) because it provides an overall spatial analysis at different scales and makes less subjective the selection of geomorphosites (Sellier 2009). This method has already been tested in France on several areas and has also been conducted in the Rondane National Park (Kerguillec 2013; Kerguillec and Sellier 2015). Furthermore, it can be used preliminary to an assessment of geomorphosites but this combination with the "selective method" remains to be tested.

The first aim of this chapter is to increase awareness of the reliefs of Dovrefjell in the perspective of their valuation in the National Park, thus justifying the registration of reliefs in its natural heritage considering that one of the objectives of the Park is to "*to protect landscape formations and distinctive geological features*" (Dovrefjell-Sunndalsfjella Official Website 2018). Its second aim is to propose a new way to select geomorphosites by applying the "deductive method" and the "selective method" jointly. Thus, this new way defines several geomorphotypes representative of the Dovrefjell ("deductive method") before to perform an assessment of sites ("selective method").

12.2 Study Area

12.2.1 Location of the Massif

Dovrefjell is situated in central Norway, within the boundaries of Oppland, Hedmark and Sør-Trondelag, near the 62nd north parallel and the 9th east meridian. It rises to 2 286 m a.s.l. at Snøhetta (62°19′20″ N; 9°16′15″ E) and is one of the highest massifs of the Scandes. The mountain area is surrounded by three other high massifs: Trollheimen in the north (Store Trolla, 1 850 m a.s.l.), Rondane in the southeast (Rondslottet, 2 178 m a.s.l.), and Jotunheimen in the southwest (Galdhøppigen, 2 469 m a.s.l.). It is bordered by the valley of Sunndalen in the north, Drivdalen in the east, Grøvudalen in the northwest and Gudbrandsdalen in the southeast. This study essentially concerns the massif of Snøhetta and its surrounding area on the fjell (Fig. 12.1).

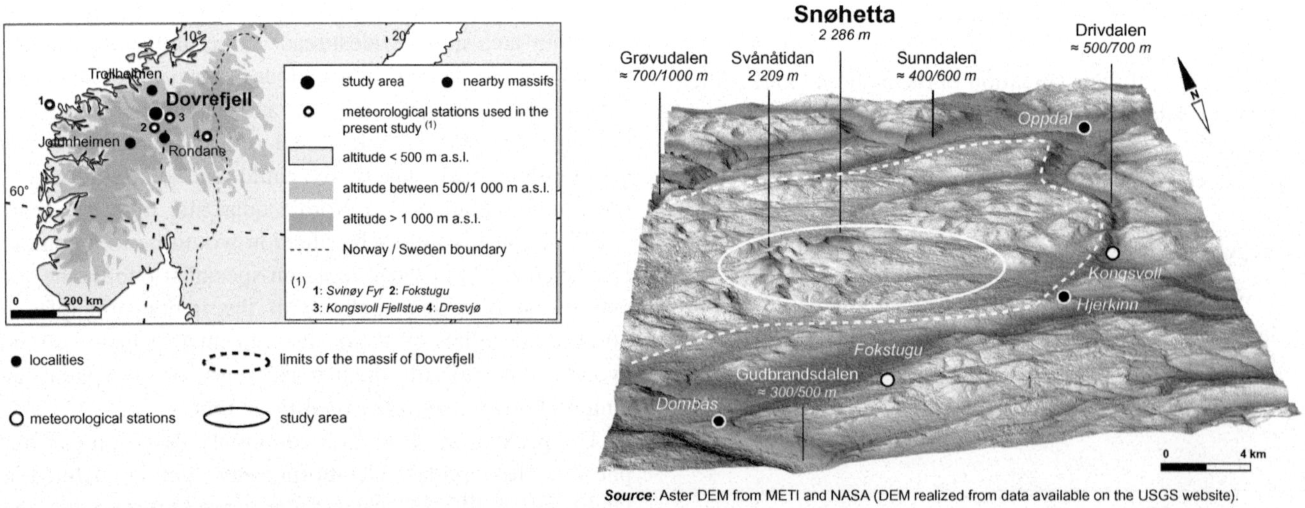

Fig. 12.1 Location map and digital elevation model of the study area

12.2.2 Description

The massif includes three major elements from the periphery to the center (Fig. 12.2a, c). The Norwegian fjell, extending from 900–1 000 m a.s.l. to 1 200–1 300 m a.s.l. and strewn with lakes and smooth landforms, is a barren highland characteristic of the action of Quaternary glacial sheets at high latitudes (Rudberg 1960; Peulvast 1985). It corresponds to a high plateau surface with an average elevation of 1 000 m a.s.l., thus located above the timberline.

The second element is a discontinuous ring of peripheral mountains (1 500–1 800 m a.s.l.) with rounded summits and wide glacial cirques (Fig. 12.2a, c). These mountains, like Tythöa in the northern part of the study area (1 773 m a.s.l.), Kolla in the eastern part (1 652 m a.s.l.) and Einövlingseggen (1 675 m a.s.l.) in the southeast, are surrounded by rectilinear external slopes, with an inclination lower than 20°. These slopes have been called "*paleïc slopes*" in previous studies because they are a remnant of the pre-Pleistocene topography (Gjessing 1967; Peulvast 1985; Sellier 2002). They are linked to the fjell by large inclined slopes with an average inclination of 5–10° (flyi).

The last element consists of six central massifs of similar size and elevation, inherited from a high erosion surface (Strøm 1945; Peulvast 1985). The central massifs of Snøhetta (2 286 m a.s.l., Fig. 12.2b) and Svånåtindan (2 209 m a.s.l.) exceed 2 000 m a.s.l., while the four others are characterized by elevations between 1 800–2 000 m a.s.l. (Fig. 12.2a). All these central massifs have in common "alpine" landscapes, not only because of their

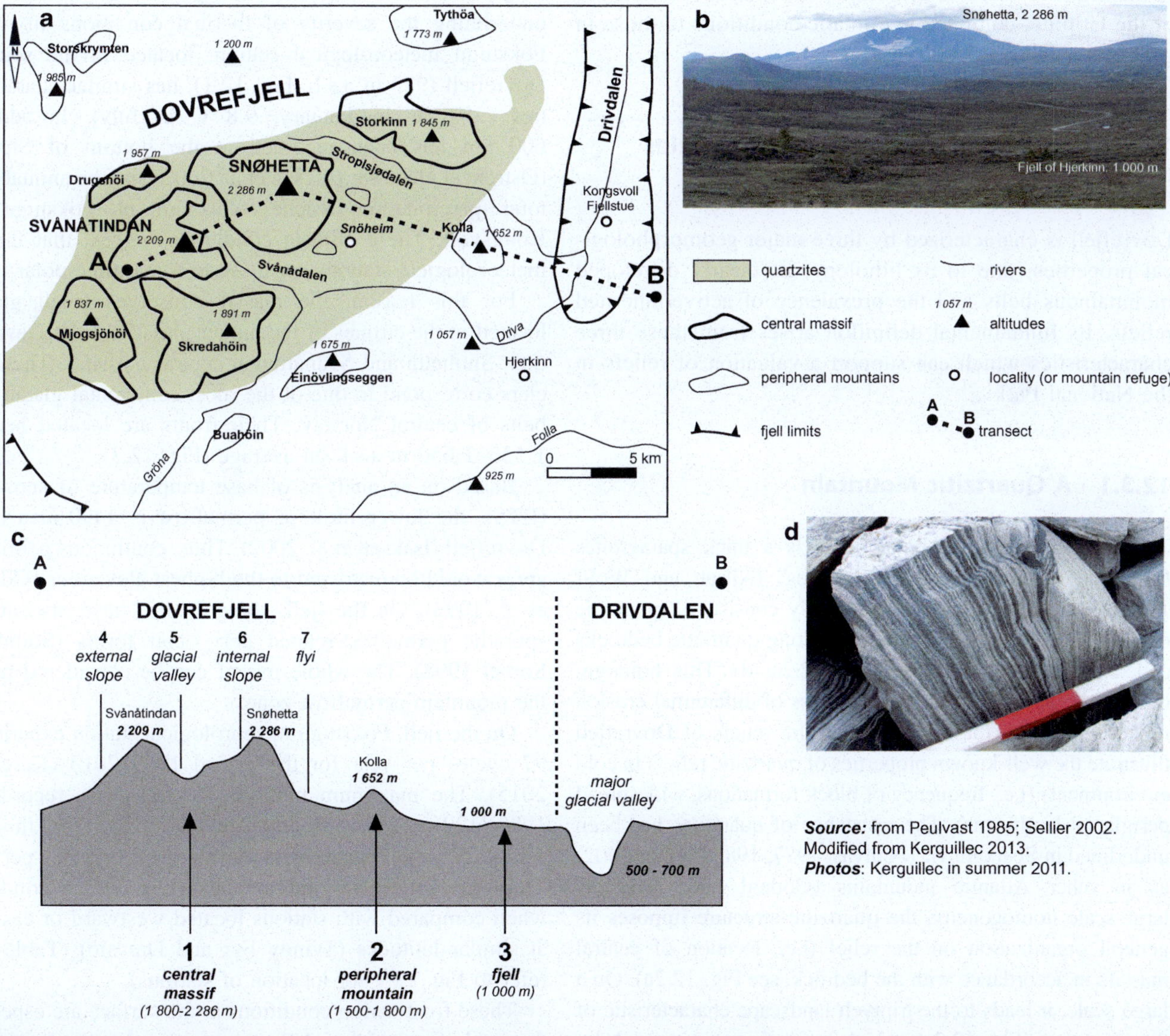

Source: from Peulvast 1985; Sellier 2002. Modified from Kerguillec 2013. ***Photos***: Kerguillec in summer 2011.

Fig. 12.2 **a** Simplified morphostructural map of Dovrefjell. **b** Photo of the massif of Snøhetta (2 286 m a.s.l.), viewed from Hjerkinn (fjell) toward the northwest. **c** Schematic northwest/southeast transect from Dovrefjell to Drivdalen. **d** Photo of a feldspathic quartzitic block. The scale measures 0.1 m per color

elevation, but also for several other reasons. They are distinguished from peripheral mountains by two types of fitted relief: first, they are wrapped by paleïc slopes that cut each other by forming sharp summits. Secondly, they are separated by internal glacial valleys (Stroplsjødalen in the north, between the massifs of Snøhetta and Storkinn, and Svånådalen in the center, separating the massifs of Svånåtindan and Skredahöin from Snøhetta). Furthermore, this appearance results from large and widespread glacial cirques as well as from their internal slopes morphology. These slopes are higher than 600–800 m. They are characterized by rocky faces and well-developed scree. The "alpine" landscape also results from the presence of small glaciers that remain in some cirques, but only in the two highest massifs (i.e., Snøhetta and Svånåtindan). Unsurprisingly, this appearance is also due to the high elevation of Dovrefjell which induces, because of the latitude, comparable climatic conditions to those in the Alps at 3 500–3 800 m a.s.l.

12.3 Main Geomorphological Properties of the Massif

Dovrefjell is characterized by three major geomorphological properties, due to its lithology, its clearly developed mountainous belts and the prevalence of active/inherited reliefs. Its fundamental definition arises from these three characteristics which can support a valuation of reliefs in the National Park.

12.3.1 A Quartzitic Mountain

The bedrock in the mountain area is a thick sparagmites formation with "feldspathic quartzites" (Nilsen and Wolff 1988). In fact, the lithology essentially consists of quartzitic units and the massif is made up of strong quartzitic bedrocks called "Dovrefjell complex" (Fig. 12.2a, d). This homogeneous lithology explains the sparseness of differential erosion reliefs within the massif. However, the reliefs of Dovrefjell illustrate the well-known properties of quartzitic reliefs in cold environments (i.e., frequency of block formations, widespread periglacial landforms). The resistance of quartzites has been underlined in Scandinavia (Peulvast 1977, 1985; Sellier 2002) as in other Atlantic mountains (Godard 1965). By its large-scale homogeneity, the quartzitic structure imposes its general organization on the relief (i.e., location of central massifs in accordance with the bedrock, see Fig. 12.2a). On a large scale, it leads to the triptych landscape characteristic of the Scandes (Fig. 12.2c) which was first recognized by Peulvast (1985). This triptych is composed of high central massifs lying on the Norwegian fjell and surrounded by large glacial valleys. This explains that the Dovrefjell has been described as a "mountain on a high surface" (Sellier 2002).

12.3.2 Mountainous Belts

The base of the massif of Dovrefjell corresponds to the annual 0 °C isotherm, which is close to 900–1 000 m a.s.l. in the region. According to the annual climate data from Kongsvoll Fjellstue meteorological station located in the eastern part of the massif (934 m a.s.l., Fig. 12.1), the MAAT is −0.4 °C. The thermal average of the coldest month is −9.4 °C (January) and the warmest month is July (9.6 °C) (Norsk Meteorologisk Institutt 2018). The station experiences 6 months with a mean air temperature above 0 °C from May to October, whereas the mean monthly air temperatures are negative during the rest of the year, underlining the severity of thermal conditions in winter. Fokstugu meteorological station, located in the south of Dovrefjell (973 m a.s.l., Fig. 12.1), has similar characteristics (−8.8 °C in January, 9.8 °C in July). In addition, 600 mm has been recorded at the bottom of Snøhetta (Østrem et al. 1988; Isaksen et al. 2002) and the annual mean total precipitation reaches 450 mm at Fokstugu and Kongsvoll. These climate conditions suggest that the two meteorological stations are close to a maritime polar type.

For this reason, the massif conserves small glaciers located in the cirques of the higher massifs of the Dovrefjell (i.e., Snøhetta and Svånåtindan central massifs). These glaciers correspond to one of the most continental glacio-nival belts of central Norway. Their fronts are located between 1 550–1 650 m a.s.l. on average (Fig. 12.3).

Based on an analysis of base temperature of snow data (BTS), the lower limit of permafrost is 1490 m a.s.l. on Dovrefjell (Isaksen et al. 2002). Thus, continuous permafrost areas would be restricted to the highest elevations (Ødegård et al. 1993). On the fjell, some studies have also noticed sporadic permafrost-related periglacial forms (Sollid and Sørbel 1998). The whole massif can be considered part of the mountain permafrost zone.

On the fjell, Fokstugu meteorological station experienced 97 cycles per year for the period 1969–2013 (Kerguillec 2015). The maximum number of cycles was recorded in 1989 (129) and the minimum in 1980 (68). Over this time series, 21 years experienced more than 100 cycles per year. Thus, Dovrefjell has intermediate frost-surface conditions when compared with stations located westward or eastward at similar latitudes (Svinøy Fyr and Drevsjø) (Table 12.1, refer to Fig. 12.1 for location of stations).

These frost-action conditions at the surface are especially favorable for periglacial features, given that the bedrock of the mountain area is suitable for frost action. The quartzitic structure is known to be prone to fragmentation and

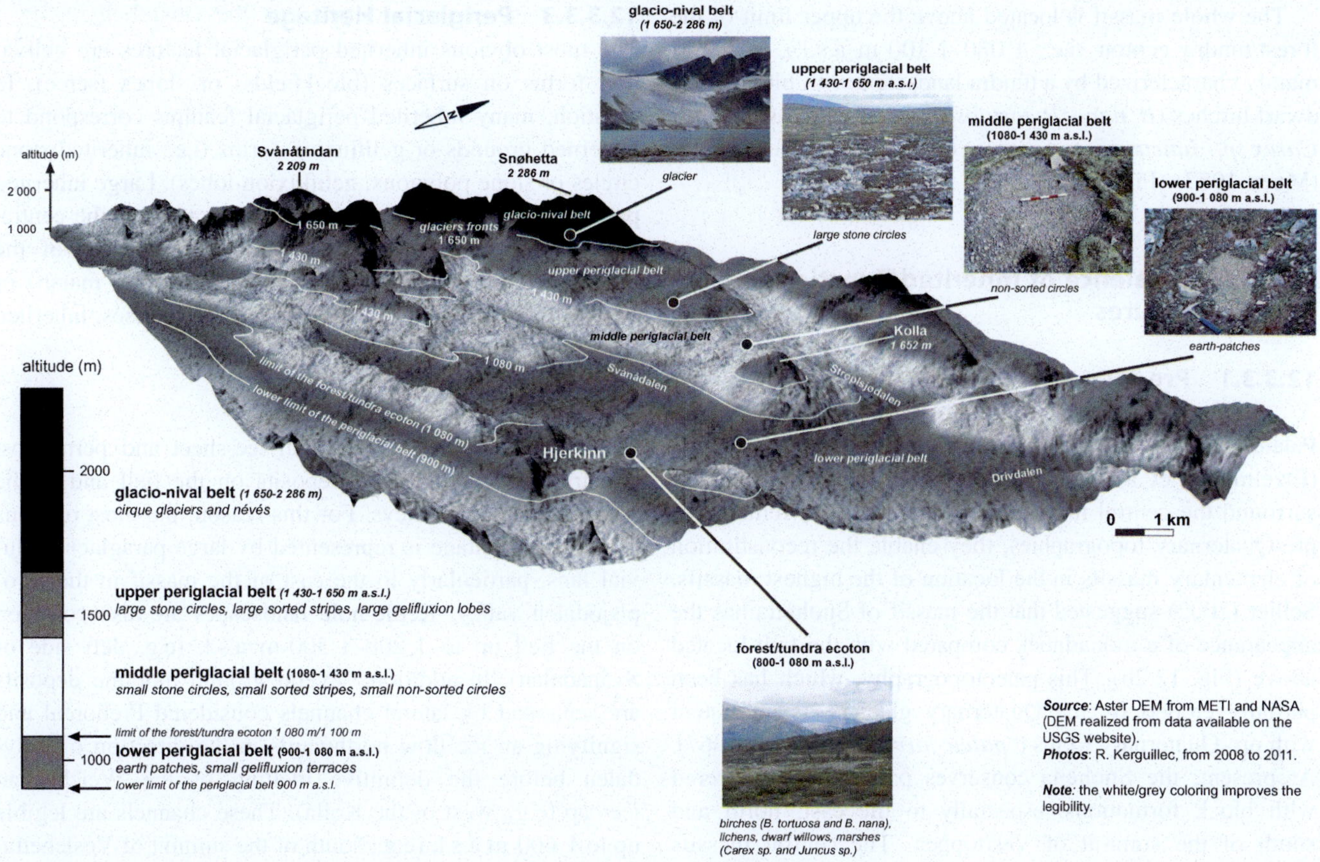

Fig. 12.3 Digital elevation model of the study area and its major mountainous belts

Table 12.1 Freeze/thaw cycle data for Svinøy Fyr, Fokstugu and Drevsjø meteorological stations (from Kerguillec 2015)

Station	Coordinates	Time series	Mean annual number of cycles *(last three decades)*	Mean annual number of cycles *(time series)*	Maximum recorded	Minimum recorded	Years > 100 cycles
Svinøy Fyr	*62° 32'94" N—5° 26'08" E alt.: 38 m*	1956–2013	23	27	52	4	0
Fokstugu	*62° 11'33" N—9° 28'62" E alt.: 973 m*	1969–2013	99	97	129	68	21
Drevsjø	*61° 88'72" N—12° 04'08" E alt.: 672 m*	1957–2013	117	114	150	77	44

two-class granulometric sorting, especially as regoliths are often mixed with feldspathic elements and clay material, which favor frost actions (Sellier 2002; Kerguillec 2013). This explains the prevalence and variety of periglacial features in Dovrefjell from the lower limit of the periglacial belt to the summits of the massif (900–2 286 m a.s.l.). In these conditions, this periglacial belt has been recognized as one of the thickest in Europe (Kerguillec 2013). From its lower limit, the functional periglacial belt can be divided into three minor belts with systematically characteristic features (Fig. 12.3), and turns out to be particularly visible in Dovrefjell (Sellier 2002; Kerguillec 2013):

- The lower periglacial belt, from 900 to 1 080 m a.s.l.;
- The middle periglacial belt, from 1 080 to 1 430 m a.s.l.;
- The upper periglacial belt, from 1 430 m a.s.l. to the summits (possibly including a supraglacial periglacial belt).

The whole massif is located above the upper limit of the forest/tundra ecoton (i.e., 1 080–1 100 m a.s.l.), which is mainly characterized by a tundra landscape with blueberries, dwarf birches (*B. nana, B. tortuosa*), dwarf willows, lichens, *Carex sp.*, *Juncus sp.* and mosses occupying the wettest sites (Moen 1987) (Fig. 12.3).

12.3.3 Prevalence of Inherited/Functional Features

12.3.3.1 Pre-quaternary Heritage ("Paleïc Heritage")

Paleïc forms are dominant in the central massifs of Norway (Etzelmüller et al. 2007). Because the paleïc slopes that surround the central massifs of Dovrefjell are a remnant of pre-Quaternary topographies, they enable the reconstitution of elementary massifs in the location of the highest massifs. Sellier (2002) suggested that the massif of Snøhetta has the appearance of a monadnock compared with the fjell located above (Fig. 12.2b). This paleotopography, which has been partially eroded by the Quaternary glaciers, is consistent with pre-Quaternary forms ("*paleïc forms*") (Gjessing 1967). At present, the Snøhetta conserves paleïc slopes covered with block formations, especially to the east, north and south of the summit of Vesttoppen. The central massifs conserve other remnants of paleotopographies, like well-developed surfaces covered with blockfields above 1 800–2 000 m a.s.l.

12.3.3.2 Glacial Heritage

The frequency of erosion forms inherited from the Quaternary glacial stages leads to one of the main characteristics of the landscape of the central massifs of Dovrefjell as well as contributing to its "alpine" appearance. In the massif of Snøhetta, the upper limit of the Weichselian glaciation was positioned at 1 850–1 950 m a.s.l. considering the upper limit of glacial landforms and the frequency of weathering evidence above 1 950 m a.s.l. (Sellier 2002). The prevalence and characteristics of glacial cirques, which already conserve small glaciers, provide narrow intersection crests that lead to a compartmented relief (e.g., crest from Vesttoppen to Snøhetta). In addition, glacial accumulation forms are widespread in the study area. These forms include several types of morainic deposits and essentially concern Little Ice Age moraines at the entrance of glacial cirques, lateral moraines that help to delineate the location of the ice during the Preboreal stage (9 000–10 000 BP) (e.g., surrounding the summit of Kolla), and numerous ground moraine deposits on the fjell (e.g., east of the massif of Snøhetta, west of the summit of Kolla).

12.3.3.3 Periglacial Heritage

The most obvious inherited periglacial features are gelivation forms on surfaces (blockfields) or slopes (scree). In addition, many inherited periglacial features correspond to patterned grounds or gelifluxion forms (i.e., inherited stone circles or stone polygons, gelifluxion lobes). Large inherited patterned grounds are common at the bottom of the central massifs, especially in the west and in the south of the Brunkollen peripheral mountain (Tvillingkollan marsh) or east of the proglacial Lake Istjønne. In most cases, inherited forms adjoin active forms of the same type.

12.3.3.4 Paraglacial Heritage

The melting of the Weichselian ice-sheet and permafrost redistributed the morainic deposits on the fjell and at the bottom of glacial valleys. For this reason, the most obvious paraglacial heritage is represented by large paraglacial alluvial fans, particularly to the east of the massif in the Stroplsjødalen valley. Kettle-hole landscapes are also common on the fjell up to 1 200–1 300 m a.s.l. (e.g., left side of Svånådalen). In addition, several ground moraine deposits are separated by lateral channels considered Preboreal and signifying an ice flow northwards in the direction of Drivdalen before the definitive melting of the Weichselian ice-cap (e.g., west of the Kolla). These channels are legible up to 1 600 m a.s.l. (e.g., south of the summit of Veslehetta) and mark the extreme limits of the ice-cap retreat in this area during the Preboreal (Sollid 1975).

12.3.3.5 Current Geomorphic Processes

At present, the most efficient geomorphic processes are periglacial processes and concern frost actions and avalanches. On slopes, functional gelivation results in widespread scree in the central massifs, especially around the glacial cirques at the bottom of rocky faces (e.g., the surrounding slopes of Lake Larsjønne, west of the Snøhetta). Some scree are sometimes supraglacial and frost processes are often assisted by runoff (snow melting and debris flow). Functional periglacial features in soils also remain particularly visible in the massif, especially around lakes (e.g., proglacial Lake Istjønne). The Dovrefjell presents in particular large areas covered by plurimetric active stones circles (e.g., Tvillingkollan marsh, proglacial Lake Istjønne).

12.4 Selection of Geomorphosites to Promote Geomorphology of Dovrefjell

The main characteristics of the massif and their associated landforms can lead to a promotion of reliefs as, using geomorphosites descriptions, they can serve as an educational tool for the comprehension of the geomorphology of the National Park. Consequently, these places of observation must be selected.

The approach proposed in the present study includes two successive stages. The first stage is analytic. It consists to apply a "deductive method" to the massif, thus to use an integrated analysis of landforms of complementary scale levels before a deductive selection of elementary units of relief representative of the Dovrefjell (geomorphotypes). The second stage is selective. It performs an assessment to select a geomorphosite for each geomorphotype previously defined during the first stage. It also leads to select the best stations to observe the selected geomorphosites (Fig. 12.4).

12.4.1 First Stage (Analysis Stage, Deductive Method)

12.4.1.1 Description

This approach has the specificity to refer to the fundamental geomorphological properties of the area concerned by the valuation of relief. It usually has three successive steps (Fig. 12.4). The first step defines the fundamental characteristics of the study area. The second step identifies the key geomorphological components in order to recognize several sub-sets with similar dimensions but different characteristics from a topographic, structural, hydrographic, or paleogeographic point of view ("*major geomorphological components*"). During the third step, each main geomorphological component is subdivided into elementary units of relief (also described as "*geomorphotypes*").

12.4.1.2 Results

The first step which aims to study the fundamental characteristics of the study area was performed in the paragraph 2 ("*Main geomorphological properties of the massif*"). This description of the fundamental properties of the Dovrefjell leads to a geomorphological map which synthesizes the main characteristics of the massif (topography, hydrography, paleïc forms, glacial forms, periglacial forms, paraglacial forms). In addition, this map helps to perform a multi-levels analysis of landforms and to characterize the geomorphic processes. It is also used in the second stage to inventory the sites (Fig. 12.5).

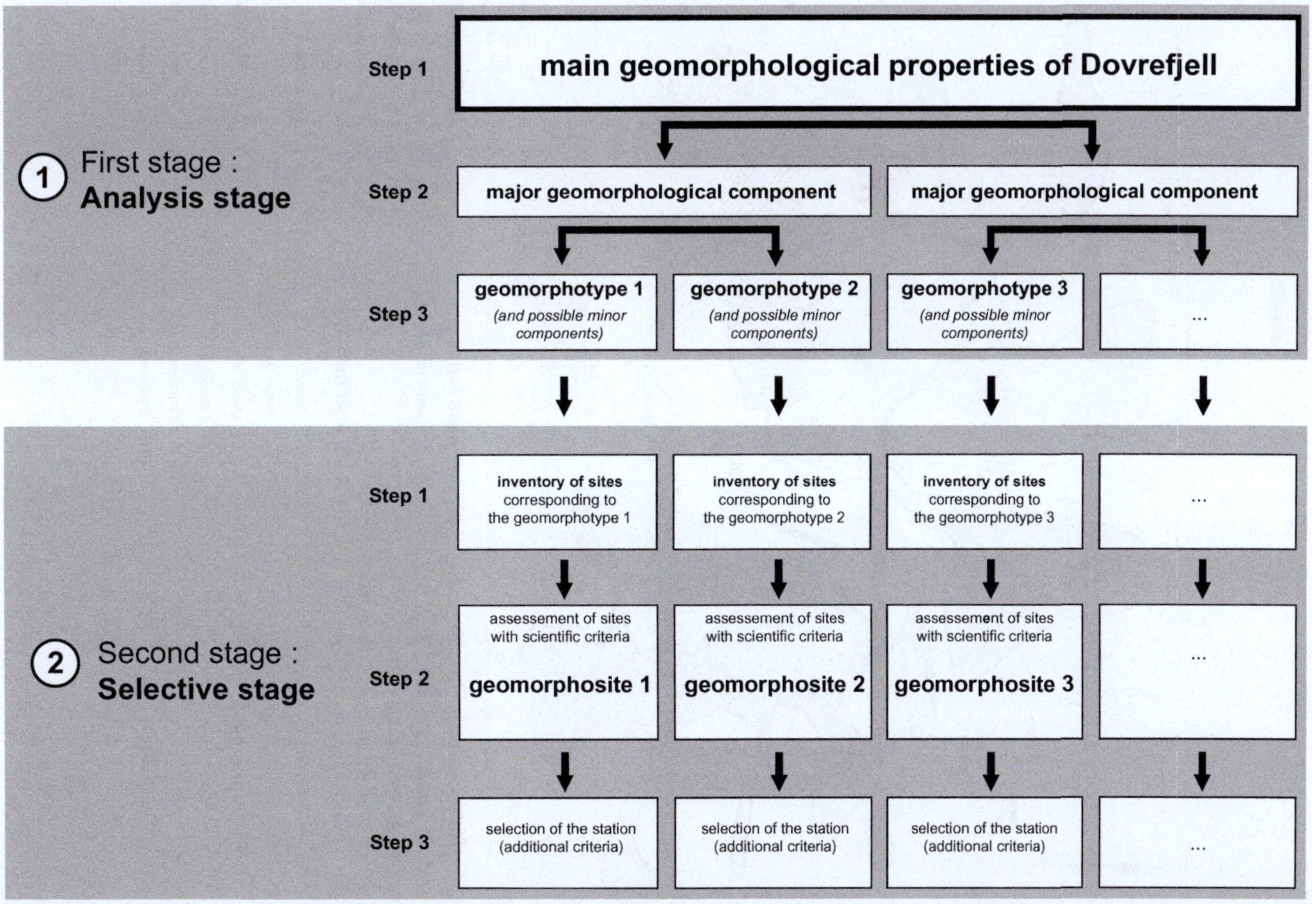

Fig. 12.4 Method for the selection of geomorphosites applied to the relief of Dovrefjell

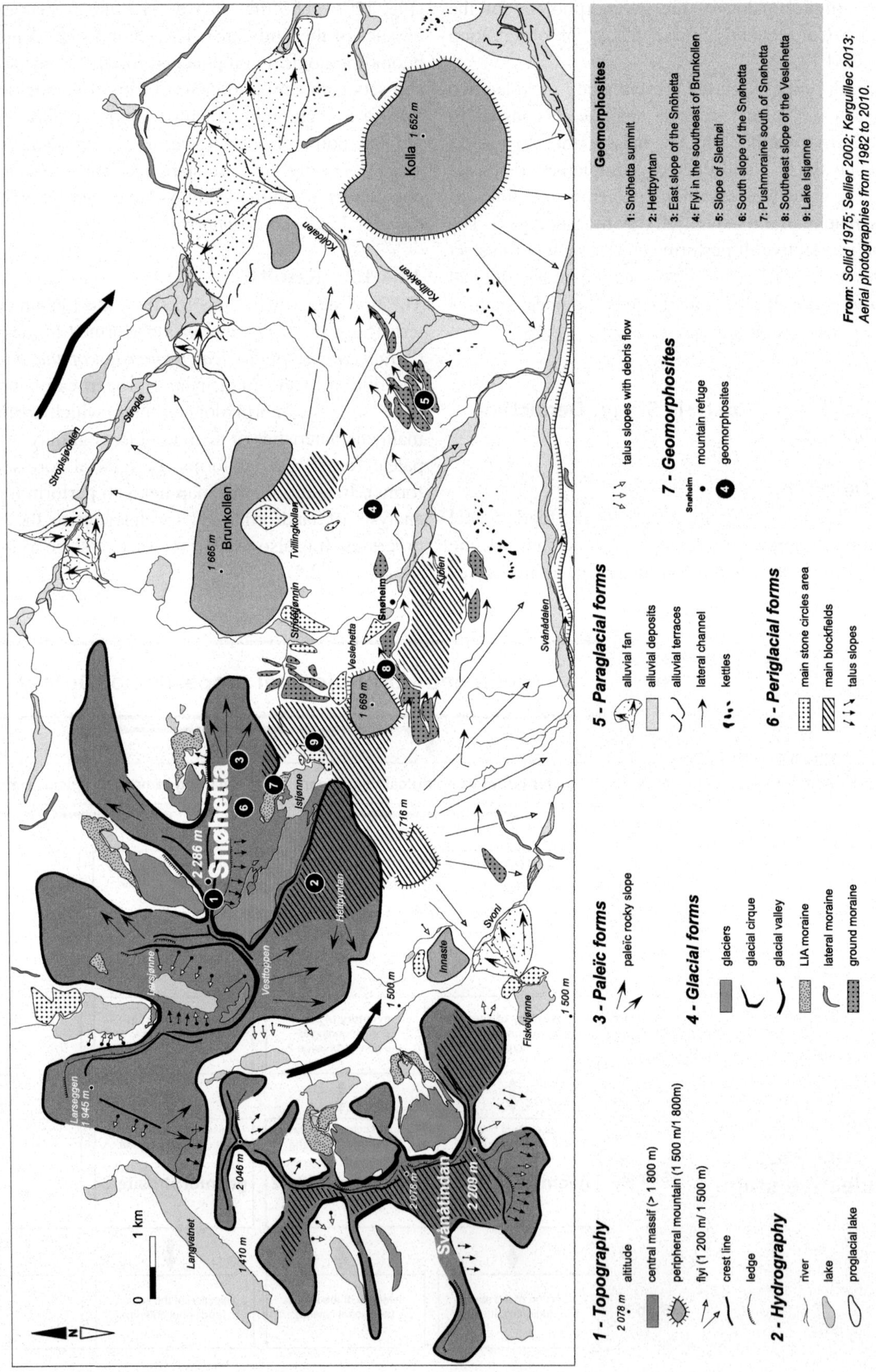

Fig. 12.5 Geomorphological map of the study area

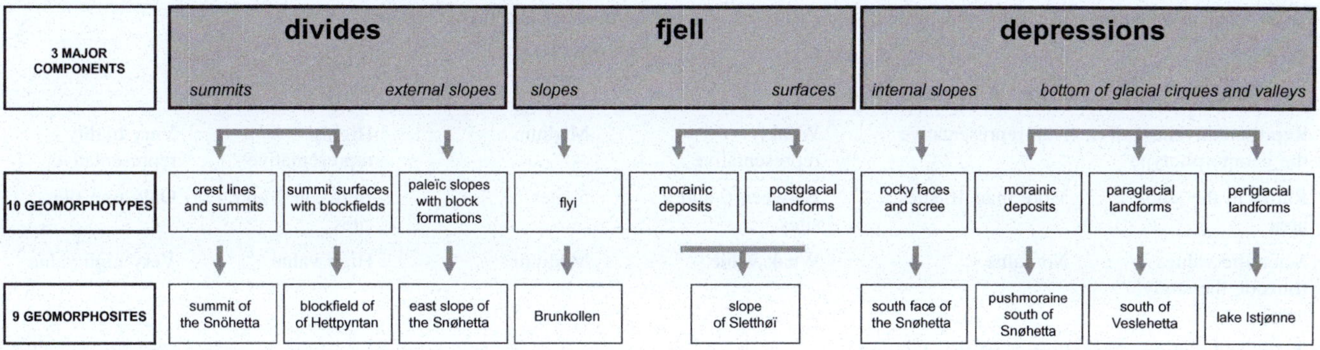

Fig. 12.6 Proposition of geomorphosites for the Dovrefjell

The second step of the deductive method aims to identify the key geomorphological components of the Dovrefjell and eventually their secondary units. In this study, the "divides", the "depressions" and the "fjell" are considered as the three major components of the massif (Fig. 12.6). The "Divides" are divided into two secondary units (i.e., "summits" and "external slopes"), as well as the "Depressions" (i.e., "bottom of glacial cirques and valleys" and "internal slopes" of glacial cirques) and the "fjell" if taking into account its detailed topography ("slopes" and "surfaces").

The third step of the method specifies elementary units of relief (i.e., "*geomorphotypes*") for each secondary unit of the major components. It leads to 10 geomorphotypes (three geomorphotypes derived from the "divides", two geomorphotypes for the "fjell" and four geomorphotypes for the "depressions"). The geomorphotypes corresponding to the "divides" are "crest lines and summits" and "summit surfaces with blockfields". "External slopes" gives the geomorphotype "paleïc slopes with blocky formations". The "fjell" leads to the three geomorphotypes "flyi" (geomorphotype "slope"), "morainic deposits" and "postglacial landforms" (i.e., paraglacial and periglacial landforms). Four geomorphotypes correspond to the major component "depressions": "internal slopes" are represented by "rocky faces and scree", "bottom of the glacial cirques and valleys" includes three different geomorphotypes ("morainic deposits", "paraglacial features", and "periglacial features").

12.4.2 Second Stage (Selective Stage, Selective Method)

12.4.2.1 Description

The *International Association of Geomorphologists (IAG)* formed a Working Group for the period 2001–2005, which focused on several topics, including the development of assessment methods for valuation and selection of geomorphosites. Researchers from the Universities of Lausanne, Cantabria, Modena, and Valladolid proposed several methods of selection (Reynard and Panizza 2005), which attribute a score to the geomorphosites by taking into account scientific criteria (e.g., exemplarity, educational interest, integrity, representativeness, paleogeographic value) and additional values (e.g., management criteria, accessibility, legibility, historical values).

In this study, a preliminary inventory of the sites was done before performing the assessment (Fig. 12.4). This first step aims to inventory all the sites corresponding to the 10 geomorphotypes resulting from the deductive method. The sites were surveyed during four field missions (2008–2011) and this inventory was completed by the analysis of the geomorphological map (Fig. 12.5) and in some cases by photointerpretation of aerial photos.

The second step consists to perform an assessment to all the sites corresponding to the 10 geomorphotypes resulting from the deductive method applied on the first stage. To carry out this assessment, a preliminary selection was made from the scientific criteria proposed by Reynard (2006), according to the recommendations of the author who suggests that the evaluation grid should be adapted according to circumstances (Table 12.2). On the other hand, several criteria were added to those proposed by Reynard, especially "the scientific value" of a site compared to other sites of the same geomorphotype located in other places in the world. Some criteria were also moved from the category "scientific criteria" to the category "additional criteria": for example, the integrity of a site is not considered here as a scientific indicator but as an additional indicator. In this study, it was associated with the value "protection".

In Dovrefjell, some criteria which are commonly used in the selective method are irrelevant or not directly applicable. Some additional criteria are not appropriate or not applicable, because either they should be evaluated by specialists or they do not concern the study area. As an example, the valuation of the ecological importance of a geomorphosite for specific ecosystems (i.e., "ecological influence") should be carried out by biologists. It is well known that Dovre has a great symbolic meaning and a high ecological and

Table 12.2 Evaluation grid used in the present study to select the geomorphosites and their stations

Score	0	0.25	0.5	0.75	1
SV = scientific value (selection of the geomorphosites)					
Representativeness of the geomorphotype	Non-representative	Weakly representative	Medium	Highly representative	Very highly representative
Rarity in the study area	More than 10 sites	Between 6 and 9 sites	5 sites	Between 2 and 4 sites	Only one site
Scientific value (outside the study area)	No value	Weak value	Medium	High value	Very high value
AV = additional value (selection of the stations)					
Accessibility	Very difficult (i.e. hike > 10 km, no marked footpath)	Difficult (i.e. hike 5/10 km, no marked footpath)	Medium (i.e. hike 5/10 km, marked footpath)	Easy (i.e. hike 2/5 km, marked footpath)	Very easy (i.e. hike < 2 km, marked footpath)
Esthetic value	No value	Weak value	Medium	High value	Very high value
Integrity and protection	Destroyed, non-protected area	Very damaged	Fairly damaged	Slightly damaged	Intact, National Park area
Intrinsic value (e.g. cultural, geomorphologic, etc.)	No value	Weak value	Medium	High value	Very high value

biological value in Norway (e.g., endemic plants, reindeer, musk ox), but the assessment of these values has been outside the scope of this study because it should be carried out by specialists. Accessibility apart, the economic and tourist value of a site should be estimated by economists or the National Park authorities.

12.4.2.2 Results

This study presents the results for 80 assessed sites previously inventoried during the first step. For each value, a score from 0 to 1 was attributed to each site at intervals of a quarter of a point: 0 (no value) 0.25 (weak value) 0.5 (mean value) 0.75 (high value) 1 (very high value).

The scientific value (SV) corresponds to the addition of all the scores of scientific criteria, and the additional value (AV) corresponds to the addition of all the scores of additional criteria. Taking into account that the additional values should not have the same weight as the scientific value for the selection of priority geomorphosites, AV was divided by two in the scoring in order to give more importance to SV (Reynard 2006). Thus, the final score of a site is attributed in the following way:

$$\text{Score} = \text{SV} + \text{AV}/2.$$

The assessment leads to the selection of nine geomorphosites which are mainly located in the massif of Snøhetta or in its vicinity (Figs. 12.4 and 12.6).

The summit of the Snøhetta (geomorphosite 1) was selected to represent the geomorphotype "crest lines and summits". The Snøhetta is one of the higher summits of Norway (2 286 m a.s.l.). It has the appearance of a large dome cut by narrow crevices and prolonged by a narrow rocky crest in the west. The Snøhetta is surrounded by well-developed paleïc rocky slopes and was recognized as a paleonunatak during the last glacial stage (Sellier 2002). It also provides a 360° panorama on the whole massif of Dovrefjell and also on its neighboring massifs (massif of Rondane in the southeast, Jotunheimen in the southwest, Trollheimen in the north).

The blockfield of Hettpyntan (geomorphosite 2) represents the geomorphotype "summit surfaces with blockfields". This blockfield of low inclination from the southeast to the northwest is located between the summit of Veslehetta and the Snøhetta glacier in the northwest (1 600–1 900 m a.s.l.). It has been considered of a glacial origin (Sellier 2002) and is a good way to show to the public the mechanical sensitivity of quartzites to glacial erosion processes and frost actions.

The east slope of the Snøhetta (geomorphosite 3) was selected to represent the geomorphotype "paleïc slopes with block formations". The paleïc slope, which extends from 1 680–1 700 m to the summit (2 286 m), reaches an elevation of 600 m and has an average inclination of 20°. It is covered by an openwork blocky formation which mainly results from frost-heaving and gelifluxion. This geomorphosite could eventually be associated to the geomorphosite 1 (i.e., summit of the Snøhetta) in an operation of a valuation of relief because of its proximity with the summit.

The gentle slope located in the east of Snøheim mountain refuge and in the southeast of Brunkollen (geomorphosite 4) represents the geomorphotype “flyi”. In addition, this site provides a general viewpoint of the central massif of the Snøhetta toward the west.

The extended morainic deposit located on the slope of Sletthøi between Snøheim mountain refuge and the summit of the Kolla (geomorphosite 5) was selected to represent the two geomorphotypes derived from the fjell (i.e., “morainic deposits” and “postglacial landforms”). This site has been recently recognized as an example of a post-Little Ice Age periglacial recovery developed in a nivation hollow (Kerguillec et al. 2015). It comprises several types of forms of different ages, with Weichselian glacial deposits (ground moraine), paraglacial landforms (lateral channel) and also widespread post-Little Ice Age active periglacial forms (large stone circles and polygons). This site could be used to explain to the public the succession of geomorphic processes in time on a same place during the postglacial period.

The south rocky face of the Snøhetta (geomorphosite 6) was selected to represent the geomorphotype “rocky faces and scree”. It is one of the most developed rocky faces in elevation considering that it extends from 1 750 m to the summit of the Snøhetta (i.e., 500–600 m in elevation).

The push moraine situated at the entrance of the glacial cirque located south of the Snøhetta (geomorphosite 7) would be a good site to represent the geomorphotype “morainic deposits”. This moraine has been reported to the Little Ice Age (Sollid 1975). It is several hundred meters long and is composed of plurimetric angular blocks which would be a good way to show the public the glacial processes of erosion and transport.

Lateral channel is common in the massif. The lateral channel located in the southeast of the peripheral summit of the Veslehetta (geomorphosite 8) which represents the geomorphotype “paraglacial landforms”, is easily accessible from Snøheim mountain refuge.

The wide large stone circles field located around Lake Istjønne (geomorphosite 9) represents the geomorphotype “periglacial landforms”. These periglacial landforms are among the largest active stone circles of the massif (4–5 m in diameter with blocks > 1 m). This site could be used to show to the public the effects of periglacial sorting in soils. It could be visited in association with geomorphosites 6 and 7.

The last step of the method used in this chapter consists in the selection of the best stations to observe the geomorphosites (Fig. 12.4). This last step uses non-scientific criteria (i.e., “additional criteria”). The role of the station is to be emphasized because it is the privileged place from which the geomorphosite will be observed by the public. Therefore, the station is the place where the knowledge is disseminated and photographs are taken. It is also chosen according to its intrinsic opportunities to valorize geomorphology (e.g., other landforms or processes, outcrop) or another thematic (Table 12.2).

12.5 Summary and Conclusions

Located in central Norway, the massif of Dovrefjell is part of the Scandes and has been a National Park since 1974. The Dovrefjell–Sunndalsfjella National Park is well known for its intact ecosystems and also for its cultural heritage, its landscape and landforms. The aim of this case study is to increase awareness of these landforms in the perspective of their valuation in the National Park, and to propose a new way to select geomorphosites.

The relief is representative of a high Atlantic mountain landscape and because of its fundamental properties, the massif has a great geomorphological interest for the natural heritage valuation into the National Park. First, its homogeneous lithology plays a major role for the relief. At a large scale, the quartzites impose its general organization on the relief. At a medium scale, this lithology is also suitable for periglacial processes on surfaces and slopes. Functional periglacial landforms are well developed in the Dovrefjell, which shows distinctive mountainous belts and one of the thickest periglacial belts in Europe after the massif of Rondane. Moreover, the massif is characterized by its numerous inherited forms of various origins, especially Pre-Quaternary, glacial, periglacial, and paraglacial heritages.

One of the main results of this study is to increase awareness of the landforms of Dovrefjell in the perspective of their valuation in the National Park and ultimately to valorize the characteristic landforms of a Nordic high mountain. The method uses the combination of a “deductive method” and the “selective method” in order to propose to the National Park authorities several geomorphosites to be presented to the visitors. The interest of this new geomorphosites selection methodology is to combine the advantages of the two methods involved. The deductive method has the advantage of being able to be used in regions with different reliefs and to deduce representative geomorphotypes from the fundamental properties of the space concerned by the promotion of relief. Thus, its strong point is to propose an approach to enhance the geomorphology of a region because the sites are connected to each other by their geomorphological relationships at complementary scales. The interest of the selective method is to give a note to the sites according to a scoring grid that can be adapted to the space concerned by the promotion of relief. This method allows taking into account a large number of sites to retain only the most representative. For all these reasons, the method developed in this chapter and applied to the Dovrefjell provides a new tool for the use of structures that want to increase awareness of the landforms of their area by selecting representative and interconnected geomorphosites.

This method is also of interest for the public because it establishes strong relationships between the reliefs at different levels of scale, by providing complementary geomorphotypes whose singularity could each time be valued. Thus, it enhances the global geomorphological value of the Dovrefjell as well as its value in detail by means of exemplary sites. Moreover, the accessibility of the Dovrefjell strengthens its interest for the valorization of the typical reliefs of a high Nordic mountain (i.e., the vehicles can penetrate in the center of the massif, which makes the stations of observation particularly easy access). Other massifs, such as the Jotunheimen or the Rondane, require hikes of several kilometers to observe the sites of interest (e.g., at least 7 km to reach the center of the massif of Rondane). The Dovrefjell also has a strong aesthetic value (e.g., presence of glaciers) and finally allows an educational and representative geomorphologic synthesis of the Norwegian relief and its main processes. Finally, the selection method presented in this chapter can also be used as a method to describe and explain the relief to the public.

References

Dovrefjell-Sunndalsfjella Official Website (2018) www.nasjonalparker.org. Accessed June 2018

Etzelmüller B, Romstad B, Fjellanger J (2007) Automatic regional classification of topography in Norway. Nor Geol Tidsskr 87:167–180

Gjessing J (1967) Norway's paleïc surface. Nor Geogr Tidsskr 21 (2):69–132

Godard A (1965) Recherches géomorphologiques en Ecosse du Nord-Ouest. Thèse d'Etat, Publications de la Faculté des Lettres de l'Université de Strasbourg, Paris

Grandgirard V (1999) L'évaluation des géotopes. Geologia Insubrica 4:59–66

Isaksen K, Hauck C, Gudevang E, Ødegård RS, Sollid JL (2002) Mountain permafrost distribution in Dovrefjell and Jotunheimen, southern Norway, based on BTS and DC resistivity tomography data. Nor Geogr Tidsskr 56:122–136

Kerguillec R (2013) Les dynamiques périglaciaires actuelles dans un milieu de haute montagne atlantique: parcs nationaux du Oppland et du Sør-Trondelag, Norvège centrale. PhD thesis, University of Nantes

Kerguillec R (2015) Seasonal distribution and variability of atmospheric freeze/thaw cycles in Norway over the last six decades (1950–2013). Boreas 44(3):526–542

Kerguillec R, Sellier D (2015) Selection and promotion of geomorphosites in the Rondane National park (Central Norway): landforms, heritages and geodynamics. Géomorphologie: relief, processus, environnement 21(2):131–144

Kerguillec R, Sellier D, Beylich A (2015) An example of a periglacial recovery: the slope of Sletthøi (Dovrefjell, central Norway). Zeitschrift für Geomorphologie 59(2):173–196

Moen A (1987) The regional vegetation of Norway: that of Central Norway in particular. Nor Geogr Tidsskr 41(4):179–226

Nilsen O, Wolff FC (1988) Røros & Sveg BerggrunnskartNorsk

Norsk Meteorologisk Institutt (2018) http://www.eklima.no. Accessed Sept 2018

Ødegård RS, Hoelzle M, Johansen KV, Sollid JL (1993) Permafrost prospecting and mapping in southern Norway. In: Ødegård RS (ed) Ground and glacier thermal regimes related to periglacial and glacial processes: cases studies from Svalbard and southern Norway, pp 79–98

Østrem G, Dale Selvig K, Tandberg K (1988) Atlas over breer in Sør-Norge. Meddelser 61, fra Hydrologisk avdeling, Norwegian Water Resources and Energy Directorate (NVE), Oslo

Panizza M (2001) Geomorphosites: concepts, methods and example of geomorphological survey. Chin Sci Bull 46:2–6

Pereira P, Pereira D (2010) Methodological guidelines for geomorphosites assessment. Géomorphologie: relief, processus, environnement 2:215–222

Peulvast J-P (1977) Le bourrelet scandinave et les Calédonides: aspects et problèmes de la géomorphologie de la Norvège. Rev Géogr Phys Géol Dynam 19(5):503–514

Peulvast J-P (1985) Relief, érosion différentielle et morphogenèse dans un bourrelet montagneux de Haute latitude: Lofoten-Vesterålen et Sogn-Jotun (Norvège). Thèse d'Etat, University of Paris I

Reynard E (2005) Géomorphosites et paysages. Géomorphologie: relief, processus, environnement 3:181–188

Reynard E (2006) Fiche d'inventaire des géomorphosites. Université de Lausanne, Institut de géographie, not published

Reynard E, Panizza M (2005) Géomorphosites: définition, évaluation et cartographie, une introduction. Géomorphologie: relief, processus, environnement 3:177–180

Reynard E, Fontana G, Kolzlik L, Scapozza C (2007) A method for assessing "scientific" and "additional values" of geomorphosites. Geographica Helvetica 62(3):148–158

Rudberg S (1960) Geology and geomorphology. In: Sømme A (ed) A geography of Norden. Cappelens Forlag, Oslo, pp 31–47

Sellier D (2002) Géomorphologie des versants quartzitiques en milieux froids: l'exemple de montagnes d'Europe du nord-ouest. Thèse d'Etat, University of Paris I

Sellier D (2009) La vulgarisation du patrimoine géomorphologique: objets, moyens, perspectives. Bulletin de l'Association de Géographes Français 86(1):67–81

Sollid JL (1975) Dovrefjell Nasjonalpark - Landskapet. Geografisk Institutt, Universitetet i Oslo

Sollid JL, Sørbel L (1998) Palsa bogs as a climate indicator—examples from Dovrefjell, Southern Norway. Ambio 27(4):287–291

Strøm KM (1945) Geomorphology of the Rondane area. Nor Geogr Tidsskr 25:360–378

Index

A. A. Beylich (ed.), *Landscapes and Landforms of Norway*, World Geomorphological Landscapes,
https://doi.org/10.1007/978-3-030-52563-7

T

U

The manufacturer's authorised representative in the EU is Springer Nature Customer Service Centre GmbH, Europaplatz 3, 69115 Heidelberg, Germany. If you have any concerns regarding our products, please contact ProductSafety@springernature.com

Printed and bound by CPI Group (UK) Ltd, Croydon, CR0 4YY
30/09/2026
02225476-0003